G. GRÄTZER

GENERAL LATTICE THEORY

GENERAL LATTICE THEORY

by

George Grätzer

Professor of Mathematics
University of Manitoba

AKADEMIE-VERLAG · BERLIN

1978

Erschienen im Akademie-Verlag, 108 Berlin, Leipziger Straße 3–4

Lizenznummer: 202 · 100/557/78
Gesamtherstellung: IV/2/14 VEB Druckerei »Gottfried Wilhelm Leibniz«,
445 Gräfenhainichen · 4742
Bestellnummer: 762 291 5 (6334) · LSV 1025
Printed in GDR
DDR 72,– M

To my family, Cathy, Tom, and David,
and to the memory of my father, József

CONTENTS

PREFACE AND ACKNOWLEDGEMENTS

A book that is more than twelve years in the making has a long history and its final form is shaped by many. It all started in my formative years as a mathematician, 1955–1961, when I worked with E. T. Schmidt. We often commented upon the need for a two– or three–volume work on lattice theory that would treat the subject in depth. We felt, however, that the time was not ripe for such a project. For instance, no such work would be complete without presenting at least one example of a nondistributive uniquely complemented lattice. We did not know how to do it without reproducing the almost thirty pages of the famous proof of R. P. Dilworth. We also thought that much more had to be learned about free lattices and equational classes of lattices before the project could be attempted.

In 1962, I wrote a proposal for a volume on lattice theory that would attempt to survey the whole field in depth. Apart from doing the research necessary for the proposal, no writing was done on this book. M. H. Stone offered to publish a book on universal algebra in the D. Van Nostrand University Series in Higher Mathematics and I concentrated on that book until the end of 1967.

Maybe because mathematicians in general (or I, in particular) are like hobbits (according to J. R. R. Tolkien [1954]: "Hobbits delighted in such things if they were accurate: they liked to have books filled with things they already knew, set out fair and square with no contradictions.") or maybe because I felt that the need for an in depth book on lattice theory had not yet been satisfied, I started in 1968 on this book. In the academic year 1968–1969, I gave a course on lattice theory and I wrote a set of lecture notes. The present first two chapters are based on those notes.

This material was augmented by a chapter on pseudocomplemented distributive lattices and published under the title "Lattice Theory: First Concepts and Distributive Lattices" in 1971. (This book will be referred to as FC.) The Introduction of this book promised a companion volume on general lattices.

A number of research breakthroughs in the sixties now supplied me with the material (including almost all of Chapters V and VI) I needed to complete the project. But then it became apparent that a complete revision of my plans was in order.

While back in the late fifties it seemed reasonable to try to give a complete picture of lattice theory, this became patently unfeasible in the seventies. For instance, in 1958 there was one paper on Stone algebras; by 1974 there were more than fifty papers on Stone algebras and related problems. A number of books have appeared dealing with specialized aspects of lattice theory and with various applications.

Besides, my experience with the writing of Universal Algebra taught me not to stray too far away from my research interests. Thus it was decided that while I try to include all the basic material and research methods, the illustrations will be chosen, as far as possible, from fields in which I have some personal interest.

Another change took place in the publishing field. For the second volume it became desirable to choose a publisher with a greater interest in monographs. The new arrangement made it necessary to produce a volume that does not depend on the previous publication. That is why most of the first two chapters of the former book are reproduced here (augmented by a new section, several new exercises, with updated Further Topics and References, and with a new set of Problems), thus giving the reader a self-contained book.

The work on this new book started in 1972 and then continued with an advanced course on lattice theory at the University of Manitoba in 1973–1974. The lecture notes of this course form the basis of most of Chapters III–VI.

I am grateful to my students who took the courses in 1968–1969 or in 1973–1974 and also to my colleagues who attended for their helpful criticisms and for many simplified proofs. A corrected version of the first set of notes was read by R. Balbes, P. Burmeister, M. I. Gould, J. H. Hoffman, K. M. Koh, H. Lakser, S. M. Lee, R. Padmanabhan, P. Penner, and C. R. Platt. B. Jónsson, reading the manuscript of FC for the publisher, offered many useful suggestions. The first part was proofread by R. Antonius, J. A. Gerhard, K. M. Koh, W. A. Lampe, R. W. Quackenbush, and I. Rival.

Many readers, in particular D. D. Miller, sent me corrections to FC; this made it possible for me to improve some parts of FC that are being reproduced here.

The second set of notes was distributed widely and I am grateful to all who offered corrections, in particular, to K. A. Baker, C. C. Chen, M. I. Gould, D. Haley, K. M. Koh, V. B. Lender, G. H. Wenzel, and B. Wolk.

In the proofreading of the present volume I was assisted by M. E. Adams, R. Beazer, K. A. Baker, J. Berman, B. A. Davey, J. A. Gerhard, M. I. Gould, D. Haley, D. Kelly, C. R. Platt, and G. H. Wenzel.

A great deal of organizational work was necessary in the distribution of manuscripts and the collation of corrections; this was faithfully carried out by R. Padmanabhan.

M. E. Adams undertook the arduous task of getting the manuscript ready for the publisher.

I received help from various individuals in specific areas, including M. Doob (matroids), I. Rival (exercises on combinatorial topics), R. Venkataraman (partially ordered vector spaces), B. Wolk (projective geometry).

Thanks are due to the National Research Council of Canada for sponsoring much of the original research that has gone into this book and to Professor N. S. Mendelsohn for creating a very good environment for work. Mrs. M. McTavish did an excellent job of typing and retyping the manuscript.

Finally, I would like to thank the members and the many visitors of my seminar who, over a period of eight years, have been lecturing an average of four hours a week, 52 weeks a year, in an attempt to teach me lattice theory. Without their help I could not even have tried.

Despite the improvements so generously offered by so many, I am sure my original work can still be recognized: all the remaining mistakes are my own.

INTRODUCTION

In the first half of the nineteenth century, George Boole's attempt to formalize propositional logic led to the concept of Boolean algebras. While investigating the axiomatics of Boolean algebras at the end of the nineteenth century, Charles S. Peirce and Ernst Schröder found it useful to introduce the lattice concept. Independently, Richard Dedekind's research on ideals of algebraic numbers led to the same discovery. In fact, Dedekind also introduced modularity, a weakened form of distributivity. Although some of the early results of these mathematicians and of Edward V. Huntington are very elegant and far from trivial, they did not attract the attention of the mathematical community.

It was Garrett Birkhoff's work in the mid-thirties that started the general development of lattice theory. In a brilliant series of papers he demonstrated the importance of lattice theory and showed that it provides a unifying framework for hitherto unrelated developments in many mathematical disciplines. Birkhoff himself, Valère Glivenko, Karl Menger, John von Neumann, Oystein Ore, and others had developed enough of this new field for Birkhoff to attempt to "sell" it to the general mathematical community, which he did with astonishing success in the first edition of his *Lattice Theory*. The further development of the subject matter can best be followed by comparing the first, second, and third editions of his book (G. Birkhoff [1940], [1948], and [1967]).

The goal of the present volume can be stated very simply: to discuss in depth the basics of general lattice theory. In other words, I tried to include what I consider the most important results and research methods of all of lattice theory. To treat the rudimentary results in depth and still keep the size of the volume from getting out of hand, I had to omit a great deal. I excluded many important chapters of lattice theory that have grown into research fields on their own. Ordered algebraic systems and other applications were also excluded. The reader will find appropriate references to these throughout this book. It is hoped that even those whose main interest lies in areas not treated here in detail will find this volume useful by obtaining from this book the background in lattice theory so necessary in allied fields.

In my view, distributive lattices have played a many faceted role in the development of lattice theory. Historically, lattice theory started with (Boolean) distributive lattices; as a result, the theory of distributive lattices is one of the most extensive and most satisfying chapters of lattice theory. Distributive lattices have provided the motivation for many results in general lattice theory. Many conditions on lattices and on elements and ideals of lattices are weakened forms of distributivity. Therefore, a thorough knowledge of distributive lattices is indispensable for work in lattice theory. Finally, in many applications the condition of distributivity is imposed on lattices arising in various areas of mathematics, especially algebra.

This viewpoint moved me to break with the traditional approach to lattice theory, which proceeds from partially ordered sets to general lattices, semimodular lattices, modular lattices, and, finally, to distributive lattices. That is why distributive lattices are treated as a first priority in this book. This approach has the added advantage that the reader (or the student in the classroom) reaches interesting and deep results early in the book.

Chapter I gives a concise development of the basic concepts of lattice theory. Diagrams are emphasized because I believe that an important part of learning lattice theory is the acquisition of skill in drawing diagrams. This point of view is stressed throughout the book by about 130 diagrams (heeding Alice's advice: "and what is the use of a book without pictures or conversations", L. Carroll [1865]); the reader would be well advised to draw many times more while reading the book. A special feature of this chapter is a detailed development of free lattices generated by a partial lattice over an arbitrary equational class; this is one of the most important research tools of lattice theory.

Chapter II develops distributive lattices including representation theorems, congruences, congruence lattices of general lattices, Boolean algebras, and topological representations. The last section is a brief introduction to the theory of distributive lattices with pseudocomplementation. While the theory of distributive lattices is developed in detail, the reader should keep in mind that the purpose of this chapter is, basically, to serve as a model for the rest of lattice theory.

In Chapter III we discuss congruences and ideals of general lattices. The various types of ideals discussed all imitate to some extent the behaviour of ideals in distributive lattices.

After giving the basic facts concerning modular and semimodular lattices, Chapter IV investigates in detail the connection between lattice theory and geometry. We develop the theory of geometric lattices, in particular direct decompositions and geometric lattices arising out of geometries and graphs. As an important example, we investigate partition lattices. The last section deals with complemented modular lattices and projective geometries.

Chapters V and VI deal with two new areas of investigation. Equational classes of lattices is one of the most promising new fields. In Chapter V most of the basic facts are presented along with some more specialized methods. Chapter VI grew out of an investigation of free lattices. It intends to prove that almost all the results on free lattices can be obtained within the framework of free products of lattices. In addition, free products can be used to construct interesting examples of lattices.

The exercises, which number almost 900, form an integral part of the book. The Bibliography contains over 750 entries; it is not, however, a comprehensive bibliography of this field. With a few exceptions, it contains only items referred to in the text. The 193 research problems, the "Further Topics and References" at the end of each chapter, and the Concluding Remarks should be of help to those who are interested in further reading and research in lattice theory.

The abandonment of the traditional structure of a lattice theory book means that concepts and notations are more evenly introduced throughout. A very detailed Index and the Table of Notation should help the reader in finding where a concept or notation is first introduced.

Finally, the reader will note that the symbol ● is placed at the end of a proof; if a theorem or lemma contains more than one statement, the proof of a part is ended with ◑. The abbreviation "iff" stands for "if and only if". More difficult exercises are marked by *. "Theorem 10" refers to Theorem 10 of the same section, "Theorem 5.10" refers to Theorem 10 of Section 5 of the same chapter, whereas "Theorem I. 5.10" refers to Theorem 10 of Section 5 in Chapter I. Similarly, "Exercise III. 2.6" means Exercise 6 of Section III. 2. References to the Bibliography are given in the form "J. Jakubik [1957]", which refers to a paper (or book) by J. Jakubik published in 1957. Such references as "[1957 a]" and "[1957 b]" indicate that the Bibliography contains more than one work by the author published in that year. "R. Wille [a]" refers to a paper by R. Wille that had not appeared in print at the time the manuscript of this book was submitted for publication. In the list of problems, Problem I. 29 is Problem 29 of Chapter I; I. 31 (FC 17) signifies that Problem 31 of Chapter I is a repetition of, or is closely related to, Problem 17 of FC.

Winnipeg, Manitoba
September 1975

George Grätzer

CHAPTER I

FIRST CONCEPTS

1. Two Definitions of Lattices

Whereas the arithmetical properties of the set of reals R can be expressed in terms of addition and multiplication, the order theoretic, and thus the topological, properties are expressed in terms of the ordering relation $\leq$. The basic properties of this relation are as follows.

For all $a, b, c \in R$ we have:

(P1)	***Reflexivity*:**	$a \leq a$.
(P2)	***Antisymmetry*:**	$a \leq b$ **and** $b \leq a$ **imply that** $a = b$.
(P3)	***Transitivity*:**	$a \leq b$ **and** $b \leq c$ **imply that** $a \leq c$.
(P4)	***Linearity*:**	$a \leq b$ **or** $b \leq a$.

There are many examples of binary relations sharing properties (P1)–(P4) with the ordering relation of reals, and there are even more enjoying (P1)–(P3). This fact, by itself, would not justify the introduction of a new concept. However, it has been shown that many basic concepts and results about the reals depend only on (P1)–(P3), and these can be profitably used whenever we have a relation satisfying (P1)–(P3). Relations satisfying (P1)–(P3) are called *partial ordering relations*, and sets equipped with such relations are called *partially ordered sets* or *posets*.

To make the definitions formal, let us start with two sets A, B and form the set $A \times B$ of all ordered pairs $\langle a, b\rangle$ with $a \in A, b \in B$. If $A = B$, we write A^2 for $A \times A$. Then a *binary relation* ϱ on A can simply be defined as a subset of A^2. The elements a, b $(a, b \in A)$ are *in relation* with respect to ϱ iff $\langle a, b\rangle \in \varrho$. For $\langle a, b\rangle \in \varrho$, we will also write $a \varrho b$, or $a \equiv b\ (\varrho)$. Binary relations will be denoted by small Greek letters or by special symbols.

This formal definition can be compared with the intuitive one: A binary relation ϱ on A is a "rule" that decides whether or not $a \varrho b$ for any given $a, b \in A$. Of course, any such rule will determine the set $\{\langle a, b\rangle \mid a \varrho b,\ a, b \in A\}$, and this set determines ϱ, so we might as well regard ϱ as being identical with this set.

A *partially ordered set* $\langle A; \varrho\rangle$ consists of a nonvoid set A and a binary relation ϱ on A, such that ϱ satisfies properties (P1)–(P3). Note that these can be stated as follows: For all $a, b, c \in A$, $\langle a, a\rangle \in \varrho$; $\langle a, b\rangle, \langle b, a\rangle \in \varrho$ imply that $a = b$; $\langle a, b\rangle, \langle b, c\rangle \in \varrho$ imply that $\langle a, c\rangle \in \varrho$.

If ϱ satisfies (P1)–(P3), ϱ is a *partial ordering relation*, and it will usually be denoted by $\leq$. Also, $a \geq b$ will mean $b \leq a$. Sometimes we will say that A (rather than $\langle A;\ \leq\rangle$)

is a poset, meaning that the partial ordering is understood. This is an ambiguous although widely accepted practice.

A poset $\langle A; \leq \rangle$ that also satisfies (P4) is called a *chain* (also called *fully ordered set, linearly ordered set*, and so on).

If $\langle A; \leq \rangle$ is a poset, $a, b \in A$, then a and b are *comparable* if $a \leq b$ or $b \leq a$. Otherwise, a and b are *incomparable*, in notation, $a \parallel b$. A chain is, therefore, a poset in which there are no incomparable elements.

Let $\langle A; \leq \rangle$ be a poset and let B be a nonvoid subset of A. Then there is a natural partial order $\leq_B$ on B induced by $\leq$: for $a, b \in B$, $a \leq_B b$ iff $a \leq b$; we call $\langle B; \leq_B \rangle$ (or, simply, $\langle B; \leq \rangle$) a *subposet* of $\langle A; \leq \rangle$.

An *unordered poset* is one in which $a \parallel b$ for all $a \neq b$.

A *chain C in a poset P* is a nonvoid subset which, as a subposet, is a chain. An *antichain C in a poset P* is a nonvoid subset which, as a subposet, is unordered.

The *length*, $l(C)$, of a finite chain is $|C| - 1$. A poset P is said to be *of length n* (in formula, $l(P) = n$), where n is a natural number, iff there is a chain in P of length n and all chains in P are of length $\leq n$. A *poset P is of finite length* iff it is of length n for some natural number n.

The *width* of a poset P is n, where n is a natural number, iff there is an antichain in P of n elements and all antichains in P have $\leq n$ elements.

Next we define inf and sup in an arbitrary poset P (that is, $\langle P; \leq \rangle$), similarly as it is done for reals.

Let $H \subseteq P$ and $a \in P$. Then a is an *upper bound* of H iff $h \leq a$ for all $h \in H$. An upper bound a of H is the *least upper bound* of H or *supremum* of H iff, for any upper bound b of H, we have $a \leq b$. We shall write $a = \sup H$ or $a = \bigvee H$. (The notations $a = \text{l.u.b.}\ H$, $a = \Sigma H$ are also common in the literature.) This notation can be justified only if we show the uniqueness of the supremum. Indeed, if a_0 and a_1 are both suprema of H, then $a_0 \leq a_1$, since a_1 is an upper bound, and a_0 is a supremum. Similarly, $a_1 \leq a_0$; thus $a_0 = a_1$ by (P2).

Let $\varnothing$ be the empty set. Let a be the least upper bound of $\varnothing$, $a = \sup \varnothing$. For every $b \in P$, we have $h \leq b$ for all $h \in \varnothing$ (since there is no such h) and so every $b \in P$ is an upper bound of $\varnothing$. Thus $a \leq b$ for all $b \in P$. We conclude that $\sup \varnothing$ exists iff P has a smallest element. We call $\sup \varnothing$ the *zero* of P and denote it by 0.

The concepts of *lower bound* and *greatest lower bound* or *infimum* are similarly defined; the latter is denoted by $\inf H$ or $\bigwedge H$. (The notations g.l.b. H, ΠH are also used in the literature.) The uniqueness is proved as in the last but one paragraph. Observe that $\inf \varnothing$ exists iff P has a largest element. We call $\inf \varnothing$ the *unit* (or *identity*) and denote it by 1.

The adverb "similarly" in the paragraph introducing infima can be given a very concrete meaning. Let $\langle P; \leq \rangle$ be a poset. The notation $a \geq b$ (meaning $b \leq a$) can also be regarded as a definition of a binary relation on P. This binary relation $\geq$ satisfies (P1)–(P3); as an example, let us check (P2). If $a \geq b$ and $b \geq a$, then by the definition of $\geq$ we have $b \leq a$ and $a \leq b$; using (P2) for $\leq$ we conclude that $a = b$. (P1) and (P3) are equally trivial. Thus $\langle P; \geq \rangle$ is also a poset, called the *dual* of $\langle P; \leq \rangle$. Now, if Φ is a "statement" about posets, and if in Φ we replace all occurrences of $\leq$ by $\geq$, we get the *dual* of Φ.

Duality Principle. *If a statement Φ is true in all posets, then its dual is also true in all posets.*

This is true simply because Φ holds for $\langle P; \leq \rangle$ iff the dual of Φ holds for $\langle P; \geq \rangle$, which is also a poset.

As an example take for Φ the statement: "If sup H exists it is unique." We get as its dual: "If inf H exists it is unique." The dual of "$\langle P; \leq \rangle$ has a zero" is "$\langle P; \geq \rangle$ has a unit".

It is hard to imagine that anything as trivial as the Duality Principle could yield anything profound, and it does not; but it can save a lot of work.

A poset $\langle L; \leq \rangle$ is a *lattice* if $\sup \{a, b\}$ and $\inf \{a, b\}$ exist for all $a, b \in L$.

In other words, lattice theory singles out a special type of poset for detailed investigation. To make such a definition worthwhile, it must be shown that this class of posets is a very useful class, that there are many such posets in various branches of mathematics (analysis, topology, logic, algebra, geometry), and that a general study of these posets will lead to a better understanding of the behavior of the examples. This was done in the first edition of G. Birkhoff's *Lattice Theory* [1940]. As we go along we shall see many examples, most of them in the exercises. For a general survey of lattices in mathematics, see G. Birkhoff [1967] and H. H. Crapo and G.-C. Rota [1970].

We will take the usefulness of lattice theory for granted (this is a less touchy subject now than it formerly was) and hope that the reader will like it for its intrinsic beauty.

To make the definition of a lattice less arbitrary, we note that an equivalent definition is the following:

A poset $\langle L; \leq \rangle$ is a lattice iff $\sup H$ *and* $\inf H$ *exist for any finite nonvoid subset* H *of* L.

Proof. It is enough if we prove that the first definition implies the second. So let $\langle L; \leq \rangle$ satisfy the first definition and let $H \subseteq L$ be nonvoid and finite. If $H = \{a\}$, then $\sup H = \inf H = a$ follows from the reflexivity of $\leq$ and the definitions of sup and inf. Let $H = \{a, b, c\}$. To show that $\sup H$ exists, set $d = \sup \{a, b\}$, $e = \sup \{d, c\}$; we claim that $e = \sup H$. First of all, $a \leq d$, $b \leq d$, and $d \leq e$, $c \leq e$; therefore (by transitivity) $x \leq e$, for all $x \in H$. Secondly, if f is an upper bound of H, then $a \leq f$, $b \leq f$, and thus $d \leq f$; also $c \leq f$, so that $c, d \leq f$; therefore $e \leq f$, since $e = \sup \{d, c\}$. Thus e is the supremum of H.

If $H = \{a_0, \ldots, a_{n-1}\}$, $n \geq 1$, then

$$\sup \{\ldots \sup \{\sup \{a_0, a_1\}, a_2\} \ldots, a_{n-1}\}$$

is the supremum of H, by an inductive argument whose steps are analogous to those in the preceding paragraph.

By duality (in other words, by applying the Duality Principle), we conclude that $\inf H$ exists. ●

A lattice need not have zero or unit so $\inf \varnothing$ and $\sup \varnothing$ need not exist.

This simple proof can be varied to yield a large number of equally trivial statements about lattices and partially ordered sets in general. Some of these will be stated as exercises and used later. To make the use of the Duality Principle legitimate for lattices, note:

If $\langle P; \leq \rangle$ is a lattice, so is its dual $\langle P; \geq \rangle$.

Thus the Duality Principle applies to lattices.

We will use the notations

$$a \wedge b = \inf \{a, b\}$$

and

$$a \vee b = \sup \{a, b\}$$

and call $\wedge$ the *meet* and $\vee$ the *join*. In lattices, they are both *binary operations*, which means that they can be applied to a pair of elements a, b of L to yield again an element of L. Thus $\wedge$ is a map of L^2 into L and so is $\vee$, a remark that might fail to be very illuminating at this point.

The previous proof yields that

$$(\cdots ((a_0 \vee a_1) \vee a_2) \cdots) \vee a_{n-1} = \sup \{a_0, \ldots, a_{n-1}\}$$

and there is a similar formula for inf. Now observe that the right-hand side does not depend on the way the elements a_i are listed. Thus $\wedge$ and $\vee$ are idempotent, commutative, and associative—that is, they satisfy the following:

(L1) *Idempotency*: $a \wedge a = a$, $a \vee a = a$.
(L2) *Commutativity*: $a \wedge b = b \wedge a$, $a \vee b = b \vee a$.
(L3) *Associativity*: $(a \wedge b) \wedge c = a \wedge (b \wedge c)$,
$(a \vee b) \vee c = a \vee (b \vee c)$.

These properties of the operations are also called the *idempotent identities, commutative identities*, and *associative identities*, respectively.

As always in algebra, associativity makes it possible to write

$$a_0 \wedge a_1 \wedge \cdots \wedge a_{n-1}$$

without using parentheses (and the same for $\vee$).

There is another pair of rules that connect $\wedge$ and $\vee$. To derive them, note that if $a \leq b$, then $\inf \{a, b\} = a$; that is, $a \wedge b = a$, and conversely. Thus

$$a \leq b \text{ iff } a \wedge b = a.$$

By duality (and by interchanging a and b) we have

$$a \leq b \text{ iff } a \vee b = b.$$

Applying the "only if" part of the first rule to a and $a \vee b$, and that of the second rule to $a \wedge b$ and a, we get

(L4) *Absorption identities*: $a \wedge (a \vee b) = a$, $a \vee (a \wedge b) = a$.

Now we are faced with the crucial question: Do we know enough about $\wedge$ and $\vee$ so that lattices can be characterized purely in terms of properties of $\wedge$ and $\vee$?

Two comments are in order. It is obvious that $\leq$ can be characterized by $\wedge$ and $\vee$ (in fact, by either of them); therefore, obtaining such a characterization is only a matter of persistence. More importantly, why should we try to get such a characterization? To rephrase the question, why should we want to characterize $\langle L; \leq\rangle$ as $\langle L; \wedge, \vee\rangle$, which is an algebra—that is, a set equipped with operations (in this case, two binary operations)? Note that $\leq$ is a subset of L^2, whereas $\wedge$ and $\vee$ are maps from L^2 into L. The answer is simple: We want such a characterization be-because if we can treat lattices as algebras, then all the concepts and methods of universal algebra will become applicable. The usefulness of treating lattices as algebras will soon become clear.

An algebra $\langle L; \wedge, \vee\rangle$ is called a *lattice* iff: L is a nonvoid set, $\wedge$ and $\vee$ are binary operations on L, both $\wedge$ and $\vee$ are idempotent, commutative, and associative, and they satisfy the two absorption identities. The following theorem states that a lattice as an algebra and a lattice as a poset are "equivalent" concepts. (The word "equivalent" will not be defined.)

Theorem 1.

(i) *Let the poset* $\mathfrak{L} = \langle L; \leq\rangle$ *be a lattice. Set* $a \wedge b = \inf\{a, b\}$, $a \vee b = \sup\{a, b\}$. *Then the algebra* $\mathfrak{L}^a = \langle L; \wedge, \vee\rangle$ *is a lattice.*

(ii) *Let the algebra* $\mathfrak{L} = \langle L; \wedge, \vee\rangle$ *be a lattice. Set* $a \leq b$ *iff* $a \wedge b = a$. *Then* $\mathfrak{L}^p = \langle L; \leq\rangle$ *is a poset, and the poset* $\mathfrak{L}^p$ *is a lattice.*

(iii) *Let the poset* $\mathfrak{L} = \langle L; \leq\rangle$ *be a lattice. Then* $(\mathfrak{L}^a)^p = \mathfrak{L}$.

(iv) *Let the algebra* $\mathfrak{L} = \langle L; \wedge, \vee\rangle$ *be a lattice. Then* $(\mathfrak{L}^p)^a = \mathfrak{L}$.

Remark. (i) and (ii) describe how we pass from poset to algebra and back, whereas (iii) and (iv) state that going there and back takes us back to where we started.

Proof.

(i) This has already been proved. ◑

(ii) We set $a \leq b$ to mean $a \wedge b = a$. Now $\leq$ is reflexive since $\wedge$ is idempotent; $\leq$ is antisymmetric since $a \leq b$ and $b \leq a$ mean that $a \wedge b = a$ and $b \wedge a = b$, which, by the commutativity of $\wedge$, imply that $a = a \wedge b = b \wedge a = b$; $\leq$ is transitive, since if $a \leq b$ and $b \leq c$, then $a = a \wedge b$ and $b = b \wedge c$, and so

$$a = a \wedge b = a \wedge (b \wedge c) = (\wedge \text{ is associative}) = (a \wedge b) \wedge c = a \wedge c,$$

that is, $a \leq c$. Thus $\langle L; \leq\rangle$ is a poset. To prove that $\langle L; \leq\rangle$ is a lattice we will verify that $a \wedge b = \inf\{a, b\}$ and $a \vee b = \sup\{a, b\}$. (This is not a definition.) Indeed, $a \wedge b \leq a$, since

$$(a \wedge b) \wedge a = a \wedge (b \wedge a) = a \wedge (a \wedge b) = (a \wedge a) \wedge b = a \wedge b,$$

using associativity, commutativity, and idempotency of $\wedge$; similarly, $a \wedge b \leq b$. Now if $c \leq a$ and $c \leq b$, that is, $c \wedge a = c$ and $c \wedge b = c$, then $c \wedge (a \wedge b) = (c \wedge a) \wedge b = c \wedge b = c$; thus $a \wedge b = \inf\{a, b\}$. Finally, $a \leq a \vee b$ and $b \leq a \vee b$, because $a = a \wedge (a \vee b)$ and $b = b \wedge (a \vee b)$ by the first absorption identity; if $a \leq c$ and $b \leq c$, that is, $a = a \wedge c$ and $b = b \wedge c$, then $a \vee c = (a \wedge c) \vee c = c$, and $b \vee c = c$ by the second absorption identity. Thus

$$(a \vee b) \wedge c = (a \vee b) \wedge (a \vee c) = (a \vee b) \wedge (a \vee (b \vee c))$$
$$= (a \vee b) \wedge ((a \vee b) \vee c) = a \vee b,$$

that is, $a \vee b \leq c$, completing the proof of $a \vee b = \sup \{a, b\}$. ◑

(iii) It is enough to observe that the partial orderings of $\mathfrak{L}$ and $(\mathfrak{L}^a)^p$ are identical to get (iii). ◑

(iv) The proof of (iv) is similar to the proof of (iii). ●

The proof of Theorem 1, and even the statement of Theorem 1, are subject to criticism. To begin with, in the definition of a lattice as an algebra, idempotency is redundant. The last step of the proof of (ii) can be made neater by first proving that

$$a = a \wedge b \text{ iff } b = a \vee b.$$

Theorem 1 should be preceded by a similar theorem for "semilattices." All these questions will be dealt with in the exercises that follow this section.

Finally, note that for lattices as algebras, the Duality Principle takes on the following very simple form.

Let Φ be a statement about lattices expressed in terms of $\wedge$ and $\vee$. The dual of Φ is the statement we get from Φ by interchanging $\wedge$ and $\vee$. If Φ is true for all lattices, then the dual of Φ is also true for all lattices.

To prove this we have only to observe that if $\mathfrak{L} = \langle L; \wedge, \vee \rangle$, then the dual of $\mathfrak{L}^p$ is $(\langle L; \vee, \wedge \rangle)^p$.

Most of the time Φ involves $\leq$ and maybe 0 and 1 in addition to $\vee$ and $\wedge$. When dualizing such a Φ we interchange $\wedge$ and $\vee$, replace $\leq$ by $\geq$, and interchange 0 and 1.

Exercises

Posets

1. Define $x < y$ to mean $x \leq y$ and $x \neq y$. Prove that, in a partially ordered set, $x < x$ for no x, and that $x < y$, $y < z$ imply that $x < z$.
2. Let the binary relation $<$ satisfy the conditions of Exercise 1. Define $x \leq y$ to mean $x < y$ or $x = y$. Then show that $\leq$ is a partial ordering.
3. Prove the following extension of antisymmetry: If $x_0 \leq x_1 \leq \cdots \leq x_{n-1} \leq x_0$, then $x_0 = x_1 = \cdots = x_{n-1}$.
4. Enumerate all partial orderings on a five-element set.
5. Let $\leq$ be a partial ordering on A and let B be a subset of A. For $a, b \in B$, set $a \leq_B b$ iff $a \leq b$. Prove that $\leq_B$ is a partial ordering on B.
6. Let A be a set and let P be the set of all partial orderings on A. For $\varrho, \sigma \in P$, set $\varrho \leq \sigma$ iff $a \varrho b$ implies that $a \sigma b$ $(a, b \in A)$. Prove that $\langle P; \leq \rangle$ is a poset.
7. Find an example of a poset in which $\inf \varnothing$ does not exist.
8. Prove that if $\inf H$ exists for all nonempty subsets H of a poset P, then $\sup \varnothing$ also exists in P.

Lattices as Posets

9. Prove that the following are examples of posets:
   (i) Let $A = P(X)$, the set of all subsets of a set X; let $X_0 \leq X_1$ mean $X_0 \subseteq X_1$ (set inclusion).
   (ii) Let A be the set of all real valued functions defined on X; for $f, g \in A$, set $f \leq g$ iff $f(x) \leq g(x)$ for all $x \in X$.
   (iii) Let A be the set of all continuous concave real valued functions defined on the real interval; define $f \leq g$ as in (ii).
   (iv) Let A be the set of all open sets of a topological space; define $\leq$ as in (i).
   (v) Let A be the set of all human beings; $a < b$ means that a is a descendant of b.
10. Which of the examples in Exercise 9 are lattices? For those that are lattices, compute $a \wedge b$ and $a \vee b$.
11. Show that every chain is a lattice.
12. Let A be the set of all subgroups (normal subgroups) of a group G; for $X, Y \in A$, set $X \leq Y$ to mean $X \subseteq Y$. Prove that $\langle A; \leq \rangle$ is a lattice; compute $X \wedge Y$ and $X \vee Y$.
13. Let $\langle P; \leq \rangle$ be a poset in which $\inf H$ exists for *all* $H \subseteq P$. Show that $\langle P; \leq \rangle$ is a lattice. (Hint: For $a, b \in P$, let H be the set of all upper bounds of $\{a, b\}$. Prove that $\sup \{a, b\} = \inf H$.) Relate this to Exercise 12.

Semilattices as Posets

14. A poset is a *join-semilattice* (dually, *meet-semilattice*) iff $\sup \{a, b\}$ (dually, $\inf \{a, b\}$) exists for any two elements a and b. Prove that the dual of a join-semilattice is a meet-semilattice, and conversely.
15. Let A be the set of finitely generated subgroups of a group G, partially ordered under set inclusion (as in Exercise 12). Prove that $\langle A; \subseteq \rangle$ is a join-semilattice, but not necessarily a lattice.
16. Let C be the set of all continuous strictly convex real valued functions defined on the real interval $[0, 1]$. For $f, g \in C$, set $f \leq g$ iff $f(x) \leq g(x)$ for all $x \in [0, 1]$. Prove that $\langle C; \leq \rangle$ is a meet-semilattice, but not a join-semilattice.
17. Show that the poset $\langle P; \leq \rangle$ is a lattice iff it is both a join- and meet-semilattice.

Semilattices as Algebras

18. Let $\langle A; \circ \rangle$ be an algebra with one binary operation $\circ$. The algebra $\langle A; \circ \rangle$ is a *semilattice* iff $\circ$ is idempotent, commutative, and associative. Let $\langle P; \leq \rangle$ be a join-semilattice. Show that the algebra $\langle P; \vee \rangle$ is a semilattice where $a \vee b = \sup \{a, b\}$. State the analogous result for meet-semilattices.
19. Let the algebra $\langle A; \circ \rangle$ be a semilattice. Define the binary relations $\leq_\wedge$ and $\leq_\vee$ on A as follows: $a \leq_\wedge b$ iff $a = a \circ b$; $a \leq_\vee b$ iff $b = a \circ b$. Prove that $\langle A; \leq_\wedge \rangle$ is a poset, as a poset it is a meet-semilattice, and $a \wedge b = a \circ b$; that $\langle A; \leq_\vee \rangle$ is a poset, as a poset it is a join-semilattice, and $a \vee b = a \circ b$; and that the dual of $\langle A; \leq_\wedge \rangle$ is $\langle A; \leq_\vee \rangle$.

Theorem 1 for Semilattices

20. Prove the following statements:
   (i) Let the poset $\mathfrak{A} = \langle A; \leq \rangle$ be a join-semilattice. Set $a \vee b = \sup \{a, b\}$. Then the algebra $\mathfrak{A}^a = \langle A; \vee \rangle$ is a semilattice.
   (ii) Let the algebra $\mathfrak{A} = \langle A; \circ \rangle$ be a semilattice. Set $a \leq b$ iff $a \circ b = b$. Then $\mathfrak{A}^p = \langle A; \leq \rangle$ is a poset, and the poset $\mathfrak{A}^p$ is a join-semilattice.

(iii) Let the poset $\mathfrak{A} = \langle A; \leq \rangle$ be a join-semilattice. Then $(\mathfrak{A}^a)^p = \mathfrak{A}$.
(iv) Let the algebra $\mathfrak{A} = \langle A; \circ \rangle$ be a semilattice. Then $(\mathfrak{A}^p)^a = \mathfrak{A}$.

21. Formulate and prove Theorem 1 for meet-semilattices.

Lattices as Algebras

22. Prove that the absorption identities imply the idempotency of $\wedge$ and $\vee$. (Hint: simplify $a\vee(a\wedge(a\vee a))$ in two ways to yield $a = a\vee a$.)
23. Let the algebra $\langle A; \wedge, \vee \rangle$ be a lattice. Define $a \leq_\wedge b$ iff $a\wedge b = a$; $a \leq_\vee b$ iff $a\vee b = b$. Prove that $a \leq_\wedge b$ iff $a \leq_\vee b$.
24. Prove that the algebra $\langle A; \wedge, \vee \rangle$ is a lattice iff $\langle A; \wedge \rangle$ and $\langle A; \vee \rangle$ are semilattices and $a = a\wedge b$ is equivalent to $b = a\vee b$. Verify that if $\langle A; \wedge, \vee_1 \rangle$ and $\langle A; \wedge, \vee_2 \rangle$ are both lattices, then $\vee_1$ is the same as $\vee_2$.
25. Are the three identities defining semilattices independent (that is, none follows from the others)?
*26. Prove that an algebra $\langle A; \wedge, \vee \rangle$ is a lattice iff it satisfies

$$a = (b\wedge a)\vee a$$

and

$$(((a\wedge b)\wedge c)\vee d)\vee e = (((b\wedge c)\wedge a)\vee e)\vee((f\vee d)\wedge d).$$

(See J. A. Kalman [1968]. The first definition of lattices by two identities was found by R. Padmanabhan in 1967, *Notices Amer. Math. Soc.* 14, No. 67T-468, and published in full in R. Padmanabhan [1969]. J. A. Kalman's two identities are slightly improved versions of those of R. Padmanabhan. R. N. McKenzie [1970] proved the existence of a single identity characterizing lattices.)

Miscellany

27. Let ϱ be a binary relation on the set A. The *transitive extension of* ϱ is a binary relation $\hat{\varrho}$ defined by the following rule: $a\hat{\varrho}b$ iff there exists a sequence $a = x_0, x_1, \ldots, x_n = b$ with $x_i\varrho x_{i+1}$ for $i = 0, \ldots, n-1$. Show that for a reflexive relation ϱ, the transitive extension $\hat{\varrho}$ is a partial ordering relation iff $\hat{\varrho}$ is antisymmetric. Express this condition in terms of ϱ.
28. Let ϱ be a reflexive and transitive binary relation (*quasi-ordering relation*) on the nonvoid set A. Call $B \subseteq A$ a *block* iff $B \neq \varnothing$, $a\varrho b$ and $b\varrho a$ for any $a, b \in B$, and for $a \in B$, $b \in A$, $a\varrho b$ and $b\varrho a$ imply that $b \in B$. Let P be the set of all blocks and set $B_1 \leq B_2$ ($B_1, B_2 \in P$) iff $b_1\varrho b_2$ for some (thus for all) $b_1 \in B_1$ and $b_2 \in B_2$. Prove that $\langle P; \leq \rangle$ is a poset.
29. Let A be a set of sets. Let $a\varrho b$ mean that there is a one-to-one map from a into b. What is $\langle P; \leq \rangle$? (Notation is that of Exercise 28.)
30. Let the binary operation $\circ$ on the set A be associative. Give a rigorous proof of the statement that any meaningful bracketing of $a_0 \circ \cdots \circ a_{n-1}$ will yield the same element.
31. Suppose that in a poset $b\vee c$, $a\vee(b\vee c)$, and $a\vee b$ exist. Prove that $(a\vee b)\vee c$ exists and that $a\vee(b\vee c) = (a\vee b)\vee c$.
32. Prove that if $a\wedge b$ exists in a poset, so does $a\vee(a\wedge b)$.
33. Let H and K be subsets of a poset. Suppose that $\sup H$, $\sup K$, and $\sup (H \cup K)$ exist. ($H \cup K$ is the set union of H and K.) Under these conditions verify that $(\sup H)\vee(\sup K)$ exists and equals $\sup (H \cup K)$.

34. In a poset P define the *comparability relation* γ: For $a, b \in P$, $a\gamma b$ iff $a \leq b$ or $b \leq a$. Take a sequence $a_1, \ldots, a_k$ of elements of P satisfying the following conditions:

(i) $a_i \neq a_{i+1}$, $a_i \gamma a_{i+1}$ for $i = 1, \ldots, k-1$; $a_k \neq a_1$, $a_k \gamma a_1$.

(ii) For no elements $a, b \in P$, and $i, j < k$, $i \neq j$ is $a = a_i = a_j$, $b = a_{i+1} = a_{j+1}$, or $a = a_i = a_k$, $b = a_{i+1} = a_1$.

(iii) For no $1 \leq i \leq k-2$ is $a_i \gamma a_{i+2}$, and neither $a_{k-1} \gamma a_1$ nor $a_k \gamma a_2$ holds.

Prove that k is even.

***35.** Prove that a binary relation γ on a set A is the comparability relation of some poset $\langle A; \leq \rangle$ iff γ satisfies the condition of Exercise 34 (A. Ghouilà-Houri [1962]; see also P. C. Gilmore and A. J. Hoffman [1964] and M. Aigner [1969]).

2. How to Describe Lattices

To illustrate results and to refute conjectures we shall have to describe a large number of examples of lattices. This can be done by basing the examples on known mathematical structures, as illustrated in the exercises of Section 1. In this section we list a few other methods.

A finite lattice can always be described by a *meet table* and a *join table*. For example: let $L = \{0, a, b, 1\}$.

$\wedge$	0	a	b	1
0	0	0	0	0
a	0	a	0	a
b	0	0	b	b
1	0	a	b	1

$\vee$	0	a	b	1
0	0	a	b	1
a	a	a	1	1
b	b	1	b	1
1	1	1	1	1

We see that most of the information provided by the tables is redundant. Since both operations are commutative, the tables are symmetric with respect to the diagonal. Furthermore, $x \wedge x = x$, $x \vee x = x$; thus the diagonals themselves do not provide new information. Therefore, the two tables can be condensed into one.

$\wedge$	0	a	b	1
$\vee$				
0		0	0	0
a	a		0	a
b	b	1		b
1	1	1	1	

It should be emphasized again that the part above the diagonal determines the part below the diagonal since either determines the partial ordering. To show that this table defines a lattice, we have only to check the associative identities and the absorption identities.

An alternative way is to describe the partial ordering, that is, all pairs $\langle x, y \rangle$ with $x \leq y$. In the preceding example we get $\leq = \{\langle 0, 0\rangle, \langle 0, a\rangle, \langle 0, b\rangle, \langle 0, 1\rangle, \langle a, a\rangle, \langle a, 1\rangle, \langle b, b\rangle, \langle b, 1\rangle, \langle 1, 1\rangle\}$.

Obviously, all pairs of the form $\langle x, x\rangle$ can be omitted from the list, since we know that $x \leq x$. Also, if $x \leq y$ and $y \leq z$, then $x \leq z$. For instance, when we know that $0 \leq a$ and $a \leq 1$, we do not have to be told that $0 \leq 1$. To make this idea more precise, let us say that in the poset $\langle P; \leq\rangle$, *a covers b* or *b is covered by a*, in notation, $a \succ b$ or $b \prec a$ iff $a > b$ and for *no* x, $a > x > b$. The covering relation of the preceding example is simply:

$$\prec = \{\langle 0, a\rangle, \langle 0, b\rangle, \langle a, 1\rangle, \langle b, 1\rangle\}.$$

Does the covering relation determine the partial ordering? The following lemma shows that in the finite case it does.

Lemma 1. *Let $\langle P; \leq\rangle$ be a finite poset. Then $a \leq b$ iff $a = b$ or there exists a finite sequence of elements $x_0, \ldots, x_{n-1}$ such that $x_0 = a$, $x_{n-1} = b$, and $x_i \prec x_{i+1}$, for $0 \leq i < n-1$.*

Proof. If there is such a finite sequence, then $a = x_0 \leq x_1 \leq \cdots \leq x_{n-1} = b$, and a trivial induction on n yields $a \leq b$. Thus it suffices to prove that if $a < b$, then there is such a sequence. Fix $a, b \in P$, $a < b$, and take all subsets H of P such that H is a chain in P, a is the smallest element of H, and b is the largest element of H. There are such subsets: $\{a, b\}$, for example. Choose such an H with the largest possible number of elements, say, with m elements. (P is finite.) Then $H = \{x_0, \ldots, x_{m-1}\}$, and we can assume that $x_0 < x_1 < \cdots < x_{m-1}$. We claim that in $\langle P; \leq\rangle$ we have $a = x_0 \prec x_1 \prec \cdots \prec x_{m-1}$. Indeed, $x_i < x_{i+1}$ by assumption. Thus if $x_i \prec x_{i+1}$ does not hold, then $x_i < x < x_{i+1}$ for some $x \in P$, and $H \cup \{x\}$ will be a chain of $m+1$ elements between a and b, contrary to the maximality of the number of elements of H. ●

The *diagram* of a poset $\langle P; \leq\rangle$ represents the elements with small circles ∘; the circles representing two elements x, y are connected by a straight line iff one covers the other; if x covers y, then the circle representing x is higher than the circle representing y. The diagram of the lattice discussed previously is shown in Figure 1.

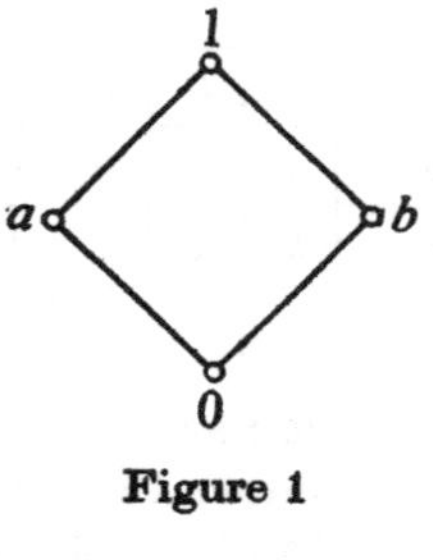

Figure 1

Sometimes the "diagram" of an infinite poset is drawn. Such diagrams are always accompanied by explanations in the text. Note that in a diagram the intersection of two lines does not indicate an element. A diagram is *planar* if no two lines intersect. A diagram is *optimal* if the number of pairs of intersecting lines is minimal. Figures 1,

2, and 3 show planar diagrams; Figure 4 is an optimal but not a planar diagram; Figure 6 is not optimal. As a rule, optimal diagrams are the most practical to use.

The methods we will use will be combinations of the previous ones. The lattice $\mathfrak{N}_5$ of Figure 2 has five elements: o, a, b, c, i, and $b < a$, $c \vee b = i$, $a \wedge c = o$. This description

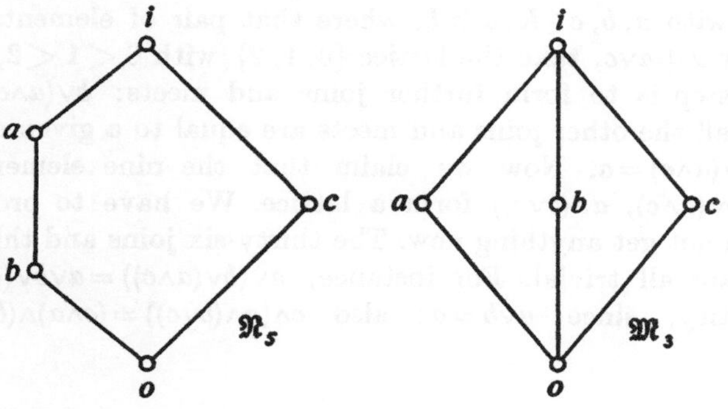

Figure 2

is complete—that is, all the relations follow from the ones given. $\mathfrak{M}_3$ has five elements: o, a, b, c, i and $a \wedge b = a \wedge c = b \wedge c = o$, $a \vee b = a \vee c = b \vee c = i$.

In contrast, we can start with some elements (say, a, b, c), with some relations (say, $b < a$), and ask for the "most general" lattice that can be formed, *without* specifying the elements to be used. (The exact meaning of "most general" will be given in Section 5.) In this case we continue to form meets and joins until we get a lattice. A meet (or join) formed is identified with an element that we already have only if this identification is forced by the lattice axioms or by the given relations. The lattice we get from a, b, c and satisfying $b < a$ is shown in Figure 3.

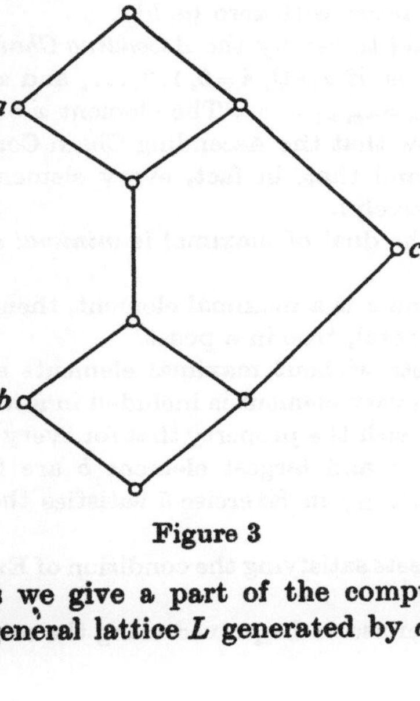

Figure 3

To illustrate these ideas we give a part of the computation that goes into the construction of the most general lattice L generated by a, b, c with $b < a$. We start

by constructing joins and meets $a\vee c$, $b\vee c$, $a\wedge c$, $b\wedge c$; note that $a\vee b\vee c=(a\vee b)\vee c$ $=a\vee c$ since $a\vee b=a$; similarly, $a\wedge b\wedge c=b\wedge c$. Next we have to show that the seven elements (a, b, c, $a\vee c$, $b\vee c$, $a\wedge c$, $b\wedge c$) that we already have are all distinct. Remember that two were equal if such equality would follow from the relation ($b<a$) and the lattice axioms. Therefore, to show a pair of them distinct, it is enough to find a lattice K with $a, b, c\in K$, $a>b$, where that pair of elements is distinct. For instance, to show $a\neq a\vee c$, take the lattice $\{0, 1, 2\}$ with $0<1<2$, and $b=0$, $a=1$, $c=2$. The next step is to form further joins and meets: $b\vee(a\wedge c)$, $a\wedge(b\vee c)$. It is easy to see that all the other joins and meets are equal to a given one—for example, $b\wedge(a\wedge c)=b\wedge c$, $a\vee(a\wedge c)=a$. Now we claim that the nine elements (a, b, c, $a\vee c$, $b\vee c$, $a\wedge c$, $b\wedge c$, $b\vee(a\wedge c)$, $a\wedge(b\vee c)$) form a lattice. We have to prove that by joins and meets we cannot get anything new. The thirty-six joins and thirty-six meets we have to check are all trivial. For instance, $a\vee(b\vee(a\wedge c))=a\vee b\vee(a\wedge c)=a$, by the absorption identity, since $a\vee b=a$; also $c\wedge(a\wedge(b\vee c))=(c\wedge a)\wedge(b\vee c)=a\wedge c$, since $c\wedge a\leq b\vee c$.

Exercises

1. Give the join and meet table of the lattice in Exercise 1.9(a) for one-, two-, and three-element sets X.
2. Give the set $\leq$ for the lattices in Exercise 1. Which is simpler: the meet and join table or $\leq$?
3. Describe a practical method of checking associativity in a join (meet) table.
4. Relate Lemma 1 to Exercise 1.27.
5. Let $\langle P;\ \leq\rangle$ be a poset, $a, b\in P$, $a<b$, and let C denote the set of all chains H in P with smallest element a, largest element b. Let $H_0\leq H_1$ mean $H_0\subseteq H_1$ for $H_0, H_1\in C$. Show that $\langle C;\ \leq\rangle$ is a poset with zero $\{a, b\}$.
6. The poset $\langle Q;\ \leq\rangle$ is said to satisfy the *Ascending Chain Condition* iff all increasing chains terminate; that is, if $x_i\in Q$, $i=0, 1, 2, \ldots$, and $x_0\leq x_1\leq\cdots\leq x_n\leq\cdots$, then for some m we have $x_m=x_{m+1}=\cdots$. The element x of Q is *maximal* if $x\leq y$ ($y\in Q$) implies that $x=y$. Show that the Ascending Chain Condition implies the existence of maximal elements and that, in fact, every element is included in a maximal element (but not conversely).
7. Dualize Exercise 6. (The dual of maximal is *minimal* and the dual of ascending is *descending*.)
8. If $\langle Q;\ \leq\rangle$ is a lattice and x is a maximal element, then x is the unit. Show that this statement is not, in general, true in a poset.
9. Give examples of posets without maximal elements and of posets with maximal elements in which not every element is included in a maximal element.
10. Let $\langle P;\ \leq\rangle$ be a poset with the property that for every $a, b\in P$, $a<b$, all chains in P with smallest element a and largest element b are finite. Show that the poset $\langle C;\ \leq\rangle$ formed from $\langle P;\ \leq\rangle$ in Exercise 5 satisfies the Ascending Chain Condition (see Exercise 6).
11. Extend Lemma 1 to posets satisfying the condition of Exercise 10 (combine Exercises 6 and 10).
12. Could the result of Exercise 11 be proved using the reasoning of Lemma 1?

13. **Is the result of Exercise 11 the best possible? (No.)**
14. **Describe a method of finding the meet and join table of a lattice given by a diagram.**
15. **Are the posets of Figures 4 and 5 lattices?**

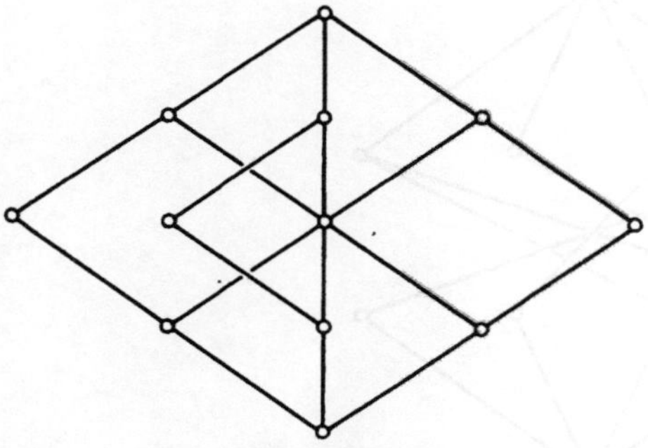

Figure 4

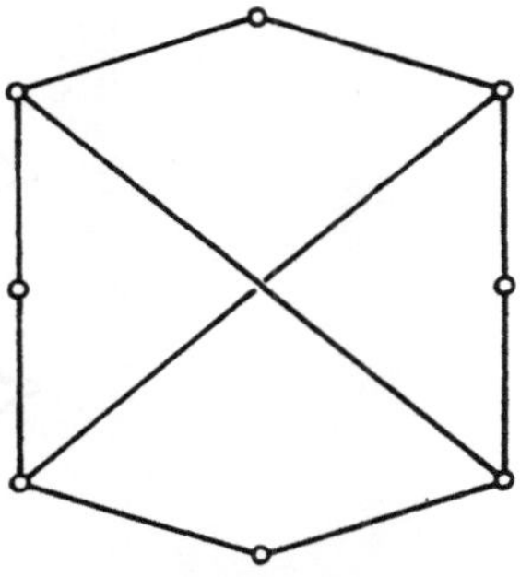

Figure 5

16. **Show that Figures 6 and 7 represent the same lattice.**

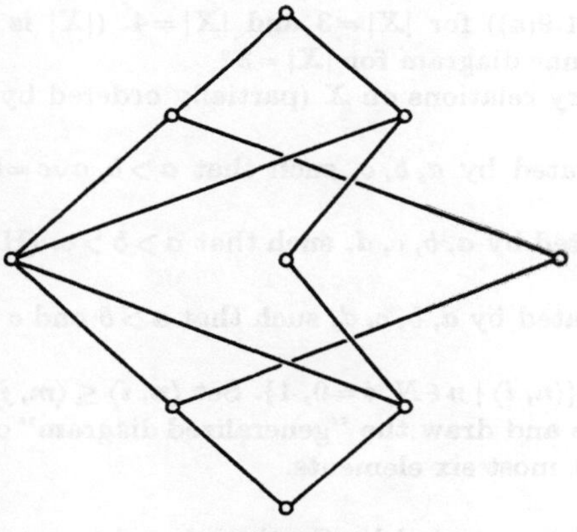

Figure 6

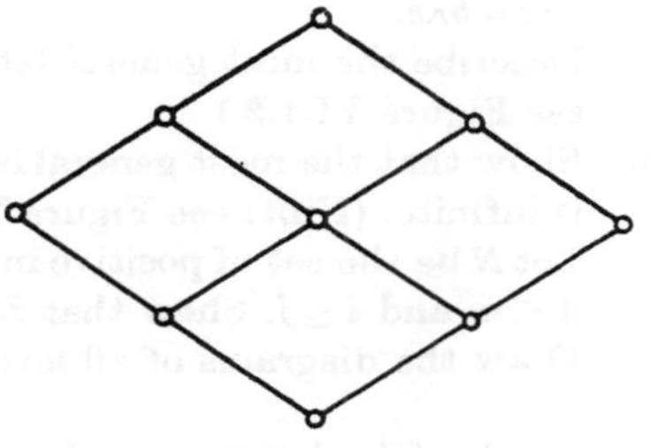

Figure 7

17. **Simplify Figure 8.**

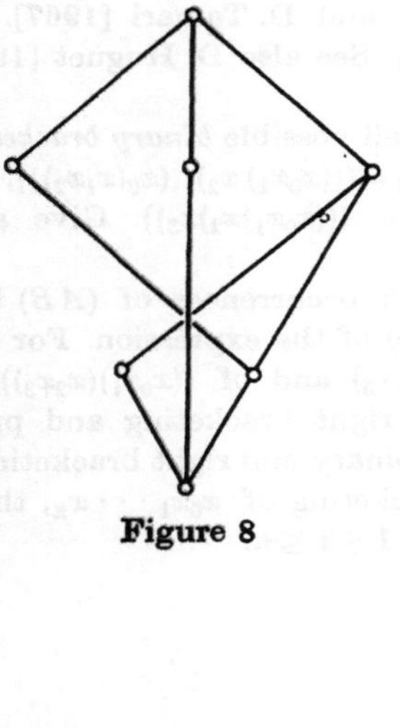

Figure 8

18. Simplify Figure 9. What is the number of pairs of intersecting lines in an optimal diagram? (Zero.)

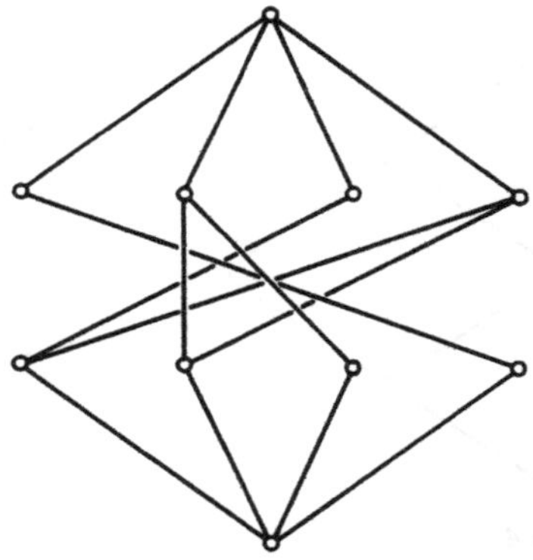

Figure 9

19. Draw the diagrams of $P(X)$ (Exercise 1.9(a)) for $|X| = 3$ and $|X| = 4$. ($|X|$ is the cardinality of X.) Does $P(X)$ have a planar diagram for $|X| = 3$?
20. Draw the diagrams of the lattice of binary relations on X (partially ordered by set inclusion) for $|X| \leq 4$.
21. Describe the most general lattice generated by a, b, c, such that $a > b$, $a \vee c = b \vee c$, $a \wedge c = b \wedge c$.
*22. Describe the most general lattice generated by a, b, c, d, such that $a > b > c$. (Hint: see Figure VI.1.2.)
*23. Show that the most general lattice generated by a, b, c, d, such that $a > b$ and $c > d$, is infinite. (Hint: see Figure VI.1.6.)
24. Let N be the set of positive integers, $L = \{\langle n, i \rangle \mid n \in N, i = 0, 1\}$. Set $\langle n, i \rangle \leq \langle m, j \rangle$ iff $n \leq m$ and $i \leq j$. Show that L is a lattice and draw the "generalized diagram" of L.
25. Draw the diagrams of all lattices with at most six elements.

To dispel a false impression that may have been created by Sections 1 and 2, namely, that the proof that a poset is a lattice is always trivial, we present Exercises 26–36 showing that the poset T_n defined in Exercise 34 is a lattice. This is a result of D. Tamari [1951], first published in H. Friedman and D. Tamari [1967]. The present proof is based on S. Huang and D. Tamari [1972]. See also D. Huguet [1975].

26. Let T_n denote the set of all possible *binary bracketings* of $x_0 x_1 \cdots x_n$; for instance, $T_0 = \{x_0\}$, $T_1 = \{(x_0 x_1)\}$, $T_2 = \{((x_0 x_1) x_2), (x_0(x_1 x_2))\}$, $T_3 = \{(x_0(x_1(x_2 x_3))), (x_0((x_1 x_2) x_3)), ((x_0 x_1)(x_2 x_3)), ((x_0(x_1 x_2))x_3), (((x_0 x_1)x_2)x_3)\}$. Give a formal (inductive) definition of T_n.
27. Replacing consistently all occurrences of (AB) by $A(B)$ in a binary bracketing, we get the *right bracketing* of the expression. For instance, the right bracketing of $((x_0(x_1 x_2))x_3)$ is $x_0(x_1(x_2))(x_3)$ and of $((x_0 x_1)(x_2 x_3))$ is $x_0(x_1)(x_2(x_3))$. Give a formal (inductive) definition of right bracketing and prove that there is a one-to-one correspondence between binary and right bracketings.
28. Show that in a right bracketing of $x_0 x_1 \cdots x_n$, there is one and only one opening bracket preceding any x_i, $1 \leq i \leq n$.

29. Associate with a right bracketing of $x_0 x_1 \cdots x_n$ a *bracketing function* $E: \{1, \ldots, n\} \to \{1, \ldots, n\}$ defined as follows: For $1 \leq i \leq n$ there is, by Exercise 28, an opening bracket before x_i; let this bracket close following x_j; set $E(i) = j$. Show that E has the following properties: (i) $i \leq E(i)$ for $1 \leq i \leq n$; (ii) $i \leq j \leq E(i)$ imply that $E(j) \leq E(i)$ for $1 \leq i \leq j \leq n$.

***30.** Show that (i) and (ii) of Exercise 29 characterize bracketing functions.

31. Let E_n denote the set of all bracketing functions defined on $\{1, \ldots, n\}$. For $E, F \in E_n$, set $E \leq F$ iff $E(i) \leq F(i)$ for all i, $1 \leq i \leq n$. Show that $\leq$ is a partial ordering, E_n as a poset is a lattice, and $(E \wedge F)(i) = \inf \{E(i), F(i)\}$.

32. The *semiassociative identity* applied at the place i is a map $\sigma_i: T_n \to T_n$ defined as follows: If $E = \cdots (A(BC)) \cdots$, where the first variable in B and C is x_i and x_j, respectively, then $E\sigma_i = \cdots ((AB)C) \cdots$; if E is not of such form, then $E\sigma_i = E$. Let X and Y denote the bracketing functions associated with E and $E\sigma_i$, respectively. Show that $X(k) = Y(k)$ for $k \neq i$, $i \leq k \leq n$, and $Y(i) = j - 1$. Conclude that $X > Y$.

33. Show the converse of Exercise 32.

34. For $E, F \in T_n$ define $E < F$ to mean the existence of a sequence $E = X_0, X_1, \ldots, X_k = F$, $X_i \in T_n$, $0 \leq i \leq k$ such that X_{i+1} can be gotten from X_i by some application of the semiassociative law for $0 \leq i < k$. Let $E \leq F$ mean $E = F$ or $E < F$. Show that $\leq$ is a partial ordering. Verify that Figure 10 is the diagram of T_3 and T_4. Is the diagram of T_4 optimal? (No.)

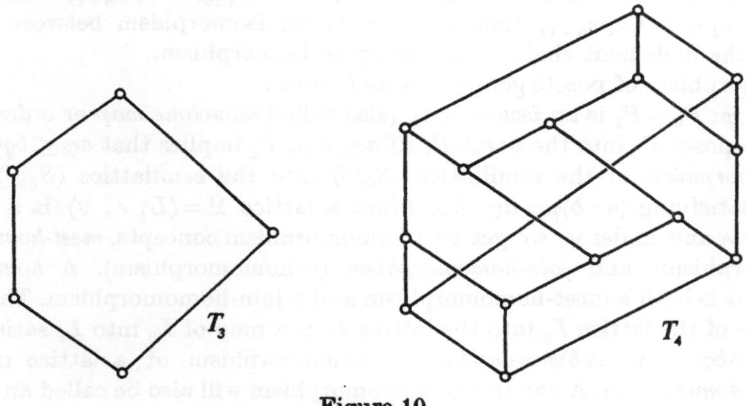

Figure 10

***35.** Let $X, Y \in E_n$ and $X \succ Y$. Let E and F be the binary bracketings associated with X and Y, respectively. Show that $F = E\sigma_i$ for some i.

36. Show that T_n is a lattice for each $n \geq 0$. (In fact, $T_n \cong E_n$.)

3. Some Algebraic Concepts

The purpose of any algebraic theory is the investigation of algebras up to isomorphism. We can introduce two concepts of isomorphism for lattices.

The lattices $\mathfrak{L}_0 = \langle L_0; \leq \rangle$ and $\mathfrak{L}_1 = \langle L_1; \leq \rangle$ are *isomorphic* (in symbol, $\mathfrak{L}_0 \cong \mathfrak{L}_1$), and the map $\varphi: L_0 \to L_1$ is an *isomorphism* iff φ is one-to-one and onto, and

$$a \leq b \text{ in } \mathfrak{L}_0 \text{ iff } a\varphi \leq b\varphi \text{ in } \mathfrak{L}_1.$$

The lattices $\mathfrak{L}_0 = \langle L_0;\ \wedge,\ \vee\rangle$ and $\mathfrak{L}_1 = \langle L_1;\ \wedge,\ \vee\rangle$ are *isomorphic* (in symbol, $\mathfrak{L}_0 \cong \mathfrak{L}_1$), and the map $\varphi\colon L_0 \to L_1$ is an *isomorphism* iff φ is one-to-one and onto, and

$$(a \wedge b)\varphi = a\varphi \wedge b\varphi,$$
$$(a \vee b)\varphi = a\varphi \vee b\varphi.$$

An isomorphism of a lattice with itself is called an *automorphism.*

It is easy to see that the two isomorphism concepts coincide under the equivalence of Theorem 1.1. However, when we generalize these to homomorphism concepts, we get various new nonequivalent notions. In order to avoid confusion, each will be given a different name.

From now on we will abandon the precise notation $\langle L;\ \wedge,\ \vee\rangle$ and $\langle L;\ \leq\rangle$ for lattices and posets; we will simply write italic capitals, indicating the underlying sets, unless for some reason we want to be more exact.

Note that the first definition of isomorphism can be applied to any two posets L_0, L_1, thus yielding an isomorphism concept for arbitrary posets. Having this concept of isomorphism, we can restate the content of Lemma 2.1: The diagram of a finite poset determines the poset up to isomorphism.

Let $\mathfrak{C}_n$ denote the set $\{0, \ldots, n-1\}$ ordered by $0 < 1 < 2 < \cdots < n-1$. Then $\mathfrak{C}_n$ is an n-element chain. Observe that $l(\mathfrak{C}_n) = n-1$. If $C = \{x_0, \ldots, x_{n-1}\}$ is an n-element chain, $x_0 < x_1 < \cdots < x_{n-1}$, then $\varphi\colon i \to x_i$ is an isomorphism between $\mathfrak{C}_n$ and C. Therefore, the n-element chain is unique up to isomorphism.

The isomorphism of posets generalizes as follows:

The map $\varphi\colon P_0 \to P_1$ is an *isotone map* (also called *monotone map* or *order-preserving map*) of the poset P_0 into the poset P_1 iff $a \leq b$ in P_0 implies that $a\varphi \leq b\varphi$ in P_1.

A *homomorphism* of the semilattice $\langle S_0;\ \cdot\rangle$ into the semilattice $\langle S_1;\ \cdot\rangle$ is a map $\varphi\colon S_0 \to S_1$ satisfying $(a \cdot b)\varphi = a\varphi \cdot b\varphi$. Since a lattice $\mathfrak{L} = \langle L;\ \wedge,\ \vee\rangle$ is a semilattice both under $\wedge$ and under $\vee$, we get two homomorphism concepts, *meet-homomorphism* ($\wedge$-homomorphism) and *join-homomorphism* ($\vee$-homomorphism). A *homomorphism* is a map that is both a meet-homomorphism and a join-homomorphism. Thus a homomorphism φ of the lattice L_0 into the lattice L_1 is a map of L_0 into L_1 satisfying both $(a \wedge b)\varphi = a\varphi \wedge b\varphi$ and $(a \vee b)\varphi = a\varphi \vee b\varphi$. A homomorphism of a lattice into itself is called an *endomorphism.* A one-to-one homomorphism will also be called an *embedding.* (The list of homomorphism concepts will be further extended in Section 6.)

Note that meet-homomorphisms, join-homomorphisms, and (lattice) homomorphisms are all isotone. Let us prove this statement for meet-homomorphisms. If $\varphi\colon L_0 \to L_1$, $(a \wedge b)\varphi = a\varphi \wedge b\varphi$ for all $a, b \in L_0$, and if $x, y \in L_0$, $x \leq y$ in L_0, then $x = x \wedge y$; thus $x\varphi = (x \wedge y)\varphi = x\varphi \wedge y\varphi$, and $x\varphi \leq y\varphi$ in L_1. Note that the converse does not hold, nor is there any connection between meet- and join-homomorphisms.

Figures 1–3 are three maps of the four-element lattice L of Figure 2.1 into the three-element chain $\mathfrak{C}_3$. The map of Figure 1 is isotone but is neither a meet- nor a join-homomorphism. The map of Figure 2 is a join-homomorphism but is not a meet-homomorphism, thus not a homomorphism. The map of Figure 3 is a homomorphism.

The second basic algebraic concept is that of a subalgebra:

A *sublattice* $\mathfrak{K} = \langle K;\ \wedge,\ \vee\rangle$ of the lattice $\mathfrak{L} = \langle L;\ \wedge,\ \vee\rangle$ is defined on a nonvoid subset K of L with the property that $a, b \in K$ implies that $a \wedge b$, $a \vee b \in K$ ($\wedge$, $\vee$ taken in $\mathfrak{L}$), and the $\wedge$ and the $\vee$ of $\mathfrak{K}$ are restrictions to K of the $\wedge$ and the $\vee$ of $\mathfrak{L}$.

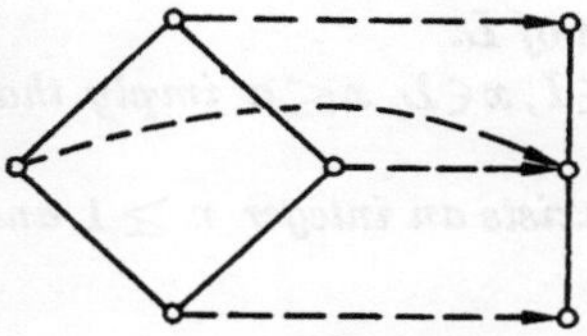

Figure 1

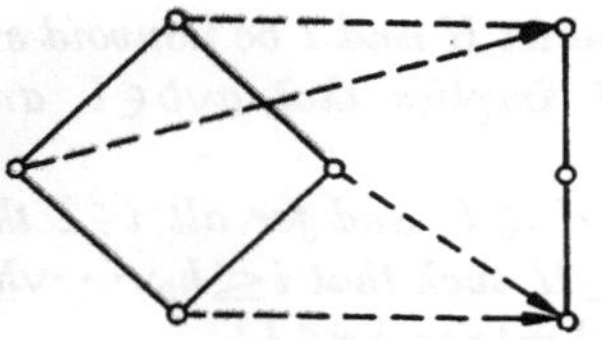

Figure 2

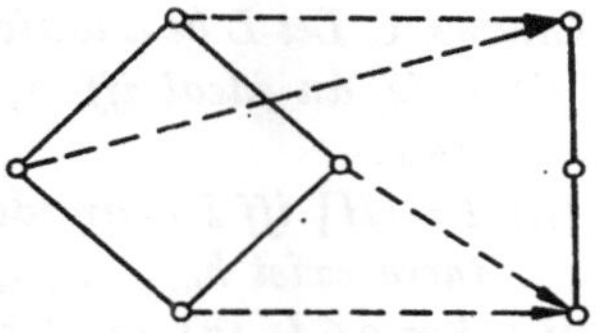

Figure 3

To put this in simpler language, we take a subset K of a lattice L such that K is closed under $\wedge$ and $\vee$. Under the same $\wedge$ and $\vee$, K is a lattice; this is a sublattice of L.

The concept of a lattice as a poset would suggest the following sublattice concept: Take a nonvoid subset K of the lattice L; if the subposet K is a lattice, call K a sublattice of L. This concept is different from the previous one and we will not use it at all; recently some use was made of this in D. Kelly and I. Rival [1975a].

Let A_λ, $\lambda \in \Lambda$, be sublattices of L. Then $\bigcap(A_\lambda \mid \lambda \in \Lambda)$ (the set theoretic intersection of A_λ, $\lambda \in \Lambda$) is also closed under $\wedge$ and $\vee$; thus for every $H \subseteq L$, $H \neq \varnothing$, there is a smallest $[H] \subseteq L$ containing H and closed under $\wedge$ and $\vee$. The sublattice $[H]$ is called the *sublattice of L generated by H*, and H is a *generating set* of $[H]$. If $H = \{a, b, c, \ldots\}$, then we write $[a, b, c, \ldots]$ for $[H]$.

The subset K of the lattice L is called *convex* iff $a, b \in K$, $c \in L$, and $a \leq c \leq b$ imply that $c \in K$. For $a, b \in L$, $a \leq b$, the *interval* $[a, b] = \{x \mid a \leq x \leq b\}$ is an important example of a convex sublattice. For a chain C, $a, b \in C$, $a \leq b$, we can also define the *half-open intervals*: $(a, b] = \{x \mid a < x \leq b\}$ and $[a, b) = \{x \mid a \leq x < b\}$, and the *open interval*: $(a, b) = \{x \mid a < x < b\}$. These are also, whenever nonvoid, examples of convex sublattices. A sublattice I of L is an *ideal* iff $i \in I$ and $a \in L$ imply that $a \wedge i \in I$. An ideal I of L is *proper* iff $I \neq L$. A proper ideal I of L is *prime* iff $a, b \in L$ and $a \wedge b \in I$ imply that $a \in I$ or $b \in I$. In Figure 4, I is an ideal and P is a prime ideal; note that I is not prime.

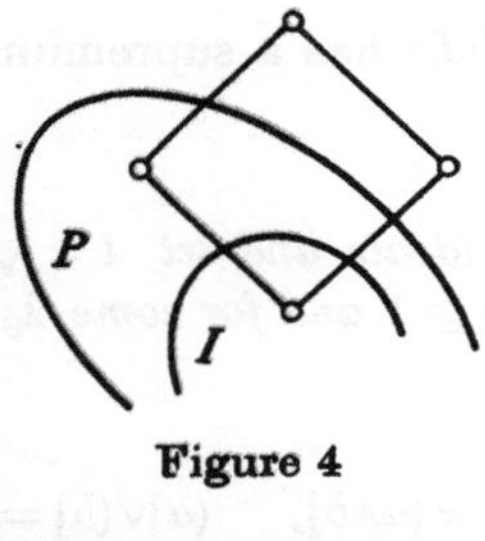

Figure 4

The concept of a convex sublattice is a typical example of the interplay between the algebraic and order theoretic concepts. We will now examine these concepts more closely.

Since the intersection of any number of convex sublattices (ideals) is a convex sublattice (ideal) unless void, we can define the *convex sublattice generated by a subset H*, and the *ideal generated by a subset H* of the lattice L, provided that $H \neq \varnothing$. The ideal generated by a subset H will be denoted by $(H]$, and if $H = \{a\}$, we write $(a]$ for $(\{a\}]$; we shall call $(a]$ a *principal ideal*.

Lemma 1. *Let L be a lattice and let H and I be nonvoid subsets of L.*

(i) *I is an ideal iff $a, b \in I$ implies that $a \vee b \in I$, and $a \in I, x \in L, x \leq a$ imply that $x \in I$.*

(ii) *$I = (H]$ iff I is an ideal, $H \subseteq I$, and for all $i \in I$ there exists an integer $n \geq 1$ and there exist $h_0, \ldots, h_{n-1} \in H$ such that $i \leq h_0 \vee \cdots \vee h_{n-1}$.*

(iii) *For $a \in L$, $(a] = \{x \mid x \leq a\} = \{x \wedge a \mid x \in L\}$.*

Proof.

(i) Let I be an ideal. Then $a, b \in I$ implies that $a \vee b \in I$, since I is a sublattice. If $x \leq a \in I$, then $x = x \wedge a \in I$, and the condition of (i) is verified. Conversely, let I satisfy the condition in (i). Let $a, b \in I$. Then $a \vee b \in I$ by definition, and, since $a \wedge b \leq a \in I$, we also have $a \wedge b \in I$; thus I is a sublattice. Finally, if $x \in L$ and $a \in I$, then $a \wedge x \leq a \in I$, thus $a \wedge x \in I$, proving that I is an ideal. ◑

(ii) Set $I_0 = \{i \mid i \leq h_0 \vee \cdots \vee h_{n-1}$ for $h_0, \ldots, h_{n-1} \in H$ and for some integer $n \geq 1\}$. Using (i), it is clear that I_0 is an ideal, and obviously $H \subseteq I_0$. Finally, if $H \subseteq J$ and J is an ideal, then $I_0 \subseteq J$, and thus I_0 is the smallest ideal containing H; that is, $I = I_0$. ◑

(iii) This proof is obvious directly, or by application of (ii). ●

Let $I(L)$ denote the set of all ideals of L and let $I_0(L) = I(L) \cup \{\varnothing\}$. We call $I(L)$ the *ideal lattice* and $I_0(L)$ the *augmented ideal lattice* of L.

Corollary 2. *$I(L)$ and $I_0(L)$ are posets under set inclusion, and as posets they are lattices.*

In fact, $I \vee J = (I \cup J]$, if we agree that $(\varnothing] = \varnothing$. From (ii) of Lemma 1 we see that for $I, J \in I(L)$, $x \in I \vee J$ iff $x \leq i \vee j$ for some $i \in I, j \in J$. It does not matter that we consider only two ideals in this formula. In general,[1]

$$\bigvee(I_\lambda \mid \lambda \in \Lambda) = (\bigcup(I_\lambda \mid \lambda \in \Lambda)];$$

that is, any nonvoid subset of $I(L)$ has a supremum. Combining this formula with Lemma 1(ii), we have:

Corollary 3. *Let I_λ, $\lambda \in \Lambda$, be ideals and let $I = \bigvee(I_\lambda \mid \lambda \in \Lambda)$. Then $i \in I$ iff $i \leq j_{\lambda_0} \vee \cdots \vee j_{\lambda_{n-1}}$, for some integer $n \geq 1$ and for some $\lambda_0, \ldots, \lambda_{n-1} \in \Lambda$, $j_{\lambda_i} \in I_{\lambda_i}$.*

Now observe the formulas:

$$(a] \wedge (b] = (a \wedge b], \qquad (a] \vee (b] = (a \vee b].$$

Since $a \neq b$ implies that $(a] \neq (b]$, these formulas yield:

Corollary 4. *L can be embedded in $I(L)$ and also in $I_0(L)$, and $a \to (a]$ is an embedding.*

---

[1] The notation $(I_\lambda \mid \lambda \in \Lambda)$ stands for a *family* (of ideals) indexed by Λ. The operations $\bigcap$, $\bigcup$ for families of sets and $\bigwedge$, $\bigvee$ for families of elements of a poset are defined the same way as they are defined for sets of sets and for subsets of posets, respectively.

Let us connect homomorphisms and ideals (recall that $\mathfrak{C}_2$ denotes the two-element chain with elements 0 and 1).

Lemma 5.
(i) *I is a proper ideal of L iff there is a join-homomorphism φ of L onto $\mathfrak{C}_2$ such that $I = 0\varphi^{-1}$ (the complete inverse image of 0, that is, $I = \{x \mid x\varphi = 0\}$).*
(ii) *I is a prime ideal of L iff there is a homomorphism φ of L onto $\mathfrak{C}_2$ with $I = 0\varphi^{-1}$.*

Proof.
(i) Let I be a proper ideal and define φ by $x\varphi = 0$ if $x \in I$, $x\varphi = 1$ if $x \notin I$; obviously, this φ is a join-homomorphism. Conversely, if φ is a join-homomorphism of L onto $\mathfrak{C}_2$ and $I = 0\varphi^{-1}$, then for $a, b \in I$, we have $a\varphi = b\varphi = 0$; thus $(a \vee b)\varphi = a\varphi \vee b\varphi = 0 \vee 0 = 0$, that is, $a \vee b \in I$. If $a \in I$, $x \in L$, $x \leq a$, then $x\varphi \leq a\varphi = 0$, that is, $x\varphi = 0$; thus $x \in I$. Finally, φ is onto, therefore $I \neq L$. ◑
(ii) If I is prime, take the φ constructed in the proof of (i) and note that φ can violate the property of being a homomorphism only with $a, b \notin I$. However, since I is prime, $a \wedge b \notin I$; consequently $(a \wedge b)\varphi = 1 = a\varphi \wedge b\varphi$, and so φ is a homomorphism. Conversely, let φ be a homomorphism of L onto $\mathfrak{C}_2$ and let $I = 0\varphi^{-1}$. If $a, b \notin I$, then $a\varphi = b\varphi = 1$, thus $(a \wedge b)\varphi = a\varphi \wedge b\varphi = 1$, and therefore $a \wedge b \notin I$; I is prime. ●

By dualizing we get the concepts of *dual ideal* (also called *filter*[1]), *principal dual ideal*, $[a)$ (*principal filter*), the *dual ideal* $[H)$ *generated by* H, *proper dual ideal*, *prime dual ideal* (*prime filter*, or *ultrafilter*), the lattice $\mathcal{D}(L)$ of dual ideals ordered by set inclusion, and $\mathcal{D}_0(L) = \mathcal{D}(L) \cup \{\varnothing\}$ ordered by set inclusion. Note that in $\mathcal{D}(L)$ (and $\mathcal{D}_0(L)$) the largest element is L; if L has 0 and 1, then $L = [0)$ is the largest and $\{1\} = [1)$ is the smallest element of $\mathcal{D}(L)$. Furthermore, for $a, b \in L$ we have

$$[a) \wedge [b) = [a \vee b), \quad \text{and} \quad [a) \vee [b) = [a \wedge b).$$

Lemma 6. *Let I be an ideal and let D be a dual ideal. If $I \cap D \neq \varnothing$, then $I \cap D$ is a convex sublattice, and every convex sublattice can be expressed in this form in one and only one way.*

Proof. The first statement is obvious. To prove the second, let C be a convex sublattice and set $I = (C]$, $D = [C)$. Then $C \subseteq I \cap D$. If $t \in I \cap D$, then $t \in I$, and thus by (ii) of Lemma 1, $t \leq c$ for some $c \in C$; also, $t \in D$; therefore, by the dual of (ii) of Lemma 1, $t \geq d$ for some $d \in C$. This implies that $t \in C$ since C is convex, and so $C = I \cap D$.

Suppose now that C has another representation, $C = I_1 \cap D_1$. Since $C \subseteq I_1$, we have $(C] \subseteq I_1$. Let $a \in I_1$ and let c be an arbitrary element of C. Then $a \vee c \in I_1$ and $a \vee c \geq c \in D_1$, so $a \vee c \in D_1$, thus $a \vee c \in I_1 \cap D_1 = C$. Finally, $a \leq a \vee c \in C$; therefore, $a \in (C]$.

[1] In the literature, filter means one of the following four concepts: (i) dual ideal; (ii) proper dual ideal; (iii) dual ideal of the lattice of all subsets of a set; (iv) proper dual ideal of the lattice of all subsets a set. Further variants allow the empty set as a filter under (i) or (ii).

This shows that $I_1 = (C]$. The dual argument shows that $D_1 = [C)$. Hence the uniqueness of such representations. ●

An *equivalence relation* Θ (that is, a reflexive, symmetric, and transitive binary relation) on a lattice L is called a *congruence relation* of L iff $a_0 \equiv b_0\ (\Theta)$ and $a_1 \equiv b_1\ (\Theta)$ imply that $a_0 \wedge a_1 \equiv b_0 \wedge b_1\ (\Theta)$ and $a_0 \vee a_1 \equiv b_0 \vee b_1\ (\Theta)$ (*Substitution Property*). Trivial examples are ω, ι, defined by $x \equiv y\ (\omega)$ iff $x = y$; $x \equiv y\ (\iota)$ for all x and y. (ω is the Greek o and stands for 0; ι is the Greek i and stands for identity and 1.) For $a \in L$, we write $[a]\Theta$ for the *congruence class* containing a, that is, $[a]\Theta = \{x \mid x \equiv a\ (\Theta)\}$.

Lemma 7. *Let Θ be a congruence relation of L. Then for every $a \in L$, $[a]\Theta$ is a convex sublattice.*

Proof. Let $x, y \in [a]\Theta$; then $x \equiv a\ (\Theta)$ and $y \equiv a\ (\Theta)$. Therefore, $x \wedge y \equiv a \wedge a = a\ (\Theta)$, and $x \vee y \equiv a \vee a = a\ (\Theta)$, proving that $[a]\Theta$ is a sublattice. If $x \leq t \leq y$, $x, y \in [a]\Theta$, then $x \equiv a\ (\Theta)$ and $y \equiv a\ (\Theta)$. Therefore, $t = t \wedge y \equiv t \wedge a\ (\Theta)$, and $t = t \vee x \equiv (t \wedge a) \vee x \equiv (t \wedge a) \vee a = a\ (\Theta)$, proving that $[a]\Theta$ is convex. ●

Sometimes a long computation is required to prove that a given binary relation is a congruence relation. Such computations are often facilitated by the following lemma (G. Grätzer and E. T. Schmidt [1958d] and F. Maeda [1958]):

Lemma 8. *A reflexive binary relation Θ on a lattice L is a congruence relation iff the following three properties are satisfied for $x, y, z, t \in L$:*

(i) *$x \equiv y\ (\Theta)$ iff $x \wedge y \equiv x \vee y\ (\Theta)$.*

(ii) *$x \leq y \leq z$, $x \equiv y\ (\Theta)$, and $y \equiv z\ (\Theta)$ imply that $x \equiv z\ (\Theta)$.*

(iii) *$x \leq y$ and $x \equiv y\ (\Theta)$ imply that $x \wedge t \equiv y \wedge t\ (\Theta)$ and $x \vee t \equiv y \vee t\ (\Theta)$.*

Proof. The "only if" part being trivial, assume now that a reflexive binary relation Θ satisfies conditions (i)–(iii). Let $b, c \in [a, d]$ and $a \equiv d\ (\Theta)$; we claim that $b \equiv c\ (\Theta)$. Indeed, $a \equiv d\ (\Theta)$ and $a \leq d$ imply by (iii) that $b \wedge c = a \vee (b \wedge c) \equiv d \vee (b \wedge c) = d\ (\Theta)$. Now $b \wedge c \leq d$ and (iii) imply that $b \wedge c = (b \wedge c) \wedge (b \vee c) \equiv d \wedge (b \vee c) = b \vee c\ (\Theta)$; thus by (i), $b \equiv c\ (\Theta)$.

To prove that Θ is transitive, let $x \equiv y\ (\Theta)$ and $y \equiv z\ (\Theta)$. Then by (i), $x \wedge y \equiv x \vee y\ (\Theta)$, and by (iii), $y \vee z = (y \vee z) \vee (y \wedge x) \equiv (y \vee z) \vee (y \vee x) = x \vee y \vee z\ (\Theta)$, and similarly, $x \wedge y \wedge z \equiv y \wedge z\ (\Theta)$. Therefore, $x \wedge y \wedge z \equiv y \wedge z \equiv y \vee z \equiv x \vee y \vee z\ (\Theta)$ and $x \wedge y \wedge z \leq y \wedge z \leq y \vee z \leq x \vee y \vee z$. Thus, applying (ii) twice, we get $x \wedge y \wedge z \equiv x \vee y \vee z\ (\Theta)$. Now we apply the statement of the previous paragraph with $a = x \wedge y \wedge z$, $b = x$, $c = z$, $d = x \vee y \vee z$ to conclude that $x \equiv z\ (\Theta)$.

Let $x \equiv y\ (\Theta)$; we claim that $x \vee t \equiv y \vee t\ (\Theta)$. Indeed, $x \wedge y \equiv x \vee y\ (\Theta)$ by (i); thus by (iii), $(x \wedge y) \vee t \equiv x \vee y \vee t\ (\Theta)$. Since $x \vee t, y \vee t \in [(x \wedge y) \vee t, x \vee y \vee t]$, we conclude, by the first paragraph of the proof, that $x \vee t \equiv y \vee t\ (\Theta)$.

To prove the Substitution Property for $\vee$, let $x_0 \equiv y_0\ (\Theta)$ and $x_1 \equiv y_1\ (\Theta)$. Then $x_0 \vee x_1 \equiv x_0 \vee y_1 \equiv y_0 \vee y_1\ (\Theta)$, implying that $x_0 \vee x_1 \equiv y_0 \vee y_1\ (\Theta)$ since Θ is transitive. The Substitution Property for $\wedge$ is similarly proved. ●

Let $C(L)$ denote the set of all congruence relations on L partially ordered by set inclusion (remember that every $\Theta \in C(L)$ is a subset of L^2). As a first application of Lemma 8 we prove

Theorem 9. *$C(L)$ is a lattice. For $\Theta, \Phi \in C(L)$, $\Theta \wedge \Phi = \Theta \cap \Phi$. The join, $\Theta \vee \Phi$, can be described as follows:*

$x \equiv y\ (\Theta \vee \Phi)$ iff there is a sequence $z_0 = x \wedge y, z_1, \ldots, z_{n-1} = x \vee y$ of elements of L such that $z_0 \leq z_1 \leq \cdots \leq z_{n-1}$ and for each i, $0 \leq i < n-1$, $z_i \equiv z_{i+1}\ (\Theta)$ or $z_i \equiv z_{i+1}\ (\Phi)$.

Remark. $C(L)$ is called the *congruence lattice* of L. Observe that $C(L)$ is a sublattice of Part(L) (Exercises 45, 46); that is, the join and meet of congruence relations as congruence relations and as equivalence relations (partitions) coincide.

Proof. $\Theta \wedge \Phi = \Theta \cap \Phi$ is obvious. To prove the statement for the join, let Ψ be the binary relation described in Theorem 9. Then $\Theta \subseteq \Psi$ and $\Phi \subseteq \Psi$ are obvious. If Γ is a congruence relation, $\Theta \subseteq \Gamma$, $\Phi \subseteq \Gamma$, and $x \equiv y\ (\Psi)$, then for each i, either $z_i \equiv z_{i+1}\ (\Theta)$ or $z_i \equiv z_{i+1}\ (\Phi)$; thus $z_i \equiv z_{i+1}\ (\Gamma)$. By the transitivity of Γ, $x \wedge y \equiv x \vee y\ (\Gamma)$; thus $x \equiv y\ (\Gamma)$. Therefore, $\Psi \subseteq \Gamma$. This shows that if Ψ is a congruence relation, then $\Psi = \Theta \vee \Phi$.

Ψ is obviously reflexive and satisfies Lemma 8(i). If $x \leq y \leq z$, $x \equiv y\ (\Psi)$, and $y \equiv z\ (\Psi)$, then $x \equiv z\ (\Psi)$ is established by putting together the sequences showing $x \equiv y\ (\Psi)$ and $y \equiv z\ (\Psi)$; this verifies Lemma 8(ii). To show Lemma 8(iii), let $x \equiv y\ (\Psi)$, $x \leq y$, with $z_0, \ldots, z_{n-1}$ establishing this, and $t \in L$. Then $x \wedge t \equiv y \wedge t\ (\Psi)$ and $x \vee t \equiv y \vee t\ (\Psi)$ can be shown with the sequences $z_i \wedge t$, $0 \leq i < n$, $z_i \vee t$, $0 \leq i < n$, respectively. Thus the hypotheses of Lemma 8 hold for Ψ, and we conclude that Ψ is a congruence relation. ●

Homomorphisms and congruence relations express two sides of the same phenomenon. To establish this fact we first define quotient lattices (also called *factor lattices*). Let L be a lattice and let Θ be a congruence relation on L. Let L/Θ denote the set of blocks of the partition of L induced by Θ, that is,

$$L/\Theta = \{[a]\Theta \mid a \in L\}.$$

Set

$$[a]\Theta \wedge [b]\Theta = [a \wedge b]\Theta$$

and

$$[a]\Theta \vee [b]\Theta = [a \vee b]\Theta.$$

This defines $\wedge$ and $\vee$ on L/Θ. Indeed, if $[a]\Theta = [a_1]\Theta$ and $[b]\Theta = [b_1]\Theta$, then $a \equiv a_1\ (\Theta)$ and $b \equiv b_1\ (\Theta)$; therefore, $a \wedge b \equiv a_1 \wedge b_1\ (\Theta)$, that is, $[a \wedge b]\Theta = [a_1 \wedge b_1]\Theta$. Thus $\wedge$ and (dually) $\vee$ are well defined on L/Θ. The lattice axioms are easily verified. The lattice L/Θ is the *quotient lattice* of L modulo Θ.

Lemma 10. *The map*

$$\varphi_\Theta : x \to [x]\Theta \qquad (x \in L)$$

is a homomorphism of L onto L/Θ.

Remark. The lattice K is a *homomorphic image* of the lattice L iff there is a homomorphism of L *onto* K. Lemma 10 states that any quotient lattice is a homomorphic image.

Proof. The proof is trivial. ●

Theorem 11 (The Homomorphism Theorem). *Every homomorphic image of a lattice L is isomorphic to a suitable quotient lattice of L. In fact, if $\varphi: L \to L_1$ is a homomorphism of L onto L_1 and if Θ is the congruence relation of L defined by $x \equiv y\ (\Theta)$ iff $x\varphi = y\varphi$, then*

$$L/\Theta \cong L_1;$$

an isomorphism (see Figure 5) *is given by*

$$\psi: [x]\Theta \to x\varphi, \qquad x \in L.$$

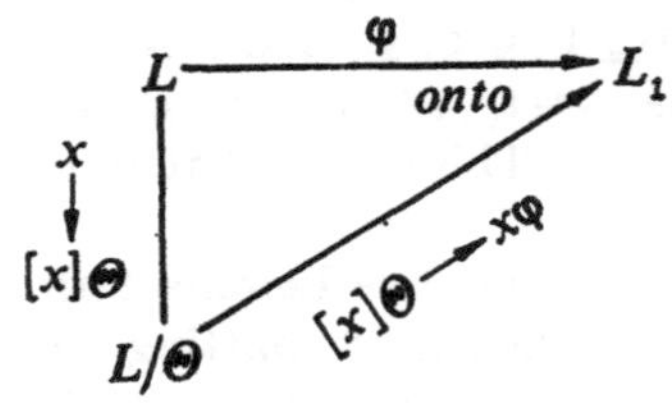

Figure 5

Proof. It is easy to check that Θ is a congruence relation. To prove that ψ is an isomorphism we have to check that ψ (i) is well defined, (ii) is one-to-one, (iii) is onto, and (iv) preserves the operations.

(i) Let $[x]\Theta = [y]\Theta$. Then $x \equiv y\ (\Theta)$; thus $x\varphi = y\varphi$, that is,

$$([x]\Theta)\psi = ([y]\Theta)\psi.$$

(ii) Let $([x]\Theta)\psi = ([y]\Theta)\psi$, that is, $x\varphi = y\varphi$. Then $x \equiv y\ (\Theta)$, and so $[x]\Theta = [y]\Theta$.
(iii) Let $a \in L_1$. Since φ is onto, there is an $x \in L$ with $x\varphi = a$. Thus $([x]\Theta)\psi = a$.
(iv) $([x]\Theta \wedge [y]\Theta)\psi = ([x \wedge y]\Theta)\psi = (x \wedge y)\varphi = x\varphi \wedge y\varphi = ([x]\Theta)\psi \wedge ([y]\Theta)\psi$. The computation for $\vee$ is identical. ●

For a homomorphism $\varphi: L \to L_1$ (not necessarily onto), the relation Θ described in Theorem 11 is called the *congruence kernel of the homomorphism* φ; it will be denoted by $\mathrm{Ker}(\varphi)$. If L_1 has a zero, 0, then $0\varphi^{-1}$ is an ideal of L, called the *ideal kernel of the homomorphism* φ.

Let L be a lattice and let Θ be a congruence relation of L. If L/Θ has a zero, $[a]\Theta$, then $[a]\Theta$ as a subset of L is an ideal, called the *ideal kernel of the congruence relation* Θ.

In contrast with group or ring theory all three kernel concepts are useful in lattice theory. Note the obvious connections, for instance, if I is the ideal kernel of φ, then it is the ideal kernel of $\mathrm{Ker}(\varphi)$.

The final algebraic concept introduced in this section is that of direct product. Let L, K be lattices and form the set $L \times K$ of all ordered pairs $\langle a, b\rangle$ with $a \in L$, $b \in K$.

Define $\wedge$ and $\vee$ in $L \times K$ "componentwise":

$$\langle a, b\rangle \wedge \langle a_1, b_1\rangle = \langle a \wedge a_1, b \wedge b_1\rangle,$$
$$\langle a, b\rangle \vee \langle a_1, b_1\rangle = \langle a \vee a_1, b \vee b_1\rangle.$$

This makes $L \times K$ into a lattice, called the *direct product* of L and K (for an example, see Figure 6).

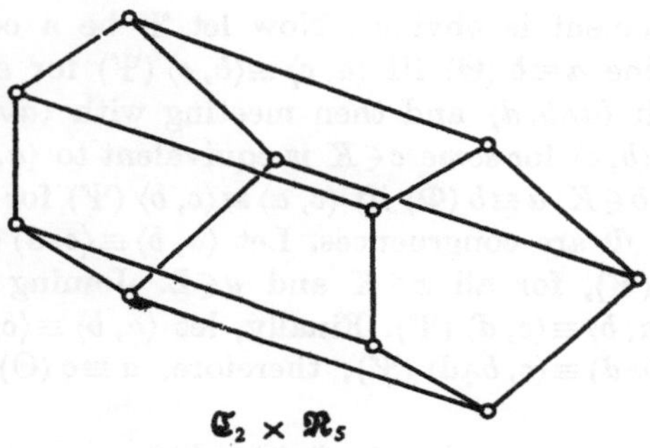

$\mathfrak{C}_2 \times \mathfrak{N}_5$

Figure 6

Lemma 12. *Let L, L_1, K, K_1 be lattices, $L \cong L_1$, $K \cong K_1$. Then*

$$L \times K \cong L_1 \times K_1 \cong K_1 \times L_1.$$

Remark. This means that $L \times K$ is determined up to isomorphism if we know L and K up to isomorphism, and the direct product is determined up to isomorphism by the factors, the order in which they are given being irrelevant.

Proof. Let $\varphi: L \to L_1$ and $\psi: K \to K_1$ be isomorphisms; for $a \in L$ and $b \in K$, define $\langle a, b\rangle\chi = \langle a\varphi, b\psi\rangle$. Then $\chi: L \times K \to L_1 \times K_1$ is an isomorphism. Of course, $L_1 \times K_1 \cong K_1 \times L_1$ is proved by showing that $\langle a, b\rangle \to \langle b, a\rangle$ $(a \in L_1, b \in K_1)$ is an isomorphism. ●

If L_i, $i \in I$, is a family of lattices, again we first form the *Cartesian product* of the sets, $\Pi(L_i \mid i \in I)$, which is defined as the set of all functions $f: I \to \bigcup(L_i \mid i \in I)$ such that $f(i) \in L_i$, for all $i \in I$. We then define $\wedge$ and $\vee$ "componentwise"; that is, $f \wedge g = h$, $f \vee g = k$ means:

$$f(i) \wedge g(i) = h(i), \qquad f(i) \vee g(i) = k(i), \qquad \text{for all } i \in I.$$

The resulting lattice is the *direct product* $\Pi(L_i \mid i \in I)$. If $L_i = L$ for all $i \in I$, we get the *direct power* L^I. Letting n denote the set $\{0, \dots, n-1\}$, L^n is $\underbrace{((L \times L) \times \cdots) \times L}_{n\text{-times}}$ (at least up to isomorphism). In particular, if we identify $f: 2 \to L$ with $\langle f(0), f(1)\rangle$, then we get $L^2 = L \times L$.

We shall use the convention that a one-element lattice (respectively, algebra) is the direct product of the empty family of lattices (respectively, algebras).

A very important property of direct products is:

Theorem 13. *Let L and K be lattices, let Θ be a congruence relation of L, and let Φ be a congruence relation of K. Define the relation $\Theta \times \Phi$ on $L \times K$ by*

$$\langle a, b\rangle \equiv \langle c, d\rangle \ (\Theta \times \Phi) \ \textit{iff} \ a \equiv c \ (\Theta) \ \textit{and} \ b \equiv d \ (\Phi).$$

Then $\Theta \times \Phi$ is a congruence relation on $L \times K$. Conversely, every congruence relation of $L \times K$ is of this form.

Proof. The first statement is obvious. Now let Ψ be a congruence relation on $L \times K$. For $a, b \in L$ define $a \equiv b \ (\Theta)$ iff $\langle a, c\rangle \equiv \langle b, c\rangle \ (\Psi)$ for some $c \in K$. Let $d \in K$. Joining both sides with $\langle a \wedge b, d\rangle$ and then meeting with $\langle a \vee b, d\rangle$, we get $\langle a, d\rangle \equiv \langle b, d\rangle \ (\Psi)$; thus $\langle a, c\rangle \equiv \langle b, c\rangle$ for *some* $c \in K$ is equivalent to $\langle a, c\rangle \equiv \langle b, c\rangle$ for *all* $c \in K$. Similarly, define for $a, b \in K$, $a \equiv b \ (\Phi)$ iff $\langle c, a\rangle \equiv \langle c, b\rangle \ (\Psi)$ for any/for all $c \in L$. It is easily seen that Θ and Φ are congruences. Let $\langle a, b\rangle \equiv \langle c, d\rangle \ (\Theta \times \Phi)$; then $\langle a, x\rangle \equiv \langle c, x\rangle \ (\Psi)$, $\langle y, b\rangle \equiv \langle y, d\rangle \ (\Psi)$, for all $x \in K$ and $y \in L$. Joining the two with $y = a \wedge c$ and $x = b \wedge d$, we get $\langle a, b\rangle \equiv \langle c, d\rangle \ (\Psi)$. Finally, let $\langle a, b\rangle \equiv \langle c, d\rangle \ (\Psi)$. Meeting with $\langle a \vee c, b \wedge d\rangle$, we get $\langle a, b \wedge d) \equiv \langle c, b \wedge d\rangle \ (\Psi)$; therefore, $a \equiv c \ (\Theta)$. Similarly, $b \equiv d \ (\Phi)$, and so

$$\langle a, b\rangle \equiv \langle c, d\rangle \ (\Theta \times \Phi),$$

proving that $\Psi = \Theta \times \Phi$. ●

In conclusion, we introduce a nonalgebraic concept to balance the false impression that might have been created in this section that all lattice theoretic concepts are algebraic.

A lattice L is called *complete* if $\bigwedge H$ and $\bigvee H$ exist for *any subset* $H \subseteq L$. The concept is self-dual, and half of the hypothesis is redundant.

Lemma 14. *Let P be a poset in which $\bigwedge H$ exists for all $H \subseteq P$. Then P is a complete lattice.*

Proof. For $H \subseteq P$, let K be the set of all upper bounds of H. By hypothesis, $\bigwedge K$ exists; set $a = \bigwedge K$. If $h \in H$, then $h \leq k$ for all $k \in K$; therefore $h \leq a$, and $a \in K$. Thus a is the smallest member of K, that is, $a = \bigvee H$. ●

Lemma 14 can be applied to $I_0(L)$ and $C(L)$. For a further application, let $\mathrm{Sub}(L)$ denote the set of all subsets A of L closed under $\wedge$ and $\vee$ partially ordered under set inclusion. In other words, if $A \in \mathrm{Sub}(L)$ and $A \neq \varnothing$, then A is a sublattice of L. Obviously, $\mathrm{Sub}(L)$ is closed under arbitrary intersections.

Corollary 15. *$I_0(L)$ and $C(L)$ are complete lattices. If L has a smallest element, $I(L)$ is a complete lattice. The lattice* $\mathrm{Sub}(L)$ *is a complete lattice.*

Exercises

1. Formalize and prove the equivalence of the two isomorphism concepts.
2. Let $\varphi: L_0 \to L_1$, $\psi: L_1 \to L_0$ be (lattice) homomorphisms. Show that if $\varphi\psi$ is the identity map on L_0 and $\psi\varphi$ is the identity map on L_1, then φ is an isomorphism and $\psi = \varphi^{-1}$. Furthermore, prove that if $\varphi: L_0 \to L_1$, $\psi: L_1 \to L_2$ are isomorphisms, then so are φ^{-1} (the *inverse map*, $\varphi^{-1}: L_1 \to L_0$) and $\varphi\psi$.
3. A one-to-one and onto homomorphism is an isomorphism. Is a one-to-one and onto isotone map an isomorphism?
4. Find a general construction of meet-homomorphisms and join-homomorphisms that are not homomorphisms.
5. Find a subset H of a lattice L such that H is not a sublattice of L but H is a lattice under the partial ordering of L restricted to H.
6. Show that a lattice L is a chain iff every nonvoid subset of L is a sublattice.
7. Prove that a sublattice generated by two distinct elements has two or four elements.
*8. Find an infinite lattice generated by three elements.
9. Verify that a nonempty subset I of a lattice L is an ideal iff, for $a, b \in L$, $a \vee b \in I$ is equivalent to $a, b \in I$.
10. Prove that if L is finite, then L and $I(L)$ are isomorphic. How about $I_0(L)$?
*11. Is there an infinite lattice L such that $L \cong I(L)$, but not every ideal is principal? (There is no such lattice; see D. Higgs [1971].)
12. Prove the completeness of $I_0(L)$ without using Lemma 14.
13. Prove that $C(L)$ is complete by showing the following description of infinite joins: Let $H \subseteq C(L)$, $\Phi = \bigvee H$. Then $x \equiv y\ (\Phi)$ iff there exists a finite sequence $x \wedge y = z_0, z_1, \ldots, z_{n-1} = x \vee y$, such that $z_0 \leq z_1 \leq \cdots \leq z_{n-1}$, and $z_i \equiv z_{i+1}\ (\Theta_i)$ for $0 \leq i < n-1$ and for some $\Theta_i \in H$.
14. Let $\varphi: L \to K$ be an onto homomorphism, let I be an ideal of L, and let J be an ideal of K. Show that $I\varphi$ is an ideal of K, and $J\varphi^{-1} = \{a \mid a \in L,\ a\varphi \in J\}$ is an ideal of L.
15. Is the image $P\varphi$ of a prime ideal under a homomorphism φ prime again?
16. Show that the complete inverse image $P\varphi^{-1}$ of a prime ideal P under an onto lattice homomorphism φ is prime again.
17. Show that an ideal P is a prime ideal of L iff $L - P$ is a dual ideal—in fact, a prime dual ideal.
18. Let H be a nonvoid subset of the lattice L such that $a, b \in H$ implies that $a \vee b \in H$. Then
$$(H] = \{t \mid t \leq h \text{ for some } h \in H\};$$
that is,
$$(H] = \bigcup((h] \mid h \in H).$$
19. Show that if K is a sublattice of L, then K is isomorphic to a sublattice of $I(L)$.
*20. Verify that the converse of Exercise 19 is false even for some finite K.
21. Prove that if $\mathfrak{N}_5$ (see Figure 2.2) is isomorphic to a sublattice of $I(L)$, then $\mathfrak{N}_5$ is isomorphic to a sublattice of L.
22. Find a lattice L and a convex sublattice C of L that cannot be represented as $[a]\Theta$ for any congruence relation Θ of L.
23. State and prove an analogue of Lemma 8 for *join-congruence relations*, that is, for equivalence relations on a lattice satisfying the Substitution Property for joins.
24. Describe $\Theta \vee \Phi$ for join-congruences.

*25. Find a lattice L such that $L \cong L/\Theta$ for all $\Theta \neq \iota$, and there are infinitely many such Θ.
26. Describe the congruence lattice of $\mathfrak{N}_5$.
27. Describe the congruence lattice of an n-element chain.
28. Describe the congruence lattice of the lattice of Figure 2.3, and list all quotient lattices.
29. Construct a lattice that has exactly three congruence relations.
30. Construct infinitely many lattices L such that each lattice is isomorphic to its congruence lattice.
*31. Can an L in Exercise 30 be infinite?
32. Generalize Lemma 12 to the direct product of more than two lattices.
33. Show that $\mathfrak{N}_5 \cong L \times K$ implies that L or K has only one element.
34. Show that $I(L)$ is *conditionally complete*: Every nonvoid set H with an upper bound has a supremum, and dually. $I(L)$ is complete iff L has a zero.
35. Show that the ideal kernel of a homomorphism is an ideal of L.
36. Find an ideal that is the ideal kernel of no homomorphism.
37. Find an ideal that is the kernel of more than one (infinitely many) homomorphisms (congruence relations).
38. Prove that every ideal of a lattice L is prime iff L is a chain.
39. Under what conditions does $L \times K$ have a planar diagram?
40. Let L and K be lattices and let $\varphi: L \to K$ be one-to-one and onto satisfying $\{(a \wedge b)\varphi, (a \vee b)\varphi\} = \{a\varphi \wedge b\varphi, a\varphi \vee b\varphi\}$ for all $a, b \in L$. Let $A \in \mathrm{Sub}(L)$. Show that $A\varphi \in \mathrm{Sub}(K)$; that in fact $A \to A\varphi$ is a (lattice) isomorphism between $\mathrm{Sub}(L)$ and $\mathrm{Sub}(K)$.
41. Prove the converse of Exercise 40.
42. Generalize Theorem 13 to finitely many lattices.
43. Show that the first part of Theorem 13 holds for any number of lattices but that the second part does not.
44. Show that the second statement of Theorem 13 fails for (Abelian) groups.
45. For a set X, let $\mathrm{Part}(X)$ denote the set of all equivalence relations on X partially ordered under set inclusion. Show that $\mathrm{Part}(X)$ is a (complete) lattice. (Equivalence relations and partitions can be identified as explained in Section IV.4 and there we use the notation $\mathrm{Part}(X)$ for the set of equivalence relations on X.)
46. Show that $C(L)$ is a sublattice of $\mathrm{Part}(L)$.
47. *The First Isomorphism Theorem*: Let L be a lattice, let Θ be a congruence relation on L, and let L_1 be a sublattice of L. If for every $a \in L$ there exists exactly one $b \in L_1$ satisfying $a \equiv b\ (\Theta)$, then $L/\Theta \cong L_1$.
48. Let L_0 and L_1 be lattices and let $\pi_i: \langle x_0, x_1 \rangle \to x_i$ be the projection map of $L_0 \times L_1$ onto L_i, $i = 0, 1$. Prove that if A is a lattice and φ_i is a homomorphism of A into L_i, $i = 0, 1$, then there is a unique homomorphism $\psi: A \to L_0 \times L_1$ satisfying $\psi\pi_i = \varphi_i$, $i = 0, 1$.
49. In what way does Exercise 48 characterize $L_0 \times L_1$?
50. Characterize $\Pi(L_i \mid i \in I)$ by projection maps and homomorphisms.

4. Polynomials, Identities, and Inequalities

From variables $x_0, x_1, \ldots, x_n, \ldots$, we can form polynomials in the usual manner using $\wedge, \vee$, and, of course, parentheses. Examples of polynomials are: x_0, x_3, $x_0 \vee x_0$, $(x_0 \wedge x_2) \vee (x_3 \wedge x_0)$, $(x_0 \wedge x_1) \vee (x_0 \wedge x_2) \vee (x_1 \wedge x_2)$. A formal definition is:

Definition 1. *The set* $\mathbf{P}^{(n)}$ *of* n-ary lattice polynomials *is the smallest set satisfying* (i) *and* (ii):

(i) $x_i \in \mathbf{P}^{(n)}, 0 \leq i < n$.

(ii) *If* $p, q \in \mathbf{P}^{(n)}$, *then* $(p \wedge q), (p \vee q) \in \mathbf{P}^{(n)}$.

Remark. We shall omit the outside parentheses and write $p_1 \vee \cdots \vee p_n$ for $(\cdots(p_1 \vee p_2) \cdots \vee p_n)$, and the same for $\wedge$. Thus we write $x_0 \vee x_1$ for $(x_0 \vee x_1)$ and $x_0 \vee x_1 \vee x_2$ for $((x_0 \vee x_1) \vee x_2)$. Note that if $n < m$ then $\mathbf{P}^{(n)} \subset \mathbf{P}^{(m)}$. It is easy to define formally what it means to substitute in a polynomial p all occurrences of x_i by x_j; for instance, substituting x_4 for x_1, $p = (x_1 \vee x_2) \wedge (x_1 \wedge x_3)$ becomes $q = (x_4 \vee x_2) \wedge (x_4 \wedge x_3)$; observe that $p \in \mathbf{P}^{(4)}$ but $q \notin \mathbf{P}^{(4)}$. In general, for $p = p(x_0 \ldots, x_{n-1}) \in \mathbf{P}^{(n)}$, we have $p(x_{i_0}, \ldots, x_{i_{n-1}}) \in \mathbf{P}^{(m)}$ for any $m > i_0, \ldots, i_{n-1}$.

By Definition 1, a polynomial is just a sequence of symbols. It is defined because in terms of such a sequence of symbols we can define a function on any lattice:

Definition 2. *An n-ary polynomial p defines a function in n variables* (a polynomial function, *or simply, a* polynomial) *on a lattice L by the following rules* $(a_0, \ldots, a_{n-1} \in L)$:

(i) *If* $p = x_i$, $p(a_0, \ldots, a_{n-1}) = a_i$, $0 \leq i < n$.

(ii) *If* $p(a_0, \ldots, a_{n-1}) = a$, $q(a_0, \ldots, a_{n-1}) = b$, *and* $p \wedge q = r$, $p \vee q = t$, *then* $r(a_0, \ldots, a_{n-1}) = a \wedge b$ *and* $t(a_0, \ldots, a_{n-1}) = a \vee b$.

Thus if $p = (x_0 \wedge x_1) \vee (x_2 \vee x_1)$, then $p(a, b, c) = (a \wedge b) \vee (c \vee b) = b \vee c$. Definitions 1 and 2 are quite formal but their meaning is very simple.

A *polynomial* is an n-ary polynomial for some n. We will also use $x, y, z, \ldots$ instead of the x_i.

Note that if p is a *unary* $(n = 1)$ lattice polynomial, then $p(a) = a$ for any $a \in L$. If p is *binary*, then for all $a, b \in L$, $p(a, b) = a$; or for all $a, b \in L$, $p(a, b) = b$; or for all $a, b \in L$, $p(a, b) = a \wedge b$; or for all $a, b \in L$, $p(a, b) = a \vee b$.

We shall prove statements on polynomials by induction on the *rank*. The rank of x_i is 1; that of $p \wedge q$ (and of $p \vee q$) is the sum of the ranks of p and q.

Now we are in a position to describe $[H]$, the sublattice generated by $H \neq \varnothing$:

Lemma 3. $a \in [H]$ *iff* $a = p(h_0, \ldots, h_{n-1})$ *for some integer* $n \geq 1$, *for some n-ary polynomial p, and for some* $h_0, \ldots, h_{n-1} \in H$.

Proof. First we must show that if $a = p(h_0, \ldots, h_{n-1})$, $h_0, \ldots, h_{n-1} \in H$, then $a \in [H]$, which can be easily accomplished by induction on the rank of p. Then we form the set $\{a \mid a = p(h_0, \ldots, h_{n-1}), n \geq 1, h_0, \ldots, h_{n-1} \in H\}$; observe that it contains H and that it is closed under $\wedge$ and $\vee$. Since it is contained in $[H]$, it has to equal $[H]$. ●

Corollary 4. $|[H]| \leq |H| + \aleph_0$.

Proof. By Lemma 3, every element of $[H]$ can be associated with a finite sequence of elements of $H \cup \{(,), \wedge, \vee\}$, and there are no more than $|H| + \aleph_0$ such sequences. ●

Definition 5. *A* lattice identity (inequality) *is an expression of the form* $p=q$ $(p\leq q)$, *where* p *and* q *are polynomials. An* identity, $p=q$ (inequality, $p\leq q$), holds in the lattice L *iff* $p(a_0, a_1, \ldots)=q(a_0, a_1, \ldots)$ $(p(a_0, a_1, \ldots)\leq q(a_0, a_1, \ldots))$ *holds for any* $a_0, a_1, \ldots \in L$.

An identity $p=q$ is equivalent to the two inequalities $p\leq q$ and $q\leq p$, and the inequality $p\leq q$ is equivalent to the identity $p\vee q=q$ (and to $p\wedge q=p$). Frequently, the validity of identities is shown by verifying that the two inequalities hold.

One of the most basic properties of polynomials is:

Lemma 6. *A polynomial (function)* p *is isotone; that is, if* $a_0\leq b_0$, $a_1\leq b_1, \ldots$, *then* $p(a_0, a_1, \ldots)\leq p(b_0, b_1, \ldots)$. *Furthermore,*

$$x_0\wedge\cdots\wedge x_{n-1}\leq p(x_0, \ldots, x_{n-1})\leq x_0\vee\cdots\vee x_{n-1}.$$

Proof. We prove the first statement by induction on the rank of p. The first statement is certainly true for $p=x_i$. Suppose that it is true for q and r and that $p=q\wedge r$. Then

$$\begin{aligned}p(a_0, \ldots)\wedge p(b_0, \ldots) &= (q(a_0, \ldots)\wedge r(a_0, \ldots))\wedge(q(b_0, \ldots)\wedge r(b_0, \ldots))\\ &= (q(a_0, \ldots)\wedge q(b_0, \ldots))\wedge(r(a_0, \ldots)\wedge r(b_0, \ldots))\\ &= q(a_0, \ldots)\wedge r(a_0, \ldots)=p(a_0, \ldots);\end{aligned}$$

thus $p(a_0, \ldots)\leq p(b_0, \ldots)$. The proof is similar for $p=q\vee r$. Since $x_0\wedge\cdots\wedge x_{n-1}\leq x_i\leq x_0\vee\cdots\vee x_{n-1}$ for $0\leq i\leq n-1$, using the idempotency of $\wedge$ and $\vee$, we obtain:

$$\begin{aligned}x_0\wedge\cdots\wedge x_{n-1} &= p(x_0\wedge\cdots\wedge x_{n-1}, \ldots, x_0\wedge\cdots\wedge x_{n-1})\leq p(x_0, x_1, \ldots, x_{n-1})\\ &\leq p(x_0\vee\cdots\vee x_{n-1}, \ldots, x_0\vee\cdots\vee x_{n-1})=x_0\vee x_1\vee\cdots\vee x_{n-1},\end{aligned}$$

proving the second statement. ●

A simple application is:

Lemma 7. *Let* $p_i=q_i$, $0\leq i<n$, *be lattice identities. Then there is a single identity* $p=q$ *such that all* $p_i=q_i$, $0\leq i<n$, *hold in a lattice* L *iff* $p=q$ *holds in* L.

Proof. Let us take two identities, $p_0=q_0$ and $p_1=q_1$. Suppose that all polynomials are n-ary and consider the identity (we use now $2n$-ary polynomials formed by substitution, see Remark to Definition 1):

(N) $p_0(x_0, \ldots, x_{n-1})\wedge p_1(x_n, \ldots, x_{2n-1})=q_0(x_0, \ldots, x_{n-1})\wedge q_1(x_n, \ldots, x_{2n-1})$.

It is obvious that if $p_0=q_0$ and $p_1=q_1$ hold in L, then (N) holds in L. Now let (N) hold in L and let $a_0, \ldots, a_{n-1}\in L$. Substitute $x_0=a_0, \ldots, x_{n-1}=a_{n-1}$, $x_n=\cdots=x_{2n-1}=a_0\vee\cdots\vee a_{n-1}=a$. By Lemma 6,

$$p_0(a_0, \ldots, a_{n-1})\leq p_0(a, \ldots, a)=a=p_1(a, \ldots, a),$$

whence $p_0(a_0, \ldots, a_{n-1}) = p_0(a_0, \ldots, a_{n-1}) \wedge p_1(a, \ldots, a)$, and similarly for q_0, q_1; thus (N) yields $p_0(a_0, \ldots, a_{n-1}) = q_0(a_0, \ldots, a_{n-1})$. The second identity is derived similarly from (N). The general proof is similar. ●

The most important (and, in fact, characteristic) properties of identities are given by

Lemma 8. *Identities are preserved under the formation of sublattices, homomorphic images, direct products, and ideal lattices.*

Proof. Let the polynomials p and q both be n-ary and let $p = q$ hold in L. If L_1 is a sublattice of L, then $p = q$ obviously holds in L_1. Let $\varphi: L \to K$ be an onto homomorphism. A simple induction shows that

$$p(a_0, \ldots, a_{n-1})\varphi = p(a_0\varphi, \ldots, a_{n-1}\varphi)$$

and proves the similar formula for q. Therefore,

$$p(a_0\varphi, \ldots, a_{n-1}\varphi) = p(a_0, \ldots, a_{n-1})\varphi = q(a_0, \ldots, a_{n-1})\varphi = q(a_0\varphi, \ldots, a_{n-1}\varphi),$$

and so $p = q$ holds in K. The statement for direct products is also obvious.

The last statement is an easy corollary of the following formula. Let p be an n-ary polynomial and let $I_0, \ldots, I_{n-1}$ be ideals of L. Then $I_0, \ldots, I_{n-1} \in I(L)$; thus we can substitute the I_j into $p: p(I_0, \ldots, I_{n-1})$ is also in $I(L)$, that is, an ideal of L. This ideal can be described by a simple formula:

$$p(I_0, \ldots, I_{n-1}) = \{x \mid x \leq p(i_0, \ldots, i_{n-1}), \text{ for some } i_0 \in I_0, \ldots, i_{n-1} \in I_{n-1}\}.$$

This follows easily from Lemma 6 and the formula in Section 3 describing $I \vee J$, by induction on the rank of p. ●

Now we list a few important inequalities:

Lemma 9. *The following inequalities hold in any lattice:*

(i) $(x \wedge y) \vee (x \wedge z) \leq x \wedge (y \vee z)$.

(ii) $x \vee (y \wedge z) \leq (x \vee y) \wedge (x \vee z)$.

(iii) $(x \wedge y) \vee (y \wedge z) \vee (z \wedge x) \leq (x \vee y) \wedge (y \vee z) \wedge (z \vee x)$.

(iv) $(x \wedge y) \vee (x \wedge z) \leq x \wedge (y \vee (x \wedge z))$.

Remark. (i)–(iii) are called *distributive inequalities*, and (iv) is the *modular inequality*.

Proof. We prove (iv) as an example and leave the rest to the reader. Since $x \wedge y \leq x$, $x \wedge z \leq x$, we obtain $(x \wedge y) \vee (x \wedge z) \leq x$; $x \wedge y \leq y \leq y \vee (x \wedge z)$ and $x \wedge z \leq y \vee (x \wedge z)$; therefore, $(x \wedge y) \vee (x \wedge z) \leq y \vee (x \wedge z)$. The meet of the two inequalities $(x \wedge y) \vee (x \wedge z) \leq x$ and $(x \wedge y) \vee (x \wedge z) \leq y \vee (x \wedge z)$ yields (iv). ●

Lemma 10. *Consider the following two identities and inequality:*
(i) $(x\wedge y)\vee(x\wedge z)=x\wedge(y\vee z)$.
(ii) $(x\vee y)\wedge(x\vee z)=x\vee(y\wedge z)$.
(iii) $(x\vee y)\wedge z\leq x\vee(y\wedge z)$.
Then (i), (ii), *and* (iii) *are equivalent in any lattice* L.

Remark. A lattice satisfying identities (i) or (ii) is called *distributive*. Note that (i) and (ii) are *not* equivalent for fixed elements; that is, (i) can hold for three elements $a, b, c\in L$, whereas (ii) does not.

Proof. Let (i) hold in L and let $a, b, c\in L$; then, using (i) with $x=a\vee b$, $y=a$, $z=c$,

$$(a\vee b)\wedge(a\vee c)=((a\vee b)\wedge a)\vee((a\vee b)\wedge c)$$

and using $a=(a\vee b)\wedge a$ and (i) for $x=c$, $y=a$, $z=b$,

$$=a\vee(a\wedge c)\vee(b\wedge c)=a\vee(b\wedge c),$$

verifying (ii).

The proof that (ii) implies (i) is the dual of the preceding paragraph.

Let (ii) hold in L; then $x\vee(y\wedge z)=(x\vee y)\wedge(x\vee z)\geq(x\vee y)\wedge z$, since $x\vee z\geq z$, verifying (iii).

Let (iii) hold in L. Then, with $x=a$, $y=b$, $z=a\vee c$ in (iii),

$$(a\vee b)\wedge(a\vee c)\leq a\vee(b\wedge(a\vee c))=a\vee((a\vee c)\wedge b)$$

and with $x=a$, $y=c$, $z=b$ in (iii)

$$\leq a\vee(a\vee(c\wedge b))=a\vee(c\wedge b).$$

This, combined with the dual of Lemma 9(i), gives (ii). ●

Corollary 11. *The dual of a distributive lattice is distributive.*

Lemma 12. *The identity*

$$(x\wedge y)\vee(x\wedge z)=x\wedge(y\vee(x\wedge z))$$

is equivalent to the condition:

$$x\geq z \quad \textit{implies that} \quad (x\wedge y)\vee z=x\wedge(y\vee z).$$

Remark. A lattice satisfying either condition is called *modular*.

Proof. If $x\geq z$, then $z=x\wedge z$; thus the implication follows from the identity. Conversely, if the implication holds, then since $x\geq x\wedge z$, we have $(x\wedge y)\vee(x\wedge z)=x\wedge(y\vee(x\wedge z))$. ●

Exercises

1. Give a formal definition of substituting all occurrences of x_i by x_j in a polynomial p. Prove formally the statement of the last sentence of the Remark to Definition 1 and the statement of the first sentence of the proof of Lemma 3.
2. Let $H=\{h_0, \ldots, h_{n-1}\}$. Prove that $a \in [H]$ iff $a=p(h_0, \ldots, h_{n-1})$ for some n-ary polynomial p.
3. Show that the upper bound in Corollary 4 is best possible if $|H| \geq 3$. Give the best estimates for $|H| \leq 2$.
4. Give a formal proof of Lemma 8.
5. Prove (without reference to Lemma 8) that if L is distributive (modular), then so is $I(L)$.
6. Show that the dual of a modular lattice is modular.
7. Prove that L is distributive iff the identity

$$(x \wedge y) \vee (y \wedge z) \vee (z \wedge x) = (x \vee y) \wedge (y \vee z) \wedge (z \vee x)$$

holds in L.
8. Prove that every distributive lattice is modular, but not conversely. Find the smallest modular but nondistributive lattice.
9. Find an identity $p=q$ characterizing distributive lattices such that neither $p \leq q$ nor $q \leq p$ holds in a general lattice.
10. Show that in any lattice

$$\bigvee(\bigwedge(x_{ij} \mid 0 \leq i < m) \mid 0 \leq j < n) \leq \bigwedge(\bigvee(x_{ij} \mid 0 \leq j < n) \mid 0 \leq i < m).$$

11. Prove that the following identity holds in a distributive lattice:

$$\bigvee(\bigwedge(x_{ij} \mid 0 \leq j < n) \mid 0 \leq i < m) = \bigwedge(\bigvee(x_{if(i)} \mid 0 \leq i < m) \mid f \in F),$$

where F is the set of all functions from $\{0, 1, \ldots, m-1\}$ into $\{0, 1, \ldots, n-1\}$.
12. Derive (i)–(iii) of Lemma 9 from Exercise 10.
13. Verify that any chain is a distributive lattice.
14. Let L be a lattice with more than one element, let $L' = L \cup \{0\}$, and let $0 < x$ for all $x \in L$. Show that L' is then a lattice and that an identity $p=q$ holds in L iff it holds in L'.
15. Show that if the identity $x_0 = p(x_0, x_1, \ldots)$ holds in the two-element lattice, then it holds in every lattice.
16. Prove that $P(X)$ is a distributive lattice.
17. Show that Part(X) is distributive iff $|X| \leq 2$ and modular iff $|X| \leq 3$. (See Exercise 3.45.)
18. Find an identity that holds in $\mathfrak{C}_2$ but not in $\mathfrak{N}_5$.
19. Find an identity that holds in $\mathfrak{M}_3$ but not in $\mathfrak{N}_5$.
20. Let K and L be modular lattices, let D be a dual ideal of K, and let I be an ideal of L such that there exists an isomorphism φ of D with I. Set $A = K \cup (L - I)$ (disjoint union) and define $x \leq y$, for $x, y \in A$, as follows: For $x, y \in K$ or $x, y \in L - I$, $x \leq y$ retains its meaning; for $x \in K$, $y \in L - I$, let $x \leq y$ mean the existence of a $z \in D$ such that $x \leq z$, $z\varphi \leq y$; $x \leq y$ for no $x \in L - I$, $y \in K$. Is $\langle A; \leq \rangle$ then a modular lattice? (Ph. M. Whitman [1943] and R. P. Dilworth and M. Hall [1944].)
21. If K and L of Exercise 20 are distributive, is A distributive?
22. In Exercise 20, put $I = D = \{a\}$ and show that any identity holding in K and L holds in A as well.

23. Examine the statements of Lemma 8 for properties of the form, "If $p_0 = q_0$, then $p_1 = q_1$."

*24. Show that $\langle L; \wedge, \vee \rangle$ is a lattice satisfying the identity $p = q$ iff it satisfies the identities
$$(w \wedge x) \vee x = x$$
and
$$(((x \wedge p) \wedge z) \vee u) \vee v = (((q \wedge z) \wedge x) \vee v) \vee ((t \vee u) \wedge u),$$
where x, z, u, v, and w are variables that do not occur in p or q (R. Padmanabhan [1968]).

*25. Show that the result of Exercise 24 is best possible: If $p = q$ is an identity *not* satisfied by some lattice, then the two identities of Exercise 24 cannot be replaced by one (R. N. McKenzie [1970]).

26. Show that the lattice L is modular iff the inequality $x \wedge (y \vee z) \leq y \vee ((x \vee y) \wedge z)$ holds in L.

5. Free Lattices

Though it would be quite easy to develop a feeling for the most general lattice (generated by a set of elements and satisfying some relation) of Section 2 by way of some examples, a general definition seems to be hard to accept. So we ask the reader to withhold judgment on whether Definition 2 expresses his intuitive feelings until the theory is developed and further examples are presented.

The most general lattice will be called free. Since we might be interested, for instance, in the most general distributive lattice generated by a, b, c, satisfying $b < a$, it seems desirable to define freeness with respect to a class **K** of lattices.

DEFINITION 1. *Let $p_i = q_i$ be identities for $i \in I$. The class* **K** *of all lattices satisfying all identities $p_i = q_i$, $i \in I$, is called an* equational class of lattices. *An equational class is* trivial *iff it contains one-element lattices only.*

The class **L** of all lattices, the class **D** of all distributive lattices, and the class **M** of all modular lattices are examples of equational classes of lattices.

Next we have to agree on what kinds of relations to allow in a generating set. Can we prescribe only relations of the form $b < a$, or do we allow relations of the form $a \wedge b = c$ or $d \vee e = f$? Lemma 9 can be rephrased to furnish an example in which the four generators a, b, c, d are required to satisfy $a \vee b = a \vee c = b \vee c = d$. Let us therefore agree that for a generating set we take a poset P, and for relations we take all $a \leq b$ that hold in P, all $a \wedge b = c$, where $\inf\{a, b\} = c$ in P, and all $a \vee b = c$, where $\sup\{a, b\} = c$ in P. (A more liberal approach will be presented later in this section.)

Now we are ready to formulate our basic concept:

DEFINITION 2. *Let P be a poset and let* **K** *be an equational class of lattices. A lattice $F_{\mathbf{K}}(P)$ is called a* free lattice over **K** generated by P *iff the following conditions are satisfied:*

(i) $F_{\mathbf{K}}(P) \in \mathbf{K}$.

(ii) $P \subseteq F_{\mathbf{K}}(P)$, *and for* $a, b, c \in P$, $\inf\{a, b\} = c$ *in* P *iff* $a \wedge b = c$ *in* $F_{\mathbf{K}}(P)$, *and* $\sup\{a, b\} = c$ *in* P *iff* $a \vee b = c$ *in* $F_{\mathbf{K}}(P)$.
(iii) $[P] = F_{\mathbf{K}}(P)$.
(iv) *Let* $L \in \mathbf{K}$ *and let* $\varphi: P \rightarrow L$ *be an isotone map with the properties that if* $a, b, c \in P$, $\inf\{a, b\} = c$ *in* P, *then* $a\varphi \wedge b\varphi = c\varphi$ *in* L, *and if* $\sup\{a, b\} = c$ *in* P, *then* $a\varphi \vee b\varphi = c\varphi$ *in* L. *Then there exists a (lattice) homomorphism* $\psi: F_{\mathbf{K}}(P) \rightarrow L$ *extending* φ (*that is,* $a\varphi = a\psi$ *for all* $a \in P$).

Let ε denote the identity map on P; then the crucial condition (iv) can be expressed by Figure 1. In that and in all such similar *diagrams* (not in the sense the word is

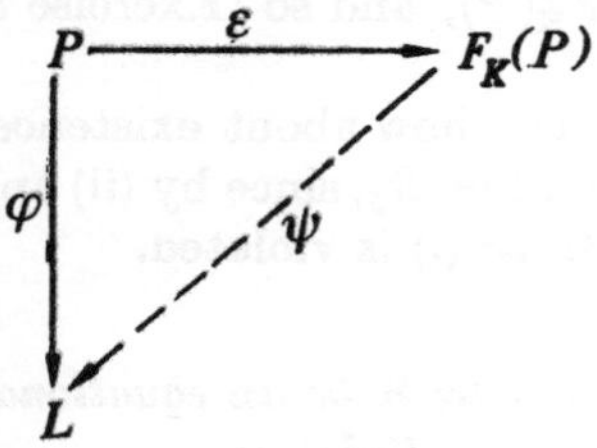

Figure 1

used in Section 2), the capital letters represent lattices or posets, and the arrows indicate homomorphisms, or maps with certain properties. Such diagrams are usually assumed to be *commutative*, which, in this particular case, means $\varepsilon\psi = \varphi$, which is (iv). Note that (ii) is also included in the diagram if the arrows are supposed to represent maps as required in (iv). In fact, we could have required in (ii) that the identity map on P be an embedding of P into $F_{\mathbf{K}}(P)$ in the sense of (iv).

If P is an unordered set, $|P| = \mathfrak{m}$, we shall write $F_{\mathbf{K}}(\mathfrak{m})$ for $F_{\mathbf{K}}(P)$ and call it a *free lattice on* $\mathfrak{m}$ *generators* over $\mathbf{K}$. In case $\mathbf{K} = \mathbf{L}$, we shall always omit "over $\mathbf{L}$"; thus "*free lattice generated by P*" shall mean "free lattice over $\mathbf{L}$ generated by P"–in notation, $F(P)$.

It should be noted that if $b \in F_{\mathbf{K}}(P)$, then by (iii) and by Lemma 4.3, $b = p(a_0, \ldots, a_{n-1})$, where p is a polynomial and $a_0, \ldots, a_{n-1} \in P$. Thus if the ψ of (iv) exists (in fact, if φ is any mapping of P into L and if $\psi: F_{\mathbf{K}}(P) \rightarrow L$ is any homomorphism extending φ), then we must have

$$\begin{aligned} b\psi &= p(a_0, \ldots, a_{n-1})\psi \quad \text{(and since } \psi \text{ is to be a homomorphism)} \\ &= p(a_0\psi, \ldots, a_{n-1}\psi) = p(a_0\varphi, \ldots, a_{n-1}\varphi), \end{aligned}$$

since $a_i\psi = a_i(\varepsilon\psi) = a_i\varphi$.

From this we conclude that there is at most one homomorphism $\psi: F_{\mathbf{K}}(P) \rightarrow L$ extending φ, whence:

Corollary 3. *The homomorphism* ψ *in* (iv) *is unique.*

This corollary is used to prove:

Corollary 4. *Let both $F_{\mathbf{K}}(P)$ and $F_{\mathbf{K}}^{*}(P)$ satisfy the conditions of Definition 2. Then there exists an isomorphism χ: $F_{\mathbf{K}}(P)\to F_{\mathbf{K}}^{*}(P)$, and χ can be chosen so that $a\chi=a$ for all $a\in P$. In other words, free lattices (over* **K** *generated by P) are unique up to isomorphism.*

Proof. Let us use Figure 1 with $L=F_{\mathbf{K}}(P)$ and $\varphi=\varepsilon$. Then there exist ψ_1: $F_{\mathbf{K}}(P)\to F_{\mathbf{K}}^{*}(P)$ and ψ_2: $F_{\mathbf{K}}^{*}(P)\to F_{\mathbf{K}}(P)$ such that $\varepsilon\psi_1=\varepsilon$ and $\varepsilon\psi_2=\varepsilon$. Thus $\psi_1\psi_2$: $F_{\mathbf{K}}(P)\to F_{\mathbf{K}}(P)$ is the identity map ε on P. By the statement preceding Corollary 3, ε has a unique extension to a homomorphism $F_{\mathbf{K}}(P)\to F_{\mathbf{K}}(P)$; the identity map on $F_{\mathbf{K}}(P)$ is one such extension. Therefore, $\psi_1\psi_2$ is the identity map on $F_{\mathbf{K}}(P)$. Similarly, $\psi_2\psi_1$ is the identity map on $F_{\mathbf{K}}^{*}(P)$, and so (Exercise 3.2) ψ_1 is the required isomorphism. ●

This settles the uniqueness, but how about existence? Naturally, $F_{\mathbf{K}}(P)$ need not exist. For instance, $F_{\mathbf{D}}(\mathfrak{N}_5)$ should be $\mathfrak{N}_5$, since by (ii) and (iii) of Definition 2, $F_{\mathbf{K}}(P)=P$ if P is a lattice; but $\mathfrak{N}_5\notin\mathbf{D}$, so (i) is violated.

Theorem 5. *Let P be a poset and let* **K** *be an equational class of lattices. Then $F_{\mathbf{K}}(P)$ exists iff the following condition is satisfied:*

(E) *There exists a lattice L in* **K** *such that $P\subseteq L$ and for $a, b, c\in P$,* inf $\{a, b\}=c$ *in P iff $a\wedge b=c$ in L, and* sup $\{a, b\}=c$ *in P iff $a\vee b=c$ in L.*

Proof. Condition (E) is obviously necessary for the existence of $F_{\mathbf{K}}(P)$; indeed, if $F_{\mathbf{K}}(P)$ exists, (E) can always be satisfied with $L=F_{\mathbf{K}}(P)$ by (i) and (ii) of Definition 2.

Now assume that (E) is satisfied. In this proof a map φ: $P\to N$ ($N\in\mathbf{K}$) will be called a *homomorphism* iff it satisfies the conditions set forth in Definition 2(iv).

Obviously, in Definition 2(iv) it suffices to consider L with $L=[P\varphi]$. Let $\langle N, \varphi\rangle$ denote this situation—that is, $N\in\mathbf{K}$, φ: $P\to N$ is a homomorphism and $N=[P\varphi]$. Then $F_{\mathbf{K}}(P)$ or, more precisely, $\langle F_{\mathbf{K}}(P), \varepsilon\rangle$ has the property that for any $\langle L, \varphi\rangle$ there exists a ψ: $F_{\mathbf{K}}(P)\to L$ with $\varphi=\varepsilon\psi$. To construct $F_{\mathbf{K}}(P)$ we have to construct a lattice having this property for *all* $\langle L, \varphi\rangle$. How would we construct such a lattice for two?

Let $\langle L_1, \varphi_1\rangle$ and $\langle L_2, \varphi_2\rangle$ be given. Form $L_1\times L_2$ and define a map φ: $P\to L_1\times L_2$ by $p\varphi=\langle p\varphi_1, p\varphi_2\rangle$; set $N=[P\varphi]$. The fact that φ is a homomorphism is easy to check. A simple example is illustrated in Figures 2–4. Now we define ψ_i: $\langle x_1, x_2\rangle\to x_i$. Obviously, for $p\in P$, $p\varphi\psi_i=p\varphi_i$ and ψ_i: $N\to L_i$.

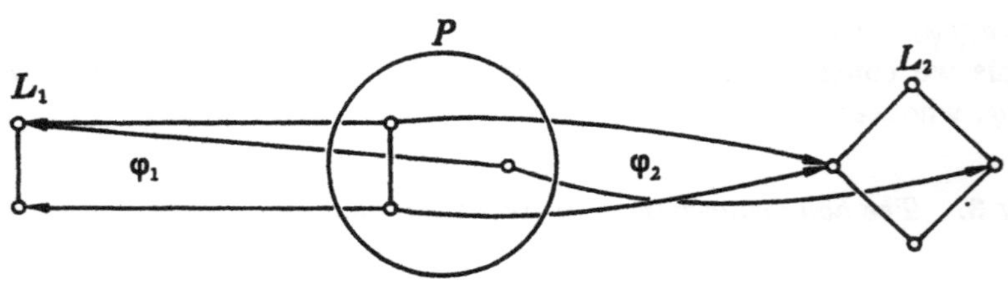

Figure 2

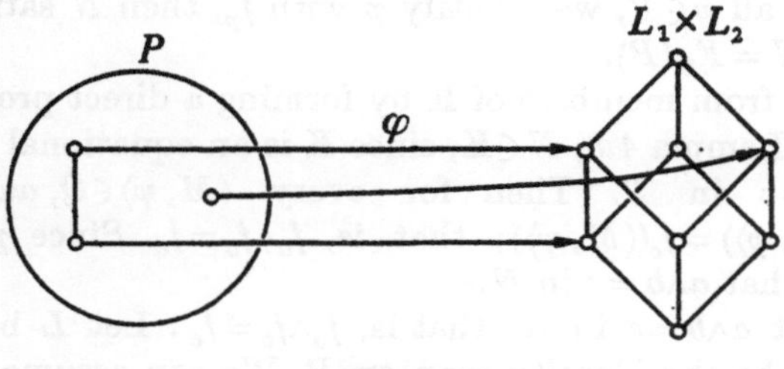

Figure 3

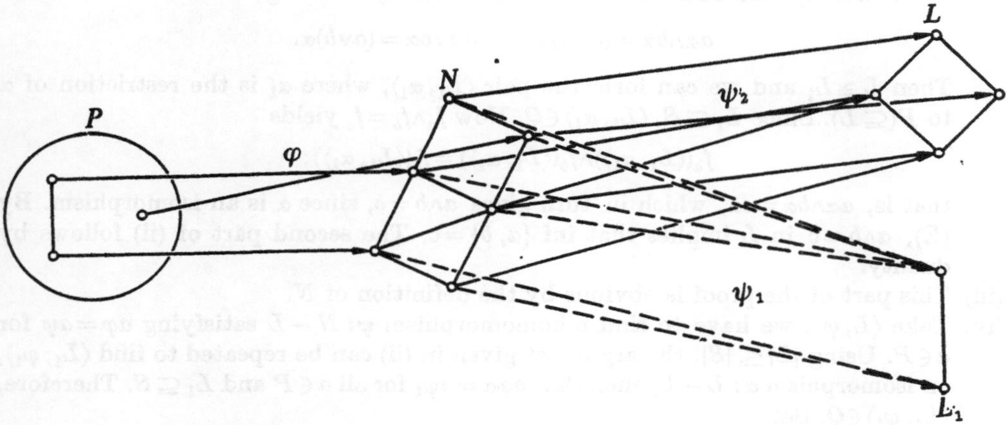

Figure 4

If we are given any number of $\langle L_i, \varphi_i \rangle$, $i \in I$, we can proceed as before and get $\langle N, \varphi \rangle$; if one of the $\langle L_i, \varphi_i \rangle$ is the $\langle L, \varepsilon \rangle$ given by (E), then (ii) of Definition 2 will also be satisfied. There is only one problem: All the pairs $\langle L_i, \varphi_i \rangle$ do not form a set, so their direct product cannot be formed. The $\langle L_i, \varphi_i \rangle$ do not form a set because a lattice and all its isomorphic copies do not form a set; therefore, if we can somehow restrict taking too many isomorphic copies, the previous procedure can be followed. Observe that by Corollary 4.4, in every pair $\langle L_i, \varphi_i \rangle$ we have

$$|L_i| \leq |P\varphi_i| + \aleph_0 \leq |P| + \aleph_0.$$

Thus, by choosing a large enough set S and taking only those $\langle L_i, \varphi_i \rangle$ that satisfy $L_i \subseteq S$, we can solve our problem.

Now we are ready to proceed with the formal proof. Choose a set S satisfying $|P| + \aleph_0 = |S|$. Let Q be the set of all pairs $\langle M, \psi \rangle$ where $M \subseteq S$. Form $A = \Pi(M \mid \langle M, \psi \rangle \in Q)$, and for each $p \in P$ let $f_p \in A$ be defined by

$$f_p(\langle M, \psi \rangle) = p\psi.$$

Finally, set

$$N = [\{f_p \mid p \in P\}].$$

We claim that if, for all $p \in P$, we identify p with f_p, then N satisfies (i)–(iv) of Definition 2, and thus $N = F_{\mathbf{K}}(P)$.

(i) N is constructed from members of $\mathbf{K}$ by forming a direct product and by taking a sublattice. By Lemma 4.8, $N \in \mathbf{K}$, since $\mathbf{K}$ is an equational class.

(ii) Let $\inf\{a, b\} = c$ in P. Then for every $\langle M, \psi \rangle \in Q$, $a\psi \wedge b\psi = c\psi$, so that $f_a(\langle M, \psi \rangle) \wedge f_b(\langle M, \psi \rangle) = f_c(\langle M, \psi \rangle)$; that is, $f_a \wedge f_b = f_c$. Since p is identified with f_p, we conclude that $a \wedge b = c$ in N.

Conversely, let $a \wedge b = c$ in N, that is, $f_a \wedge f_b = f_c$. Let L be the lattice given by (E) and let ε be the identity map on P. We can assume that $L = [P]$; thus we can form $\langle L, \varepsilon \rangle$. By Corollary 4.4, $|L| \leq |S|$, so there is a one-to-one map α: $L \to S$. Let $L_1 = L\alpha$ and make L_1 into a lattice by defining

$$a\alpha \wedge b\alpha = (a \wedge b)\alpha, \qquad a\alpha \vee b\alpha = (a \vee b)\alpha.$$

Then $L \cong L_1$ and we can form the pair $\langle L_1, \alpha_1 \rangle$, where α_1 is the restriction of α to $P (\subseteq L)$. Since $L_1 \subseteq S$, $\langle L_1, \alpha_1 \rangle \in Q$. Now $f_a \wedge f_b = f_c$ yields

$$f_a(\langle L_1, \alpha_1 \rangle) \wedge f_b(\langle L_1, \alpha_1 \rangle) = f_c(\langle L_1, \alpha_1 \rangle);$$

that is, $a\alpha \wedge b\alpha = c\alpha$, which in turn gives $a \wedge b = c$, since α is an isomorphism. By (E), $a \wedge b = c$ in L implies that $\inf\{a, b\} = c$. The second part of (ii) follows by duality.

(iii) This part of the proof is obvious by the definition of N.

(iv) Take $\langle L, \varphi \rangle$; we have to find a homomorphism ψ: $N \to L$ satisfying $a\varphi = a\psi$ for $a \in P$. Using $|L| \leq |S|$, the argument given in (ii) can be repeated to find $\langle L_1, \varphi_1 \rangle$, an isomorphism α: $L \to L_1$ such that $a\varphi\alpha = a\varphi_1$ for all $a \in P$ and $L_1 \subseteq S$. Therefore, $\langle L_1, \varphi_1 \rangle \in Q$. Set

$$\psi_1 : f \to f(\langle L_1, \varphi_1 \rangle), \quad f \in N.$$

Then for $a \in P$,

$$a\psi_1 = f_a\psi_1 = f_a(\langle L_1, \varphi_1 \rangle) = a\varphi_1 = a\varphi\alpha.$$

Thus the homomorphism $\psi = \psi_1\alpha^{-1}$: $N \to L$ will satisfy the requirement of (iv). ●

Two consequences of this theorem are very important:

Corollary 6. *For any nontrivial equational class* $\mathbf{K}$ *and for any cardinal* $\mathfrak{m}$, *a free lattice over* $\mathbf{K}$ *with* $\mathfrak{m}$ *generators*, $F_{\mathbf{K}}(\mathfrak{m})$, *exists.*

Proof. It suffices to find an $L \in \mathbf{K}$, $X \subseteq L$ such that $|X| = \mathfrak{m}$, and for $x, y \in X$, $x \neq y$, x and y are incomparable. This is easily done. Since $\mathbf{K}$ is nontrivial, there exists an $N \in \mathbf{K}$, $|N| > 1$; thus, $\mathfrak{C}_2$ is a sublattice of N. By Lemma 4.8, $\mathfrak{C}_2 \in \mathbf{K}$; by Lemma 4.8, $(\mathfrak{C}_2)^I \in \mathbf{K}$ for any set I. Let $|I| = \mathfrak{m}$, let $L = (\mathfrak{C}_2)^I$; for $i \in I$, define $f_i \in L$ by $f_i(i) = 1$, $f_i(j) = 0$ for $i \neq j$, and set $X = \{f_i \mid i \in I\}$. Obviously, X satisfies the condition. ●

Corollary 7. *For any poset* P, *a free lattice (over* $\mathbf{L}$*) generated by the poset* P *exists.*

Proof. Take a poset P and define $I_0(P)$ to be the set of all subsets I of P satisfying the condition: $\sup\{a, b\} \in I$ iff $a, b \in I$. Partially ordering $I_0(P)$ by set inclusion makes $I_0(P)$ a lattice. Identifying a with $\{x \mid x \leq a\}$, we see that $I_0(P)$ contains P and satisfies (E) of Theorem 5. The detailed computation is almost the same as that for Theorem 20, so it will be omitted. ●

The argument given in the proof of Corollary 3 shows that whenever L is generated by P, any homomorphism φ of P has at most one extension to L, and if there is one, it is given by

$$\psi: p(a_0, \ldots, a_{n-1}) \rightarrow p(a_0\varphi, \ldots, a_{n-1}\varphi).$$

This formula gives a homomorphism iff ψ is well defined; in other words, iff $p(a_0, \ldots, a_{n-1}) = q(b_0, \ldots, b_{m-1})$ implies that $p(a_0\varphi, \ldots, a_{n-1}\varphi) = q(b_0\varphi, \ldots, b_{m-1}\varphi)$, for any $a_0, \ldots, a_{n-1}, b_0, \ldots, b_{m-1} \in P$ and $\varphi: P \rightarrow N \in \mathbf{K}$.

This yields a very practical method of finding free lattices and verifying their freeness.

Theorem 8. *In the definition of $F_{\mathbf{K}}(P)$, (iv) can be replaced by the following condition:*
If $b \in F_{\mathbf{K}}(P)$ has two representations, $b = p(a_0, \ldots, a_{n-1})$ and $b = q(b_0, \ldots, b_{m-1})$ $(a_0, \ldots, a_{n-1}, b_0, \ldots, b_{m-1} \in P)$, then

$$p(a_0, \ldots, a_{n-1}) = q(b_0, \ldots, b_{m-1})$$

can be derived from the identities defining $\mathbf{K}$ *and the relations of P of the form $a \wedge b = c$ and $a \vee b = c$.*

Remark. Thus, in proving $p = q$, we can use only the $\wedge$ and $\vee$ table of P, but we cannot use $a \nparallel b$ or $a \nparallel b \vee c$, and so on.

We illustrate Theorem 8 first by determining $F_{\mathbf{D}}(3)$. The following simple observation will be useful:

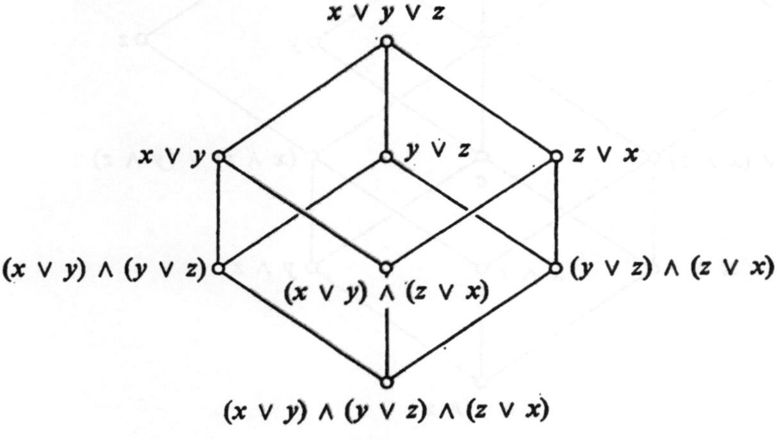

Figure 5

Lemma 9. *Let x, y, and z be elements of a lattice L and let $x \vee y$, $y \vee z$, $z \vee x$ be pairwise incomparable. Then $\{x \vee y, y \vee z, z \vee x\}$ generates a sublattice of L isomorphic to $(\mathfrak{C}_2)^3$* (see Figure 5).

Proof. Almost all the meets and joins are obvious; by symmetry, the nonobvious ones are typified by $((x \vee y) \wedge (y \vee z)) \vee ((x \vee y) \wedge (z \vee x)) = x \vee y$ and $((x \vee y) \wedge (y \vee z)) \vee (z \vee x) = x \vee y \vee z$. Since $y \leq (x \vee y) \wedge (y \vee z)$ and $x \leq (x \vee y) \wedge (z \vee x)$, we get $x \vee y \leq ((x \vee y) \wedge (y \vee z)) \vee ((x \vee y) \wedge (z \vee x))$, and $\geq$ is trivial. The second equality follows from $y \leq (x \vee y) \wedge (y \vee z)$. Note that, for example, $(x \vee y) \wedge (y \vee z) = (x \vee y) \wedge (y \vee z) \wedge (z \vee x)$ would imply, by joining both sides with $z \vee x$, that $x \vee y \vee z = z \vee x$; thus $x \vee y \leq z \vee x$, a contradiction. Therefore, all eight elements are distinct. ●

Theorem 10. *A free distributive lattice on three generators, $F_{\mathbf{D}}(3)$, has eighteen elements* (see Figure 6).

$$a = (x \vee y) \wedge (x \vee z) \wedge (y \vee z) = (x \wedge y) \vee (x \wedge z) \vee (y \wedge z)$$

$$b = (x \vee y) \wedge (y \vee z)$$

$$c = (x \wedge y) \vee (y \wedge z)$$

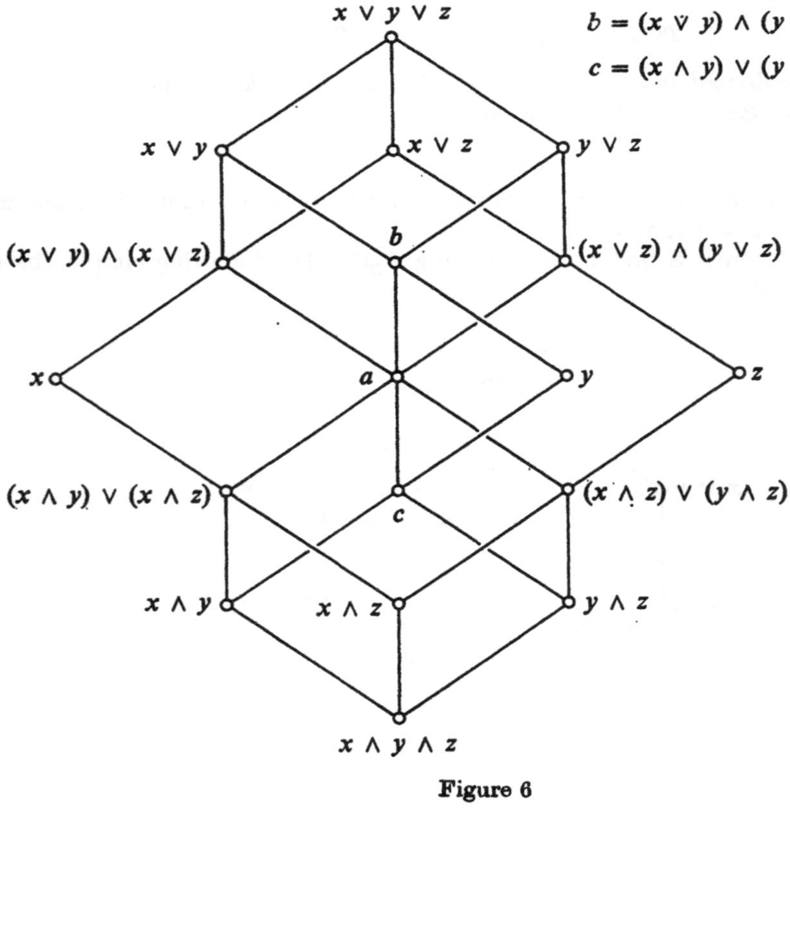

Figure 6

Proof. Let x, y, and z be the free generators. The top eight and the bottom eight elements form sublattices by Lemma 9 and its dual; note that $(x\wedge y)\vee(y\wedge z)\vee(z\wedge x) = (x\vee y)\wedge(y\vee z)\wedge(z\vee x)$ by Exercise 4.7.

According to Theorem 8, we have only to verify that the lattice L of Figure 6 is a distributive lattice, and that if p, q, r are polynomials representing elements of L and $p\wedge q=r$ in L, then $p\wedge q=r$ in every distributive lattice and similarly for $\vee$. The first statement is easily proved by representing L by sets (see Exercise 13). The second statement requires a complete listing of all triples p, q, r with $p\wedge q=r$. If p, q, r belong to the top eight or bottom eight elements, the statement follows from Lemma 9. The remaining cases are all trivial except when p or q is x, y, or z. By symmetry, only $p=x$, $q=y\vee z$, $r=(x\wedge y)\vee(x\wedge z)$ is left to discuss, but then $p\wedge q=r$ is the distributive law. ●

Theorem 11. (R. Dedekind [1900]). *A free modular lattice on three generators, $F_M(3)$, has twenty-eight elements* (see Figure 7).

$u = (x \wedge y) \vee (y \wedge z) \vee (x \wedge z)$

$v = (x \vee y) \wedge (y \vee z) \wedge (x \vee z)$

$x_1 = (x \wedge v) \vee u$

$y_1 = (y \wedge v) \vee u$

$z_1 = (z \wedge v) \vee u$

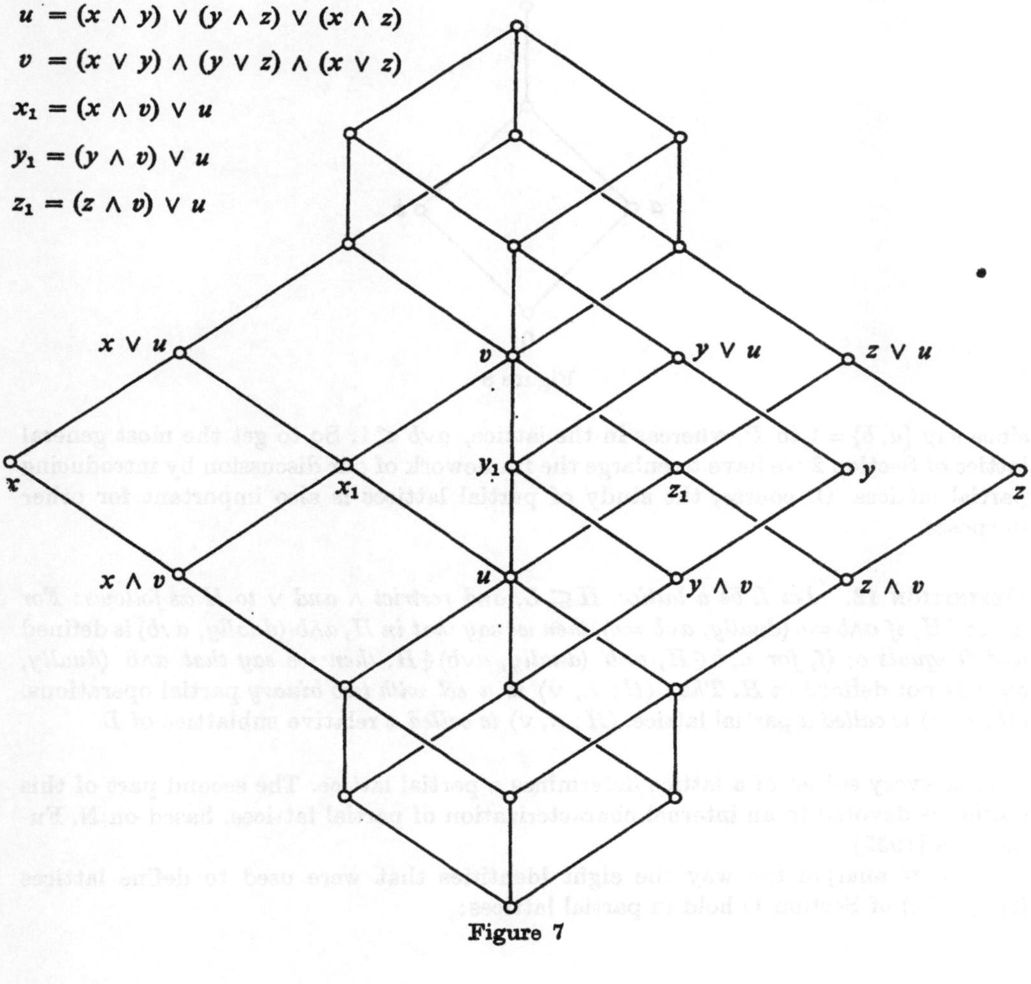

Figure 7

Proof. Let x, y, and z be the free generators. Again modularity is proved by a representation (see Exercise 16). Theorem 10 takes care of most meets and joins not involving x_1, y_1, z_1. Of the rest, only one relation (and the symmetric and the dual cases) is difficult to prove: $x_1 \wedge y_1 = u$. This we do now, leaving the rest to the reader. Compute:

$$\begin{aligned} x_1 \wedge y_1 &= ((x \wedge v) \vee u) \wedge ((y \wedge v) \vee u) \quad \text{(since } u \leq (y \wedge v) \vee u) \\ &= ((x \wedge v) \wedge ((y \wedge v) \vee u)) \vee u \\ &= ((x \wedge v) \wedge (y \vee u) \wedge v) \vee u \quad \text{(substitute } u \text{ and } v) \\ &= (x \wedge (y \vee z) \wedge (y \vee (x \wedge z))) \vee u \\ &= (x \wedge y) \vee (x \wedge z) \vee u = u. \quad ● \end{aligned}$$

Consider the lattice represented by Figure 8. We would like to say that it is freely generated by $\{0, a, b, 1\} = P$, but this is clearly not the case according to Definition 2,

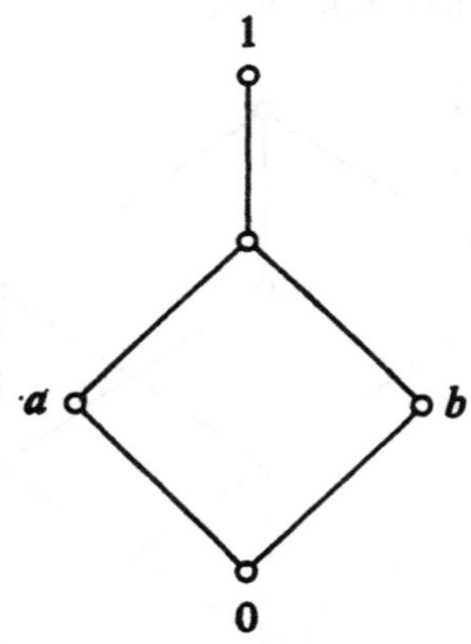

Figure 8

since $\sup \{a, b\} = 1$ in P, whereas in the lattice, $a \vee b < 1$. So to get the most general lattice of Section 2 we have to enlarge the framework of our discussion by introducing partial lattices. Of course, the study of partial lattices is also important for other purposes.

Definition 12. *Let L be a lattice, $H \subseteq L$, and restrict $\wedge$ and $\vee$ to H as follows: For $a, b, c \in H$, if $a \wedge b = c$ (dually, $a \vee b = c$), then we say that in H, $a \wedge b$ (dually, $a \vee b$)* is defined *and it equals c; if, for $a, b \in H$, $a \wedge b$ (dually, $a \vee b$) $\notin H$, then we say that $a \wedge b$ (dually, $a \vee b$)* is not defined *in H. Thus $\langle H; \wedge, \vee \rangle$ is a set with two binary* partial operations. *$\langle H; \wedge, \vee \rangle$ is called a* partial lattice. *$\langle H; \wedge, \vee \rangle$ is called a* relative sublattice *of L.*

Thus every subset of a lattice determines a partial lattice. The second part of this section is devoted to an internal characterization of partial lattices, based on N. Funayama [1953].

We now analyze the way the eight identities that were used to define lattices ((L1)–(L4) of Section 1) hold in partial lattices:

Lemma 13. *Let $\langle H; \wedge, \vee\rangle$ be a partial lattice, $a, b, c \in H$.*
(i) *$a\wedge a$ exists and $a\wedge a = a$.*
(ii) *If $a\wedge b$ exists, then $b\wedge a$ exists, and $a\wedge b = b\wedge a$.*
(iii) *If $a\wedge b$, $(a\wedge b)\wedge c$, $b\wedge c$ exist, then $a\wedge(b\wedge c)$ exists, and $(a\wedge b)\wedge c = a\wedge(b\wedge c)$. If $b\wedge c$, $a\wedge(b\wedge c)$, $a\wedge b$ exist, then $(a\wedge b)\wedge c$ exists, and $(a\wedge b)\wedge c = a\wedge(b\wedge c)$.*
(iv) *If $a\wedge b$ exists, then $a\vee(a\wedge b)$ exists, and $a = a\vee(a\wedge b)$.*

Proof. As an illustration let us prove the first statement of (iii). Let $H \subseteq L$ as in Definition 12. The assumption that $a\wedge b$, $(a\wedge b)\wedge c$, $b\wedge c$ exist in H means that in L, $a, b, c, a\wedge b, (a\wedge b)\wedge c, b\wedge c \in H$. But in L, $(a\wedge b)\wedge c = a\wedge(b\wedge c)$, and so $a\wedge(b\wedge c) \in H$; that is, $a\wedge(b\wedge c)$ exists in H, and, of course, $(a\wedge b)\wedge c = a\wedge(b\wedge c)$. ●

Lemma 13′. *Let* (i′)–(iv′) *denote the statements we get from* (i)–(iv) *of Lemma 13 by interchanging $\wedge$ and $\vee$. Then* (i′)–(iv′) *hold in any partial lattice.*

Proof. This proof is trivial by duality. ●

Statements (i)–(iv) and (i′)–(iv′) give the required interpretation of the eight identities for partial lattices.

Definition 14. *A* weak partial lattice *is a set with two binary partial operations satisfying* (i)–(iv) *and* (i′)–(iv′).

Corollary 15. *Every partial lattice is a weak partial lattice.*

Based on Figure 9, we give the following example of a weak partial lattice that is not a partial lattice.

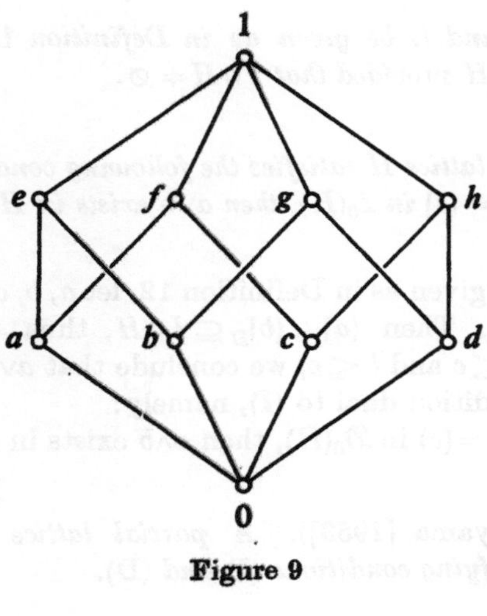

Figure 9

Let $H=\{0, a, b, c, d, e, f, g, h, 1\}$. Consider the lattice $\mathfrak{H}$ of Figure 9. Define $x\wedge y=z$ in H iff $x\wedge y=z$ in $\mathfrak{H}$. Define $x\vee y=z$ in H iff either $x\leq y$ in $\mathfrak{H}$ and $y=z$, or $y\leq x$ in $\mathfrak{H}$ and $x=z$; or if $\{x, y\}=\{a, c\}$ and $z=f$; or $\{x, y\}=\{b, d\}$ and $z=g$; or $\{x, y\}=\{f, g\}$ and $z=1$. Then $\langle H; \wedge, \vee\rangle$ is a weak partial lattice (check the axioms). Now suppose that there exists a lattice L, $H\subseteq L$, such that $\langle H; \wedge, \vee\rangle$ is a relative sublattice of L. Then $1=(a\vee c)\vee(b\vee d)$ in L, and thus $1=\sup\{a, b, c, d\}$. Since $e\geq a, b$, and $h\geq c, d$ in L, and $1\geq e, h$, we get $1=\sup\{e, h\}$ in L. The fact that $e, h, 1\in H$ implies that $e\vee h$ is defined in H (and equals 1), contrary to the definition of $\langle H; \wedge, \vee\rangle$. (Compare this with Lemma 19.)

To avoid such anomalies we shall introduce two further conditions. To prepare for them we prove:

Lemma 16. *Let $\langle H; \wedge, \vee\rangle$ be a weak partial lattice. Then we define a partial ordering relation $\leq$ on H by "$a\leq b$ iff $a\wedge b$ exists and $a\wedge b=a$." If $a\vee b$ exists, then $a\vee b=\sup\{a, b\}$. If $a\wedge b$ exists, then $a\wedge b=\inf\{a, b\}$. Also, $a\leq b$ iff $a\vee b=b$.*

Proof. This proof is the same as the proof of the corresponding parts of Theorem 1.1, except that the arguments are a bit longer. ●

Note that in a partial lattice $\sup\{a, b\}$ may exist but $a\vee b$ does not. For instance, let L be the lattice of Figure 9, $H=\{0, a, b, 1\}$. Then $\sup\{a, b\}=1$ in H, but $a\vee b$ is not defined in H because $a\vee b\notin H$.

Definition 17. *An* ideal *of a weak partial lattice H is a nonvoid subset I of H such that if $a, b\in I$ and $a\vee b$ exists, then $a\vee b\in I$, and $x\leq a\in I$ implies that $x\in I$. Again we set $\{x \mid x\leq a\}=(a]$.* Dual ideal *and* $[a)$ *are defined dually. $I_0(H)$ is the lattice consisting of $\varnothing$ and all ideals of H (partial ordering is $\subseteq$), $\mathcal{D}_0(H)$ is the lattice consisting of $\varnothing$ and all dual ideals of H (partial ordering is $\subseteq$). For $K\subseteq H$, $(K]$ is the ideal and $[K)$ is the dual ideal generated by K.*

Corollary 18. *Let H and L be given as in Definition 12. Let I be an ideal of L. Then $I\cap H$ is an ideal of H provided that $I\cap H\neq\varnothing$.*

Lemma 19. *Any partial lattice H satisfies the following condition:*

(I) *If $(a]\vee(b]=(c]$ in $I_0(H)$, then $a\vee b$ exists in H and equals c.*

Proof. Let H and L be given as in Definition 12, let $a, b, c\in H$, and let $(a]\vee(b]=(c]$ in $I_0(H)$. Set $I=(a\vee b]_L$. Then $(a]_H, (b]_H\subseteq I\cap H$, thus $(c]_H=(a]_H\vee(b]_H\subseteq(a\vee b]_L$; that is, $c\leq a\vee b$. Since $a\leq c$ and $b\leq c$, we conclude that $a\vee b=c$. ●

Let (D) denote the condition dual to (I), namely:

(D) If $[a)\vee[b)=[c)$ in $\mathcal{D}_0(H)$, then $a\wedge b$ exists in H and equals c.

Theorem 20. (N. Funayama [1953]). *A partial lattice can be characterized as a weak partial lattice satisfying conditions* (I) *and* (D).

Proof. Corollary 15, Lemma 19, and its dual prove that a partial lattice is a weak partial lattice satisfying (I) and (D). Conversely, let $\langle H; \wedge, \vee\rangle$ be a weak partial lattice satisfying (I) and (D). Consider the map

$$\varphi: x \to \langle (x], [x)\rangle,$$

sending H into $I_0(H) \times \tilde{\mathcal{D}}_0(H)$, where $\tilde{\mathcal{D}}_0(H)$ is the dual of $\mathcal{D}_0(H)$. This map φ is one-to-one. If $x \vee y = z$, then $(x] \vee (y] = (z]$ in $I_0(H)$ and $[x) \vee [y) = [z)$ in $\tilde{\mathcal{D}}_0(H)$, thus $x\varphi \vee y\varphi = (x \vee y)\varphi$. Conversely, if $x\varphi \vee y\varphi = z\varphi$, then $(x] \vee (y] = (z]$ in $I_0(H)$. Therefore, by (I), $x \vee y$ exists and equals z, so $x\varphi \vee y\varphi = z\varphi$ implies that $x \vee y = z$. A similar argument shows that $x \wedge y = z$ iff $x\varphi \wedge y\varphi = z\varphi$. Thus we can identify x with $x\varphi$, getting $H \subseteq L = I_0(H) \times \tilde{\mathcal{D}}_0(H)$. We have just proved that $\langle H; \wedge, \vee\rangle$ is a relative sublattice of L. ●

Let $\langle P; \leq\rangle$ be a poset. We make P into a partial lattice as follows: $a \wedge b$ is defined iff $\inf\{a, b\}$ exists and $a \wedge b = \inf\{a, b\}$, and similarly for $a \vee b$.

Lemma 21. *$\langle P; \wedge, \vee\rangle$ is a partial lattice.*

Proof. It is easy to verify that (I), (D), (i)–(iv) of Lemma 13, and (i′)–(iv′) of Lemma 13′ hold. ●

We need some further definitions.

Definition 22. *Let $\langle A; \wedge, \vee\rangle$, $\langle B; \wedge, \vee\rangle$ be weak partial lattices, $\varphi: A \to B$. We call φ a* homomorphism *iff whenever $a \wedge b$ exists for $a, b \in A$, then $a\varphi \wedge b\varphi$ exists and $(a \wedge b)\varphi = a\varphi \wedge b\varphi$, and the similar condition holds for $\vee$. A one-to-one homomorphism φ is an* embedding *provided that $a \wedge b$ exists iff $a\varphi \wedge b\varphi$ exists, and the similar condition holds for $\vee$. If φ is onto and φ is an embedding, then φ is an* isomorphism.

Now we are ready again to define the most general lattices of Section 2.

Definition 23. *Let $\mathfrak{A} = \langle A; \wedge, \vee\rangle$ be a partial lattice and let $\mathbf{K}$ be an equational class of lattices. The lattice $F_{\mathbf{K}}(\mathfrak{A})$ (or, simply, $F_{\mathbf{K}}(A)$) is a* free lattice over $\mathbf{K}$ generated by $\mathfrak{A}$, *iff the following conditions are satisfied:*

(i) *$F_{\mathbf{K}}(A) \in \mathbf{K}$.*
(ii) *$A \subseteq F_{\mathbf{K}}(A)$, and A is a relative sublattice of $F_{\mathbf{K}}(A)$.*
(iii) *$[A] = F_{\mathbf{K}}(A)$.*
(iv) *If $L \in \mathbf{K}$ and $\varphi: A \to L$ is a homomorphism, then there exists a homomorphism ψ: $F_{\mathbf{K}}(A) \to L$ extending φ (that is, $a\varphi = a\psi$ for $a \in A$).*

If P is a poset, then $F_{\mathbf{K}}(P)$ is a free lattice over $\mathbf{K}$ generated by P, where P is considered a partial lattice as in Lemma 21; thus Definition 23 contains Definition 2 as a special case. The general theory can be developed exactly as it was in the first part of this section. The final result is:

Theorem 24. *Let $\mathfrak{A} = \langle A; \wedge, \vee \rangle$ be a partial lattice and let* **K** *be an equational class. Then $F_{\mathbf{K}}(\mathfrak{A})$ exists iff there exists a lattice L in* **K** *such that $\mathfrak{A}$ is a relative sublattice of L.*

As an application we prove the existence of the lattice completely freely generated by a poset. Let P be a poset; we define a partial lattice P^m on P as follows:

$x \wedge y = z$ in P^m iff x and y are comparable and $z = \inf\{x, y\}$;

$x \vee y = z$ in P^m iff x and y are comparable and $z = \sup\{x, y\}$.

Definition 25. $F(P^m)$ $(= F_{\mathbf{L}}(P^m))$ *is called a* lattice completely freely generated by P.

Theorem 26. *For any poset P, a lattice completely freely generated by P exists.*

Proof. A subset $A \subseteq P$ is called *hereditary* iff $x \in A$ and $y \leq x$ imply that $y \in A$. Let $H(P)$ be the set of all hereditary subsets of P partially ordered under set inclusion. Let $P_1 = H(P)$ and $\tilde{P}_2 = H(\tilde{P}_1)$, where $\tilde{P}_i$ is the dual of P_i. Identifying $p \in P$ with $(p]$, we get $P \subseteq P_1$; identifying $p \in P_1$ with $[p)$, we get $P_1 \subseteq P_2$, thus $P \subseteq P_2$. Obviously, P_2 is a lattice. Let $a, b, c \in P$ and $a \vee b = c$ in P_2. If a and b are incomparable, then $(a] \cup (b]$ $(\neq (c])$ is an upper bound for a and b, thus $a \vee b < c$ in P_2. A similar argument works for $a \wedge b = c$. Therefore, $P_2 \supseteq P$ satisfies the condition of Theorem 24. ●

$CF(P)$ is the usual notation for $F(P^m)$ in the literature.

Exercises

1. Show that an equational class is closed under the formation of sublattices, homomorphic images, and direct products.
2. Show that the class **T** of all one-element lattices is an equational class. For any equational class **K**, prove that $\mathbf{K} \supseteq \mathbf{T}$.
3. Let $\mathbf{K}_i$, $i \in I$, be equational classes. Show that $\bigcap(\mathbf{K}_i \mid i \in I)$ is again an equational class.
4. Find an equational class **A** with $\mathbf{A} \subseteq \mathbf{M}$, $\mathbf{D} \subseteq \mathbf{A}$, and $\mathbf{A} \neq \mathbf{M}$, $\mathbf{A} \neq \mathbf{D}$. (Hint: Let **A** be defined by

$$(x \vee y) \wedge (x \vee z) \wedge (x \vee u) = x \vee ((x \vee y) \wedge z \wedge u) \vee ((x \vee z) \wedge y \wedge u) \vee ((x \vee u) \wedge y \wedge z).)$$

5. Let **K** be a nontrivial equational class. Show that **K** contains arbitrarily large lattices.
6. Let $\mathbf{P} = \{0, a, b\}$, $\inf\{a, b\} = 0$, and let **K** be a nontrivial equational class. Show that $F_{\mathbf{K}}(P) \cong F(2)$.
7. Find an example of an isomorphism $\varphi: F_{\mathbf{K}}(P) \to F_{\mathbf{K}}(P)$ that is *not* the identity map on P.

8. Let P be a poset, let $\mathbf{K}$, $\mathbf{N}$ be equational classes, and assume that $F_{\mathbf{K}}(P)$ and $F_{\mathbf{N}}(P)$ exist. Prove that if $\mathbf{K} \supseteq \mathbf{N}$, then there exists a homomorphism φ from $F_{\mathbf{K}}(P)$ onto $F_{\mathbf{N}}(P)$ such that φ is the identity map on P.
9. Work out a proof of the existence of $F(3)$ without any reference to Theorem 5.
10. Prove Theorem 8.
11. Formulate and prove the form of Theorem 8 that is used in the proofs of Theorems 10 and 11.
12. Prove that $F_{\mathbf{M}}(4)$ is infinite (G. Birkhoff [1933]). (Hint: Let R be the set of real numbers and let L be the lattice of vector subspaces of R^3. Set

$$a = \{\langle x, 0, x\rangle \mid x \in R\}, \quad b = \{\langle 0, x, x\rangle \mid x \in R\},$$
$$c = \{\langle 0, 0, x\rangle \mid x \in R\}, \quad d = \{\langle x, x, x\rangle \mid x \in R\}.$$

Then $[\{a, b, c, d\}]$ is infinite.)
13. Let A and B be disjoint three-element sets. Let L be the set of the following subsets of $A \cup B$: all $X \subseteq A$, all $A \cup Y$, $Y \subseteq B$, all three-element sets Z with $|Z \cap A| = 2$. Prove that $\langle L; \subseteq\rangle$ is a lattice, that $\wedge$ and $\vee$ are intersection and union, and that thus L is distributive. Figure 6 is the diagram of L (A. D. Campbell [1943]).
14. Represent the lattice of Figure 6 as a sublattice of $(\mathfrak{C}_2)^6$.
15. Prove that the exponent six is best possible in Exercise 14.
16. Represent the lattice of Figure 7 as a sublattice of $L \times \mathfrak{M}_3$, where L is the lattice of Exercise 13 and $\mathfrak{M}_3$ is given in Figure 2.2.
17. Show that condition (E) of Theorem 5 is equivalent to the following: For any $a, b, c \in P$ not satisfying $\inf\{a, b\} = c$, there exist a lattice L in $\mathbf{K}$ and a homomorphism $\varphi\colon P \to L$ with $a\varphi \wedge b\varphi \neq c\varphi$, and dually.
18. The statement "$\mathfrak{A} = \langle A; \wedge, \vee\rangle$ is a *partial algebra*" means that A is a nonvoid set and that $\wedge$, $\vee$ are partial binary operations on A. For an n-ary polynomial p, $a_0, \ldots, a_{n-1} \in A$, interpret $p(a_0, \ldots, a_{n-1})$. (When is it defined and what is its value?)
19. In the weak partial lattice represented by Figure 9, $p = x \wedge ((x \vee y) \vee (x \vee z))$ is not defined for $x = 0$, $y = e$, $z = h$. Verify that in every lattice $p = q$, where $q = x$, and q is defined in every partial lattice.
20. An identity $p = q$ *holds in the partial algebra* $\mathfrak{A} = \langle A; \wedge, \vee\rangle$ iff the following three conditions are satisfied:
    (i) If $p(a_0, \ldots)$, $q(a_0, \ldots)$ are defined $(a_0, \ldots \in A)$, then $p(a_0, \ldots) = q(a_0, \ldots)$.
    (ii) If $p(a_0, \ldots)$ is defined, $q = q_0 * q_1$, where $*$ is $\wedge$ or $\vee$, and $q_0(a_0, \ldots)$, $q_1(a_0, \ldots)$ are defined, then $q_0(a_0, \ldots) * q_1(a_0, \ldots)$ is defined.
    (iii) This condition is the same as (ii) with p and q interchanged.
    Check that Lemmas 13 and 13′ give this interpretation to the lattice axioms.
21. Let $p = (((x \vee z) \vee (y \vee u)) \vee v) \vee w$ and let $q = ((v \vee x) \vee (v \vee y)) \vee ((w \vee z) \vee (w \vee u))$. Show that $p = q$ in any lattice. Show that $p = q$ does not hold in the weak partial lattice defined in connection with Figure 9.
22. Let I_0 and I_1 be ideals of a weak partial lattice. Set $J_0 = I_0 \cup I_1$, $J_n = \{x \mid x \leq y \vee z,\ y, z \in J_{n-1}\}$, $n = 1, 2, \ldots$, and $J = \bigcup(J_i \mid i = 0, 1, 2, \ldots)$. Show that $J = I_0 \vee I_1$.
23. Let the weak partial lattice L violate (I); that is, $(a] \vee (b] = (c]$, but $a \vee b$ is undefined. Let $I_0 = (a]$, $I_1 = (b]$, and $c \in J_n$ (see Exercise 22). Generalizing Exercise 21, find an identity $p = q$ that holds in any lattice but not in L. (Exercise 21 is the special case in which $n = 2$.)
24. Prove that a partial algebra $\mathfrak{A} = \langle A; \wedge, \vee\rangle$ is a partial lattice iff every identity $p = q$ holding in any lattice also holds in $\mathfrak{A}$.
25. A *homomorphism* of a partial algebra $\mathfrak{A} = \langle A; \wedge, \vee\rangle$ into a lattice L is a map φ:

$A \to L$ such that $(a \wedge b)\varphi = a\varphi \wedge b\varphi$ whenever $a \wedge b$ exists, and the same for $\vee$. Prove that there exists a one-to-one homomorphism of $\mathfrak{A}$ into some lattice L iff there exists a partial ordering $\leq$ on A satisfying $a \wedge b = \inf \{a, b\}$, whenever $a \wedge b$ is defined in $\mathfrak{A}$, and $a \vee b = \sup \{a, b\}$, whenever $a \vee b$ is defined in $\mathfrak{A}$.

26. Show that every weak partial lattice satisfies the condition of Exercise 25, but not conversely.
27. Let $A = \{0, a, b, c, 1\}$, $0 \leq a, b, c \leq 1$. For $x \leq y$ define
$$x \wedge y = y \wedge x = x, \quad x \vee y = y \vee x = y,$$
and define
$$a \wedge b = b \wedge a = 0, \quad a \vee b = b \vee a = 1.$$
Show that $\langle A; \wedge, \vee \rangle$ is a partial lattice.
28. Let $A = \{0, a, b, c, d, 1\}$, $0 \leq a, b, c, d \leq 1$. For $x \leq y$ define
$$x \wedge y = y \wedge x = x, \quad x \vee y = y \vee x = y,$$
and define
$$a \wedge b = b \wedge a = c \wedge d = d \wedge c = 0,$$
$$a \vee b = b \vee a = c \vee d = d \vee c = 1.$$
Show that $\langle A; \wedge, \vee \rangle$ is a partial lattice.
29. Let L and K be lattices and let $L \cap K$ be a sublattice of L and of K. For $x, y \in L \cup K$ define $x \wedge y = z$ iff $x, y, z \in L$ and $x \wedge y = z$ in L, or $x, y, z \in K$ and $x \wedge y = z$ in K; define $x \vee y$ similarly. Is $\langle L \cup K; \wedge, \vee \rangle$ a partial lattice?
30. Are the eight axioms of a weak partial lattice independent?
31. Are the ten axioms of a partial lattice independent?
32. Define weak partial semilattice and partial semilattice; prove the analogue of Theorem 20 for partial semilattices.
33. Let A be a weak partial lattice in which $a \wedge b$ exists for all $a, b \in A$. Then A is a partial lattice iff $a \to (a]$ is an embedding of A into $I_0(A)$.
34. Let $\mathbf{T}$ be the equational class of all one-element lattices. Show that $F_{\mathbf{T}}(A)$ exists iff $|A| = 1$.
35. Let $P = (\mathfrak{C}_2)^2$. Show that for any nontrivial equational class $\mathbf{K}$, $F_{\mathbf{K}}(P)$ and $F_{\mathbf{K}}(P^m)$ exist and that always $F_{\mathbf{K}}(P) \ncong F_{\mathbf{K}}(P^m)$.
36. Determine $F(P)$, $F_{\mathbf{M}}(P)$, and $F_{\mathbf{D}}(P)$, where $P = \{a, b, c\}$ and $a < b$.
37. Discuss the set of all weak partial lattices on A inducing a given partial ordering on A.
38. Repeat Exercise 37 for partial lattices.
39. Show that a one-to-one-homomorphism of weak partial lattices need not be an embedding.
40. Show that in Theorem 24, the condition "there exists a lattice L in $\mathbf{K}$ such that $\mathfrak{A}$ is a relative sublattice of L" can be replaced by the following condition:
For all $a, b, c \in A$ for which $a \wedge b = c$ does not hold, there exists a lattice L in $\mathbf{K}$, and a homomorphism φ of A into L such that $a\varphi \wedge b\varphi \neq c\varphi$, and the same condition holds for $\vee$.
41. Let $\langle A; \wedge, \vee \rangle$ be a partial algebra, let $\mathbf{K}$ be an equational class of lattices, let $L \in \mathbf{K}$, and let M be a relative sublattice of L. Then M is called a ***maximal homomorphic image of A in*** $\mathbf{K}$ iff there is a homomorphism φ of A onto M such that whenever ψ is a homomorphism of A into $N \in \mathbf{K}$, then there is a homomorphism α: $M \to N$ such that $\varphi\alpha = \psi$. Prove that a maximal homomorphic image is unique up to isomorphism provided it exists.

42. Starting with an arbitrary partial algebra $\langle A; \wedge, \vee\rangle$, carry out the construction of Theorem 5 (Theorem 24).
43. Are there posets $P_0 \subset P$ such that $F(P_0^m) = F(P^m)$?
44. After M. M. Gluhov [1960], a finite partial lattice $\mathfrak{A}$ is a *basis* of a lattice L iff $L = F(\mathfrak{A})$, but $L = F(\mathfrak{A}_0)$ for no $A_0 \subset A$. Show that the lattice $L = F(\mathfrak{A})$ has more than one basis where $\mathfrak{A}$ is defined as follows: Let $\langle A; \leq\rangle$ be the poset given by Figure 10; for $x \leq y$ let $x \wedge y = y \wedge x = x$; furthermore, the join of any two elements is defined in $\{0, a, b, c, d, e, f, 1\}$ as supremum, and $1 \vee g = g \vee 1 = 1$. (This example, which is due to C. Herrmann, contradicts M. M. Gluhov's result.)

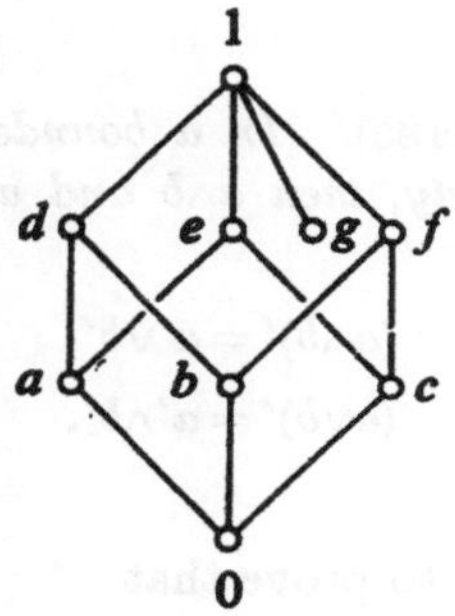

Figure 10

45. Let L and L_1 be lattices, let $L = [A]$, and let φ be a map of A into L_1. Then φ can be extended to a homomorphism ψ of L into L_1 iff $p(a_0, \ldots, a_{n-1}) = q(a_0, \ldots, a_{n-1})$ implies that $p(a_0\varphi, \ldots, a_{n-1}\varphi) = q(a_0\varphi, \ldots, a_{n-1}\varphi)$ holds for any $a_0, \ldots, a_{n-1} \in A$ and for any integer n, and pair p, q of n-ary polynomials.
46. Under the hypotheses of Exercise 45 describe ψ.
47. Characterize $\mathbf{K}$-free lattices using Exercise 45.
48. Under the hypotheses of Exercise 45, show that if A is finite and φ is isotone, then φ can always be extended to an isotone map of L into L_1.

6. Special Elements

A *bounded* poset is one that has both 0 and 1. A $\{0, 1\}$-*homomorphism* (of a bounded lattice into another one) is a homomorphism taking zero into zero and unit into unit. A $\{0, 1\}$-*sublattice* of a bounded lattice L is a sublattice containing the 0 and 1 of L. Similarly, we can define $\{0\}$-*homomorphism*, and so on, for lattices and semilattices.

In a bounded lattice L, a is a *complement* of b iff $a \wedge b = 0$ and $a \vee b = 1$.

Lemma 1. *In a bounded distributive lattice, an element can have only one complement.*

Proof. If b_0 and b_1 are both complements of a, then $b_0 = b_0 \wedge 1 = b_0 \wedge (a \vee b_1) = (b_0 \wedge a) \vee (b_0 \wedge b_1) = 0 \vee (b_0 \wedge b_1) = b_0 \wedge b_1$; similarly, $b_1 = b_0 \wedge b_1$, thus $b_0 = b_1$. ●

Let $a \in [b, c]$; x is a *relative complement of a in* $[b, c]$ iff $a \wedge x = b$ and $a \vee x = c$.

Lemma 2. *In a bounded distributive lattice, if a has a complement, then it also has a relative complement in any interval containing it.*

Proof. Let d be the complement of a. Then $x=(d\vee b)\wedge c$ is the relative complement of a in $[b, c]$, provided that $b\leq a\leq c$. Indeed,

$$a\wedge x=a\wedge(d\vee b)\wedge c=((a\wedge d)\vee(a\wedge b))\wedge c=(0\vee b)\wedge c=b,$$

and $a\vee x=c$ by duality. ●

Lemma 3 (De Morgan's Identities). *In a bounded distributive lattice, if a, b have complements a' and b', respectively, then $a\wedge b$ and $a\vee b$ have complements $(a\wedge b)'$ and $(a\vee b)'$, respectively, and*

$$(a\wedge b)'=a'\vee b',$$
$$(a\vee b)'=a'\wedge b'.$$

Proof. By Lemma 1 it suffices to prove that

$$(a\wedge b)\wedge(a'\vee b')=0, \qquad (a\wedge b)\vee(a'\vee b')=1$$

to verify the first identity; the second is dual. Compute:

$$(a\wedge b)\wedge(a'\vee b')=(a\wedge b\wedge a')\vee(a\wedge b\wedge b')=0\vee 0=0$$

and

$$(a\wedge b)\vee(a'\vee b')=(a\vee a'\vee b')\wedge(b\vee a'\vee b')=1\wedge 1=1. \quad ●$$

A *complemented lattice* is a bounded lattice in which every element has a complement. A *relatively complemented lattice* is a lattice in which every element has a relative complement in any interval containing it. A *Boolean lattice* is a complemented distributive lattice. Thus, in a Boolean lattice B, every element a has a unique complement, and B is also relatively complemented.

A *Boolean algebra* is a Boolean lattice in which 0, 1, and $'$ (complementation) are also considered to be operations. Thus a Boolean algebra is a system: $\langle B; \wedge, \vee, ', 0, 1\rangle$, where $\wedge$, $\vee$ are binary operations, $'$ is a unary operation, and 0, 1 are nullary operations. (A nullary operation picks out an element of B.) A *homomorphism* φ preserves 0, 1 and $'$; that is, it is a $\{0, 1\}$-homomorphism satisfying $(x\varphi)'=x'\varphi$. A *subalgebra* is a $\{0, 1\}$-sublattice closed under $'$. $\mathfrak{B}_2$ will denote the two-element Boolean algebra.

Note that in a bounded distributive lattice L, if b is a complement of a, then b is the largest element x of L with $a\wedge x=0$. More generally, let L be a lattice with 0; an element a^* is a *pseudocomplement* of $a(\in L)$ iff $a\wedge a^*=0$, and $a\wedge x=0$ implies that $x\leq a^*$. An element can have at most one pseudocomplement. A *pseudocomplemented lattice* is one in which every element has a pseudocomplement.

The concept of pseudocomplement involves only the meet operation. Thus we can also define *pseudocomplemented semilattices*.

Theorem 4. *Let L be a pseudocomplemented meet-semilattice, $S(L) = \{a^* \mid a \in L\}$. Then the partial ordering of L partially orders $S(L)$ and makes $S(L)$ into a Boolean lattice. For $a, b \in S(L)$ we have $a \wedge b \in S(L)$, and the join in $S(L)$ is described by*

$$a \vee b = (a^* \wedge b^*)^*.$$

Remark. Even if L is a lattice, the join in L need not be the same as the join in $S(L)$. This result was proved for complete distributive lattices by V. Glivenko [1929] and published in its full generality for the first time by O. Frink [1962]. Both proofs used special axiomatizations of Boolean algebras to get around the difficulty of proving distributivity. The present proof is direct.

Proof. We start with the following observations:

$$a \leq a^{**}. \tag{1}$$

$$a \leq b \text{ implies that } a^* \geq b^*. \tag{2}$$

$$a^* = a^{***}. \tag{3}$$

$$a \in S(L) \text{ iff } a = a^{**} \tag{4}$$

$$a, b \in S(L) \text{ implies that } a \wedge b \in S(L). \tag{5}$$

$$\text{For } a, b \in S(L), \sup_{S(L)}\{a, b\} = (a^* \wedge b^*)^*. \tag{6}$$

Formulas (1) and (2) follow from the definitions.

Formulas (1) and (2) yield $a^* \geq a^{***}$, and by (1) $a^* \leq a^{***}$, thus (3).

If $a \in S(L)$, then $a = b^*$; therefore, by (3), $a^{**} = b^{***} = b^* = a$. Conversely, if $a = a^{**}$, then $a = b^*$ with $b = a^*$; thus $a \in S(L)$, proving (4).

If $a, b \in S(L)$, then $a = a^{**}$ and $b = b^{**}$, and so $a \geq (a \wedge b)^{**}$ and $b \geq (a \wedge b)^{**}$, thus $a \wedge b \geq (a \wedge b)^{**}$; by (1), $a \wedge b = (a \wedge b)^{**}$, thus $a \wedge b \in S(L)$. If $x \in S(L)$, $x \leq a$, and $x \leq b$, then $x \leq a \wedge b$; therefore, $a \wedge b = \inf_{S(L)}\{a, b\}$, proving (5).

$a^* \geq a^* \wedge b^*$, thus by (2) and (4), $a \leq (a^* \wedge b^*)^*$; similarly, $b \leq (a^* \wedge b^*)^*$. If $a \leq x$ and $b \leq x$ ($x \in S(L)$), then $a^* \geq x^*$ and $b^* \geq x^*$ by (2); thus by (2) and (4), $(a^* \wedge b^*)^* \leq x$, proving (6).

For $a, b \in S(L)$, define $a \vee b = (a^* \wedge b^*)^*$. By Formulas (5) and (6), $\langle S(L); \wedge, \vee \rangle$ is a bounded lattice. Since for $a \in S(L)$, $a \vee a^* = (a^* \wedge a^{**})^* = 0^* = 1$ and $a \wedge a^* = 0$, $S(L)$ is a complemented lattice. Now we need only prove that $S(L)$ is distributive. For $x, y, z \in S(L)$, $x \wedge z \leq x \vee (y \wedge z)$ and $y \wedge z \leq x \vee (y \wedge z)$; therefore, $x \wedge z \wedge (x \vee (y \wedge z))^* = 0$ and $y \wedge z \wedge (x \vee (y \wedge z))^* = 0$. Thus $z \wedge (x \vee (y \wedge z))^* \leq x^*, y^*$, and so $z \wedge (x \vee (y \wedge z))^* \leq x^* \wedge y^*$. Consequently, $z \wedge (x \vee (y \wedge z))^* \wedge (x^* \wedge y^*)^* = 0$, which implies that $z \wedge (x^* \wedge y^*)^* \leq (x \vee (y \wedge z))^{**}$. Now the left-hand side is $z \wedge (x \vee y)$ by Formula (6), and the right-hand side is $x \vee (y \wedge z)$ by Formula (4). Thus we get $z \wedge (x \vee y) \leq x \vee (y \wedge z)$, which is distributivity by Lemma 4.10. ●

Other types of special elements: An element a is an *atom* if $a \succ 0$ and a *dual atom* if $a \prec 1$; it is *join-irreducible* if $a = b \vee c$ implies that $a = b$ or $a = c$; it is *meet-irreducible* if $a = b \wedge c$ implies that $a = b$ or $a = c$. An element which is both join- and meet-irreducible is called *doubly irreducible*. Examples are given in the following exercises.

Exercises

1. Find a homomorphism of bounded lattices that is not a $\{0, 1\}$-homomorphism and a sublattice that is not a $\{0, 1\}$-sublattice.
2. Find a modular lattice in which every element $x \neq 0, 1$ has exactly $\mathfrak{m}$ complements.
3. Let L be a distributive lattice, $a, b \in L$. Prove that if $a \vee b$ and $a \wedge b$ have complements, so do a and b.
4. Show that Lemma 2 holds in any modular lattice.
5. In a bounded lattice L, let x be a relative complement of a in $[b, c]$; let y be a relative complement of c in $[x, 1]$; let z be a relative complement of b in $[0, x]$; and let t be a relative complement of x in $[z, y]$. Verify that t is a complement of a.
6. Let B_0, B_1 be Boolean algebras and let φ be a $\{0, 1\}$-(lattice) homomorphism of B_0 into B_1. Show that φ is a homomorphism of the Boolean algebras.
7. Let L be a distributive lattice with 0. Show that $I(L)$ is pseudocomplemented.
8. Is the converse of Exercise 7 true?
9. Prove that if in Theorem 4 L is a lattice, then $a \mathbf{v} b$ in $S(L)$ can be described by $a \mathbf{v} b = (a \vee b)^{**}$.
10. Let L be a pseudocomplemented lattice. Show that $a^{**} \mathbf{v} b^{**} = (a \vee b)^{**}$.
11. Find arbitrarily large pseudocomplemented lattices in which $S(L) = \{0, 1\}$.
12. Prove that in a Boolean lattice, $x \neq 0$ is join-irreducible iff x is an atom.
13. Show that in a finite lattice every element is the join of join-irreducible elements.
14. Verify that "finite lattice" in Exercise 13 can be replaced by "lattice satisfying the Descending Chain Condition." (A lattice L or, in general, a poset, satisfies the *Descending Chain Condition* iff $x_0, x_1, x_2, \ldots \in L$, $x_0 \geq x_1 \geq x_2 \geq \cdots$ implies that $x_n = x_{n+1} = \cdots$ for some n; see Exercise 2.7).
15. Show that some form of the Axiom of Choice must be used to verify Exercise 14.
16. Prove that if a lattice satisfies the Descending Chain Condition, then every nonzero element contains an atom.
17. The dual of the Descending Chain Condition is the Ascending Chain Condition (see Exercise 2.6). Dualize Exercises 13–16.
18. Show that a lattice satisfies the Ascending Chain Condition and the Descending Chain Condition iff all chains are finite.
19. Find a lattice in which all chains are finite but that contains a chain of n elements for every natural number n.
20. Find a lattice in which there are no join- or meet-irreducible elements.
21. Prove that the Ascending Chain Condition (Descending Chain Condition) holds in a lattice L iff every ideal (dual ideal) of L is principal.
22. Let L be a lattice and let C be a chain with $|C| \leq \aleph_0$. Prove that if there is a homomorphism of L onto C, then L contains an isomorphic copy of C. How does this result extend to countable lattices C? Does this result hold for the chain of real numbers? (No.)
23. Let $\wedge$ be a binary operation on L, let $*$ be a unary operation on L (that is, for every $a \in L$, $a^* \in L$), and let 0 be a nullary operation (that is, $0 \in L$). Let us assume that the following hold for all $a, b, c \in L$: $a \wedge b = b \wedge a$, $(a \wedge b) \wedge c = a \wedge (b \wedge c)$, $a \wedge a = a$, $0 \wedge a = 0$, $a \wedge (a \wedge b)^* = a \wedge b^*$, $a \wedge 0^* = a$, $(0^*)^* = 0$. Show that $\langle L; \wedge \rangle$ is a meet-semilattice with 0 as zero, and for all $a \in L$, a^* is the pseudocomplement of a (R. Balbes and A. Horn [1970]).
24. Let L be a pseudocomplemented meet-semilattice and let $a, b \in L$. Verify the formulas $(a \wedge b)^* = (a^{**} \wedge b)^* = (a^{**} \wedge b^{**})^*$.

25. Let L be a meet-semilattice and let $a, b \in L$. The *pseudocomplement of a relative to b* is an element $a * b$ of L satisfying $a \wedge x \leq b$ iff $x \leq a * b$. Show that $a * b$ is unique if it exists; show that $a * a$ exists iff L has a unit.
26. Let L be a *relatively pseudocomplemented meet-semilattice* (that is, L is a meet-semilattice and $a * b$ exists for all $a, b \in L$). Show that L has 1; for all $a, b, c \in L$: $a * (b * c) = (a \wedge b) * c$ and $a * (b * c) = (a * b) * (a * c)$. Furthermore, if L is a lattice, then L is distributive.
27. Let L be a pseudocomplemented distributive lattice. Prove that for each $a \in L$, $(a]$ is a pseudocomplemented distributive lattice; in fact, the pseudocomplement of $x \in (a]$ in $(a]$ is $x^* \wedge a$.
*28. Using the notation of Exercise 27, let $S(a)$ denote the elements of the form $x^* \wedge a$, $x \leq a$. Then $S(a)$ is a Boolean algebra by Theorem 4. Let $\vee_a$ denote the join in $S(a)$. Show that if $x, y \in S(a)$ and $x, y \in S(b)$ $(a, b \in L)$, then $x \vee_a y = x \vee_b y$.
*29. Let $b \in S(a)$. Prove that $S(b) \subseteq S(a)$. (The results of Exercises 28 and 29 first appeared in FC.)
30. Show that T_n (see Exercise 2.36) is complemented.
31. Show that in T_n every interval is pseudocomplemented. (Exercises 30 and 31 are due to H. Lakser.)
32. For a finite lattice L, let $\mathrm{Irr}(L)$ denote the set of all doubly irreducible elements. Prove that[1] $|L| \geq 2\,(l(L)+1) - |\mathrm{Irr}(L)|$.
33. Show that for a finite lattice L,

$$l(\mathrm{Sub}(L)) = |\mathrm{Irr}(L)| + l(\mathrm{Sub}(L - \mathrm{Irr}(L))).$$

34. A finite lattice L of n elements is *dismantlable* iff there is a chain $L_1 \subset L_2 \subset \cdots \subset L_n = L$ of sublattices satisfying $|L_i| = i$. Show that every lattice with at most seven elements in dismantlable. (Hint: use Exercise 32.)
35. Show that for every integer $n \geq 8$ there is a lattice of n elements which is not dismantlable.
36. A finite lattice is *planar* iff it has a planar diagram. Show that every finite planar lattice is dismantlable.
37. For every integer $n \geq 9$, construct an n-element dismantlable lattice which is not planar.[2]
38. If a dismantlable lattice L is not a chain, then it contains two incomparable doubly irreducible elements.
39. Prove that every sublattice and homomorphic image of a dismantlable lattice is dismantlable.[3] Is this also true for planar lattices?

***

Let P and Q be posets and let $\alpha: P \to Q$ and $\beta: Q \to P$ satisfy the following conditions.
(G_1) For $x, y \in P$, if $x \leq y$, then $x\alpha \leq y\alpha$.
(G_2) For $x, y \in Q$, if $x \leq y$, then $x\beta \leq y\beta$.
(G_3) For $x \in P$, $x\alpha\beta \geq x$.
(G_4) For $x \in Q$, $x\beta\alpha \geq x$.
Then $\langle \alpha, \beta \rangle$ is a *Galois connection* between P and Q.
This concept was introduced in G. Birkhoff [1940]. The results presented in the following exercises are based on G. Birkhoff [1940] and O. Ore [1944]. They are

---

[1] Exercises 32–35 are based on I. Rival [1974a].
[2] For Exercises 36–37, see K. A. Baker, P. C. Fishburn, and F. S. Roberts [1970].
[3] Exercises 38 and 39 are from D. Kelly and I. Rival [1974].

modelled after the correspondence between subgroups of the Galois group and subfields of a separable field extension which is the subject of the classical Fundamental Theorem of Galois Theory.

40. Let L be a pseudocomplemented meet-semilattice. Let $P=Q=L$ and, for $x \in L$, let $x\alpha = x\beta = x^*$. Show that $\langle \alpha, \beta \rangle$ is a Galois connection.

41. Let X_0 and X_1 be sets and $r \subseteq X_0 \times X_1$. For $A \subseteq X_0$, set

$$A^1 = \{x_1 \mid x_1 \in X_1 \text{ and } \langle x_0, x_1 \rangle \in r \text{ for some } x_0 \in A\};$$

for $B \subseteq X_1$, set

$$B^0 = \{x_0 \mid x_0 \in X_0 \text{ and } \langle x_0, x_1 \rangle \in r \text{ for some } x_1 \in B\}.$$

$A \to A^1$ and $B \to B^0$ set up a Galois connection between $P(X_0)$ and $P(X_1)$.

42. Let $\langle \alpha, \beta \rangle$ be a Galois connection between the posets P and Q. Show that $\alpha\beta$ is a closure map on P and $\beta\alpha$ is a closure map on Q. (A *closure map* χ on a poset is an isotone map of the poset into itself satisfying $x \leq x\chi$, $x\chi\chi = x\chi$. If $x\chi = x$, then x is called *closed*.)

43. Let P_C and Q_C be the set of closed elements in P and Q, respectively (P, Q, α, β as in Exercise 42). Prove that if P and Q are complete lattices, then so are P_C and Q_C, α is an isomorphism of P_C with the dual of Q_C, and β is the inverse of α.

44. How much of Theorem 4 can be derived from Exercises 40–43?

Further Topics and References

Many of the concepts and results discussed in Chapter 1 are special cases of universal algebraic concepts and results. To see this, the reader needs the definition of a universal algebra. An *n-ary operation* f on a nonvoid set A is a map from A^n into A; in other words, if $a_1, \ldots, a_n \in A$, then $f(a_1, \ldots, a_n) \in A$. If $n=1$, f is called *unary*; if $n=2$, f is called *binary*. Since $A^0 = \{\varnothing\}$, a *nullary operation* ($n=0$) is determined by $f(\varnothing) \in A$, and f is sometimes identified with $f(\varnothing)$. A *universal algebra*, or simply *algebra*, consists of a nonvoid set A and a set F of operations; each $f \in F$ is an n-ary operation for some n (depending on f). We denote this algebra by $\mathfrak{A}$ or $\langle A; F \rangle$. Many of the results of Sections 3–5 can be formulated and proved for arbitrary universal algebras. For more details, see Chapters 1–4 of the author's book [1968].

In every poset we can introduce (as suggested by the real line) a ternary relation r called *betweenness*: $r(a, b, c)$ iff $a \leq b \leq c$ or $c \leq b \leq a$. M. Altwegg [1950] proves that the partial ordering can be defined in terms of betweenness (naturally, up to duality); see also J. M. Cibulskis [1969]. In lattices, several betweenness relations are known, see E. Pitcher and M. F. Smiley [1942]. For applications of these relations, see M. F. Smiley and W. R. Transcue [1943] and R. Padmanabhan [1966a].

Lattices and Boolean algebras can be defined by identities in innumerable ways. Of the eight identities we used to define lattices ((L1)–(L4) of Section 1), two ((L1)) can be dropped. Ju. I. Sorkin [1951] reduces six to four, R. Padmanabhan [1969] to two, and finally in R. N. McKenzie [1970] a single identity is found characterizing lattices. Ju. I. Sorkin's identities use only three variables; the others use more. More recently, R. Padmanabhan [1972] has found two identities in three variables characterizing lattices. It is easy to see that two variables would not suffice. Take the lattice of Figure 5.8 and redefine the join of the two atoms to be 1; otherwise keep all the joins and meets. The resulting algebra is not a lattice, but every subalgebra generated by two elements is a lattice. Therefore, lattices cannot be defined by identities in two variables. By means of a more complicated construction, A. H. Diamond and J. C. C. McKinsey [1947] derive the same conclusion for Boolean algebras. Finite algebras in which every two-generated subalgebra is Boolean have been investigated in R. W. Quackenbush [1974].

A result of A. Tarski [1968] states that, given any integer n, there exists a set of n identities defining lattices such that no identity can be dropped from the set.

By Exercise 4.24, modular and distributive lattices can be defined by two identities. Nicer sets of identities for these cases can be found in M. Kolibiar [1956] ($(a \vee (b \wedge b)) \wedge b = b$ and $((a \wedge b) \wedge c) \vee (a \wedge d) = ((d \wedge a) \vee (c \wedge b)) \wedge a$ characterize modular lattices) and M. Sholander [1951] ($a \wedge (a \vee b) = a$ and $a \wedge (b \vee c) = (c \wedge a) \vee (b \wedge a)$ characterize distributive lattices). See also J. Riečan [1958].

One of the most useful axiomatizations of Boolean algebras is that of E. V. Huntington [1904] (observe that a proof of Huntington's result is implicit in the proof of Theorem 6.4) according to which a Boolean algebra is a complemented lattice in which the complementation is pseudocomplementation (that is, $a \wedge x = 0$ implies that $x \leq a'$). Contrast this with Corollary VI. 3.8.

One of the briefest axiom systems of Boolean algebras in terms of $\wedge$ and $'$ is due to L. Byrne [1946]: $a \wedge b = b \wedge a$, $a \wedge (b \wedge c) = (a \wedge b) \wedge c$, $a \wedge b' = c \wedge c'$ iff $a \wedge b = a$. Characterization by identities is usually longer. R. N. McKenzie, A. Tarski, and the author observed independently that Boolean algebras can be defined by a single identity (see A. Tarski [1968], and G. Grätzer and R. N. McKenzie [1967]). A thorough survey of the axiom systems of Boolean algebras is given in S. Rudeanu [1963]; see also F. M. Sioson [1964], and S. Rudeanu [1974]. The only known irredundant selfdual axiom system can be found in R. Padmanabhan [1974a].

Boolean algebras originated as an algebraic formalization of propositional logic. Most introductory logic texts give satisfactory expositions of these ideas; see also P. R. Halmos [1963]. There is a similar relationship between Boolean algebras and the theory of switching circuits; see M. A. Harrison [1965].

P. Ribenboim [1949] first pointed out that pseudocomplementation can be described by identities (involving *). A. Monteiro [1955] accomplished the same for relative pseudocomplementation; see also R. Balbes and A. Horn [1970a] and R. Balbes and Ph. Dwinger [1974]. This fact is applied in G. Grätzer [1969].

The examples of lattice identities we have seen so far seem to suggest that all such identities are selfdual. The identity $(a \wedge b) \vee (a \wedge c) = a$, where $a = x \wedge ((x \wedge y) \vee (y \wedge z) \vee (z \wedge x))$, $b = (x \wedge y) \vee (y \wedge z)$, $c = (x \wedge z) \vee (y \wedge z)$, is an example of a nonselfdual identity. This identity holds in a lattice L iff L does not have a sublattice isomorphic to the dual of the lattice of Figure V. 2.7; see H. F. J. Löwig [1943].

A diagram of a finite lattice is a graph; however, very little work has been done in lattice theory from a graph theoretic point of view. It is known that a finite distributive lattice is planar iff no element is covered by more than two elements. Planar modular lattices are characterized in R. Wille [1974]. It is pointed out in K. A. Baker, P. C. Fishburn, and F. S. Roberts [1970] that a finite lattice is planar iff it has order dimension one or two (a poset $\langle P; \leq \rangle$ is of *order dimension* n iff the relation $\leq$ is the set intersection of n, but not of fewer, linear orders on P), a result they attribute to J. Zilber; they also point out that the corresponding result for posets does not hold. A graph theoretic characterization of planar lattices is given in C. R. Platt [1976]. A characterization by exclusions of an infinite list of lattices as subposets can be found in D. Kelly and I. Rival [1975a]. (D. Kelly and I. Rival [1974] investigate dismantlable lattices, a generalization of planar lattices.) For a general review of planar lattices see R. W. Quackenbush [1973]. Pairs of lattices whose diagrams are isomorphic as graphs are investigated by J. Jakubík [1954], [1954a]. See also J. Jakubík and M. Kolibiar [1954].

Automorphism groups of planar graphs are described in L. Babai [1972] and [1973]. L. Babai used these results to characterize automorphism groups of planar lattices (unpublished).

Sublattices suggest a number of problems: For an equational class $\mathbf{K}$ of lattices, let $f_{\mathbf{K}}(k, n)$ denote the smallest integer such that any lattice in $\mathbf{K}$ having at least $f_{\mathbf{K}}(k, n)$ elements has at least k sublattices of exactly n elements; write $f(k, n)$ for $f_{\mathbf{L}}(k, n)$. The function $f(1, n)$ has been investigated: $f(1, t) = t$ for $t \leq 6$ (I. Kaplansky); $f(1, 7) = 9, f(1, 8) = 8, f(1, 9) \geq 17, f(1, n) \geq n + 2$ for $n \geq 10$ (I. Rival [c]); $f(1, n) \leq n^{4n-3}$ (R. Wille, unpublished); and $f(1, n) \leq n^{3^n}$ (G. Havas and M. Ward [1969]); see also M. Curzio [1953]. Related questions concerning the *spectrum* of a finite lattice L ($= \{k \mid L$ has a k element sublattice$\}$), the size of maximal sublattices, dismantlability, etc. have been investigated in a number of papers; see I. Rival [1974a], [c] and D. Kelly and I. Rival [1974] and [1975a].

Infinite lattices have, as a rule, very many large sublattices. T. P. Whaley [1969] proves that if $|L|$ is an infinite regular cardinal, then the lattice L has an infinite chain of sublattices of cardinality $|L|$.

The intersection $\Phi(L)$ of all maximal proper sublattices of a lattice is called the *Frattini sublattice* (a misnomer, since $\Phi(L) = \varnothing$ is possible). If the lattice L satisfies the Ascending or the Descending Chain Condition, then $\Phi(L) = \varnothing$ iff L is a chain; furthermore, every lattice K can be represented as $\Phi(L)$ for a suitable lattice K; see K. M. Koh [1971]. For planar modular lattices and distributive lattices, the Frattini sublattices are investigated in K. M. Koh [1973a] and C. C. Chen, K. M. Koh, and S. K. Tan [1973], [1975]. It is proved in M. E. Adams [1973] that every distributive lattice is the Frattini sublattice of a suitable distributive lattice.

The lattice of sublattices Sub(L) and the lattice of subsemilattices of a semilattice are investigated in N. D. Filippov [1966] and L. N. Ševrin [1964], respectively. They give necessary and sufficient conditions for $\mathrm{Sub}(L) \cong \mathrm{Sub}(L_1)$. Filippov's results show that if L is distributive (Boolean) and $\mathrm{Sub}(L) \cong \mathrm{Sub}(L_1)$, then L_1 is distributive (Boolean): see also I. Rival [1972]. The lattice of subalgebras of a Boolean algebra has been characterized in D. Sachs [1962]; see also G. Grätzer, K. M. Koh, and M. Makkai [1972].

Sub(L) has few of the properties discussed in this book. Sub(L) is modular iff L is a chain (K. M. Koh [1973]) and it is semimodular (in the sense of Section IV.2) iff L does not contain a sublattice isomorphic to $\mathfrak{C}_2 \times \mathfrak{C}_3$ (H. Lakser [1973b]).

A new area of study is the investigation of finite sublattices of lattices to prove that many properties can be characterized by the existence of finite sublattices of a certain type. Of course, results of this sort are well known for identities; see Theorem II.1.1 and the discussion of splitting lattices in Section V.2. Recently, similar results were found for properties of many different kinds: three-generated lattices (B. A. Davey and I. Rival [a]), ($SD_\wedge$) and ($SD_\vee$) of Section VI.1 (B. A. Davey, W. Poguntke, and I. Rival [1975], (W) of Section VI.1 in the presence of ($SD_\wedge$) and ($SD_\vee$) (R. Antonius and I. Rival [1974]). A free lattice also has very many finite sublattices; see H. S. Gaskill [a]. Apparently, an infinite lattice has many more typical finite sublattices than has been suspected before.

By Lemma I.4.8, identities are preserved when passing from L to $I(L)$. In G. Grätzer [1970] the problem is posed: which first-order properties (properties that can be formulated as a first order sentence in the sense of Exercises V.1.23 and V.1.24) are preserved when passing from L to $I(L)$, or conversely? The theory of transferable and sharply transferable lattices and semilattices that grew out of this problem will be discussed in the Concluding Remarks. See also K. A. Baker and A. W. Hales [1974] and I. Rival and B. Sands [1975].

A very important property of complete lattices is the Fixed-Point Theorem:

Any isotone map f of a complete lattice L into itself has a fixed point (that is, $f(a)=a$ for some $a \in L$); in fact, $f(a)=a$ if a is defined by

$$a=\bigvee(b \mid b \in L, b \leq f(b))$$

–see A. Tarski [1955] and B. Knaster [1928]. A. C. Davis [1955] proved that if this theorem holds in a lattice L, then L is complete. Various generalizations are given in A. Tarski [1955], E. S. Wolk [1957], A. Pelczar [1961] and [1962], S. Abian and A. B. Brown [1961], V. Devidé [1963], S. R. Kogalovskiĭ [1964], R. Demarr [1964], and B. R. Salinas [1969]. Posets of length one in which every isotone map has a fixed point are characterized in I. Rival [a].

The poset of all partial order relations on a set X has an interesting property recently discovered by B. M. Schein [1972a]: every lattice can be embedded in such a poset as a sublattice. Compare this with Whitman's result that every lattice can be embedded in a partition lattice (a stronger form of which is Theorem IV.4.4).

It is hard to overemphasize the importance of free lattices; they provide one of the most important research tools of lattice theory. Two typical applications to modular lattices can be found in G. Grätzer and E. T. Schmidt [1961] and G. Grätzer [1966]. We discuss free lattices in great detail in Section VI.2 and we use the method of constructing special lattices $F_{\mathbf{K}}(P)$ throughout the book.

Few of the topics in this chapter have been investigated for semilattices. Congruences are one exception (see G. Zacher [1952], D. Papert [1964], R. A. Dean and R. H. Oehmke [1964], R. Permutti [1964], J. Varlet [1965], E. T. Schmidt [1969], R. Freese and J. B. Nation [1973]). Congruences of partial lattices are considered in G. Grätzer and H. Lakser [1968].

Distributivity (see also Section II.5) and modularity of semilattices is investigated in J. B. Rhodes [1975] and B. M. Schein [1972], see also I. Fleischer [1975]. For some other topics on semilattices, see G. Grätzer and H. Lakser [1969b], G. Bruns and H. Lakser [1970], A. Horn and N. Kimura [1971], H. S. Gaskill [1972], K. A. Baker [1973], and K. M. Koh [1971a].

Problems

I.1 (FC 5). Characterize the comparability relations of semilattices and lattices. (See Exercises 1.34 and 1.35. R. N. McKenzie pointed out that neither of these two problems has a first-order solution.)

I.2 (FC 6). Characterize the lattices T_n and finite lattices that can be embedded in some T_n. (See Exercises 2.26–2.36, 6.30 and 6.31.)

I.3 (FC 7). Characterize the lattice Sub(L) of all sublattices of a lattice L.

I.4. Find conditions under which Sub(L) determines L up to isomorphism.

I.5 (FC 8). For what equational classes K does $L \in \mathbf{K}$ and $\mathrm{Sub}(L) \cong \mathrm{Sub}(L_1)$ imply that $L_1 \in \mathbf{K}$? (This is known for $\mathbf{K} = \mathbf{L}, \mathbf{M}, \mathbf{D}$, and $\mathbf{T}$.)

I.6. For an equational class K, let Sub(K) denote the equational class generated by $\{\mathrm{Sub}(L) \mid L \in \mathbf{K}\}$. Determine all equational classes K, for which $\mathrm{Sub}(\mathbf{K}) = \mathbf{K}$.

I.7. Determine all equational classes K, for which $\mathrm{Sub}(\mathbf{K}) = \mathbf{L}$.

I.8. Which first-order properties of lattices are preserved under projectivities of lattices, that is, under $\mathrm{Sub}(L) \cong \mathrm{Sub}(L_1)$?

I.9. Which first-order properties are preserved when passing from L to Sub(L), and conversely?

I.10. Let Csub(L) be the lattice of all convex sublattices (including $\varnothing$) of the lattice L. Solve I.3–I.9 with Csub(L) replacing Sub(L). (In I.8, projectivity, that is, $\mathrm{Sub}(L) \cong \mathrm{Sub}(L_1)$, has to be replaced by $\mathrm{Csub}(L) \cong \mathrm{Csub}(L_1)$. Csub($L$) is investigated in K. M. Koh [1972], C. C. Chen and K. M. Koh [1972], and J. W. de Bakker [1967]. See also W. D. Duthie [1942].)

I.11. Relate the automorphism groups of L, Sub(L), and Csub(L).

I.12. Characterize planar posets.

I.13. Determine the order dimensions of planar posets.[1]

I.14. Is there an analogue of the Kelly-Rival theorem for finite lattices of order dimension n ($\leq n$)?

I.15. Relate the automorphism group of a planar lattice L to the automorphism group of a planar diagram of L (as a graph).

I.16 (FC 10). Define the *complexity* of a finite lattice as the number of intersections of lines in an optimal diagram. Investigate this concept.

I.17 (FC 11). For an equational class K of lattices, determine the set of all natural numbers n, such that n occurs as the complexity of a finite $L \in \mathbf{K}$.

I.18. Determine the complexity of $(\mathfrak{C}_2)^n$ and Part(A).

I.19. Compute the complexity of $A \times B$.

I.20. Is there a connection between complexity and order dimension?

I.21. Investigate first-order properties which are preserved when passing from L to $I(L)$; and conversely. (See G. Grätzer [1970] and K. A. Baker and A. W. Hales [1974].)

I.22 (FC 14). A (semi) lattice K is called *transferable* iff whenever, for a (semi) lattice L, K has an embedding φ in $I(L)$, then K has an embedding ψ in L. Characterize transferability. (See Concluding Remarks.)

---

[1] Solved by D. Kelly, On the dimension of partially ordered sets. Manuscript.

I.23. Is a sublattice or the dual of a finite transferable lattice transferable?

I.24. If in I.22 the embedding ψ can be chosen so as to satisfy the condition: $x\psi \in y\varphi$ iff $x \leq y$, then K is called *sharply transferable*. Characterize sharply transferable lattices. (See Concluding Remarks.)[1]

I.25. Find a direct proof that a sublattice and the dual of a finite sharply transferable lattice are sharply transferable.

I.26. Is a finite transferable lattice sharply transferable? The same question for finite semilattices.

I.27. Impose conditions on a class **K** of partial lattices such that **K**-transferable and sharply **K**-transferable become meaningful and include the corresponding concepts for lattices and semilattices. Investigate these new concepts, especially in the light of Problems I.22–I.26.

I.28. For an equational class **K** of lattices, investigate **K**-transferability and sharp **K**-transferability. (Results are known for $\mathbf{K} = \mathbf{L}$ and $\mathbf{K} = \mathbf{D}$, see Concluding Remarks.)

I.29. Is there an equational class **K** of lattices, $\mathbf{K} \neq \mathbf{L}$, such that any finite lattice L which is sharply **K**-transferable is projective in **K**?

I.30 (FC 16). Describe the equational class of lattices with a single identity in three variables.

I.31 (FC 17). Find the shortest single identity characterizing lattices.

I.32 (FC 19). Define and investigate weak partial lattices and partial lattices satisfying an identity.

I.33. Which lattice identities are equivalent to one of the form $p \leq q$, where p and q are lattice polynomials and no variable occurs twice in p. (See Lemma IV.5.11.)

I.34 (FC 20). For a partial lattice A, define a congruence relation as follows: A binary relation Θ on A is a *congruence* iff there exists a lattice L and a congruence relation Φ on L such that (i) A is a relative sublattice of L and (ii) Φ restricted to A is Θ. Give an intrinsic characterization of this concept of congruence relation. (This is known for universal algebras in general; see G. Grätzer [1968], Theorem 13.3.)

I.35 (FC 21). Characterize the lattice of all congruence relations of a semilattice.

I.36 (FC 22). Let L be a meet-semilattice with 0 for which $[0, a]$ is pseudocomplemented for each $a \in L$; let $S(a)$ denote the pseudocomplements in $[0, a]$. Characterize the family of Boolean algebras $\{S(a) \mid a \in L\}$. (See Exercises 6.28 and 6.29.)

I.37 (FC 23). What is the order of magnitude of $f(k, n)$? (Note that $f(k, n)$ is defined in Further Topics and References. B. Wolk pointed out that the argument of G. Havas and M. Ward [1969] can be modified so as to prove that if L has more than n^{3^n} elements, then L has an n element sublattice in which all but at most two elements are doubly irreducible. This implies the existence of $f(k, n)$.)

I.38 (FC 24). For what equational classes **K**, do all $f_{\mathbf{K}}(k, n)$ exist?

I.39 (FC 27). Investigate the lattice of join-endomorphisms of a finite lattice. (See G. Grätzer and E. T. Schmidt [1958].)

---

[1] Solved by G. Grätzer and C. R. Platt, A characterization of sharply transferable lattices. Abstract. Notices Amer. Math. Soc., February 1977.

I.40. Determine those equational classes $\mathbf{K}$ of lattices for which every $K \in \mathbf{K}$ is the Frattini sublattice of a suitable $L \in \mathbf{K}$.

I.41. For an equational class $\mathbf{K}$, a partial $\mathbf{K}$-lattice is a relative sublattice of a lattice in $\mathbf{K}$. Examine special cases when characterizations along the lines of Theorem 5.20 can be found.

I.42. Find a class $\mathbf{K}$ of lattices such that for all pairs of ordinals $\alpha < \beta$, $\underrightarrow{L}^{\alpha}(\mathbf{K}) \subset \underrightarrow{L}^{\beta}(\mathbf{K})$. The same problem for inverse limits. Can this $\mathbf{K}$ (or the $\mathbf{K}$ for which $\underrightarrow{L}^{\alpha}(\mathbf{K}) \neq \underrightarrow{L}^{\alpha+1}(\mathbf{K})$) be constructed to satisfy some nontrivial identities? ($\underrightarrow{L}^{\alpha}(\mathbf{K})$ ($\underleftarrow{L}^{\alpha}(\mathbf{K})$) is obtained from $\mathbf{K}$ by applying direct limits (inverse limits) α times, see Problem 26 of G. Grätzer [1968]. For fixed α and β, C. R. Platt [1974] constructed a class $\mathbf{K}$ of lattices satisfying $\underrightarrow{L}^{\alpha}(\mathbf{K}) \neq \underrightarrow{L}^{\alpha+1}(\mathbf{K})$ and $\underleftarrow{L}^{\beta}(\mathbf{K}) \neq \underleftarrow{L}^{\beta+1}(\mathbf{K})$.)

CHAPTER II

DISTRIBUTIVE LATTICES

1. Characterization Theorems and Representation Theorems

The two typical examples of nondistributive lattices are $\mathfrak{M}_3$ and $\mathfrak{N}_5$, whose diagrams are given in Figure 1.

Theorem 1 is a striking and useful characterization of distributive lattices; Theorem 2 is a more detailed version of Theorem 1.

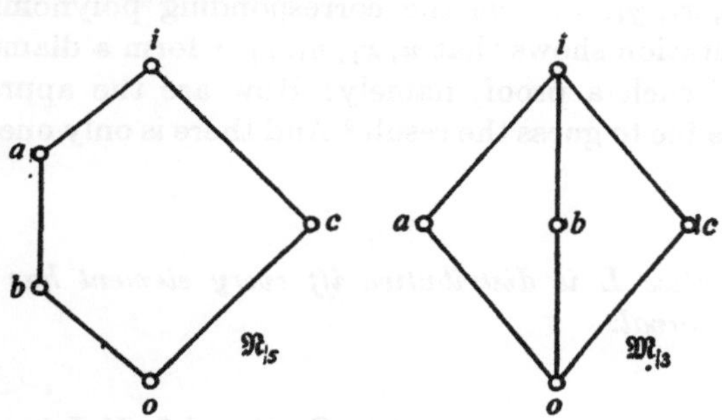

Figure 1

It is clear from Theorem 1 that sublattices isomorphic to $\mathfrak{N}_5$ or $\mathfrak{M}_3$ play an important role. We introduce special names and notation for them. A subset A of a lattice L is called a *pentagon* or *diamond* iff A is a sublattice isomorphic to $\mathfrak{N}_5$ or $\mathfrak{M}_3$, respectively. If we say that $A=\{x_0, x_1, x_2, x_3, x_4\}$ is a pentagon (respectively, diamond) we also assume that $x_0 \to o$, $x_1 \to a$, $x_2 \to b$, $x_3 \to c$, $x_4 \to i$ is an isomorphism of A and $\mathfrak{N}_5$ (respectively, $\mathfrak{M}_3$).

THEOREM 1. *A lattice L is distributive iff L does not contain a pentagon or a diamond.*

THEOREM 2.
(i) *A lattice L is modular iff it does not contain a pentagon.*
(ii) *A modular lattice L is distributive iff it does not contain a diamond.*

PROOF.
(i) If L is modular, then every sublattice of L is also modular; $\mathfrak{N}_5$ is not modular, thus it cannot be isomorphic to a sublattice of L. Conversely, let L be nonmodular,

let $a, b, c \in L$, $a \geq b$, and let $(a \wedge c) \vee b \neq a \wedge (c \vee b)$. The free lattice generated by a, b, and c with $a \geq b$ is shown in Figure I.2.3. Therefore, the sublattice of L generated by a, b, and c must be a homomorphic image of the lattice of Figure I.2.3. Observe that if any two of the five elements $a \wedge c$, $(a \wedge c) \vee b$, $a \wedge (b \vee c)$, $b \vee c$, c are identified under a homomorphism, then so are $(a \wedge c) \vee b$ and $a \wedge (b \vee c)$. Consequently, these five elements are distinct in L, and they form a pentagon. ◑

(ii) Let L be modular, but nondistributive, and choose $x, y, z \in L$ such that $x \wedge (y \vee z) \neq (x \wedge y) \vee (x \wedge z)$. The free modular lattice generated by x, y, z is shown in Figure I.5.7. By inspecting the diagram we see that u, x_1, y_1, z_1, v form a diamond. Thus in *any* modular lattice they form a sublattice isomorphic to a quotient lattice of $\mathfrak{M}_3$. But $\mathfrak{M}_3$ has only two quotient lattices: $\mathfrak{M}_3$ and the one-element lattice. In the former case we have finished the proof. In the latter case, note that if u and v collapse, then so do $x \wedge (y \vee z)$ and $(x \wedge y) \vee (x \wedge z)$, contrary to our assumption. ●

Naturally, these results could be proved without any reference to free lattices. A routine proof of (ii) runs as follows: Take $x, y, z \in L$ such that $x \wedge (y \vee z) \neq (x \wedge y) \vee (x \wedge z)$ and define u, x_1, y_1, z_1, v as the corresponding polynomials of Figure I.5.7. Then a direct computation shows that u, x_1, y_1, z_1, v form a diamond. There is a very natural objection to such a proof, namely: How are the appropriate polynomials found? How is it possible to guess the result? And there is only one answer: by working out the free lattice.

Corollary 3. *A lattice L is distributive iff every element has at most one relative complement in any interval.*

Proof. The "only if" part was proved in Section I.6. If L is nondistributive, then by Theorem 2 it contains a pentagon or a diamond, and each has an element with two relative complements in some interval. ●

Corollary 4. *A lattice L is distributive iff for any two ideals I, J of L:*

$$I \vee J = \{i \vee j \mid i \in I, j \in J\}.$$

Proof. Let L be distributive. By Lemma I.3.1(ii), if $t \in I \vee J$, then $t \leq i \vee j$ for some $i \in I$ and $j \in J$. Therefore, $t = (t \wedge i) \vee (t \wedge j)$, $t \wedge i \in I$ and $t \wedge j \in J$. Conversely, if L is nondistributive, then L contains elements a, b, c as in Figure 1. Let $I = (b]$ and $J = (c]$; observe that $a \in I \vee J$, since $a \leq b \vee c$. However, a has no representation as required by Corollary 4, since if $a = i \vee j$, $i \in I$, $j \in J$, then $j \leq a$ and $j \leq c$. Therefore, $j \leq a \wedge c < b$; thus $j \in I$, and so $a = i \vee j \in I$, a contradiction. ●

Another important property of ideals of a distributive lattice is:

Lemma 5. *Let I and J be ideals of a distributive lattice L. If $I \wedge J$ and $I \vee J$ are principal, then so are I and J.*

Proof. Let $I \wedge J = (x]$ and $I \vee J = (y]$. Then $y = i \vee j$ for some $i \in I$ and $j \in J$. Set $c = x \vee i$ and $b = x \vee j$; note that $c \in I$ and $b \in J$ (since $x \in I \wedge J = I \cap J$). We claim that $I = (c]$ and $J = (b]$. Indeed, if, for instance, $J \neq (b]$, then there is an $a > b$, $a \in J$, and $\{x, a, b, c, y\}$ form a pentagon. ●

Theorem 6. *Let L be a distributive lattice and let $a \in L$. Then the map*

$$\varphi : x \to \langle x \wedge a, x \vee a \rangle, \quad x \in L,$$

is an embedding of L into $(a] \times [a)$; it is an isomorphism if a has a complement.

Proof. φ is one-to-one, since if $x\varphi = y\varphi$, then x and y are both relative complements of a in the same interval; thus $x = y$ by Corollary 3. Distributivity also implies that φ is a homomorphism.

If a has a complement, b, and $\langle u, v \rangle \in (a] \times [a)$, then for $x = (u \vee b) \wedge v$, $x\varphi = \langle u, v \rangle$; therefore, φ is an isomorphism. ●

We start the detailed investigation of the structure of distributive lattices with the finite case.

Definition 7. *For a distributive lattice L let J(L) denote the set of all nonzero join-irreducible elements, regarded as a poset under the partial ordering of L. For $a \in L$ set*

$$r(a) = \{x \mid x \leq a, x \in J(L)\} = (a] \cap J(L).$$

Definition 8. *For a poset P, call $A \subseteq P$* hereditary *iff $x \in A$ and $y \leq x$ imply that $y \in A$. Let H(P) denote the set of all hereditary subsets partially ordered by set inclusion.*

Note that $H(P)$ is a lattice in which meet and join are intersection and union, respectively, and thus $H(P)$ is distributive. The structure of finite distributive lattices is revealed by the following result:

Theorem 9. *Let L be a finite distributive lattice. Then the map*

$$\varphi : a \to r(a)$$

is an isomorphism between L and $H(J(L))$.

Proof. Since L is finite, every element is the join of nonzero join-irreducible elements; thus

$$a = \bigvee r(a),$$

showing that φ is one-to-one. Obviously, $r(a) \cap r(b) = r(a \wedge b)$, and so $(a \wedge b)\varphi = a\varphi \wedge b\varphi$. The formula $(a \vee b)\varphi = a\varphi \vee b\varphi$ is equivalent to

$$r(a \vee b) = r(a) \cup r(b).$$

To verify this formula, note that $r(a) \cup r(b) \subseteq r(a \vee b)$ is trivial. Now let $x \in r(a \vee b)$. Then $x = x \wedge (a \vee b) = (x \wedge a) \vee (x \wedge b)$; therefore, $x = x \wedge a$ or $x = x \wedge b$, since x is join-irreducible. Thus $x \in r(a)$ or $x \in r(b)$, that is, $x \in r(a) \cup r(b)$.

Finally, we have to show that if $A \in H(J(L))$, then $a\varphi = A$ for some $a \in L$. Set $a = \bigvee A$. Then $r(a) \supseteq A$ is obvious. Let $x \in r(a)$; then $x = x \wedge a = x \wedge \bigvee A = \bigvee (x \wedge y \mid y \in A)$. So $x = x \wedge y$ for some $y \in A$, implying that $x \in A$, since A is hereditary. ●

Corollary 10. *The correspondence $L \to J(L)$ makes the class of all finite distributive lattices with more than one element correspond to the class of all finite posets; isomorphic lattices correspond to isomorphic posets, and vice versa.*

Proof. This is obvious from $J(H(P)) \cong P$ and $H(J(L)) \cong L$. ●

A subset S of $P(A)$ is called a *ring of sets* if $X, Y \in S$ implies that $X \cap Y, X \cup Y \in S$. Since $H(J(L))$ is a ring of sets, we obtain:

Corollary 11. *A finite lattice is distributive iff it is isomorphic to a ring of sets.*

If the poset Q is unordered, $H(Q) = P(Q)$; if B is Boolean, $J(B)$ is the set of all atoms or $\varnothing$ and therefore $J(B)$ is unordered. Thus we get:

Corollary 12. *A finite lattice is Boolean iff it is isomorphic to the Boolean lattice of all subsets of a finite set.*

A representation $a = x_0 \vee \cdots \vee x_{n-1}$ is *redundant* iff

$$a = x_0 \vee \cdots \vee x_{i-1} \vee x_{i+1} \vee \cdots \vee x_{n-1},$$

for some $0 \leq i < n$; otherwise it is *irredundant*.

Corollary 13. *Every element of a finite distributive lattice has a unique irredundant representation as a join of join-irreducible elements.*

Proof. The existence of such a representation is obvious. If

$$a = x_0 \vee \cdots \vee x_{n-1}$$

is an irredundant representation, then $r(a) = \bigcup (r(x_i) \mid 0 \leq i < n)$. Thus x occurs in such a representation iff x is a maximal element of $r(a)$; hence the uniqueness. ●

Corollary 14. *Every maximal chain C of the finite distributive lattice L is of length $|J(L)|$.*

Proof. For $a \in J(L)$, let $m(a)$ be the smallest member of C containing a. Then $a \to m(a)$ is a one-to-one map of $J(L)$ onto the nonzero elements of C. Indeed, if

$m(a) = m(b)$, $m(a) \succ x$, and $x \in C$, then $x \vee a = x \vee b$; therefore, $a = (a \wedge x) \vee (a \wedge b)$, implying that $a \leq x$ or $a \leq b$. But $a \leq x$ implies that $m(a) \leq x < m(a)$, a contradiction. Consequently, $a \leq b$; similarly, $b \leq a$; thus $a = b$. Let $y \in C, y \succ z, z \in C$. Then $r(y) \supset r(z)$, and so $y = m(a)$ for any $a \in r(y) - r(z)$. ●

Let $M(L)$ denote the set of meet-irreducible elements of L excluding 1. Corollary 14 and its dual yield:

$$|J(L)| = |M(L)|.$$

For the modular case see Exercise IV.5.26.

The crucial Theorem 9 and its most important consequence, Corollary 11, depend on the existence of sufficiently many join-irreducible elements. In an infinite distributive lattice there may be no join-irreducible element. Note that in a distributive lattice, $a \neq 0$ is join-irreducible iff $L - [a)$ is a prime ideal. In the infinite case the role of join-irreducible elements is taken over by prime ideals. The crucial result is the existence of sufficiently many prime ideals (as illustrated in Figure 2).

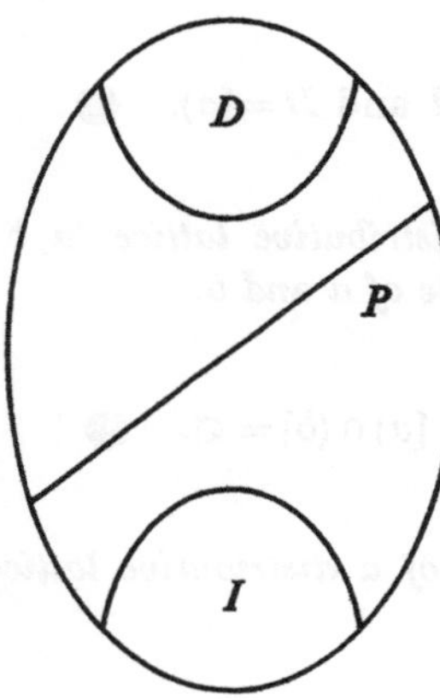

Figure 2

Theorem 15 (M. H. Stone [1936]). *Let L be a distributive lattice, let I be an ideal, let D be a dual ideal of L, and let $I \cap D = \varnothing$. Then there exists a prime ideal P of L such that $P \supseteq I$ and $P \cap D = \varnothing$.*

Proof. Some form of the Axiom of Choice is needed to prove this statement. The most convenient form for this proof is:

Zorn's Lemma. *Let A be a set and let $\mathscr{X}$ be a nonvoid subset of $P(A)$. Let us assume that $\mathscr{X}$ has the following property: If C is a chain in $\langle \mathscr{X}; \subseteq \rangle$, then $\bigcup(X \mid X \in C) \in \mathscr{X}$. Then $\mathscr{X}$ has a maximal member.*

Let $\mathscr{X}$ be the set of all ideals of L that contain I and are disjoint from D. We verify that $\mathscr{X}$ satisfies the hypothesis of Zorn's Lemma. Since $I \in \mathscr{X}$, we conclude that $\mathscr{X}$ is nonvoid. Let C be a chain in $\mathscr{X}$ and let $M = \bigcup(X \mid X \in C)$. If $a, b \in M$, then $a \in X$ and $b \in Y$ for some $X, Y \in C$; since C is a chain, either $X \subseteq Y$ or $Y \subseteq X$; if, say, $X \subseteq Y$, then $a, b \in Y$, and so $a \vee b \in Y \subseteq M$, since Y is an ideal. Also, if $b \leq a \in M$, then $a \in X$; since X is an ideal, $b \in X \subseteq M$. Thus M is an ideal. It is obvious that $M \supseteq I$ and $M \cap D = \varnothing$, verifying that $M \in \mathscr{X}$. Therefore, by Zorn's Lemma, $\mathscr{X}$ has a maximal element P. We claim that P is a prime ideal. Indeed, if P is not prime, there exist $a, b \in L$ such that $a, b \notin P$ but $a \wedge b \in P$. The maximality of P yields $(P \vee (a]) \cap D \neq \varnothing$ and $(P \vee (b]) \cap D \neq \varnothing$. Thus there are $p, q \in P$ such that $p \vee a \in D$ and $q \vee b \in D$. Then $x = (p \vee a) \wedge (q \vee b) \in D$, since D is a dual ideal. Also,

$$x = (p \wedge q) \vee (p \wedge b) \vee (a \wedge q) \vee (a \wedge b) \in P;$$

thus $P \cap D \neq \varnothing$, a contradiction. ●

Corollary 16. *Let L be a distributive lattice, let I be an ideal of L, and let $a \in L$ and $a \notin I$. Then there is a prime ideal P such that $P \supseteq I$ and $a \notin P$.*

Proof. Apply Theorem 15 to I and $D = [a)$. ●

Corollary 17. *Let L be a distributive lattice, $a, b \in L$ and $a \neq b$. Then there is a prime ideal containing exactly one of a and b.*

Proof. Either $(a] \cap [b) = \varnothing$, or $[a) \cap (b] = \varnothing$. ●

Corollary 18. *Every ideal I of a distributive lattice is the intersection of all prime ideals containing it.*

Proof. Let $I_1 = \bigcap(P \mid P \supseteq I,\ P$ is a prime ideal of $L)$. If $I \neq I_1$, then there is an $a \in I_1 - I$, and so by Corollary 16 there is a prime ideal P, with $P \supseteq I$ and $a \notin P$. But then $a \notin P \supseteq I_1$ is a contradiction. ●

Theorem 19 (G. Birkhoff [1933] and M. H. Stone [1936]). *A lattice is distributive iff it is isomorphic to a ring of sets.*

Proof. Let L be a distributive lattice and let X be the set of all prime ideals of L. For $a \in L$ set

$$r(a) = \{P \mid a \notin P,\ P \in X\}.$$

Then the $r(a)$ form a ring of sets, and $a \to r(a)$ is an isomorphism. The details are similar to the proof of Theorem 9, except for the first step, which now uses Corollary 17. ●

This result has a very useful corollary.

Corollary 20. *Let L be a distributive lattice with more than one element. An identity holds in L iff it holds in the two-element chain, $\mathfrak{C}_2$.*

Proof. Let $p = q$ hold in L. Since $|L| > 1$, $\mathfrak{C}_2$ is a sublattice of L, and $p = q$ holds in $\mathfrak{C}_2$. Conversely, let $p = q$ hold in $\mathfrak{C}_2$. Note that $\mathfrak{C}_2 = P(X)$, with $|X| = 1$, and that $P(A)$ is isomorphic to the direct power $P(X)^{|A|}$. Therefore, $p = q$ holds in any $P(A)$. By Theorem 19, L is a sublattice of some $P(A)$; thus $p = q$ holds in L. ●

We can reformulate Theorem 19 using the concept of a field of sets: A *field of sets* is a ring of sets closed under set complementation.

Corollary 21 (M. H. Stone [1936]). *A lattice is Boolean iff it is isomorphic to a field of sets.*

Proof. Use the representation of Theorem 19. Obviously $r(a') = X - r(a)$, and thus complements are also preserved. ●

For a distributive lattice L with more than one element, let $\mathcal{P}(L)$ denote the set of all prime ideals of L, regarded as a poset under $\subseteq$. The importance of $\mathcal{P}(L)$ should be clear from the previous results. A topology on $\mathcal{P}(L)$ will be discussed in Section 5.

Some interesting properties of L are reflected in $\mathcal{P}(L)$. An important result of this type is the following theorem (see also L. Rieger [1949]):

Theorem 22 (L. Nachbin [1947]). *Let L be a distributive lattice with $0 \neq 1$. Then L is a Boolean lattice iff $\mathcal{P}(L)$ is unordered.*

Proof. Let L be Boolean, $P, Q \in \mathcal{P}(L)$, and $P \subset Q$. Choose $a \in Q - P$. Since $a \in Q$, we have $a' \notin Q$, and thus $a' \notin P$. Thus $a, a' \notin P$, but $a \wedge a' = 0 \in P$, a contradiction, showing that $\mathcal{P}(L)$ is unordered. This proof in fact verifies that in a Boolean algebra every prime ideal is maximal.

Now let $\mathcal{P}(L)$ be unordered and $a \in L$, and let us assume that a has no complement. Set $D = \{x \mid a \vee x = 1\}$. Then D is a dual ideal. Take $D_1 = D \vee [a) = \{x \mid x \geq d \wedge a$ for some $d \in D\}$. The dual ideal D_1 does not contain 0, since $0 = d \wedge a$, $a \vee d = 1$ would mean that d is a complement of a. Thus there exists a prime ideal P disjoint to D_1. Note that $1 \notin (a] \vee P$, otherwise $1 = a \vee p$ for some $p \in P$, contradicting $P \cap D = \varnothing$. Thus some prime ideal Q contains $(a] \vee P$; and so $P \subset Q$, which is impossible since $P(L)$ is unordered. ●

According to Corollary 18, every ideal is an intersection of prime ideals. When is this representation unique?

Theorem 23 (J. Hashimoto [1952]). *Let L be a distributive lattice with $0 \neq 1$. Every ideal has a unique representation as an intersection of prime ideals iff L is a finite Boolean lattice.*

PROOF. If L is a finite Boolean lattice, then P is a prime ideal iff $P=(a]$, where a is a dual atom; the uniqueness follows from Corollary 13 (or is obvious by direct computation).

Now let every ideal of L have a unique representation as a meet of prime ideals. We claim that $I(L)$ is Boolean. Let $I \in I(L)$; define

$$J = \bigcap(P \mid P \in \mathcal{P}(L),\ P \not\supseteq I).$$

Then $I \wedge J = \bigcap(P \mid P \in \mathcal{P}(L)) = (0]$. If $L \neq I \vee J$, then there is a prime ideal $P_0 \supseteq I \vee J$, and consequently J has two representations:

$$\bigcap(P \mid P \not\supseteq I) = P_0 \cap \bigcap(P \mid P \not\supseteq I).$$

Thus $L = I \vee J$, and J is a complement of I in $I(L)$.

By Lemma 5, every ideal of L is principal; therefore $L \cong I(L)$, and so L is Boolean. By Exercise I.6.21, L satisfies the Ascending Chain Condition; thus every element of L other than the unit is contained in a dual atom. Since the complement of a dual atom is an atom, by taking complements we find that every nonzero element of L contains an atom.

If $p_0, p_1, \ldots, p_n, \ldots$ are distinct atoms in L, then the ascending chain p_0, $p_0 \vee p_1, \ldots, p_0 \vee p_1 \vee \cdots \vee p_n, \ldots$ does not terminate, and thus L has only finitely many atoms, $p_0, \ldots, p_{n-1}$. Let $a = p_0 \vee \cdots \vee p_{n-1}$. If $a' \neq 0$, then a' has to contain an atom, which is impossible. Therefore, $a'=0$, $a=1$, and $L \cong P(X)$, with $|X|=n$. ●

Exercises

1. Work out a direct proof of Theorem 2(i).
2. Work out a direct proof of Theorem 2(ii).
3. Let K be a five-element distributive lattice. Is there an identity $p=q$ such that $p=q$ holds in a lattice L iff L has no sublattice isomorphic to K?
4. Does the property stated in Lemma 5 characterize distributive lattices?
5. Let L be a distributive lattice with 0 and 1. Prove that the direct decompositions $L \cong L_0 \times L_1$ of L are in one-to-one correspondence with the complemented elements of L.
6. Prove that the complemented elements of a distributive lattice form a sublattice.
7. Let L be a distributive lattice with 0 and 1. Let $L \cong L_0 \times L_1 \cong K_0 \times K_1$. Show that there is a direct decomposition $L \cong A_0 \times A_1 \times A_2 \times A_3$ such that $A_0 \times A_1 \cong L_0$, $A_2 \times A_3 \cong L_1$, $A_0 \times A_2 \cong K_0$, $A_1 \times A_3 \cong K_1$.
8. Extend Theorem 9 to distributive lattices satisfying the Descending Chain Condition (see Exercise I.6.14).
9. Extend Corollary 10 to distributive lattices satisfying the Descending Chain Condition.
10. Can Exercises 8 and 9 be further sharpened?
11. Let L be a distributive lattice with 0 and 1. Let C_0 and C_1 be finite chains in L. Show that there exist chains $D_0 \supseteq C_0$, $D_1 \supseteq C_1$, such that $|D_0| = |D_1|$.
12. Derive from Exercise 11 the result that all maximal chains of a finite distributive lattice have the same length.

13. Find examples showing that Exercise 11 is not valid if "finite" is omitted.

14. Prove the theorem "L is modular iff $I(L)$ is modular" by showing that "L contains a pentagon iff $I(L)$ contains a pentagon."

***15.** Is the second statement of Exercise 14 true for the diamond rather than for the pentagon?

16. Let L be a distributive lattice, $a, b, c \in L$ and $a \leq b$. Is it true that $[a, b]$ is Boolean iff $[a \wedge c, b \wedge c]$ and $[a \vee c, b \vee c]$ are Boolean?

17. For a poset P, let $H_F(P)$ denote the poset lattices of the form of all subsets of the form $(a_0] \cup (a_1] \cup \cdots \cup (a_{n-1}]$, where n is an arbitrary natural number. Show that Theorem 9 holds for $H_F(P)$.

18. Prove that if P is a finite poset, then $H_F(P) = H(P)$.

19. Is it true that if Corollary 13 holds for a lattice L with more than one element, then $L \cong H_F(P)$ for some poset P?

20. Show that the Ascending Chain Condition implies the Descending Chain Condition for Boolean algebras.

21. Show that Exercise 20 fails to hold for *generalized Boolean algebras* (that is, relatively complemented distributive lattices with 0).

22. Use Exercise 20 to simplify the proof of Theorem 23.

23. Let L be a lattice, let P be a prime ideal of L, and let $a, b, c \in L$. Prove that if $a \vee (b \wedge c) \in P$, then $(a \vee b) \wedge (a \vee c) \in P$.

24. Using Exercise 23, show that the lattice L is distributive iff, for all $x, y \in L$, $x < y$, there exists a prime ideal P with $x \in P$ and $y \notin P$.

25. Verify the statement of Exercise 24 using Theorem 6.

26. Verify the statement of Exercise 24 using Theorem 1.

27. Let L be a distributive lattice. Then L is relatively complemented iff $\mathscr{P}(L)$ is unordered.

28. Prove Theorem 15 by well-ordering L, $L = \{a_\gamma \mid \gamma < \alpha\}$, and deciding one by one for each a_γ whether $a_\gamma \in P$ or $a_\gamma \notin P$.

29. Let L be a distributive lattice with 1. Show that every prime ideal P is contained in a *maximal prime ideal* Q (that is, $P \subseteq Q$, and for any prime ideal X of L, $Q \subseteq X$ implies that $Q = X$).

30. Let L be a distributive lattice with 0. Verify that every prime ideal P contains a *minimal prime ideal* Q (that is, $P \supseteq Q$, and for any prime ideal X of L, $Q \supseteq X$ implies that $Q = X$).

31. Find a distributive lattice L with no minimal and no maximal prime ideals.

32. Investigate the connections among the Ascending Chain Condition (and Descending Chain Condition) for a distributive lattice L, for $I(L)$, and for $\mathscr{P}(L)$.

33. Let L be a distributive lattice with 0 and let $I \in I(L)$. Show that the pseudocomplement of I is $I^* = \{x \mid (x] \wedge I = (0]\}$.

34. Prove that $I = I^{**}$ for any $I \in I(L)$ of a distributive lattice with 0 iff L is a generalized Boolean lattice satisfying the Descending Chain Condition.

35. The congruence relations Θ and Φ *permute* iff $a \equiv b\ (\Theta)$ and $b \equiv c\ (\Phi)$ imply that $a \equiv d\ (\Phi)$ and $d \equiv c\ (\Theta)$ for some d. Show that the congruences of a relatively complemented lattice permute.

36. Prove the converse of Exercise 35 for distributive lattices.

37. Generalize Theorem 23 to distributive lattices without 0 and 1.

***38.** Let L be a distributive lattice, let $a \in L$, let S be a sublattice of L, and let $a \notin S$. Show that there exists a prime ideal P and a prime dual ideal Q such that $a \notin P \cup Q \supseteq S$ provided that a is not the 0 or 1 of L (J. Hashimoto [1952]).

39. Let L be a relatively complemented distributive lattice. A sublattice K of L is *proper* iff $K \neq L$. Show that every proper sublattice of L can be extended to a maximal proper sublattice of L (K. Takeuchi [1951]; see also J. Hashimoto [1952] and G. Grätzer and E. T. Schmidt [1958c]).

40. Show that the statement of Exercise 39 is not valid in general if L is not relatively complemented (K. Takeuchi [1951]). (See also M. E. Adams [1973].)
41. Generalize Corollary 13 to infinite distributive lattices, claiming the unique irredundant representation of certain ideals as a meet of prime ideals.
42. If P is a prime ideal of L, then $(P]$ is a principal prime ideal of $I(L)$. Is the converse true?
43. Show that Corollary 17 characterizes distributivity.
44. A chain C in a poset P is called *maximal* iff, for any chain D in P, $C \subseteq D$ implies that $C = D$. Using Zorn's Lemma, show that every chain is contained in a maximal chain.
45. Prove that a finite distributive lattice is planar iff no element is covered by three elements.
46. Show that a finite distributive lattice is planar iff it is dismantlable.
47. Let S be a sublattice of the finite lattice L. Then[1] S can be represented in the form $$L - \bigcup([a, b] \mid a \text{ is join-irreducible and } b \text{ is meet-irreducible}).$$
48. Prove the converse of Exercise 47 for distributive lattices.

2. Polynomials and Freeness

We can introduce an equivalence relation $\equiv$ for lattice polynomials: $p \equiv q$ iff p and q define the same functions in the class $\mathbf{D}$ of distributive lattices. More formally, if p and q are lattice polynomials (see Section I.4), then $p \equiv q$ iff, for any distributive lattice L and $a_0, a_1, \ldots \in L$, we have $p(a_0, \ldots) = q(a_0, \ldots)$ (see Definitions I.4.1 and I.4.2). For an n-ary lattice polynomial p, let $[p]\mathbf{D}$ denote the set of all n-ary lattice polynomials q satisfying $p \equiv q$ and let $P_{\mathbf{D}}(n)$ denote the set of all these equivalence classes, that is, $P_{\mathbf{D}}(n) = \{[p]\mathbf{D} \mid p \in \mathbf{P}^{(n)}\}$. Observe that, for $p, p_1, q, q_1 \in \mathbf{P}^{(n)}$, if $p \equiv p_1$ and $q \equiv q_1$, then $p \wedge q \equiv p_1 \wedge q_1$ and $p \vee q \equiv p_1 \vee q_1$. Thus $[p]\mathbf{D} \wedge [q]\mathbf{D} = [p \wedge q]\mathbf{D}$ and $[p]\mathbf{D} \vee [q]\mathbf{D} = [p \vee q]\mathbf{D}$ define $\wedge$ and $\vee$ on $P_{\mathbf{D}}(n)$. It is easily seen that $P_{\mathbf{D}}(n)$ is a distributive lattice and $[p]\mathbf{D} \leq [q]\mathbf{D}$ iff the inequality $p \leq q$ holds in the class $\mathbf{D}$.

To describe the structure of $P_{\mathbf{D}}(n)$, let $Q(n)$ denote the dual of the poset of all proper nonvoid subsets of $\{0, 1, \ldots, n-1\}$. Define $H(\varnothing) = \mathfrak{C}_1$.

THEOREM 1.
(i) *$P_{\mathbf{D}}(n)$ is isomorphic with $H(Q(n))$.*
(ii) *$P_{\mathbf{D}}(n)$ is a free distributive lattice on n generators.*
(iii) $2^n - 2 \leq |P_{\mathbf{D}}(n)| \leq 2^{2^n - 2}$.
(iv) *A finitely generated distributive lattice is finite.*

---

[1] Exercises 47 and 48 are from I. Rival [1973].

Proof.

(i) A lattice polynomial p is called a *meet-polynomial* iff it is of the form $x_{i_0}\wedge\cdots\wedge x_{i_k}$. For $J\subseteq\{0,\ldots,n-1\}$, $J\neq\varnothing$, set

$$p_J=\bigwedge(x_i\mid i\in J).$$

We claim that $[p_J]\mathbf{D}\leq[p_K]\mathbf{D}$ ($J, K\subseteq\{0,\ldots,n-1\}$) iff $J\supseteq K$. The "if" part is obvious. Now let $J\not\supseteq K$; then there exists an $i\in K$ such that $i\notin J$. Consider the two-element chain $\mathfrak{C}_2$ and substitute $x_i=0$ and $x_j=1$ for $j\neq i$. Obviously, $p_J=1$ and $p_K=0$; thus the inequality $p_J\leq p_K$ fails in $\mathfrak{C}_2$ and therefore in $\mathbf{D}$.

Every lattice polynomial is equivalent to one of the form $\bigvee p_J$; because every x_i is of this form, the join of two such polynomials is of this form and the same holds for the meet in view of

$$\bigvee p_{J_i}\wedge\bigvee p_{K_j}\equiv\bigvee(p_{J_i}\wedge p_{K_j})$$

and

$$p_{J_i}\wedge p_{K_j}\equiv p_{J_i\cup K_j}.$$

Next we claim that $[p]\mathbf{D}\in J(P_\mathbf{D}(n))$ iff it is a $[p_J]\mathbf{D}$. Since every $[p]\mathbf{D}\in P_\mathbf{D}(n)$ is a join of some $[p_J]\mathbf{D}$, it suffices to prove that a $[p_J]\mathbf{D}$ is join-irreducible. Let

$$p_J\equiv\bigvee(p_{J_i}\mid i\in K, J_i\subseteq\{0,\ldots,n-1\});$$

$J\subseteq J_i$ follows from $[p_J]\mathbf{D}\geq[p_{J_i}]\mathbf{D}$. Now if $[p_J]\mathbf{D}>[p_{J_i}]\mathbf{D}$ for all i, then we have $J\subset J_i$. Choose $j_i\in J_i$, $j_i\notin J$ for all $i\in K$. In $\mathfrak{C}_2$ put $x_k=0$ for all $k=j_i$ and $x_k=1$ otherwise. Then $p_J=1$, and $\bigvee(p_{J_i}\mid i\in K)=0$, which is a contradiction.

A reference to Theorem 1.9 completes the proof of (i). ◐

(ii) Let L be a distributive lattice, $a_0,\ldots,a_{n-1}\in L$. Then the map $x_i\to a_i$ can be extended to the homomorphism

$$[p]\mathbf{D}\to p(a_0,\ldots,a_{n-1}),$$

proving (ii). ◐

(iii) This proof is obvious from (i). ◐

(iv) This proof is obvious from (iii). ●

Figure I.5.6 is a diagram of $P_\mathbf{D}(3)$.

Boolean polynomials are defined exactly as lattice polynomials except that all five operations $\wedge$, $\vee$, $'$, 0, 1 are used in the formation of polynomials. A formal definition is the same as Definition I.4.1 with two clauses added: If p is a Boolean polynomial, so is p'; 0 and 1 are Boolean polynomials. An n-ary Boolean polynomial p defines a function in n variables on any Boolean algebra B; $p(a_0,\ldots,a_{n-1})$ can be defined imitating Definition I.4.2.

For the Boolean polynomials p and q, set $p\equiv q$ iff, for any Boolean algebra B and $a_0, a_1,\ldots\in B$, we have $p(a_0,\ldots)=q(a_0,\ldots)$. Let $[p]\mathbf{B}$ denote the equivalence class containing p. Observe that $p\equiv q$ is equivalent to the identity $p=q$ holding in the class $\mathbf{B}$ of all Boolean algebras.

Let $P_\mathbf{B}(n)$ denote the set of all $[p]\mathbf{B}$, where p is an n-ary Boolean polynomial. It is easily seen that $[p]\mathbf{B}\wedge[q]\mathbf{B}=[p\wedge q]\mathbf{B}$, $[p]\mathbf{B}\vee[q]\mathbf{B}=[p\vee q]\mathbf{B}$, $([p]\mathbf{B})'=[p']\mathbf{B}$, $0=[0]\mathbf{B}$, $1=[1]\mathbf{B}$ defines the Boolean operations on $P_{\mathbf{B}}(n)$, and thus $P_{\mathbf{B}}(n)$ is a Boolean algebra.

Theorem 2.
(i) *$P_{\mathbf{B}}(n)$ is isomorphic to $(\mathfrak{B}_2)^{2^n}$.*
(ii) *$P_{\mathbf{B}}(n)$ is a free Boolean algebra on n generators.*
(iii) $|P_{\mathbf{B}}(n)|=2^{2^n}$.
(iv) *A finitely generated Boolean algebra is finite.*

Proof. A Boolean polynomial is called *atomic* iff it is of the form

$$x_0^{i_0}\wedge\cdots\wedge x_{n-1}^{i_{n-1}},$$

where $i_j=0$ or 1, x^0 denotes x and x^1 denotes x'. For every $J\subseteq\{0,\ldots,n-1\}$ there is an atomic polynomial p_J for which $i_j=0$ iff $j\in J$. The crucial statement is: $[p_{J_0}]\mathbf{B}\leq[p_{J_1}]\mathbf{B}$ iff $J_0=J_1$. Indeed, let $J_0\neq J_1$. We make the following substitution in $\mathfrak{B}_2$: $x_i=1$ if $i\in J_0$, $x_i=0$ if $i\notin J_0$; this makes $p_{J_0}=1$, $p_{J_1}=0$, contradicting $[p_{J_0}]\mathbf{B}\leq[p_{J_1}]\mathbf{B}$.

Let $B(n)$ be the set of all Boolean polynomials that are equivalent to one of the form $\bigvee(p_{J_i}\mid i\in K)$. Then $B(n)$ is closed under $\vee$ and $\wedge$, since

$$\bigvee p_{J_i}\wedge\bigvee p_{I_k}\equiv\bigvee(p_{J_i}\wedge p_{I_k}),$$

and $p_{J_i}\wedge p_{I_k}\equiv p_{J_k}$ if $J_i=I_k$, and $p_{J_i}\wedge p_{I_k}\equiv 0$ otherwise.

Now we prove by induction on n that x_i, $x_i'\in B(n)$ for $i<n$. For $n=1$, x_0, x_0' are atomic polynomials, so x_0, $x_0'\in B(1)$. By induction, $x_0\equiv\bigvee(p_{J_i}\mid i\in K)$, where the p_{J_i} are atomic $(n-1)$-ary polynomials; then

$$\begin{aligned}x_0&\equiv x_0\wedge(x_{n-1}\vee x_{n-1}')\equiv(x_0\wedge x_{n-1})\vee(x_0\wedge x_{n-1}')\\&\equiv\bigvee(p_{J_i}\wedge x_{n-1}\mid i\in K)\vee\bigvee(p_{J_i}\wedge x_{n-1}'\mid i\in K),\end{aligned}$$

and similarly for x_0'. Thus x_0, $x_0'\in B(n)$, and, by symmetry, x_i, $x_i'\in B(n)$ for all $i<n$. Since $(p_J)'\equiv\bigvee(x_i'\mid i\in J)\vee\bigvee(x_i\mid i\notin J)$, we conclude that $(p_J)'\in B(n)$; therefore $B(n)$ is closed under $'$. Thus $B(n)$ is closed under $\wedge$, $\vee$, $'$, 0, 1. Since $B(n)$ includes all x_i, $i<n$, $B(n)$ is the set of all n-ary Boolean polynomials.

Consequently, every $[p]\mathbf{B}$ is a join of atomic ones, the $[p]\mathbf{B}$ for p atomic polynomials are unordered and 2^n in number, implying (i) and (iii). The proof of (ii) is routine (same proof as that of Theorem 1(ii)), and (iv) follows trivially from (iii). ●

We can use Theorems 1 and 2 to characterize free distributive lattices and free Boolean algebras, respectively.

Theorem 3. *Let L be a distributive lattice generated by $\{a_i\mid i\in I\}$. L is freely generated by the a_i iff the validity in L of a relation of the form*

$$\bigwedge(a_i\mid i\in I_0)\leq\bigvee(a_i\mid i\in I_1)$$

implies that $I_0\cap I_1\neq\varnothing$, for I_0, I_1 finite nonvoid subsets of I.

Proof. The "only if" part can be easily verified by using substitutions in $\mathfrak{C}_2$. For the converse, let F be the distributive lattice freely generated by x_i, $i \in I$, and let φ be the homomorphism of F into (in fact, onto) L satisfying $x_i\varphi = a_i$ for $i \in I$. It suffices to prove that for the lattice polynomials p, q, $p\varphi \leq q\varphi$ implies that $[p]\mathbf{D} \leq [q]\mathbf{D}$. (We think of the elements of F as equivalence classes of polynomials in the x_i, $i \in I$.)

Let

$$p \equiv \bigvee(\bigwedge(x_i \mid i \in I_j) \mid j \in J)$$

and

$$q \equiv \bigwedge(\bigvee(x_i \mid i \in K_t) \mid t \in T).$$

Then $p\varphi \leq q\varphi$ takes the form

$$\bigvee(\bigwedge(a_i \mid i \in I_j) \mid j \in J) \leq \bigwedge(\bigvee(a_i \mid i \in K_t) \mid t \in T),$$

which is equivalent to

$$\bigwedge(a_i \mid i \in I_j) \leq \bigvee(a_i \mid i \in K_t)$$

for all $j \in J$, $t \in T$. By assumption this implies that $I_j \cap K_t \neq \varnothing$ for all $j \in J$, $t \in T$; thus

$$\bigwedge(x_i \mid i \in J_j) \leq \bigvee(x_i \mid i \in K_t)$$

for all $j \in J$, $t \in T$; implying that $[p]\mathbf{D} \leq [q]\mathbf{D}$. ●

Theorem 4. *Let B be a Boolean algebra generated by $\{a_i \mid i \in I\}$. Then B is freely generated by $\{a_i \mid i \in I\}$ iff, whenever I_0, I_1, J_0, J_1 are finite subsets of I with $I_0 \cup I_1 = J_0 \cup J_1$ and $I_0 \cap I_1 = \varnothing$, then*

$$\bigwedge(a_i \mid i \in I_0) \wedge \bigwedge(a'_i \mid i \in I_1) \leq \bigwedge(a_i \mid i \in J_0) \wedge \bigwedge(a'_i \mid i \in J_1)$$

implies that $I_0 = J_0$ and $I_1 = J_1$.

Proof. Again, the "only if" part is by substitution into $\mathfrak{B}_2$. On the other hand, clearly B is freely generated by $\{a_i \mid i \in I\}$ iff, for every finite subset K of I, the subalgebra $[\{a_i \mid i \in K\}]$ is freely generated by $\{a_i \mid i \in K\}$. By Theorem 2, the latter holds iff $[\{a_i \mid i \in K\}]$ has $2^{2^{|K|}}$ elements, which, in turn, is equivalent to $[\{a_i \mid i \in K\}]$ having $2^{|K|}$ atoms. Using the proof of Theorem 2 and the present hypothesis for $I_0 \cup I_1 = K$, we can see that the elements of the form $\bigwedge(a_i \mid i \in I_0) \wedge \bigwedge(a'_i \mid i \in I_1)$, where $I_0 \cup I_1 = K$ and $I_0 \cap I_1 = \varnothing$, are distinct atoms in $[\{a_i \mid i \in K\}]$, thus completing the proof. ●

For a simpler variant of Theorem 4 see Exercise 3.43.

Now we turn our attention to an important application of polynomials: finding homomorphisms.

Theorem 5. *Let the Boolean algebra B be generated by the subalgebra B_1 and the element a. Let B_2 be a Boolean algebra and let φ be a homomorphism of B_1 into B_2. The extensions of φ to homomorphisms of B into B_2 are in one-to-one correspondence with the elements p of B_2 satisfying the following conditions:*

(i) *If $x \in B_1$ and $x \leq a$, then $x\varphi \leq p$.*

(ii) *If $x \in B_1$ and $x \geq a$, then $x\varphi \geq p$.*

To prepare for the proof of this theorem we verify a simple lemma, in which $+$ denotes the *symmetric difference*; that is,

$$x+y=(x'\wedge y)\vee(x\wedge y').$$

Lemma 6. *Let the Boolean algebra B be generated by the subalgebra B_1 and the element a. Then every element x of B can be represented in the form*

$$x=(a\wedge x_0)\vee(a'\wedge x_1),\quad x_0, x_1\in B_1.$$

This representation is not unique. In fact,

$$(a\wedge x_0)\vee(a'\wedge x_1)=(a\wedge y_0)\vee(a'\wedge y_1),\quad x_0, x_1, y_0, y_1\in B_1,$$

iff

$$a\leq(x_0+y_0)' \quad \textit{and} \quad x_1+y_1\leq a.$$

Proof. Let B_0 denote the set of all elements of B having such a representation. If $x\in B_1$, then $x=(a\wedge x)\vee(a'\wedge x)$; thus $B_1\subseteq B_0$. Also $a=(a\wedge 1)\vee(a'\wedge 0)$, and so $a\in B_0$. Therefore, to show $B_0=B$, it suffices to verify that B_0 is a subalgebra, which is left as an exercise. Now note that for $p, q\in B, p=q$ iff $p\wedge a=q\wedge a$ and $p\wedge a'=q\wedge a'$; thus $(a\wedge x_0)\vee(a'\wedge x_1)=(a\wedge y_0)\vee(a'\wedge y_1)$ iff $a\wedge x_0=a\wedge y_0$ and $a'\wedge x_1=a'\wedge y_1$. However, $a\wedge x_0=a\wedge y_0$ is equivalent to $(a\wedge x_0)+(a\wedge y_0)=0$; that is, $a\wedge(x_0+y_0)=0$ (see Exercise 11), which is the same as $a\leq(x_0+y_0)'$. Similarly, $a'\wedge x_1=a'\wedge y_1$ iff $x_1+y_1\leq a$. ●

Proof of Theorem 5. Let p be an element as specified and let the map $\psi\colon B\to B_2$ be defined as follows:

$$\psi\colon (a\wedge x_0)\vee(a'\wedge x_1)\to(p\wedge x_0\varphi)\vee(p'\wedge x_1\varphi).$$

By Lemma 6, ψ is defined on B. It is well defined, because if $(a\wedge x_0)\vee(a'\wedge x_1)=(a\wedge y_0)\vee(a'\wedge y_1)$, then $x_1+y_1\leq a\leq(x_0+y_0)'$; thus $(x_1+y_1)\varphi\leq p\leq(x_0+y_0)'\varphi$, and therefore $x_1\varphi+y_1\varphi\leq p\leq(x_0\varphi+y_0\varphi)'$, implying that $(p\wedge x_0\varphi)\vee(p'\wedge x_1\varphi)=(p\wedge y_0\varphi)\vee(p'\wedge y_1\varphi)$. It is routine to check that ψ is a homomorphism. Conversely, if ψ is an extension of φ to B, then ψ is uniquely determined by $p=a\psi$, and p satisfies (i) and (ii). ●

Corollary 7. *Let us assume the conditions of Theorem 5. In addition, let B_2 be complete. Set*

$$x_0=\bigvee(x\varphi\mid x\in B_1, x\leq a) \quad \text{and} \quad x_1=\bigwedge(x\varphi\mid x\in B_1, x\geq a).$$

Then the extensions of φ to B are in one-to-one correspondence with the elements of the interval $[x_0, x_1]$. In particular, there is always at least one such extension.

Exercises

1. Get lower and upper bounds for $|P_D(n)|$ that are sharper than those given by Theorem 1(iii).
2. Work out the details of the last steps in the proof of Theorem 1.

3. Let p_J be an atomic Boolean polynomial. Show that under the substitution $x_i = 1$ for $i \in J$ and $x_i = 0$ for $i \notin J$, we get $p_J = 1$ and $p_{J_0} = 0$ for all $J_0 \neq J$.
4. Let $f: \{0, 1\}^n \to \{0, 1\}$. Prove that there is an n-ary Boolean polynomial p that defines the function f on $\mathfrak{B}_2$ (see Definition I.4.2). (Let $K = \{J \mid J \subseteq \{0, 1, \ldots, n-1\}$ such that $f(x_0, \ldots, x_{n-1}) = 1$, where $x_i = 1$ if $i \in J$ and $x_i = 0$ if $i \notin J\}$; set $p = \bigvee(p_J \mid J \in K)$ and use Exercise 3.)
5. Let B be a Boolean algebra. Prove that there is a one-to-one correspondence between n-ary Boolean polynomials over B and maps $\{0, 1\}^n \to \{0, 1\}$. In other words, all 0, 1 substitutions take 0 and 1 as values and determine the polynomial p.
6. An *algebraic function* over a Boolean algebra B is built up inductively from the variables and the elements of B using $\wedge$, $\vee$, and $'$. Show that the n-ary algebraic functions are in one-to-one correspondence with maps $\{0, 1\}^n \to B$.
7. Let p be an n-ary algebraic function over the Boolean lattice B and Θ a congruence relation on B. Show that $a_i \equiv b_i(\Theta)$, $i < n$, implies that $p(a_0, \ldots, a_{n-1}) \equiv p(b_0, \ldots, b_{n-1})(\Theta)$.
8. Show that the property described in Exercise 7 characterizes algebraic functions (G. Grätzer [1962a]).

*9. Use the property described in Exercise 7 to define *Boolean algebraic functions* over a distributive lattice. Show that for bounded distributive lattices Exercise 6 holds without any change (G. Grätzer [1964]).

10. Show that a free Boolean algebra on countably many generators has no atoms.
11. Show that $a \wedge (b + c) = (a \wedge b) + (a \wedge c)$ holds in any Boolean algebra.
12. Let B be the Boolean algebra freely generated by $\{x_i \mid i \in I\}$. Let L be the sublattice generated by $\{x_i \mid i \in I\}$. Prove that L is the free distributive lattice freely generated by $\{x_i \mid i \in I\}$.
13. Let L and L_1 be distributive lattices, let $L = [A]$, and let φ be a map of A into L_1. Show that there is a homomorphism of L into L_1 extending φ iff, for any pair of finite nonempty subsets A_1 and A_2 of A, $\bigwedge A_1 \leq \bigvee A_2$ implies that $\bigwedge A_1\varphi \leq \bigvee A_2\varphi$. (Compare this with Exercise I.5.45.)
14. State and prove Exercise 13 for Boolean algebras.
15. Interpret Lemma 6 using Exercise 14.
16. Extend the last statement of Corollary 7 to the case in which B_1 is generated by B and $a_0, \ldots, a_{n-1} \in B_1$, $n > 1$.
17. Let p and q be lattice polynomials. Since p and q can also be regarded as Boolean polynomials, $p \equiv q$ was defined in two ways in this section: with respect to $\mathbf{D}$ and with respect to $\mathbf{B}$. Show that the two definitions are equivalent for lattice polynomials.
18. Define $\equiv$ for lattice polynomials with respect to a class $\mathbf{K}$ of lattices. Show that $P_{\mathbf{K}}(n) \in \mathbf{K}$ iff the free lattice over $\mathbf{K}$ with n generators exists, in which case $P_{\mathbf{K}}(n)$ is a free lattice with n generators.

3. Congruence Relations

Let L be a lattice and let $H \subseteq L^2$. We denote by $\Theta(H)$ the smallest congruence relation such that $a \equiv b$ for all $\langle a, b\rangle \in H$, and call it the *congruence relation generated by H*.

LEMMA 1. *For any $H \subseteq L^2$, $\Theta(H)$ exists.*

PROOF. Let $\Phi = \bigwedge(\Theta \mid a \equiv b\ (\Theta)$ for all $\langle a, b\rangle \in H)$. Since in the lattice $C(L)$ the meet is intersection, it is obvious that $a \equiv b\ (\Phi)$ for all $\langle a, b\rangle \in H$; thus $\Phi = \Theta(H)$. ●

We will use special notations in two cases: If $H = \{\langle a, b\rangle\}$, we write $\Theta(a, b)$ for $\Theta(H)$. If $H = I^2$, where I is an ideal, we write $\Theta[I]$ for $\Theta(H)$. The congruence relation $\Theta(a, b)$ is called *principal*; its importance is revealed by the following formula.

LEMMA 2. $\Theta(H) = \bigvee(\Theta(a, b) \mid \langle a, b\rangle \in H)$.

PROOF. The proof is obvious. ●

Note that $\Theta(a, b)$ is the smallest congruence relation under which $a \equiv b$, whereas $\Theta[I]$ is the smallest congruence relation under which I is contained in a single class.

In general lattices, not much can be said about $\Theta(H)$. In distributive lattices, the following description of $\Theta(a, b)$ is important (G. Grätzer and E. T. Schmidt [1958d]):

THEOREM 3. *Let L be a distributive lattice, $a, b, x, y \in L$, and let $a \leq b$. Then $x \equiv y(\Theta(a, b))$ iff $x \wedge a = y \wedge a$ and $x \vee b = y \vee b$.*

REMARK. This situation is illustrated in Figure 1.

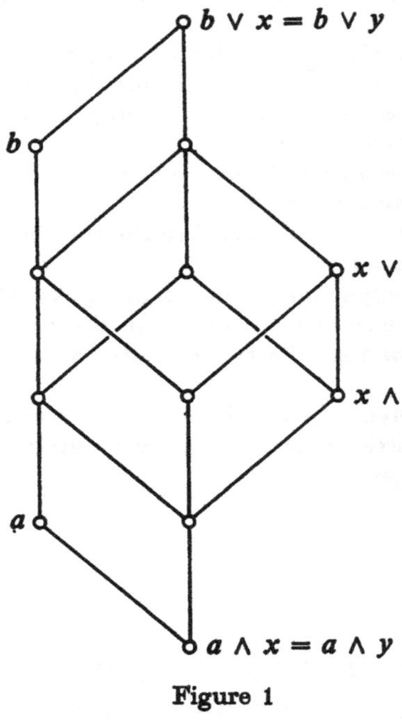

Figure 1

PROOF. Let Φ denote the binary relation under which $x \equiv y\ (\Phi)$ iff $x \wedge a = y \wedge a$ and $x \vee b = y \vee b$. Φ is obviously an equivalence relation. If $x \equiv y\ (\Phi)$ and $z \in L$, then

$(x\vee z)\wedge a=(x\wedge a)\vee(z\wedge a)=(y\wedge a)\vee(z\wedge a)=(y\vee z)\wedge a$, and $(x\vee z)\vee b=z\vee(x\vee b)=z\vee(y\vee b)=(y\vee z)\vee b$; thus $x\vee z\equiv y\vee z\ (\Phi)$. Similarly, $x\wedge z\equiv y\wedge z\ (\Phi)$, and so Φ is a congruence relation. That $a\equiv b\ (\Phi)$ is obvious. Finally, let Θ be any congruence relation such that $a\equiv b\ (\Theta)$ and let $x\equiv y\ (\Phi)$. Therefore, $x\wedge a=y\wedge a$, $x\vee b=y\vee b$, $x\vee a\equiv x\vee b\ (\Theta)$, and $x\wedge b\equiv x\wedge a\ (\Theta)$. Then, computing modulo Θ, we obtain

$$\begin{aligned} x &= x\vee(x\wedge a)=x\vee(y\wedge a)=(x\vee y)\wedge(x\vee a)\equiv(x\vee y)\wedge(x\vee b)\\ &=(x\vee y)\wedge(y\vee b)=y\vee(x\wedge b)\equiv y\vee(x\wedge a)=y\vee(y\wedge a)=y\,, \end{aligned}$$

that is, $x\equiv y\ (\Theta)$, proving that $\Phi\leq\Theta$. ●

Explanation. Since $a\equiv b$ implies that $(a\vee p)\wedge q\equiv(b\vee p)\wedge q$, we must have $x\equiv y$ $(\Theta(a,b))$ if $x\wedge y=(a\vee p)\wedge q$, $x\vee y=(b\vee p)\wedge q$. It is easy to check that the x, y satisfying the conditions of Theorem 3 are exactly the same as those for which such p, q exist. Thus Theorem 3 can be interpreted as follows: We get all pairs x, y with $x\equiv y\ (\Theta(a,b))$ and $x\leq y$ by applying the Substitution Property "twice". No further application of the Substitution Property or transitivity is needed. An equivalent form of Theorem 3 is to require $x\vee y\leq b\vee(x\wedge y)$ and $(a\vee(x\wedge y))\wedge(x\vee y)=x\wedge y$.

Some applications of Theorem 3 follow.

Corollary 4. *Let I be an ideal of the distributive lattice L. Then $x\equiv y\ (\Theta[I])$ iff $x\vee y=(x\wedge y)\vee i$ for some $i\in I$. Therefore, I is a congruence class modulo $\Theta[I]$.*

Remark. This situation is illustrated in Figure 2, in which the wavy line indicates congruence modulo $\Theta[I]$.

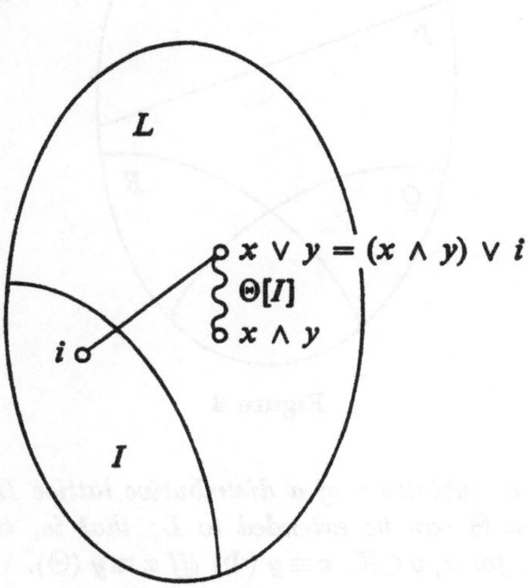

Figure 2

Proof. If $x\vee y=(x\wedge y)\vee i$, then $x\equiv y\ (\Theta(x\wedge y\wedge i,\ i))$, $x\wedge y\wedge i$, $i\in I$, and so $x\equiv y(\Theta[I])$. Conversely, $\Theta[I]=\bigvee(\Theta(u,v)\mid u,v\in I)$ by Lemma 2. However, $\Theta(u,v)\vee\Theta(u_1,v_1)\leq\Theta(u\wedge v\wedge u_1\wedge v_1,\ u\vee v\vee u_1\vee v_1)$; therefore $\Theta[I]=\bigcup(\Theta(u,v)\mid u,v\in I)$. If $x\equiv y(\Theta(u,v))$,

$u, v \in I$, $u \leq v$, then $x \vee v = y \vee v$, and so $(x \wedge y) \vee (v \wedge (x \vee y)) = x \vee y$; thus the condition of Corollary 4 is satisfied with $i = v \wedge (x \vee y) \in I$. Finally, if $a \in I$ and $a \equiv b(\Theta[I])$, then $a \vee b = (a \wedge b) \vee i$, $i \in I$, and so $a \vee b \in I$ and $b \in I$, showing that I is a full congruence class. ●

Corollary 5. *Let L be a distributive lattice, $x, y, a, b \in L$, and let $x \leq y \leq a \leq b$ or $a \leq b \leq x \leq y$. Then $x \equiv y$ $(\Theta(a, b))$ implies that $x = y$.*

A very important congruence relation has already been used in the proof of Lemma I.3.5(ii): Given a prime ideal P of the lattice L, we can construct a congruence relation that has exactly two congruence classes, P and $L - P$. This statement can be generalized as follows: Let A be a set of prime ideals of a lattice L and let us call two elements x and y congruent modulo A iff, for every $P \in A$, either $x, y \in P$ or $x, y \in L - P$; this describes a congruence relation on L. For instance, if $A = \{P, Q, R\}$, $Q \subset P$, $R \subset P$, we get five congruence classes as shown in Figure 3; the quotient lattice is shown in Figure I.5.8.

This principle will be used often. An interesting application is:

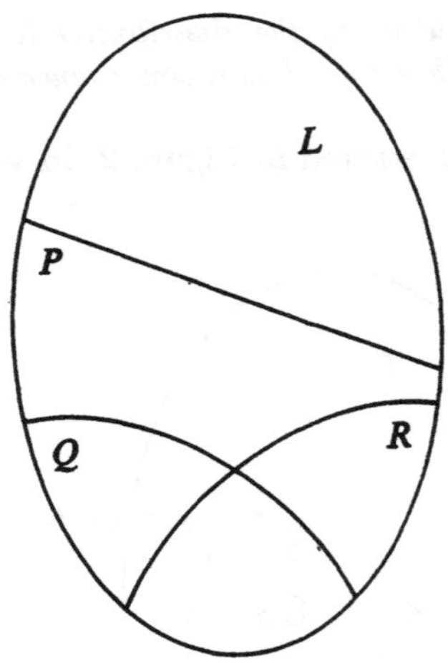

Figure 3

Theorem 6. *Let K be a sublattice of a distributive lattice L and let Θ be a congruence relation of K. Then Θ can be extended to L; that is, there exists a congruence relation Φ on L such that, for $x, y \in K$, $x \equiv y$ (Φ) iff $x \equiv y$ (Θ).*

Proof. Let φ be the natural homomorphism of K onto K/Θ, that is, $\varphi\colon x \to [x]\Theta$; then, for every prime ideal P of K/Θ, $P\varphi^{-1}$ is a prime ideal of K. Therefore, $(P\varphi^{-1}]$ is an ideal of L and $[K - P\varphi^{-1})$ is a dual ideal of L; thus by Theorem 1.15 we can choose a prime ideal P_1 of L such that $P_1 \supseteq P\varphi^{-1}$ and $P_1 \cap (K - P\varphi^{-1}) = \varnothing$. For every prime ideal P of K we choose such a prime ideal P_1 of L; let A denote the collection of all such prime ideals. Let Φ be the congruence relation associated with A as previously described. Now for $x, y \in K$, the condition $x \equiv y\ (\Theta)$ is equivalent to $x\varphi = y\varphi$, and so for every $P_1 \in A$ either $x, y \in P_1$ or $x, y \notin P_1$; thus $x \equiv y\ (\Phi)$. Conversely, if $x \equiv y\ (\Phi)$, then for every $P_1 \in A$ either $x, y \in P_1$ or $x, y \notin P_1$, and so either $x\varphi$, $y\varphi \in P$ or $x\varphi$, $y\varphi \notin P$. Since every pair of distinct elements of K/Θ is separated by a prime ideal (Corollary 1.17), we conclude that $x\varphi = y\varphi$ and thus $x \equiv y\ (\Theta)$. ●

It is well known that in rings, ideals are in a one-to-one correspondence with congruence relations. In one class of lattices the situation is exactly the same as in the class of rings.

Theorem 7. *Let L be a Boolean lattice. Then*

$$\Theta \to [0]\Theta$$

is a one-to-one correspondence between congruence relations and ideals of L.

Proof. By Corollary 4 the map is onto; therefore, we have only to prove that it is one-to-one, that is, that $[0]\Theta$ determines Θ. This fact, however, is obvious, since $a \equiv b\ (\Theta)$ iff $a \wedge b \equiv a \vee b\ (\Theta)$, which in turn is equivalent to $c \equiv 0\ (\Theta)$, where c is the relative complement of $a \wedge b$ in $[0, a \vee b]$ (see Figure 4). Thus $a \equiv b\ (\Theta)$ iff $c \in [0]\Theta$. ●

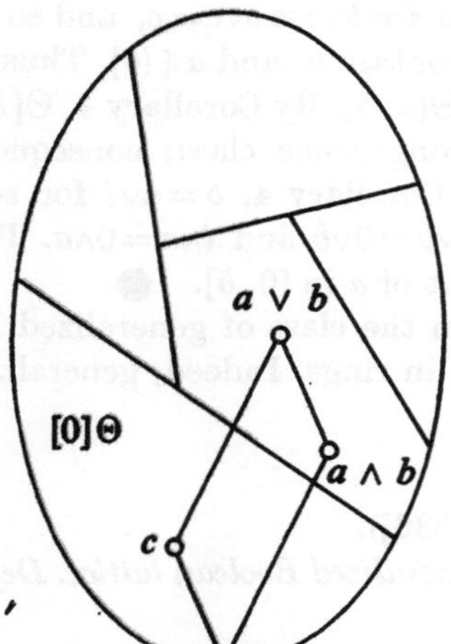

Figure 4

This proof does not make full use of the hypothesis that L is a complemented distributive lattice. In fact, all we need to make the proof work is that L has a zero and is relatively complemented. Such a distributive lattice is called a *generalized Boolean lattice.*

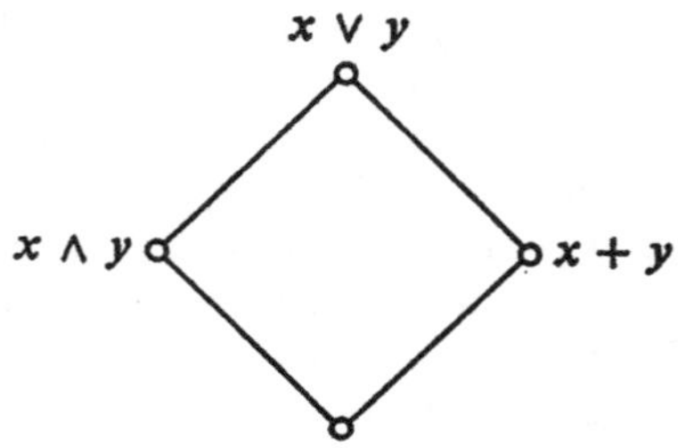

Figure 5

Theorem 8 (J. Hashimoto [1952]). *Let L be a lattice. There is a one-to-one correspondence between ideals and congruence relations of L under which the ideal corresponding to a congruence relation Θ is a whole congruence class under Θ iff L is a generalized Boolean algebra.*

Proof (G. Grätzer and E. T. Schmidt [1958d]). The proof of the "if" part is the proof of Theorem 7. We proceed with the "only if" part. The ideal corresponding to ω has to be $(0]$, and thus L has a 0. If L contains a diamond $\{o, a, b, c, i\}$, then $(a]$ cannot be a congruence class, because $a \equiv o$ implies that $i = a \vee c \equiv o \vee c = c$, and $b = b \wedge i \equiv b \wedge c = o$. But $o \in (a]$, and thus any congruence class containing $(a]$ contains b, and $b \notin (a]$. Similarly, if L contains a pentagon $\{o, a, b, c, i\}$ and a congruence class contains $(b]$, then $b \equiv o$; thus $i = b \vee c \equiv o \vee c = c$, and so $a = a \wedge i \equiv a \wedge c = o$. Therefore, this congruence class has to contain a, and $a \notin (b]$. Thus, by Theorem 1.1, L is distributive. Let $a < b$ and $I = [0]\Theta(a, b)$. By Corollary 4, $\Theta[I]$ is also a congruence relation of L having I as a whole congruence class; consequently, $\Theta[I] = \Theta(a, b)$, and so $a \equiv b$ $(\Theta[I])$. Thus, again by Corollary 4, $b = a \vee i$ for some $i \in I$, and $i \equiv 0$ $(\Theta(a, b))$. The latter is equivalent to $i \vee b = 0 \vee b$ and $i \wedge a = 0 \wedge a$. Thus $a \vee i = b$ and $a \wedge i = 0$, and so i is the relative complement of a in $[0, b]$. ●

It is no coincidence that, in the class of generalized Boolean lattices, congruences and ideals behave as they do in rings. Indeed, generalized Boolean lattices are rings in disguise.

Theorem 9 (M. H. Stone [1936]).

(i) *Let $\mathfrak{B} = \langle B; \wedge, \vee \rangle$ be a generalized Boolean lattice. Define the binary operations $\cdot$ and $+$ on B by setting*

$$x \cdot y = x \wedge y$$

and by defining $x + y$ as the relative complement of $x \wedge y$ in $[0, x \vee y]$ (see Figure 5). *Then $\mathfrak{B}^R = \langle B; +, \cdot \rangle$ is a* Boolean ring*–that is, an (associative) ring satisfying $x^2 = x$, for all $x \in B$ (and, consequently, satisfying $xy = yx$ and $x + x = 0$, for $x, y \in B$).*

(ii) *Let $\mathfrak{B} = \langle B; +, \cdot \rangle$ be a Boolean ring. Define the binary operations $\wedge$ and $\vee$ on B by*

$$x \wedge y = x \cdot y$$

and

$$x \vee y = x + y + x \cdot y.$$

Then $\mathfrak{B}^L = \langle B; \wedge, \vee \rangle$ *is a generalized Boolean lattice.*

(iii) *Let* $\mathfrak{B}$ *be a generalized Boolean lattice. Then* $(\mathfrak{B}^R)^L = \mathfrak{B}$.

(iv) *Let* $\mathfrak{B}$ *be a Boolean ring. Then* $(\mathfrak{B}^L)^R = \mathfrak{B}$.

The proof of this theorem is purely computational. Some steps will be given in the exercises. The correspondence between Boolean rings and generalized Boolean lattices preserves many algebraic properties.

Theorem 10. *Let* $\mathfrak{B}_0$ *and* $\mathfrak{B}_1$ *be generalized Boolean lattices.*

(i) *Let* $I \subseteq B_0$. *Then* I *is an ideal of* $\mathfrak{B}_0$ *iff* I *is an ideal of* $\mathfrak{B}_0^R$.

(ii) *Let* $\varphi: B_0 \to B_1$. *Then* φ *is a* $\{0\}$*-homomorphism of* $\mathfrak{B}_0$ *into* $\mathfrak{B}_1$ *iff* φ *is a homomorphism of* $\mathfrak{B}_0^R$ *into* $\mathfrak{B}_1^R$.

(iii) $\mathfrak{B}_0$ *is a* $\{0\}$*-sublattice of* $\mathfrak{B}_1$ *iff* $\mathfrak{B}_0^R$ *is a subring of* $\mathfrak{B}_1^R$.

The proof is again left to the reader.

Congruence relations on an arbitrary lattice have an interesting connection with distributive lattices:

Theorem 11 (N. Funayama and T. Nakayama [1942]). *Let* L *be an arbitrary lattice. Then* $C(L)$, *the lattice of all congruence relations of* L, *is distributive.*

Proof. Let $\Theta, \Phi, \Psi \in C(L)$. Since $\Theta \wedge (\Phi \vee \Psi) \geq (\Theta \wedge \Phi) \vee (\Theta \wedge \Psi)$, it suffices to prove that $a \equiv b \ (\Theta \wedge (\Phi \vee \Psi))$ implies that $a \equiv b \ ((\Theta \wedge \Phi) \vee (\Theta \wedge \Psi))$. So let $a \equiv b \ (\Theta \wedge (\Phi \vee \Psi))$; that is, $a \equiv b \ (\Theta)$ and $a \equiv b \ (\Phi \vee \Psi)$. By Theorem I.3.9, there exists a sequence $a \wedge b = z_0 \leq \cdots \leq z_n = a \vee b$ such that $z_i \equiv z_{i+1} \ (\Phi)$ or $z_i \equiv z_{i+1} \ (\Psi)$ for every $0 \leq i < n$. Since $a \equiv b \ (\Theta)$, we also have $a \wedge b \equiv a \vee b \ (\Theta)$, and so $z_i \equiv z_{i+1} \ (\Theta)$ for every $0 \leq i < n$. Thus for every $0 \leq i < n$, $z_i \equiv z_{i+1} \ (\Theta \wedge \Phi)$ or $z_i \equiv z_{i+1} \ (\Theta \wedge \Psi)$, implying that $a \equiv b \ ((\Theta \wedge \Phi) \vee (\Theta \wedge \Psi))$. ●

Another property of congruence lattices is given in the following definition.

Definition 12.

(i) *Let* L *be a complete lattice and let* a *be an element of* L. *Then* a *is called* compact *iff* $a \leq \bigvee X$ *for some* $X \subseteq L$ *implies that* $a \leq \bigvee X_1$ *for some finite* $X_1 \subseteq X$.

(ii) *A complete lattice is called* algebraic *iff every element is the join of compact elements.*

The name, algebraic lattice, is due to G. Birkhoff [1967], however Birkhoff does not assume completeness. In the literature, algebraic lattices are also called *compactly generated lattices.*

Just as for lattices, a nonvoid subset I of a join-semilattice F is an *ideal* iff, for $a, b \in F$, we have $a \vee b \in I$ exactly if a and $b \in I$. Again, $I(F)$ is the poset (not necessarily a lattice) of all ideals of F partially ordered under set inclusion. If F has a zero, then $I(F)$ is a lattice.

Using $I(F)$, we give a useful characterization of algebraic lattices:

Theorem 13. *A lattice L is algebraic iff it is isomorphic to the lattice of all ideals of a join-semilattice with* 0.

Proof. Let F be a join-semilattice with 0; we prove that $I(F)$ is algebraic. We know that $I(F)$ is complete. We claim that for $a \in F$, $(a]$ is a compact element of $I(F)$. Let $X \subseteq I(F)$ and let

$$(a] \subseteq \bigvee(I \mid I \in X).$$

Just as in the proof of Corollary I.3.2,

$$\bigvee(I \mid I \in X) = \{x \mid x \leq t_0 \vee \cdots \vee t_{n-1},\ t_i \in I_i,\ I_i \in X\}.$$

Therefore, $a \leq t_0 \vee \cdots \vee t_{n-1}$, $t_i \in I_i$, $I_i \in X$. Thus with $X_1 = \{I_0, \ldots, I_{n-1}\}$,

$$(a] \subseteq \bigvee(I \mid I \in X_1).$$

Since for any $I \in I(F)$ we have

$$I = \bigvee((a] \mid a \in I),$$

we see that $I(F)$ is algebraic.

Now let L be an algebraic lattice and let F be the set of compact elements of L. Obviously, $0 \in F$. Let $a, b \in F$ and $a \vee b \leq \bigvee X$ for some $X \subseteq L$. Then $a \leq a \vee b \leq \bigvee X$, and so $a \leq \bigvee X_0$, for some finite $X_0 \subseteq X$. Similarly, $b \leq \bigvee X_1$, for some finite $X_1 \subseteq X$. Thus $a \vee b \leq \bigvee(X_0 \cup X_1)$, and $X_0 \cup X_1$ is a finite subset of X. So $a \vee b \in F$.

Therefore, $\langle F; \vee \rangle$ is a join-semilattice with 0. Consider the map:

$$\varphi\colon a \to \{x \mid x \in F, x \leq a\}, \quad a \in L.$$

Obviously, φ maps L into $I(F)$. By the definition of an algebraic lattice, $a = \bigvee a\varphi$, and thus φ is one-to-one. To prove that φ is onto, let $I \in I(F)$, $a = \bigvee I$. Then $a\varphi \supseteq I$. Let $x \in a\varphi$. Then $x \leq \bigvee I$, so that by the compactness of x, $x \leq \bigvee I_1$ for some finite $I_1 \subseteq I$. Therefore $x \in I$, proving that $a\varphi \subseteq I$. Consequently, $a\varphi = I$, and so φ is onto. Thus φ is an isomorphism. ●

Now we connect the foregoing with congruence lattices.

Lemma 14. *Every principal congruence relation is compact.*

Proof. Let L be a lattice, $a, b \in L$, $X \subseteq C(L)$, and

$$\Theta(a, b) \leq \bigvee X.$$

Then $a \equiv b\ (\bigvee X)$, and thus (just as in Theorem I.3.9) there exists a sequence $a = x_0$, $x_1, \ldots, x_{n-1} = b$, in which $x_i \equiv x_{i+1}\ (\Theta_i)$ for some $\Theta_i \in X$. Therefore, $a \equiv b\ (\bigvee X_0)$, where $X_0 = \{\Theta_0, \ldots, \Theta_{n-1}\}$, and so $\Theta(a, b) \leq (\bigvee X_0)$, where X_0 is a finite subset of X. ●

Theorem 15. *Let L be an arbitrary lattice. Then $C(L)$ is an algebraic lattice.*

Proof. For every $\Theta \in C(L)$,

$$\Theta = \bigvee(\Theta(a, b) \mid a \equiv b\ (\Theta)).$$

Consequently, this theorem follows from Lemma 14 and Corollary I.3.15. ●

Combining Theorems 11 and 15 we get:

Corollary 16. *Let L be an arbitrary lattice. Then $C(L)$ is a distributive algebraic lattice.*

The converse of Corollary 16 is a long-standing conjecture of lattice theory. We shall verify the conjecture in the finite case. This was first established by R. P. Dilworth. The present proof combines a construction of G. Grätzer and E. T. Schmidt [1962] (Lemma 18) with a result of G. Grätzer and H. Lakser [1968] (Lemma 19). The reader is advised to page over the proof of Theorem 17 at the first reading of this book.

Theorem 17. *Let K be a finite distributive lattice. Then there exists a finite lattice L such that K is isomorphic to $C(L)$.*

Proof. The proof of Theorem 17 is immediate from Lemmas 18 and 19. We take $L = I(M)$ where M is given in Lemma 18. By Lemma 19, $C(L) \cong C(M)$; by Lemma 18, $C(M) \cong K$. Since M is finite, so is L. ●

Let M be a finite poset such that $\inf\{a, b\}$ exists in M for any $a, b \in M$. We define in M: $a \wedge b = \inf\{a, b\}$ and $a \vee b = \sup\{a, b\}$ whenever it exists. This makes M into a partial lattice. An equivalence relation Θ on M is a *congruence relation* iff $a_0 \equiv b_0\ (\Theta)$ and $a_1 \equiv b_1\ (\Theta)$ imply that $a_0 \wedge a_1 \equiv b_0 \wedge b_1\ (\Theta)$ and that $a_0 \vee a_1 \equiv b_0 \vee b_1\ (\Theta)$ whenever $a_0 \vee a_1$ and $b_0 \vee b_1$ exist. Then the set $C(M)$ of all congruence relations is again a lattice.

Lemma 18. *Let K be a finite distributive lattice. Then there exists a finite poset M such that* $\inf\{a, b\}$ *exists for all $a, b \in M$ and $C(M)$ is isomorphic to K.*

Proof. Take the set $M_0 = J(K) \cup \{0\}$ and make it a meet-semilattice by defining $\inf\{a, b\} = 0$ if $a \neq b$ ($J(K)$ is the set of nonzero join-irreducible elements of K; see

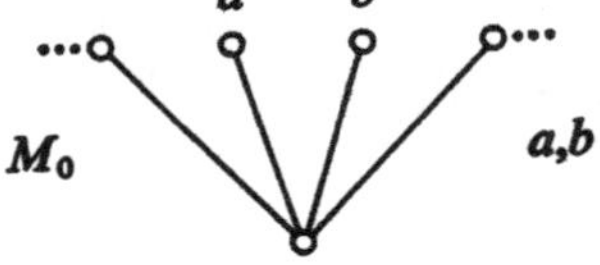

Figure 6

Section 1), as illustrated in Figure 6. Note that $a \equiv b\ (\Theta)$ and $a \neq b$ imply in M_0 that $a \equiv 0\ (\Theta)$ and $b \equiv 0\ (\Theta)$; therefore, congruence relations of M_0 are in one-to-one correspondence with subsets of $J(K)$. Thus $C(M_0)$ is a Boolean lattice whose atoms are associated with elements of $J(K)$; the congruence Φ_a associated with $a \in J(K)$ is: $x \equiv y\ (\Phi_a)$ if $\{x, y\} = \{a, 0\}$, and if $\{x, y\} \neq \{a, 0\}$, then $x \equiv y\ (\Phi_a)$ implies that $x = y$.

If $J(K)$ is unordered in K, then we are ready. However, if, say, $a, b \in J(K)$, $a > b$ in K, then we must have $\Phi_a > \Phi_b$. The simplest way to make this happen is to use the lattice $M(a, b)$ of Figure 7. Note that $M(a, b)$ has three congruence relations, namely,

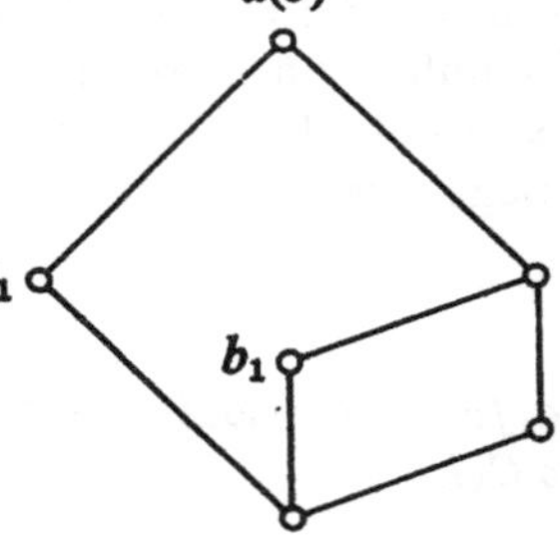

Figure 7

ω, ι, and Θ, where Θ is the congruence relation with congruence classes $\{0, b_1, b_2, b\}$, $\{a_1, a(b)\}$. Thus $\Theta(a_1, 0) = \iota$. In other words, $a_1 \equiv 0$ "implies" that $b_1 \equiv 0$, but $b_1 \equiv 0$ does not "imply" that $a_1 \equiv 0$.

We construct M by "inserting" $M(a, b)$ in M_0 whenever $a > b$ in $J(K)$. Figure 8 gives M for the three-element chain.

More precisely, M consists of four kinds of elements: (i) 0; (ii) all maximal join-irreducible elements of K (that is, all $a \in J(K)$ such that there is no $x \in J(K)$ with $a < x$ in K); (iii) for any nonmaximal join-irreducible element a of K, three elements: a, a_1, a_2; (iv) for each pair $a, b \in J(K)$ with $a > b$, a new element, $a(b)$. To simplify the notation, for each maximal join-irreducible element a, we write $a = a_1 = a_2$. For $a, b \in J(K)$ with $a > b$, we set $M(a, b) = \{0, a_1, b, b_1, b_2, a(b)\}$.

Observe that $M(a, b) \cap M(c, d) = M(a, b)$ if $a = c$ and $b = d$; $M(a, b) \cap M(c, d) = \{0, b, b_1, b_2\}$ if $a \neq c$ and $b = d$; $M(a, b) \cap M(c, d) = \{0, a_1\}$ if $a = c$ and $b \neq d$; $M(a, b) \cap M(c, d) = \{0, b_1\}$ if $b = c$; otherwise, $M(a, b) \cap M(c, d) = \{0\}$.

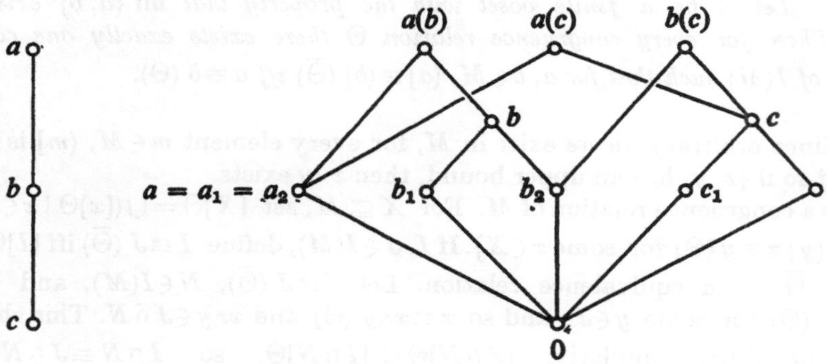

Figure 8

For $x, y \in M$, let us define $x \leq y$ to mean that for some $a, b \in J(K)$ with $a > b$, we have $x, y \in M(a, b)$ and $x \leq y$ in the lattice $M(a, b)$ as illustrated in Figure 7. It is easily seen that $x \leq y$ does not depend on the choice of a and b, and that $\leq$ is a partial ordering relation. Since, under this partial ordering, all $M(a, b)$ and $M(a, b) \cap M(c, d)$ are lattices and $x, y \in M$, $x \in M(a, b)$, and $y \leq x$ imply that $y \in M(a, b)$, we conclude that inf $\{u, v\}$ exists for all $u, v \in M$.

Now we describe $C(M)$. Let $H \in H(J(K))$ (notation of Definition 1.8). We define a binary relation Θ_H on M:

$x \equiv y\ (\Theta_H)$ iff

either $x, y \in \bigcup(M(a, b) \mid a, b \in H, a > b) \cup \bigcup(\{0, a_1, a_2, a\} \mid a \in H)$,

or $x, y \in \{a_1, a(b), a(c)\}$, where $a > b$, $a > c$, $b, c \in H$, or $x = y$.

In other words, $[0]\Theta_H$ contains all a_1, a_2, a with $a \in H$; and if $a > b$, $a, b \in H$, then it also contains $a(b)$. Outside this class the only nontrivial congruences are $a(b) \equiv a_1 \equiv a(c)$, where $a \notin H$, and $b, c \in H$, $a > b$, $a > c$.

Θ_H is obviously an equivalence relation. The fact that Θ_H restricted to any $M(a, b)$ is a congruence relation easily implies that Θ_H is a congruence relation. Given a Θ_H we get

$$H = \{a \mid a_1 \equiv 0\ (\Theta_H)\};$$

thus the map

$$\varphi: H \to \Theta_H$$

is a one-to-one order-preserving map of $H(J(K))$ into $C(M)$. To show that φ is an isomorphism, we have to show that φ is onto. So let Θ be a congruence relation of M, and

$$H = \{a \mid a_1 \equiv 0\ (\Theta)\}.$$

Since in $M(a, b)$ every congruence Θ is determined by the atoms in $[0]\Theta$, the same holds in M. Therefore, $\Theta = \Theta_H$. Thus $H(J(K)) \cong C(M)$. By Theorem 1.9, $K \cong H(J(K))$, and so $K \cong C(M)$. ●

Lemma 19. *Let M be a finite poset with the property that* $\inf\{a, b\}$ *exists for all* $a, b \in M$. *Then for every congruence relation* Θ *there exists exactly one congruence relation* $\overline{\Theta}$ *of* $I(M)$ *such that for* $a, b \in M$, $(a] \equiv (b]\ (\overline{\Theta})$ *iff* $a \equiv b\ (\Theta)$.

Proof. Since arbitrary meets exist in M, for every element $m \in M$, $(m]$ is a (finite) lattice, and so if $\{x, y\}$ has an upper bound, then $x \vee y$ exists.

Let Θ be a congruence relation of M. For $X \subseteq M$, set $[X]\Theta = \bigcup([x]\Theta \mid x \in X)$; that is, $[X]\Theta = \{y \mid x \equiv y\ (\Theta)$ for some $x \in X\}$. If $I, J \in I(M)$, define $I \equiv J\ (\overline{\Theta})$ iff $[I]\Theta = [J]\Theta$. Obviously, $\overline{\Theta}$ is an equivalence relation. Let $I \equiv J\ (\overline{\Theta})$, $N \in I(M)$, and $x \in I \cap N$. Then $x \equiv y\ (\Theta)$ for some $y \in J$, and so $x \equiv x \wedge y\ (\Theta)$ and $x \wedge y \in J \cap N$. This shows that $[I \cap N]\Theta \subseteq [J \cap N]\Theta$. Similarly, $[J \cap N]\Theta \subseteq [I \cap N]\Theta$, so $I \cap N \equiv J \cap N\ (\overline{\Theta})$. To show that $I \vee N \equiv J \vee N\ (\overline{\Theta})$, recall the description of $I \vee N$ given in Exercise I.5.22: Set $A_0 = I \cup N$ and, for $0 < n < \omega$, $A_n = \{x \mid x \leq t_0 \vee t_1$, for some $t_0, t_1 \in A_{n-1}\}$; then $I \vee N = \bigcup(A_n \mid n < \omega)$. We prove by induction on n that $A_n \subseteq [J \vee N]\Theta$. For $n = 0$, $A_0 = I \cup N \subseteq [J]\Theta \cup N \subseteq [J \vee N]\Theta$. Suppose that $A_{n-1} \subseteq [J \vee N]\Theta$ and let $x \in A_n$. Then $x \leq t_0 \vee t_1$ for some $t_0, t_1 \in A_{n-1}$. Thus $t_0 \equiv u_0\ (\Theta)$ and $t_1 \equiv u_1\ (\Theta)$ for some $u_0, u_1 \in J \vee N$, and so $t_0 \equiv t_0 \wedge u_0\ (\Theta)$ and $t_1 \equiv t_1 \wedge u_1\ (\Theta)$. Observe that $t_0 \vee t_1$ is an upper bound for $\{t_0 \wedge u_0, t_1 \wedge u_1\}$; consequently, $(t_0 \wedge u_0) \vee (t_1 \wedge u_1)$ exists. Therefore, $t_0 \vee t_1 \equiv (t_0 \wedge u_0) \vee (t_1 \wedge u_1)\ (\Theta)$. Finally, $x = x \wedge (t_0 \vee t_1) \equiv x \wedge ((t_0 \wedge u_0) \vee (t_1 \wedge u_1))\ (\Theta)$ and $x \wedge ((t_0 \wedge u_0) \vee (t_1 \wedge u_1)) \in J \vee N$.

Thus $x \in [J \vee N]\Theta$. Since $I \vee N = \bigcup(A_n \mid n < \omega)$, we conclude that $I \vee N \subseteq [J \vee N]\Theta$. Similarly, $J \vee N \subseteq [I \vee N]\Theta$, proving that $I \vee N \equiv J \vee N\ (\overline{\Theta})$. This completes the verification that $\overline{\Theta}$ is a congruence relation of $I(M)$.

If $a \equiv b\ (\Theta)$ and $x \in (a]$, then $x \equiv x \wedge b\ (\Theta)$. Thus $(a] \subseteq [(b]]\Theta$. Similarly, $(b] \subseteq [(a]]\Theta$, and so $(a] \equiv (b]\ (\overline{\Theta})$. Conversely, if $(a] \equiv (b]\ (\overline{\Theta})$, then $a \equiv b_1\ (\Theta)$ and $a_1 \equiv b\ (\Theta)$ for some $a_1 \leq a$ and $b_1 \leq b$. Forming the join of the two congruences, we get $a \equiv b\ (\Theta)$. Thus $\overline{\Theta}$ has all the properties required by Lemma 19.

To show the uniqueness, let Φ be a congruence relation of $I(M)$ satisfying $(a] \equiv (b]\ (\Phi)$ iff $a \equiv b\ (\Theta)$. Let $I, J \in I(M)$, $I \equiv J\ (\Phi)$, and $x \in I$. Then $(x] \cap I \equiv (x] \cap J\ (\Phi)$, $(x] \cap I = (x]$, and $(x] \cap J = (y]$ for some $y \in J$. Thus $(x] \equiv (y]\ (\Phi)$, and so $x \equiv y\ (\Theta)$, proving that $I \subseteq [J]\Theta$. Similarly, $J \subseteq [I]\Theta$, and so $I \equiv J\ (\overline{\Theta})$. Conversely, if $I \equiv J\ (\overline{\Theta})$, then take all congruences of the form $x \equiv y\ (\Theta)$, $x \in I$, $y \in J$. By our assumption regarding Φ, $(x] \equiv (y]\ (\Phi)$, and by our definition of $\overline{\Theta}$, the join of all these congruences yields $I \equiv J\ (\Phi)$. Thus $\Phi = \overline{\Theta}$. ●

More general results along these lines can be found in G. Grätzer and E. T. Schmidt [1962], G. Grätzer and H. Lakser [1968], and E. T. Schmidt [1968], [1969], and [1974].

Exercises

1. Let L be a distributive lattice and let $x, y, a, b \in L$. Prove that if $a \leq b$ and $x \leq y$, then $x \wedge a = y \wedge a$ and $x \vee b = y \vee b$ is equivalent to $(a \vee p) \wedge q = x$ and $(b \vee p) \wedge q = y$ for some p, q in L.

2. Verify Corollary 4 directly.
3. Let K be a sublattice of the distributive lattice L and let P be a prime ideal of K. Prove that there exists a prime ideal Q of L with $Q \cap K = P$.
4. Prove that Corollary 5 characterizes the distributivity of L.
5. Show that Theorem 6 characterizes the distributivity of L.
6. What is the smallest n such that Theorem 6 for $|K| \leq n$ characterizes the distributivity of L?
7. Let L be a *sectionally complemented lattice*; that is, L has a 0 and all intervals $[0, a]$ are complemented. Prove that $\Theta \to [0]\Theta$ is a one-to-one correspondence between congruences and *certain* ideals of L.
*8. Show that the "certain ideals" that appear in Exercise 7 form a sublattice of $I(L)$.
9. Prove that every (principal) ideal of L is of the form $[0]\Theta$ iff L is distributive.
10. Let L be a distributive lattice and let I be an ideal of L. Define a binary relation on L: $x \equiv y\ (\Phi(I))$ iff there is no $a \in L$ with $a \leq x \vee y$, $x \wedge y \wedge a \in I$, $a \notin I$. Prove that $\Phi(I)$ is the largest congruence relation of L under which I is a whole congruence class.
11. Let L be a distributive lattice with 0. Prove that there is a one-to-one correspondence between ideals and congruence relations (in the sense of Theorem 8) iff $\Theta[I] = \Phi(I)$ for all $I \in I(L)$.
12. Prove Theorem 8 using Exercises 10 and 11 (G. Ja. Areškin [1953]).
*13. Let L be a lattice and let a be an element of L. Show that every convex sublattice of L containing a is a congruence class under exactly one congruence relation iff L is distributive and all the intervals $[b, a]$ $(b \in L, b \leq a)$ and $[a, c]$ $(c \in L, a \leq c)$ are complemented (G. Grätzer and E. T. Schmidt [1958d]).
14. Derive Theorem 8 (and, also, a variant of Theorem 8) by taking $a = 0$ (arbitrary $a \in L$) in Exercise 13.
15. Let L be a relatively complemented lattice, $I, J \in I(L)$, and $I \subseteq J$. Prove that if I is an intersection of prime ideals, then so is J (J. Hashimoto [1952]).
16. Use Exercises 14 and 15 to get the following theorem: Let L be a relatively complemented lattice. Then L is distributive iff, for some element a of L, $(a]$ is an intersection of prime ideals and $[a)$ is an intersection of prime dual ideals (J. Hashimoto [1952]).
17. Prove that the verification of Theorem 9(i) can be reduced to the Boolean lattice case and that in this case $x + y = (x \wedge y') \vee (x' \wedge y)$.
18. Let B be a Boolean lattice. Verify that $x + y = (x \vee y) \wedge (x' \vee y')$.
19. Let B be a Boolean lattice. Verify that $(x+y)+z = (x \wedge y' \wedge z') \vee (x' \wedge y \wedge z') \vee (x' \wedge y' \wedge z)$ and conclude that $+$ is associative.
20. Prove that $x(y+z) = xy + xz$.
21. Prove Theorem 9(i).
22. Prove Theorem 9(ii).
23. Let $\mathfrak{B}$ be a generalized Boolean lattice. Observe that for $x, y \in B$, $x \wedge y$ is the same in $\mathfrak{B}$ as in $(\mathfrak{B}^R)^L$ (namely, $x \cdot y$); conclude that $\mathfrak{B} = (\mathfrak{B}^R)^L$.
24. Verify Theorem 9(iv).
25. Verify Theorem 10.
26. Show that, using the concept of a distributive semilattice (see Section 5), Corollary 16 can be reformulated as follows: Let L be an arbitrary lattice. Prove that there exists a distributive join-semilattice F with 0 such that $C(L)$ is isomorphic to $I(F)$.
27. Characterize the lattice of all ideals of a lattice using the concept of an algebraic lattice.
28. Characterize the lattice of all ideals of a Boolean lattice.

29. Let M be a poset such that $a \wedge b$ exists for all $a, b \in M$. Let Θ and Φ be congruences of M. Find a result for M analogous with Lemma I.3.8.

30. Let M be as in Exercise 29. Generalize Theorem I.3.9 to M.

31. Let M be as in Exercise 29. Is $C(M)$ necessarily distributive?

32. Let M be as in Exercise 29. Assume that M satisfies the following conditions:
(i) There exists $H \subseteq M$ such that for $h \in H$, $(h]$ is a lattice.
(ii) For each ideal I that belongs to the sublattice $I_P(M)$ of $I(M)$ generated by the principal ideals, there exist finite $\{h_1, \ldots, h_n\} \subseteq H$, $\{i_1, \ldots, i_n\} \subseteq I$ with $i_j \leq h_j$ and $I = (i_1] \cup \cdots \cup (i_n]$.
Under these conditions, prove Lemma 19 for M, replacing $I(M)$ with $I_P(M)$ (G. Grätzer and H. Lakser [1968]).

Exercises 33–36 are from G. Grätzer and E. T. Schmidt [1962].

***33.** Let K be a finite distributive lattice and let L be the lattice of Theorem 17 as constructed in Lemmas 18 and 19. Show that L is sectionally complemented.

34. Let L be given as in Exercise 33. Show that the congruences of L permute.

35. Let n be the length of the longest chain in K. Show that the L of Theorem 17 can be constructed so that L is of length $2n-1$ (define $a(b)$ only for $a \succ b$).

36. Let P be a poset. Generalize Theorem 17 for $K = H_F(P)$ (notation is the same as in Exercise 1.17).

Exercises 37 and 38 are from G. Grätzer and E. T. Schmidt [1958b].

***37.** Show that a chain C is the congruence lattice of a lattice iff C is algebraic.

38. Prove that a Boolean algebra B is the congruence lattice of a lattice iff B is algebraic.

39. Let L be a distributive lattice. Show that $a \rightarrow \Theta[(a]]$ embeds L into $C(L)$.

40. Let L be a bounded distributive lattice. Show that for $a, b \in L$, $a \leq b$, $\Theta(a, b)$ has a complement in $C(L)$, namely, $\Theta(0, a) \vee \Theta(b, 1)$.

41. Let L be a bounded distributive lattice. Show that the compact elements of $C(L)$ form a Boolean lattice (J. Hashimoto [1952], G. Grätzer and E. T. Schmidt [1958e]).

42. Let B be a Boolean algebra generated by X and let L be the sublattice of B generated by X. Show that B is freely generated by X iff L is freely generated by X in $\mathbf{D}$ (see Exercise 2.12).

43. Let the Boolean algebra B be generated by X. Show that B is freely generated by X iff $\wedge X_0 \leq \vee X_1$ implies that $X_0 \cap X_1 \neq \varnothing$ for any finite nonvoid $X_0, X_1 \subseteq X$ (compare this with Theorems 2.3 and 2.4).

4. Boolean Algebras *R*-generated by Distributive Lattices

The investigations of this section are based on the following result:

THEOREM 1. *Every distributive lattice can be embedded in a Boolean lattice.*

PROOF. By Theorem 1.19, every distributive lattice L is isomorphic to a ring of subsets of some set X. Obviously, L can be embedded into $P(X)$. ●

Definition 2. *Let L be a $\{0\}$-sublattice of the generalized Boolean lattice B. Then L is said to R-generate B iff L generates B as a ring, and if L has a 1 the same element is the 1 of B.*

The last condition is equivalent to the following: If $\bigvee L$ exists, then $\bigvee L = \bigvee B$.

Our goal is to show the uniqueness of the generalized Boolean lattice R-generated by L. The first result is essentially due to H. M. MacNeille [1939]:

Lemma 3. *Let B be R-generated by L. Then every $a \in B$ can be expressed in the form*

$$a_0 + a_1 + \cdots + a_{n-1}, \quad a_0 \leq a_1 \leq \cdots \leq a_{n-1}, \quad a_0, \ldots, a_{n-1} \in L.$$

Example. Let B be the Boolean lattice shown in Figure 1 and let $L = \{0, a_0, a_1, a_2\}$. Then L R-generates B.

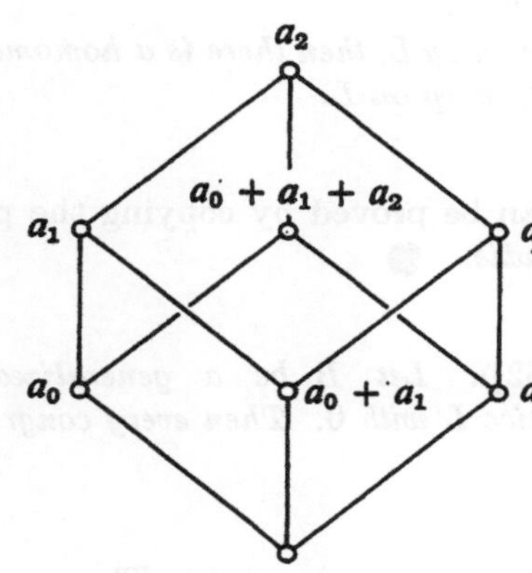

Figure 1

Proof. Let B_1 denote the set of all elements that can be represented in the form $a_0 + \cdots + a_{n-1}$, $a_0, \ldots, a_{n-1} \in L$. Then $L \subseteq B_1$, and B_1 is closed under $+$ and $-$ (since $x - y = x + y$). Furthermore,

$$(a_0 + \cdots + a_{n-1})(b_0 + \cdots + b_{m-1}) = \sum a_i b_j,$$

and each term $a_i b_j = a_i \wedge b_j \in L$, so B_1 is closed under multiplication. We conclude that $B_1 = B$.

Note that L is a sublattice of B; therefore, for $a, b \in L$, $a \vee b$ in L is the same as $a \vee b$ in B. Thus $a \vee b = a + b + ab$, and so

$$a + b = ab + (a \vee b) = (a \wedge b) + (a \vee b).$$

Take $a_0+\cdots+a_{n-1}\in B$. We prove by induction on n that the summands can be made to form an increasing sequence. For $n=1$ this is obvious. Let us assume that $a_1\leq\cdots\leq a_{n-1}$. Then

$$\begin{aligned} &a_0+a_1+\cdots+a_{n-1}\\ &\quad=(a_0\wedge a_1)+(a_0\vee a_1)+a_2+\cdots+a_{n-1}\\ &\quad=(a_0\wedge a_1)+((a_0\vee a_1)\wedge a_2)+(a_0\vee a_2)+a_3+\cdots+a_{n-1}\\ &\quad=(a_0\wedge a_1)+((a_0\vee a_1)\wedge a_2)+((a_0\vee a_2)\wedge a_3)+(a_0\vee a_3)+\cdots+a_{n-1}=\cdots\\ &\quad=(a_0\wedge a_1)+((a_0\vee a_1)\wedge a_2)+\cdots+((a_0\vee a_{n-2})\wedge a_{n-1})+(a_0\vee a_{n-1}),\end{aligned}$$

and

$$a_0\wedge a_1\leq(a_0\vee a_1)\wedge a_2\leq\cdots\leq(a_0\vee a_{n-2})\wedge a_{n-1}\leq a_0\vee a_{n-1}. \quad \bullet$$

Lemma 4. *Let L be a distributive lattice with 0. Then there exists a generalized Boolean lattice B freely R-generated by L—that is, a generalized Boolean lattice B with the following properties:*

(i) *B is R-generated by L.*

(ii) *If B_1 is R-generated by L, then there is a homomorphism φ of B onto B_1 that is the identity map on L.*

Proof. The existence of B can be proved by copying the proof of Theorem I.5.5 (Theorem I.5.24), *mutatis mutandis.* ●

Lemma 5 (J. Hashimoto [1952]). *Let B be a generalized Boolean lattice R-generated by the distributive lattice L with 0. Then every congruence relation of L has one and only one extension to B.*

Proof. The existence of an extension was proved in Theorem 3.6. By Theorems 3.7 and 3.10(i), the following statement implies the uniqueness of the extension:

If I and J are (ring) ideals of B with $I\subset J$, then there are elements $a, b\in L$, $a\neq b$, such that $a\equiv b\ (\text{mod } J)$ and $a\not\equiv b\ (\text{mod } I)$.

Indeed, let $x\in J-I$. By Lemma 3, x can be represented in the form

$$x=x_0+\cdots+x_{n-1},\quad x_0\leq\cdots\leq x_{n-1},\quad x_0,\ldots,x_{n-1}\in L.$$

If n is odd, then $x_0=x\cdot x_0\leq x\in J$, and thus $x_0\in J$; $x_0+x_1+x_2=x\cdot x_2\in J$, therefore $x_1+x_2=x_0+(x_0+x_1+x_2)\in J$. Similarly, $x_3+x_4, x_5+x_6,\ldots\in J$. Since $x_0+(x_1+x_2)+(x_3+x_4)+\cdots\in J-I$, we conclude that either $x_0\in J-I$, or $x_{2i}+x_{2i+1}\in J-I$ for some $2i<n$. If n is even, then we obtain $x_0+x_1, x_2+x_3,\ldots\in J$ (by multiplying x by $x_1, x_3,\ldots$), and we conclude that for some $2i<n$, $x_{2i}+x_{2i+1}\in J-I$.

Now if $x_{2i}+x_{2i+1}\in J-I$, then $x_{2i}\equiv x_{2i+1}\ (\text{mod } J)$, but $x_{2i}\not\equiv x_{2i+1}\ (\text{mod } I)$, x_{2i}, $x_{2i+1}\in L$. Finally, if $x_0\in J-I$, then $x_0\equiv 0\ (\text{mod } J)$ and $x_0\not\equiv 0\ (\text{mod } I)$. ●

Theorem 6. *If B_1 and B_2 are generalized Boolean lattices R-generated by a distributive lattice L with 0, then B_1 and B_2 are isomorphic.*

Remark. For a distributive lattice L with 0 we shall denote by $B(L)$ a generalized Boolean lattice R-generated by L.

Proof. Let B be a free generalized Boolean lattice R-generated by L (see Lemma 4). Let φ be a homomorphism of B onto B_1 such that φ is the identity on L (see Lemma 4(ii)). We want to show that φ is an isomorphism. Indeed, if φ is not an isomorphism, then the ideal kernel I of φ is not 0. Thus by Lemma 5, $a \equiv b \pmod{I}$ for some $a, b \in L$, $a \neq b$. This means that $a\varphi = b\varphi$, contrary to our assumptions. Similarly, there is an isomorphism ψ between B and B_2. Obviously, $\varphi^{-1}\psi$ is an isomorphism between B_1 and B_2. ●

Corollary 7. *Let L_0 and L_1 be distributive lattices with 0 and let φ be a $\{0\}$-homomorphism of L_0 onto L_1. Then φ can be extended to a homomorphism of $B(L_0)$ onto $B(L_1)$.*

Proof. Let Θ be the congruence kernel of φ, and let $\bar{\Theta}$ be the extension of Θ to $B(L_0)$ (Lemma 5). Then $B(L_0)/\bar{\Theta}$ is a generalized Boolean lattice R-generated by $L_0/\Theta \cong L_1$. Thus $B(L_0)/\bar{\Theta} \cong B(L_1)$ by Theorem 6, and using this, the proof of Corollary 7 becomes trivial. ●

Corollary 8. *Let L_0 be a $\{0\}$-sublattice of the distributive lattice L_1 with 0, and assume that if L_1 has 1, then L_0 is a $\{0, 1\}$-sublattice of L_1. Let B denote the subalgebra of $B(L_1)$ R-generated by L_0. Then $B(L_0) \cong B$.*

Proof. The proof is trivial. ●

Let L_0 and L_1 be given as in Corollary 8. It is natural to ask the conditions under which L_0 R-generates $B(L_1)$. For $H \subseteq B(L_1)$, let $[H]_R$ denote the generalized Boolean sublattice of $B(L_1)$ R-generated by H. We can answer our query by determining $[L_0]_R \cap L_1$.

Lemma 9. *Let L_0 and L_1 be given as in Corollary 8. Then $L_1 \cap [L_0]_R$ is the smallest sublattice of L_1 containing L_0 that is closed under taking relative complements in L_1. Therefore L_0 R-generates $B(L_1)$ iff the smallest sublattice of L_1 containing L_0 and closed under relative complementation in L_1 is L_1 itself.*

Proof. It is obvious that $L_0 \subseteq L_1 \cap [L_0]_R$. If $a, b, c \in L_1 \cap [L_0]_R$, $d \in L_1$, and d is a relative complement of b in $[a, c]$, then $d = a + b + c \in L_1 \cap [L_0]_R$, since (see Figure 2) d is a relative complement of $a + b$ in the interval $[0, c]$. Thus $d \in L_1 \cap [L_0]_R$. Now

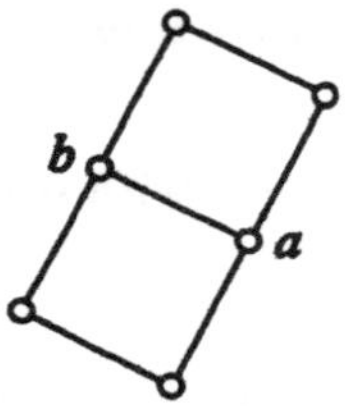

Figure 2

suppose that L is a sublattice of L_1 containing L_0 and closed under relative complementation in L_1. If $x \in L_1 \cap [L_0]_R$, then by Lemma 3 we can represent x as

$$x = a_0 + \cdots + a_{n-1}, \quad a_0, \ldots, a_{n-1} \in L_0, \quad a_0 \leq \cdots \leq a_{n-1}.$$

We prove $x \in L$ by induction on n. If $n=1$, $x=a_0 \in L_0 \subseteq L$. If $n=2$, then x is the relative complement of a_0 in $[0, a_1]$, $0, a_0, a_1 \in L_0$, thus $x \in L$. If $n=3$, then (see Figure 1) $x = a_0 + a_1 + a_2$ is the relative complement of a_1 in $[a_0, a_2]$, and so $x \in L$. Now let $n > 3$ and let $y \in L$ be proved for all $y = b_0 + \cdots + b_{k-1}$, $b_0, \ldots, b_{k-1} \in L_0$, $b_0 \leq \cdots \leq b_{k-1}$, and $k < n$. Note that $x \in L_1$ and $a_{n-3} \in L_0$ imply that $xa_{n-3} = a_0 + \cdots + a_{n-3} + a_{n-3} + a_{n-3} = a_0 + \cdots + a_{n-3} \in L_1$ and $x \vee a_{n-3} = x + a_{n-3} + xa_{n-3} = a_0 + \cdots + a_{n-1} + a_{n-3} + a_0 + \cdots + a_{n-3} = a_{n-3} + a_{n-2} + a_{n-1} \in L_1$. By the induction hypothesis, $a_0 + \cdots + a_{n-3} \in L$ and $a_{n-3} + a_{n-2} + a_{n-1} \in L$; therefore, x is the relative complement in L_1 of an element (namely, a_{n-3}) of L in an interval (namely, $[a_0 + \cdots + a_{n-3}, a_{n-3} + a_{n-2} + a_{n-1}]$) in L, and so, by assumption, $x \in L$. Thus $L_1 \cap [L_0]_R \subseteq L$. ●

In Theorem 1 we embedded L into $P(X)$, which is a complete Boolean lattice. The question arises whether we can require this embedding to be complete—that is, to preserve arbitrary meets and joins, if they exist in L.

It is easy to see that not every complete distributive lattice has a complete embedding into a complete Boolean lattice.

Lemma 10 (J. von Neumann [1936/37]). *Let B be a complete Boolean lattice. Then B satisfies the Join Infinite Distributive Identity (JID),*[1]

$$x \wedge \bigvee(x_i \mid i \in I) = \bigvee(x \wedge x_i \mid i \in I),$$

and its dual, the Meet Infinite Distributive Identity (MID).

Proof. $x \wedge x_j \leq x$ and $x \wedge x_j \leq \bigvee(x_i \mid i \in I)$ for any $j \in I$; therefore, $x \wedge \bigvee(x_i \mid i \in I)$ is an upper bound for $\{x \wedge x_i \mid i \in I\}$. Now let u be any upper bound, that is, $x \wedge x_i \leq u$ for all $i \in I$. Then

$$x_i = x_i \wedge (x \vee x') = (x_i \wedge x) \vee (x_i \wedge x') \leq u \vee x'.$$

[1] Of course, (JID) is not an identity in the sense of Section I.4 but is only an infinitary analogue of an identity.

Thus

$$x \wedge \bigvee(x_i \mid i \in I) \leq x \wedge (u \vee x') = (x \wedge u) \vee (x \wedge x') = x \wedge u \leq u,$$

showing that $x \wedge \bigvee(x_i \mid i \in I)$ is the least upper bound for $\{x \wedge x_i \mid i \in I\}$. (MID) follows by duality. ●

COROLLARY 11. ***Any complete distributive lattice that has a complete embedding into a complete Boolean lattice satisfies both (JID) and (MID).***

Easy examples show that (JID) and (MID) need not hold in all complete distributive lattices.

Our task now is to show the converse of Corollary 11 (N. Funayama [1959]). The construction depends on a property of $B(L)$ and on Theorem I.6.4.

LEMMA 12 **(V. Glivenko [1929]).** ***Let L be a distributive lattice with 0. Then $I(L)$ is a pseudocomplemented lattice in which***

$$I^* = \{x \mid x \wedge i = 0 \text{ for all } i \in I\}.$$

Let $S(I(L)) = \{I^* \mid I \in I(L)\}$. If L is a Boolean lattice, then $S(I(L))$ is a complete Boolean lattice and the map $a \to (a]$ embeds L into $S(I(L))$; this embedding preserves all existing meets and joins.

PROOF. **The first statement is trivial. Now let L be Boolean. It follows from Theorem I.6.4 that $S(I(L))$ is a Boolean lattice. Furthermore, it is easily seen that for any $X \subseteq I(L)$, the inf and sup of X in $S(I(L))$ are $\bigwedge X$ and $(\bigvee X)^{**}$, respectively, where $\bigwedge$ and $\bigvee$ are the meet and join of X in $I(L)$, respectively. Since $\bigwedge((x] \mid x \in X) = (\inf X]$, whenever $\inf X$ exists in L, the mapping $a \to (a]$ preserves all existing meets in L. Observe that, for $x, a \in L$, $x \wedge a' = 0$ iff $x \leq a$, and so $(a] = (a']^* \in S(I(L))$. Now let $a = \sup X$ in L and set $I = (X]$ $(= \bigvee((x] \mid x \in X))$. To show that $x \to (x]$ is join-preserving, we have to verify that $I^{**} = (a]$, or, equivalently, that $I^* = (a']$. Indeed, if $b \in I^*$, then $b \wedge x = 0$ for all $x \in I$, and thus $x \leq b'$. Therefore, $a = \sup X \leq b'$, proving $a' \geq b$, that is, $b \in (a']$. Conversely, let $b \in (a']$. Then $b' \geq a$; therefore, $b' \geq a = \sup X \geq x$ for all $x \in X$, and so $b \wedge x = 0$ for all $x \in X$. This shows that $b \in I^*$, proving that $I^* = (a']$.** ●

LEMMA 13. ***Let L be a complete lattice satisfying (JID) and (MID). Then the identity map is a complete embedding of L into $B(L)$.***

PROOF. **Let us write $a \in B(L)$ in the form**

$$a = a_0 + \cdots + a_{n-1}, \quad a_0 \leq a_1 \leq \cdots \leq a_{n-1}, \quad a_0, \ldots, a_{n-1} \in L.$$

If n is even let us replace a_0 by $0+a_0$; thus we can assume that n is odd. We claim that for $x\in L$ and $a\in B(L)$, we have $x\leq a$ iff

$$x\wedge a_0=x\wedge a_1,$$
$$x\wedge a_2=x\wedge a_3,\ldots,$$
$$x\wedge a_{n-3}=x\wedge a_{n-2},\quad x\leq a_{n-1}.$$

Indeed, let $x\leq a$. Then

$$xa_1=xa_1(a_0+\cdots+a_{n-1})=x(a_0+a_1+a_1+\cdots+a_1)=xa_0;$$

therefore, $x\wedge a_0=x\wedge a_1$. Thus $x(a_2+\cdots+a_{n-1})=(xa_0+xa_1)+x(a_2+\cdots+a_{n-1})=xa=x$, and so $x\leq a_2+\cdots+a_{n-1}$. Conversely, if $x\wedge a_0=x\wedge a_1$ and $x\leq a_2+\cdots+a_{n-1}$, then $xa=xa_0+xa_1+x(a_2+\cdots+a_{n-1})=x$, proving that $x\leq a$. A simple induction completes the proof of the claim.

Let $X\subseteq L$, $y=\sup X$ in L, and $a\in B(L)$. If $x\leq a$ for all $x\in X$, then the formulas of the preceding claim hold for all x and a and, by (JID), for y and a, proving that $y\leq a$. Thus $y=\sup X$ in $B(L)$. The dual argument, using (MID), completes the proof. ●

Theorem 14 (N. Funayama [1959]). *A complete lattice L has a complete embedding into a complete Boolean lattice iff L satisfies (MID) and (JID).*

Proof. Combine Lemmas 10–13. ●

The representation for $a\in B(L)$ given in Lemma 3 is not unique in general; the only exception is when L is a chain. Since this case is of special interest, we shall investigate it in detail.

Repeating the definition, a Boolean lattice B is *R-generated by a chain C* with 0 iff $B=B(C)$. This concept is due to A. Mostowski and A. Tarski [1939] and can be extended to distributive lattices as follows:

A distributive lattice L with 0 is *R-generated by a chain* $C(\subseteq L)$ with 0 iff C R-generates $B(L)$.

Lemma 15. *Let L be a distributive lattice with 0 and let C be a chain in L, $0\in C$. Then C R-generates L iff L is the smallest sublattice of itself containing C and closed under the formation of relative complements.*

Proof. Apply Lemma 9 to C. ●

An explicit representation of $B(C)$ is given as follows:

For a chain C with 0, let $B[C]$ be the set of all subsets of C of the form

$$(a_0]+(a_1]+\cdots+(a_{n-1}],\quad 0<a_0\leq a_1\leq\cdots\leq a_{n-1},\quad a_0,\ldots,a_{n-1}\in C,$$

where $+$ is the symmetric difference. We consider $B[C]$ as a poset, partially ordered by $\subseteq$. We identify $a\in C$ with $(a]$, for $a\neq 0$, and 0 with $\varnothing$. Thus $C\subseteq B[C]$.

Lemma 16. *$B[C]$ is the generalized Boolean lattice R-generated by C.*

Proof. The proof is obvious, by construction and by Theorem 6. ●

Note that every nonvoid element a of $B[C]$ can be represented in the form

$$a = (a_0] \cup (b_1, a_1] \cup \cdots \cup (b_{n-1}, a_{n-1}],$$
$$0 < a_0 < b_1 < a_1 < \cdots < b_{n-1} < a_{n-1},$$

where the union is disjoint union and the first term $(a_0]$ may be missing. An element of the form $(x, y]$ is nothing but $x + y$. Thus, $a = a_0 + b_1 + a_1 + \cdots + b_{n-1} + a_{n-1}$, and so we conclude:

Corollary 17. *In $B(C)$ every nonzero element a has a unique representation in the form*

$$a = a_0 + a_1 + \cdots + a_{n-1}, \quad 0 < a_0 < a_1 < \cdots < a_{n-1}, \quad a_0, \ldots, a_{n-1} \in C.$$

The following results show that many distributive lattices can be R-generated by chains.

Lemma 18. *Every finite Boolean lattice B can be R-generated by a chain; in fact, $B = [C]_R$ for any maximal chain C of B.*

Proof. Let B_1 be the subalgebra of B R-generated by C. Using the notation of Corollary 1.14, the length of C equals $|J(B)|$; also, the length of C equals $|J(B_1)|$; thus $|J(B)| = |J(B_1)| = n$. We conclude that both B and B_1 have 2^n elements, proving that $B = B_1$. ●

Corollary 19. *Every finite distributive lattice L can be R-generated by a chain—in fact, by any maximal chain of L.*

Proof. Let C be a maximal chain in L and let $B = B(L)$. Then $|J(L)| = |J(B)|$. By Corollary 1.14, C is maximal in B. Thus, $B = B(C) \supseteq L$. ●

Theorem 20. *Every countable distributive lattice L with 0 can be R-generated by a chain.*

Proof. Let $L = \{a_0, a_1, a_2, \ldots, a_n, \ldots\}$, $a_0 = 0$, and let L_n be the sublattice of L generated by $a_0, \ldots, a_n$. Let C_0 be a maximal chain of L_0, and, inductively, let C_n be a maximal chain of L_n containing C_{n-1}. Set $C = \bigcup(C_i \mid i < \omega)$. Obviously, $0 \in C$. We claim that C R-generates L. Take $a \in B(L)$; $a = x_0 + \cdots + x_{m-1}$, $x_0, \ldots, x_{m-1} \in L$. $L = \bigcup(L_i \mid i < \omega)$; thus for some n, $x_0, \ldots, x_{m-1} \in L_n$, and so $a \in B(L_n)$. Since L_n is finite, we get $B(L_n) = B(C_n)$; therefore $a \in B(C_n) \subseteq B(C)$, proving that $L \subseteq B(C)$. ●

Corollary 21. *The correspondence $C \to B(C)$ maps the class of countable chains with 0 onto the class of countable generalized Boolean lattices. Under this correspondence, $\{0\}$-subchains and $\{0\}$-homomorphic images correspond to $\{0\}$-subalgebras and $\{0\}$-homomorphic images.*

Note, however, that $C_0 \cong C_1$ is *not* implied by $B(C_0) \cong B(C_1)$.

Much is known about countable chains. Utilizing the previous results, such information can be used to prove results on countable generalized Boolean lattices. To help distinguish an important class of chains, we introduce the concept of prime interval. An interval $[a, b]$ is *prime* iff $a \prec b$.

Lemma 22. *Every countable chain C can be embedded in the chain Q of rational numbers. Every countable chain not containing any prime interval is isomorphic to one of the intervals $(0, 1)$, $[0, 1)$, $(0, 1]$, and $[0, 1]$ of Q.*

Proof. Let $C = \{x_0, x_1, \ldots, x_{n-1}, \ldots\}$. We define the map φ inductively as follows: Pick an arbitrary $r_0 \in Q$ and set $x_0\varphi = r_0$. If $x_0\varphi, \ldots, x_{n-1}\varphi$ have already been defined, we define $x_n\varphi$ as follows: Let

$$L_n = \bigcup((x_i\varphi] \mid x_i < x_n,\ i < n),$$
$$U_n = \bigcup([x_i\varphi) \mid x_i > x_n,\ i < n)$$

($L_n = \varnothing$ or $U_n = \varnothing$ is possible). Note that if $L_n \neq \varnothing$, then it has a greatest element l_n, and if $U_n \neq \varnothing$, then it has a smallest element u_n. If both are nonempty, then $l_n < u_n$. In any case, we can choose an $r_n \in Q$ with $r_n \notin L_n \cup U_n$. We set $x_n\varphi = r_n$. Obviously, φ is an embedding.

The second part of the proof reduces to the following statement:

Let C and D be bounded countable chains with no prime intervals. Then $C \cong D$.

To prove this, let $C = \{c_0, c_1, \ldots\}$ and $D = \{d_0, d_1, \ldots\}$. We define two maps: $\varphi\colon C \to D$ and $\psi\colon D \to C$. Let us assume that $c_0 = 0$, $c_1 = 1$, and that $d_0 = 0$, $d_1 = 1$. For each $n < \omega$, we shall define inductively finite chains C_n, D_n ($C_n \subseteq C$, $D_n \subseteq D$) and an isomorphism $\varphi_n\colon C_n \to D_n$ with inverse $\psi_n\colon D_n \to C_n$. Set $C_0 = \{c_0, c_1\} = \{0, 1\}$, $D_0 = \{d_0, d_1\} = \{0, 1\}$ and $i\varphi_0 = i$, $i\psi_0 = i$, for $i = 0, 1$. Given C_n, D_n, φ_n, ψ_n, and n even, let k be the smallest integer with $c_k \notin C_n$. Define $u_k = \bigwedge([c_k) \cap C_n)$ and $l_k = \bigvee((c_k] \cap C_n)$. Then $l_k < c_k < u_k$, and so $l_k\varphi_n < u_k\varphi_n$. Since D contains no prime intervals, we can choose a $d \in D$ satisfying $l_k\varphi_n < d < u_k\varphi_n$. Since ψ_n is isotone, $d \notin D_n$. Define $C_{n+1} = C_n \cup \{c_k\}$, $D_{n+1} = D_n \cup \{d\}$, φ_{n+1} restricted to C_n to be φ_n, and $c_k\varphi_{n+1} = d$, ψ_{n+1} restricted to D_n to be ψ_n and $d\psi_{n+1} = c_k$. If n is odd we proceed in a similar way, but we interchange the role of C and D, C_n and D_n, φ_n and ψ_n, respectively.

Finally, put $\varphi = \bigcup(\varphi_n \mid n < \omega)$. Clearly, $C = \bigcup(C_n \mid n < \omega)$, $D = \bigcup(D_n \mid n < \omega)$, and φ is the required isomorphism. ●

Corollary 23. *Up to isomorphism there is exactly one countable Boolean lattice with no atoms and exactly one countable generalized Boolean lattice with no atoms and no unit element, $B(Q)$.*

Proof. Take the rational intervals $[0, 1]$ and $[0, 1)$. The generalized Boolean lattices in question are $B([0, 1])$ and $B([0, 1))$. This follows from the observation that $[a, b]$ is a prime interval in C iff $a+b$ is an atom in $B(C)$. The results follow from Lemmas 16 and 22 and Theorem 20. ●

Theorem 24. *Let B be a countable Boolean algebra. Then B has either $\aleph_0$ or $2^{\aleph_0}$ prime ideals.*

Remark. This is obvious if we assume the Continuum Hypothesis. Interestingly enough, we can give a proof without it.

Proof. For a Boolean algebra B and an ideal I of B, we shall write B/I for $B/\Theta[I]$. If J is an ideal of B with $J \supseteq I$, then $J/I = \{[x]\Theta[I] \mid x \in J\}$ is an ideal of B/I. (This is the usual notation in ring theory.) Let B be a Boolean algebra. We define the ideals I_γ by transfinite induction. $I_0 = (0]$; I_1 is the ideal generated by the atoms of B; given I_γ, let I be the ideal of B/I_γ generated by the atoms of B/I_γ and let $\varphi: x \to x + I_\gamma$ be the homomorphism of B onto B/I_γ; we set $I_{\gamma+1} = I\varphi^{-1}$. Finally, if γ is a limit ordinal, set $I_\gamma = \bigcup(I_\delta \mid \delta < \gamma)$. The *rank* of B is defined to be the smallest ordinal α such that $I_\alpha = I_{\alpha+1}$. Obviously, the cardinality of α is at most $|B|$.

Claim 1. *Let B be countable. If $I_\alpha \neq B$, then $|\mathcal{P}(B)| = 2^{\aleph_0}$.*

Indeed, if $I_\alpha \neq B$, then B/I_α has no atoms, and thus $B/I_\alpha \cong B(C)$, where C is the rational interval $[0, 1]$. By Lemma 5 (see Exercise 27), $|\mathcal{P}(B/I_\alpha)| = |\mathcal{P}(C)| = |I(C)| = 2^{\aleph_0}$.

Claim 2. *Let B be countable. If $I_\alpha = B$, then $|\mathcal{P}(B)| = \aleph_0$.*

Indeed, for $\gamma < \alpha$, let $\mathcal{P}_\gamma(B)$ be the set of prime ideals P of B for which $I_\gamma \subseteq P$, $I_{\gamma+1} \nsubseteq P$. Since α is finite or countable, it suffices to show that $|\mathcal{P}_\gamma(B)| = \aleph_0$. If $P \in \mathcal{P}_\gamma(B)$, then, by Corollary 1.16 and Theorem 1.22, we have $P \vee I_{\gamma+1} = B$. It follows that for $P, Q \in \mathcal{P}_\gamma(B)$, $P \neq Q$, we have $P \cap I_{\gamma+1} \neq Q \cap I_{\gamma+1}$. Thus $P \to (P \cap [I_{\gamma+1}]_R)/I_\gamma$ is a one-to-one correspondence of $\mathcal{P}_\gamma(B)$ into (in fact, onto) $\mathcal{P}([I_{\gamma+1}]_R/I_\gamma)$; but $[I_{\gamma+1}]_R/I_\gamma$ is just the generalized Boolean lattice of all finite subsets of a countable set. Therefore, $|\mathcal{P}_\gamma(B)| = \aleph_0$. ●

In order to avoid giving the impression that most Boolean algebras can be R-generated by chains, we state:

Lemma 25. *Let B be a complete Boolean algebra R-generated by a chain C with 0. Then B is finite.*

Proof. Let $B=[C]_R$ and let the chain C be infinite. We can assume that C contains a subchain $\{x_0, x_1, \ldots, x_n, \ldots\}$, $0<x_0<x_1<\cdots<x_n<\cdots$ (or dually, in which case the complements form an increasing sequence). Then define $a_n=x_0+x_1+\cdots+x_{2n}+x_{2n+1}$ for each $n<\omega$. We claim that $\bigvee(a_i \mid i<\omega)$ does not exist. Indeed, let a be an upper bound for $\{a_i \mid i<\omega\}$. By the remarks immediately following Lemma 16, we can represent each a_n by the set $(x_0, x_1] \cup (x_2, x_3] \cup \cdots \cup (x_{2n}, x_{2n+1}]$, and we can represent a in the form

$$a=(b_0] \cup (b_1, b_2] \cup \cdots \cup (b_{m-2}, b_{m-1}],$$

where $0<b_0<b_1<\cdots<b_{m-1}$, $b_i \in C$ for $i<m$, and the first term, $(b_0]$, may be missing. Since a contains each a_n, there must exist an n and a $j<m$ such that both $(x_{2n}, x_{2n+1}]$ and $(x_{2n+2}, x_{2n+3}]$ are contained in $(b_{j-1}, b_j]$ or in $(b_0]$. Therefore, the interval $(x_{2n+1}, x_{2n+2}]$ can be deleted from a, and it will still contain all the a_i—that is, $a+x_{2n+1}+x_{2n+2}$ is an upper bound for $\{a_i \mid i<\omega\}$, and $a+x_{2n+1}+x_{2n+2}<a$. We conclude that $\{a_i \mid i<\omega\}$ does not have a least upper bound. ●

Next we consider which chains with 0 can be R-generating chains of a given distributive lattice.

Definition 26. *Let L be a distributive lattice with 0 and let C be a chain in L, $0 \in C$. The chain C is called* strongly maximal *in L iff, for any homomorphism φ of L onto a distributive lattice L_1, the chain $C\varphi$ is maximal in L_1.*

Lemma 27. *Let L be a distributive lattice with 0 and let C be a chain in L, $0 \in C$. If L is R-generated by C then C is maximal in L.*

Proof. If C is not maximal in L, then we can find $a \in L$, $a \notin C$, such that $C \cup \{a\}$ is a chain. Write $a=a_0+a_1+\cdots+a_{n-1}$ with $0<a_0<a_1<\cdots<a_{n-1}$ and $a_i \in C$ for $i<n$. Since $a \notin C$, $n>1$. Now $a \wedge a_0=a_0+a_0+\cdots+a_0$, which is a_0 if n is odd and 0 if n is even. But $a_0 \neq a$ and $0 \neq a$, so, since a and a_0 are comparable, $a \wedge a_0=a_0$ and n is odd. Then $a \wedge a_1=a_0+a_1+\cdots+a_1=a_0$, contradicting the comparability of a and a_1. ●

The converse of Lemma 27 is false. The following theorem settles the matter.

Theorem 28. *Let L be a distributive lattice with 0 and let C be a chain in L, $0 \in C$. Then C R-generates L iff C is strongly maximal in L.*

Proof. If C R-generates L, then, for any onto homomorphism φ, $C\varphi$ R-generates $L\varphi$. By Lemma 27, $C\varphi$ is maximal in $L\varphi$, so C is strongly maximal in L.

Next assume that C is strongly maximal in L but does not R-generate L. Without any loss of generality we can assume that L and C have a greatest element. (Otherwise, add one. Then $C \cup \{1\}$ is strongly maximal in $L \cup \{1\}$ but does not R-generate $L \cup \{1\}$.) Let $B_1=B(L)$ and let $B_0=[C]_R$. By hypothesis, $B_0 \neq B_1$, so there exists an $a \in B_1-B_0$. We claim that there exist prime ideals $P_1 \neq P_2$ of B_1 with $B_0 \cap P_1=B_0 \cap P_2$. With

$I=((a]\cap B_0]$ and $D=[a)$, we have $I\cap D=\varnothing$, so, by Theorem 1.15, there is a prime ideal P_1 such that $I\subseteq P_1$ and $P_1\cap D=\varnothing$. Then let $I_1=(a]$ and $D_1=[B_0-P_1)$. Since $(a]\cap B_0\subseteq P_1$, it follows that $I_1\cap D_1=\varnothing$. Let P_2 be a prime ideal with $I_1\subseteq P_2$, $P_2\cap D_1=\varnothing$. Then $a\in P_2-P_1$, so $P_1\neq P_2$. Because $P_2\cap(B_0-P_1)=\varnothing$, $P_2\cap B_0\subseteq P_1\cap B_0$. Then by Theorem 1.22 (prime ideals of a Boolean lattice are unordered), $P_1\cap B_0=P_2\cap B_0$, proving our claim.

Now (again by using Theorem 1.22) we can map B_1 onto $(\mathfrak{C}_2)^2$ by a homomorphism ψ: For $x\in P_1\cap P_2$, $x\psi=\langle 0,0\rangle$; for $x\in P_1-P_2$, $x\psi=\langle 0,1\rangle$; for $x\in P_2-P_1$, $x\psi=\langle 1,0\rangle$; for $x\notin P_1\cup P_2$, $x\psi=\langle 1,1\rangle$. Since $C\psi=\{\langle 0,0\rangle,\langle 1,1\rangle\}$ is not maximal, we conclude that C is not strongly maximal in L. ●

Corollary 29. *Let C and D be strongly maximal chains of the distributive lattice L with 0. Then $|C|=|D|$ and $|I(C)|=|I(D)|$.*

Proof. If L is finite, these conclusions follow from Corollary 1.14. If L is infinite, then $[C]=[D]=B(L)$, and so $|C|=|D|=|L|$. Also, by Lemma 5, $|\mathcal{P}(C)|=|\mathcal{P}(B(L))|=|\mathcal{P}(D)|$, and $\mathcal{P}(C)=I(C)$, $\mathcal{P}(D)=I(D)$; hence the second statement. ●

Corollary 29 is the strongest known extension of Corollary 1.14 to the infinite case. The second part of Corollary 29 is from G. Grätzer and E. T. Schmidt [1957].

Boolean algebras generated by chains were first investigated by A. Mostowski and A. Tarski [1939]. Theorem 20 for Boolean lattices and Theorem 24 were communicated to the author by J. R. Büchi. These results have been known for some time in topology (via the Stone topological representation theorem, see Section 5). Some of the other results appeared first in FC.

Exercises

1. Give a detailed proof of Lemma 4.
2. Try to describe the most general situation to which the idea of the proof of Theorem I.5.5 (Theorem I.5.24) could be applied.
3. Show that Lemma 5 does not remain valid if the word "generalized" is omitted.
4. Find necessary and sufficient conditions on a distributive lattice L in order that L have a Boolean extension B to which every congruence of L has exactly one extension.
5. Let L be a distributive lattice with 0 and define L_1 to be the lattice L if L has a unit element; let L_1 be L with a unit added if L does not have a unit element. The *Boolean lattice $B[L]$ R-generated by L* is defined to be $B(L_1)$. Show that if B is any Boolean lattice, containing L as a sublattice, and B is generated by L under $\wedge,\vee$, and $'$, then B is isomorphic to the Boolean lattice R-generated by L.
6. Work out Corollaries 7 and 8 for the Boolean lattice R-generated by L.

*7. The *Complete Infinite Distributive Identity* is (for $I,J\neq\varnothing$):

$$\bigwedge(\bigvee(a_{ij}\mid j\in J)\mid i\in I)=\bigvee(\bigwedge(a_{i\,i\varphi}\mid i\in I)\mid \varphi\colon I\to J).$$

Show that this holds in a complete Boolean lattice B iff it is atomic (A. Tarski [1930]). (Hint: apply the identity to $\bigwedge(a \vee a' \mid a \in B) = 1$.)

8. Prove that the Complete Infinite Distributive Identity is selfdual for Boolean lattices.
9. For a subset A of a lattice L, set

$$A^u = \{x \mid x \in L, x \text{ is an upper bound of } A\}$$
$$A^l = \{x \mid x \in L, x \text{ is a lower bound of } A\}.$$

Prove that this sets up a Galois connection and therefore $(A^u)^l \supseteq A$, $(A^l)^u \subseteq A$, $A^u = ((A^u)^l)^u$, and $A^l = ((A^l)^u)^l$.

10. Call an ideal I of a lattice L *normal* iff $I = (I^u)^l$. Show that every principal ideal is normal.
11. Let $I_N(L)$ denote the set of all normal ideals of L. Show that $I_N(L)$ is a complete lattice but that it is not necessarily a sublattice of $I_0(L)$.
12. Show that the map: $x \to (x]$ is an embedding of L into $I_N(L)$, preserving all meets and joins that exist in L. ($I_N(L)$ is called the *MacNeille completion* of L; see H. M. MacNeille [1937].)
13. Let B be a Boolean lattice and let I be an ideal of B. Show that I is normal iff $I = I^{**}$ (for the concepts, see Exercise 10 and Lemma 12).
14. Prove that the Boolean lattice $S(I(L))$ of Lemma 12 is the MacNeille completion of the Boolean lattice L.

*15. Show that the MacNeille completion of a distributive lattice need not even be modular.

16. Let L be a distributive algebraic lattice. Show that L satisfies the Join Infinite Distributive Identity. (Thus, for any lattice K, $C(K)$ satisfies (JID).)
17. Let L be a distributive lattice, $a_i, b_i \in L$ for $i < \omega$ and $[a_0, b_0] \supset [a_1, b_1] \supset \cdots$. Define

$$\Theta = \bigvee(\Theta(a_0, a_i) \vee \Theta(b_0, b_i) \mid i < \omega).$$

Show that

$$\Theta \vee \bigwedge(\Theta(a_i, b_i) \mid i < \omega) \neq \bigwedge(\Theta \vee \Theta(a_i, b_i) \mid i < \omega).$$

18. Use Exercise 17 to show that, for a distributive lattice L, the Meet Infinite Distributive Identity holds in $C(L)$ iff every interval in L is finite (G. Grätzer and E. T. Schmidt [1958b]).
19. For a chain C, introduce and describe $B(C)$ using Corollary 17.

*20. Prove the converse of Lemma 18: If every maximal chain R-generates the Boolean lattice B, then B is finite.

21. Why is it not possible to use transfinite induction to extend Theorem 20 to the uncountable case?
22. Let C be a chain with 0 and 1 and let $a \in C$. Define a new order on C: For $x, y < a$, and for $a \leq x, y$, let $x \leq y$ retain its meaning; for $x < a \leq y$, let $y < x$; let C_1 be the set C with the new order. Then C_1 is a chain, and $B(C) \cong B(C_1)$, but in general $C \cong C_1$ does not hold.
23. Describe a countable family of pairwise nonisomorphic countable Boolean algebras.
24. Give an example of a bounded distributive lattice L with a maximal chain C such that C is not maximal in $B(L)$ (G. W. Day [1970]).
25. Let L_0 be the [0, 1] rational interval and let L_1 be the [0, 1] real interval. Let $C = \{\langle x, x\rangle \mid 0 \leq x \leq 1, x \text{ rational}\}$. Then C is a maximal chain in $L_0 \times L_1$. Show that C is not strongly maximal (G. Grätzer and E. T. Schmidt [1957]).

26. Find in $L_0 \times L_1$ of Exercise 25 maximal chains of cardinality $\aleph_0$ and $2^{\aleph_0}$. What are the cardinalities of strongly maximal chains?
27. Let L be a distributive lattice with 0 and let $B = B(L)$. Show that $P \to P \cap L$ for $P \in \mathcal{P}(B)$ is a one-to-one correspondence between the prime ideals of L and B (use Lemma 5).
*28. Let A be an infinite set, $B = P(A)$. Prove that B has maximal chains of cardinality $|A|$ and $2^{|A|}$.
29. Construct an example in which the sequence of ideals I_γ of Theorem 24 does not terminate in finitely many steps.
30. Let C be the $[0, 1]$ interval of the rational numbers. Show that $B(C)$ is $F_B(\aleph_0)$.

5. Topological Representation

The poset $\mathcal{P}(L)$ of prime ideals does give a great deal of information about the distributive lattice L, but obviously it does not characterize L. For instance, for a countably infinite Boolean algebra L, $\mathcal{P}(L)$ is an unordered set of cardinality $\aleph_0$ or $2^{\aleph_0}$, whereas there are surely more than two such Boolean algebras up to isomorphism.

Therefore, it is necessary to endow $\mathcal{P}(L)$ with more structure if we want it to characterize L. M. H. Stone [1937] endowed $\mathcal{P}(L)$ with a topology (see also L. Rieger [1949]). In this section we shall discuss his approach in a slightly more general but, in our opinion, more natural framework.

A join-semilattice L is called *distributive* iff $a \leq b_0 \vee b_1$ ($a, b_0, b_1 \in L$) implies the existence of $a_0, a_1 \in L$, $a_i \leq b_i$, $i = 0, 1$, with $a = a_0 \vee a_1$ (see Figure 1). Note that a_0 and a_1 need not be unique.

Some elementary properties of a distributive join-semilattice are as follows:

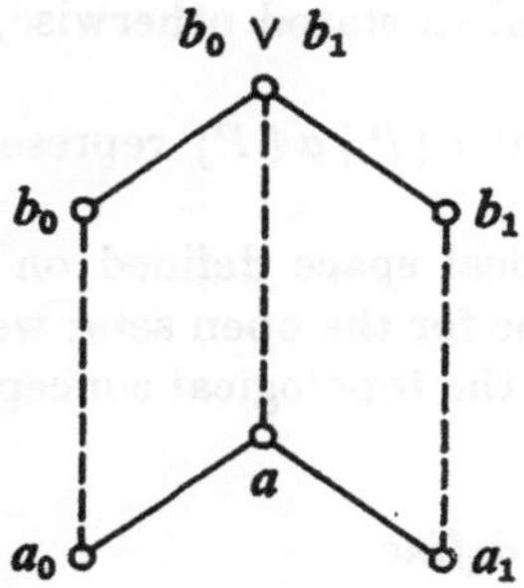

Figure 1

Lemma 1.

(i) *If $\langle L; \wedge, \vee \rangle$ is a lattice, then the join-semilattice $\langle L; \vee \rangle$ is distributive iff the lattice $\langle L; \wedge, \vee \rangle$ is distributive.*

(ii) *If a join-semilattice L is distributive, then for any $a, b \in L$ there is a $d \in L$ with $d \leq a$ and $d \leq b$. Consequently, $I(L)$ is a lattice.*

(iii) *A join-semilattice L is distributive iff $I(L)$, as a lattice, is distributive.*

Proof.

(i) If $\langle L; \wedge, \vee\rangle$ is distributive, and $a \leq b_0 \vee b_1$, then with $a_i = a \wedge b_i$, $i = 0, 1$, $a = a_0 \vee a_1$. Conversely, if $\langle L; \vee\rangle$ is distributive, and the lattice L contains a diamond or a pentagon $\{0, a, b, c, i\}$, then $a \leq b \vee c$, but a cannot be represented as $a = a_0 \vee a_1$ with $a_0 \leq b$ and $a_1 \leq c$, a contradiction. ◑

(ii) $a \leq a \vee b$, thus $a = a_0 \vee b_0$, where $a_0 \leq a$, $b_0 \leq b$. Since, in addition, $b_0 \leq a$, b_0 is a lower bound for a and b. ◑

(iii) First we observe that, for $I, J \in I(L)$,

$$I \vee J = \{i \vee j \mid i \in I, j \in J\}$$

follows from the assumption that the join-semilattice L is distributive. Therefore, the distributivity of $I(L)$ can be easily proved. Conversely, if $I(L)$ is distributive and $a \leq b_0 \vee b_1$, then

$$(a] = (a] \wedge ((b_0] \vee (b_1]) = ((a] \wedge (b_0]) \vee ((a] \wedge (b_1]),$$

and so $a = a_0 \vee a_1$, $a_0 \in (b_0]$, $a_1 \in (b_1]$, which is distributivity for L. ●

A subset D of a join-semilattice L is called a *dual ideal* iff $a \in D$ and $x \geq a$ imply that $x \in D$, and $a, b \in D$ implies that there exists a lower bound d of $\{a, b\}$ such that $d \in D$. An ideal I of L is *prime* iff $I \neq L$ and $L - I$ is a dual ideal. Again, let $\mathscr{P}(L)$ denote the set of all prime ideals of L.

Lemma 2. *Let I be an ideal and let D be a nonvoid dual ideal of a distributive join-semilattice L. If $I \cap D = \varnothing$, then there exists a prime ideal P of L with $P \supseteq I$ and $P \cap D = \varnothing$.*

Proof. The proof is a routine modification of the proof of Theorem 1.15. ●

In the rest of this section, unless stated otherwise, let L stand for a distributive join-semilattice with zero.

In $\mathscr{P}(L)$, sets of the form $r(a) = \{P \mid a \notin P\}$ represent the elements of L. We will make all these sets open.

Let $\mathscr{S}(L)$ denote the topological space defined on $\mathscr{P}(L)$ by postulating that the sets of the form $r(a)$ be a subbase for the open sets; we shall call $\mathscr{S}(L)$ the *Stone space* of L. (Exercises 1–22 review all the topological concepts that are used in this section.)

Lemma 3. *For an ideal I of L, define*

$$r(I) = \{P \mid P \in \mathscr{S}(L), P \not\supseteq I\}.$$

Then $r(I)$ is open in $\mathscr{S}(L)$. Conversely, every open set U of $\mathscr{S}(L)$ can be uniquely represented as $r(I)$ for some ideal I of L.

Proof. We simply observe that $r(I) \cap r(J) = r(I \wedge J)$,

$$r(\bigvee(I_j \mid j \in K)) = \bigcup(r(I_j) \mid j \in K),$$

and $r((a]) = r(a)$, from which it follows that the $r(I)$ form the smallest collection of sets closed under finite intersection and arbitrary union containing all the $r(a)$. Observe that $a \in I$ iff $r(a) \subseteq r(I)$. Thus $r(I) = r(J)$ iff $a \in I$ is equivalent to $a \in J$, that is, iff $I = J$. ●

Lemma 4. *The subsets of $\mathscr{S}(L)$ of the form $r(a)$ can be characterized as compact open sets.*

Proof. Indeed, if a family of open sets $\{r(I_k) \mid k \in K\}$ is a cover for $r(a)$, that is, $r(a) \subseteq \bigcup(r(I_k) \mid k \in K) = r(\bigvee(I_k \mid k \in K))$, then $a \in \bigvee(I_k \mid k \in K)$. This implies that $a \in \bigvee(I_k \mid k \in K_0)$ for some finite $K_0 \subseteq K$, proving that $r(a) \subseteq \bigcup(r(I_k) \mid k \in K_0)$. Thus $r(a)$ is compact. Conversely, if I is not principal, then $r(I) \subseteq \bigcup(r(a) \mid a \in I)$, but $r(I) \nsubseteq \bigcup(r(a) \mid a \in I_0)$ for any finite $I_0 \subseteq I$. ●

From Lemma 4 we immediately conclude:

Theorem 5. *The Stone space $\mathscr{S}(L)$ determines L up to isomorphism.*

If we want to use Stone spaces to construct new ones in order that new distributive lattices can be constructed from given ones, then we have to know what Stone spaces look like. Stone spaces are characterized in Theorem 8. To prepare for the proof of Theorem 8, we prove Lemma 6.

Let P be a prime ideal of L. Then P is represented as an element of $\mathscr{S}(L)$ and also by $r(P)$. The connection between P and $r(P)$ is given in Lemma 6 and is illustrated by Figure 2.

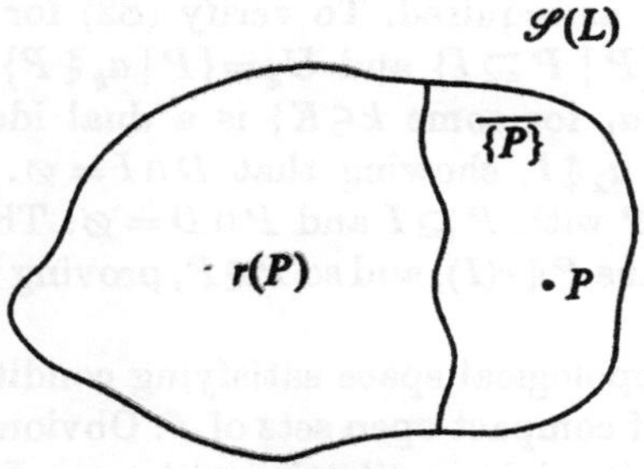

Figure 2

Lemma 6. *For every prime ideal P of L, $\overline{\{P\}} = \mathscr{S}(L) - r(P)$, where $\overline{\{P\}}$ is the topological closure of $\{P\}$.*

Proof. By definition of closure,

$$\overline{\{P\}} = \{Q \mid Q \in r(a) \text{ implies that } P \in r(a)\} = \{Q \mid Q \supseteq P\}$$
$$= \mathscr{S}(L) - \{Q \mid Q \nsupseteq P\} = \mathscr{S}(L) - r(P). \quad ●$$

Corollary 7. *If $P \neq Q$, then $\overline{\{P\}} \neq \overline{\{Q\}}$; in other words, $\mathscr{S}(L)$ is a T_0-space.*

Proof. Combine Lemmas 3 and 6. ●

Lemma 6 also shows that if P is a prime ideal, then $\mathscr{S}(L) - r(P)$ must be the closure of a singleton. In other words:

(C) If U is an open set with the property that, for the compact open sets U_0 and U_1, $U_0 \cap U_1 \subseteq U$ implies that $U_0 \subseteq U$ or $U_1 \subseteq U$, then $\mathscr{S}(L) - U = \overline{\{P\}}$ for some element P.

Now we can state the characterization theorem:

Theorem 8. *The Stone space $\mathscr{S}$ of a distributive join-semilattice with zero can be characterized (up to homeomorphism) by the following two properties:*

(S1) *$\mathscr{S}$ is a T_0-space in which the compact open sets form a base for the open sets.*

(S2) *If F is a closed set in $\mathscr{S}$, $\{U_k \mid k \in K\}$ is a dually directed[1] family of compact open sets of $\mathscr{S}$, and $U_k \cap F \neq \varnothing$, then $\bigcap(U_k \mid k \in K) \cap F \neq \varnothing$.*

Remark. The meaning of condition (S1) is clear. Condition (S2) is a complicated way of ensuring that (C) holds and that Lemma 2 holds for the join-semilattice of compact open sets of $\mathscr{S}(L)$.

Proof. To show that (S1) holds, we have to verify that the $r(a)$, $a \in L$, form a base (not only a subbase) for the open sets of $\mathscr{S}(L)$. In other words, for $a, b \in L$ and $P \in r(a) \cap r(b)$, we have to find a $c \in L$ with $P \in r(c)$ and $r(c) \subseteq r(a) \cap r(b)$. By assumption, $a \notin P$ and $b \notin P$; thus, since P is prime, there exists a $c \in L$, $c \notin P$, $c \leq a$, $c \leq b$. Then $P \in r(c)$, $r(c) \subseteq r(a)$, and $r(c) \subseteq r(b)$, as required. To verify (S2) for $\mathscr{S}(L)$, let $F = \mathscr{S}(L) - r(I)$ and $U_k = r(a_k)$. Thus $F = \{P \mid P \supseteq I\}$ and $U_k = \{P \mid a_k \notin P\}$. The assumptions on the a_k mean that $D = \{x \mid x \geq a_k \text{ for some } k \in K\}$ is a dual ideal; since $U_k \cap F \neq \varnothing$, we have $r(a_k) \nsubseteq r(I)$; that is, $a_k \notin I$, showing that $D \cap I = \varnothing$. Therefore, by Lemma 2, there exists a prime ideal P with $P \supseteq I$ and $P \cap D = \varnothing$. Then $a_k \notin P$, and so $P \in r(a_k)$ for all $k \in K$. Also $P \supseteq I$, thus $P \notin r(I)$, and so $P \in F$, proving that $P \in F \cap \bigcap(U_k \mid k \in K)$, verifying (S2).

Conversely, let $\mathscr{S}$ be a topological space satisfying conditions (S1) and (S2) of the theorem. Let L be the set of compact open sets of $\mathscr{S}$. Obviously, $\varnothing \in L$ and if $A, B \in L$, then $A \cup B \in L$, and thus L is a join-semilattice with zero. Let

$$A \subseteq B_0 \cup B_1, \quad \text{with} \quad A, B_0, B_1 \in L.$$

Then $A \cap B_i$ is open, and therefore $A \cap B_i = \bigcup(A^i_j \mid j \in J_i)$, $i = 0, 1$, where the A^i_j are compact open sets. Since $A = (A \cap B_0) \cup (A \cap B_1) \subseteq \bigcup(A^i_j \mid j \in J_0 \cup J_1, i = 0, 1)$, by the compactness of A we get $A \subseteq \bigcup(A^i_j \mid j \in \bar{J}_0 \text{ or } j \in \bar{J}_1)$, where $\bar{J}_i$ is a finite subset of J_i, $i = 0, 1$. Set

$$A_i = \bigcup(A^i_j \mid j \in \bar{J}_i), \qquad i = 0, 1.$$

[1] $K \neq \varnothing$ and for $k, l \in K$, there exists a $t \in K$ such that $U_t \subseteq U_k \cap U_l$.

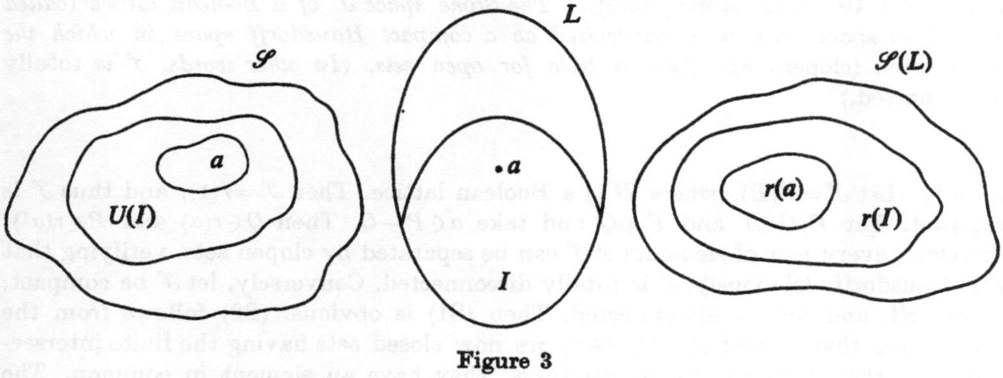

Figure 3

Then $A_0, A_1 \in L$, $A = A_0 \cup A_1$, and $A_0 \subseteq B_0$, $A_1 \subseteq B_1$, showing that L is distributive.

It follows from (S1) that the open sets of $\mathscr{S}$ are uniquely associated with ideals of L: For an ideal I of L, let $U(I) = \bigcup(a \mid a \in I)$ (keep in mind that an $a \in L$ is a subset of $\mathscr{S}$, as illustrated by Figure 3). Note that for $a \in L$, $a \in I$ iff $a \subseteq U(I)$.

Now let P be a prime ideal of L, $F = \mathscr{S} - U(P)$, and let $\{U_k \mid k \in K\}$ be the set of all compact open sets of $\mathscr{S}$ that have nonvoid intersections with F. Thus, the U_k are exactly those elements of L that are not in P. Therefore, by the definition of a prime ideal, given $k, l \in K$, there exists $t \in K$ with $U_t \subseteq U_k$, $U_t \subseteq U_l$, proving that F and $\{U_k \mid k \in K\}$ satisfy the hypothesis of condition (S2). By (S2) we conclude that there exists a $p \in F \cap \bigcap(U_k \mid k \in K)$. If $q \in F$, then for every compact open set U with $q \in U$, we have $U \cap F \neq \varnothing$; thus $p \in U$, proving that $\overline{\{p\}} = F$. Note that $\mathscr{S}$ is T_0; therefore p is unique.

Conversely, if $p \in \mathscr{S}$, let $I = \{a \mid a \in L, a \subseteq \mathscr{S} - \overline{\{p\}}\}$. Then I is an ideal of L, and $\mathscr{S} - \overline{\{p\}} = U(I)$. We claim that I is prime. Indeed, if $U, V \in L$, $U \notin I$, $V \notin I$, then $U \cap \overline{\{p\}} \neq \varnothing$, $V \cap \overline{\{p\}} \neq \varnothing$, and therefore $p \in U$, $p \in V$. Thus, $p \in U \cap V$ and so $U \cap V \nsubseteq U(I)$. By (S1) there exists a $W \in L$ with $W \subseteq U \cap V$ and $W \nsubseteq U(I)$. Therefore $W \notin I$, and so I is prime.

Summing up, the map $\varphi\colon P \to p$ is one-to-one and onto between $\mathscr{S}(L)$ and $\mathscr{S}$. To show that φ is a homeomorphism, it suffices to show that U is open in $\mathscr{S}(L)$ iff $U\varphi$ is open in $\mathscr{S}$. Since a typical open set in $\mathscr{S}(L)$ is of the form $r(I)$ $(I \in I(L))$, and an open set of $\mathscr{S}$ is of the form $U(I)$, we need only prove that $r(I)\varphi = U(I)$ and $U(I)\varphi^{-1} = r(I)$–in other words, that $P \in r(I)$ iff $(P\varphi =)\ p \in U(I)$. Indeed, $P \in r(I)$ means that $P \nsupseteq I$, which is equivalent to $U(P) \nsupseteq U(I)$; this, in turn, is the same as $U(I) \cap (\mathscr{S} - U(P)) \neq \varnothing$. Since $\mathscr{S} - U(P) = \overline{\{p\}}$ with $p = P\varphi$, the last condition means that $U(I) \cap \overline{\{p\}} \neq \varnothing$, which holds iff $p \in U(I)$. (Indeed, if $p \notin U(I)$, then $U(I) \subseteq U(P)$, and so $U(I) \cap \overline{\{p\}} = \varnothing$.) ●

Corollary 9 (M. H. Stone [1937]). *The Stone space of a distributive lattice is characterized by* (S1), (S2), *and*

(S3) *The intersection of two compact open sets is compact.*

Corollary 10 (M. H. Stone [1936]). *The Stone space $\mathscr{S}$ of a Boolean lattice (called a* Boolean space) *can be characterized as a compact Hausdorff space in which the closed open* (clopen) *sets form a base for open sets.* (*In other words, $\mathscr{S}$ is* totally disconnected.)

Proof. Let $\mathscr{S}=\mathscr{S}(B)$, where B is a Boolean lattice. Then $\mathscr{S}=r(1)$, and thus $\mathscr{S}$ is compact. Let $P, Q \in \mathscr{S}$ and $P \neq Q$ and take $a \in P-Q$. Then $Q \in r(a)$ and $P \in r(a')$; therefore, every pair of elements of $\mathscr{S}$ can be separated by clopen sets, verifying that $\mathscr{S}$ is Hausdorff. Obviously, $\mathscr{S}$ is totally disconnected. Conversely, let $\mathscr{S}$ be compact, Hausdorff, and totally disconnected. Then ($\check{\mathrm{S}}1$) is obvious. (S2) follows from the observation that F and the U_i, $i \in I$, are now closed sets having the finite intersection property; therefore, by compactness, they have an element in common. The compact open sets of $\mathscr{S}$ form a Boolean lattice B, and thus $\mathscr{S}$ is homeomorphic to $\mathscr{S}(B)$. ●

As an interesting application we prove:

Theorem 11. *Let B be an infinite Boolean lattice. Then $|\mathscr{P}(B)| \geq |B|$.*

Proof. Let $\mathscr{S}$ be a totally disconnected compact Hausdorff space. For $a, b \in \mathscr{S}$ with $a \neq b$, fix a pair of clopen sets $U_{a,b}$ and $U_{b,a}$ such that $a \in U_{a,b}$, $b \in U_{b,a}$, and $U_{a,b} \cap U_{b,a} = \varnothing$. Now let U be clopen and $a \in U$. Then $\mathscr{S}-U \subseteq \bigcup(U_{b,a} \mid b \in \mathscr{S}-U)$, and so, by the compactness of $\mathscr{S}-U$, $\mathscr{S}-U \subseteq \bigcup(U_{b,a} \mid b \in B)$ for some finite $B \subseteq \mathscr{S}-U$. Then $V_a = \bigcap(U_{a,b} \mid b \in B)$ is open and $a \in V_a \subseteq U$. Thus, $U \subseteq \bigcup(V_a \mid a \in U)$, so by the compactness of U, $U \subseteq \bigcup(V_a \mid a \in A)$ for some finite $A \subseteq U$. Therefore, $U = \bigcup(V_a \mid a \in A)$. Thus every clopen set is a finite union of finite intersections of $U_{a,b}$, and so there are no more clopen sets than there are finite sequences of elements of $\mathscr{S}$; this cardinality is $|\mathscr{S}|$ provided that $|\mathscr{S}|$ is infinite. ●

It might be illuminating to compare this to an algebraic proof; see Exercise 36.

Theorem 8 and its corollaries provide topological representations for distributive join-semilattices, distributive lattices, and Boolean lattices, respectively. It is possible to give a topological representation for homomorphisms. We do it here only for $\{0, 1\}$-homomorphisms of bounded distributive lattices.

Lemma 12. *Let L_0 and L_1 be bounded distributive lattices and let φ be a $\{0, 1\}$-homomorphism of L_0 into L_1. Then*

$$\mathscr{S}(\varphi)\colon P \to P\varphi^{-1}$$

maps $\mathscr{S}(L_1)$ into $\mathscr{S}(L_0)$; $\mathscr{S}(\varphi)$ is a continuous function with the property that if U is compact open in $\mathscr{S}(L_0)$, then $U(\mathscr{S}(\varphi))^{-1}$ is compact in $\mathscr{S}(L_1)$. Conversely, if $\psi\colon \mathscr{S}(L_1) \to \mathscr{S}(L_0)$ has these properties, then $\psi = \mathscr{S}(\varphi)$ for exactly one $\varphi\colon L_0 \to L_1$.

Proof. If $U = r(a)$, $a \in L_0$, then

$$\begin{aligned} U\mathscr{S}(\varphi)^{-1} &= \{P \mid P \in \mathscr{S}(L_1),\ P\varphi^{-1} \in r(a)\} \\ &= \{P \mid P \in \mathscr{S}(L_1),\ a \notin P\varphi^{-1}\} \\ &= \{P \mid P \in \mathscr{S}(L_1),\ a\varphi \notin P\} = r(a\varphi), \end{aligned}$$

and so $\mathscr{S}(\varphi)$ is continuous, having the desired property.

Conversely, if such a ψ is given and $U = r(a)$, $a \in L_0$, then $U\psi^{-1}$ is compact open, and so $U\psi^{-1} = r(b)$ for a unique $b \in L_1$. The map $\varphi\colon a \to b$ is a $\{0, 1\}$-homomorphism, and $\psi = \mathscr{S}(\varphi)$. ●

The following interpretation of (S2) will be useful. Let $\mathscr{S}$ be a topological space. The *Booleanization of* $\mathscr{S}$ is a topological space $\mathscr{S}^B$ on $\mathscr{S}$ that has the compact open sets of $\mathscr{S}$ and their complements as a subbase for open sets.

Lemma 13. *A compact topological space $\mathscr{S}$ satisfies* (S1), (S2), *and* (S3) *iff $\mathscr{S}^B$ is a Boolean space.*

Proof. Let $\mathscr{S}$ satisfy (S1), (S2), and (S3). Then $\mathscr{S}^B$ is obviously Hausdorff and totally disconnected. To verify the compactness of $\mathscr{S}^B$, let $\mathscr{F}_0$ be a collection of compact open sets of $\mathscr{S}$, and let $\mathscr{F}_1$ be a collection of complements of compact open sets of $\mathscr{S}$ such that in $\mathscr{F} = \mathscr{F}_0 \cup \mathscr{F}_1$ no finite intersection is void. Because of (S3), we can assume that $\mathscr{F}_0$ is closed under finite intersection. Since members of $\mathscr{F}_1$ are closed in $\mathscr{S}$ and $\mathscr{S}$ is compact, $\bigcap(X \mid X \in \mathscr{F}_1) = F$ is nonvoid. Also, for any $U \in \mathscr{F}_0$ and $X \in \mathscr{F}_1$, $U \cap X$ is closed in U, and thus $U \cap F = \bigcap(U \cap X \mid X \in \mathscr{F}_1) \neq \varnothing$. Applying (S2) to F and $\mathscr{F}_0$, we conclude that $\bigcap \mathscr{F} \neq \varnothing$, which, by Alexander's Theorem (see Exercise 14), proves compactness.

Conversely, if $\mathscr{S}^B$ is Boolean, the compact open sets of $\mathscr{S}^B$ form a Boolean lattice L. We can easily verify that the compact open sets of $\mathscr{S}$ form a sublattice L_1 of L. Thus L_1 is a distributive lattice, and so, by Corollary 9, $\mathscr{S} = \mathscr{S}(L_1)$ satisfies (S1), (S2), and (S3). ●

We close this section with an interesting application.

Let L_i, $i \in I$, be pairwise disjoint distributive lattices. Then $Q = \bigcup(L_i \mid i \in I)$ is a partial lattice. A free lattice generated by Q over the class $\mathbf{D}$ of all distributive lattices is called a *free distributive product* of the L_i, $i \in I$. To prove the existence of free distributive products, it suffices by Theorem I.5.24 to show that there exists a distributive lattice L containing Q as a relative sublattice. This is easily done: Let L be the direct product of the $L_i \cup \{0\}$, $i \in I$, where 0 is a new zero element of L_i. Identify $x \in L_i$ with $f \in L$ defined by $f(i) = x$, $f(j) = 0$ for $j \neq i$. Then Q becomes a relative sublattice of L.

An equivalent definition is:

Definition 14. *Let* $\mathbf{K}$ *be a class of lattices and let* L_i, $i \in I$, *be lattices in* $\mathbf{K}$. *A lattice* L *in* $\mathbf{K}$ *is called a* free $\mathbf{K}$-product *of the* L_i, $i \in I$, *iff every* L_i *has an embedding* ε_i *into* L *such that*:

(i) *L is generated by* $\bigcup(L_i\varepsilon_i \mid i \in I)$.

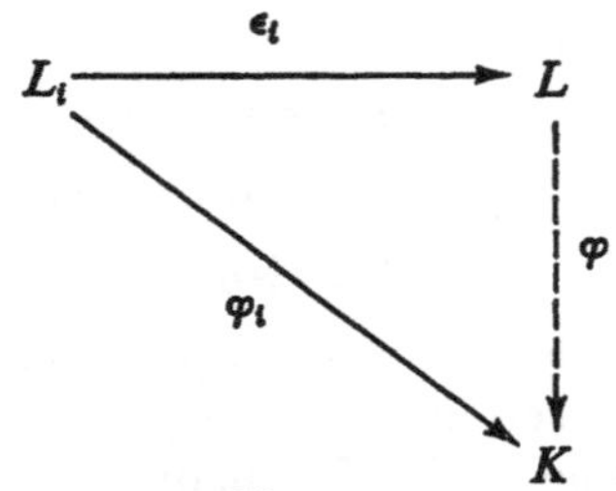

Figure 4

(ii) *If K is any lattice in* **K** *and φ_i is a homomorphism of L_i into K, for $i \in I$, then there exists a homomorphism φ of L into K satisfying $\varphi_i = \varepsilon_i \varphi$ for all $i \in I$* (see Figure 4).

For distributive lattices, this is equivalent to the first definition (see Exercise 55). In most cases we will assume that each L_i is a sublattice of L and that ε_i is the inclusion map; then (ii) will simply state that the φ_i have a common extension. Note that in all the cases we shall consider, (i) can be replaced by the requirement that the φ in (ii) be unique.

If, in Definition 14, **K** is a class of bounded lattices and all homomorphisms are assumed to be $\{0, 1\}$-homomorphisms, we get the concept of a *free* **K** *$\{0, 1\}$-product*. In particular if $\mathbf{K} = \mathbf{L}$ we get the concept of *free $\{0, 1\}$-product*, see Section VI.3, and if $\mathbf{K} = \mathbf{D}$ we obtain the concept of *free $\{0, 1\}$-distributive product*.

Our final result is the existence and description of a free $\{0, 1\}$-distributive product of a family of bounded distributive lattices.

THEOREM 15 (A. Nerode [1959]). *Let L_i, $i \in I$, be distributive lattices with* 0 *and* 1. *Let $\mathcal{S} = \Pi(\mathcal{S}(L_i) \mid i \in I)$* (see Exercise 15). *Then $\mathcal{S}$ is a Stone space, and thus $\mathcal{S} \cong \mathcal{S}(L)$ for some distributive lattice L. Any such lattice L is a free $\{0, 1\}$-distributive product of the L_i, $i \in I$.*

The proof of Theorem 15 will be preceded by two lemmas. In these two lemmas a *Stone space* is a topological space satisfying (S1), (S2), and (S3).

LEMMA 16. *Let $\mathcal{S}_i$, $i \in I$, be compact Stone spaces. Then*

$$\Pi(\mathcal{S}_i^B \mid i \in I) = (\Pi(\mathcal{S}_i \mid i \in I))^B.$$

PROOF. For $U \subseteq \mathcal{S}_j$, let $E(U) = \{f \mid f \in \Pi \mathcal{S}_i, f(j) \in U\}$ (see Exercise 15). The compact open sets form a base for open sets in $\mathcal{S}_j$; therefore, $\{E(U) \mid U$ compact open in some $\mathcal{S}_j\}$ is a subbase for open sets in $\Pi(\mathcal{S}_i \mid i \in I)$. Note that all the sets $E(U)$ are compact open in $\Pi \mathcal{S}_i$; therefore, $V \subseteq \Pi \mathcal{S}_i$ is compact open iff it is a finite union of finite intersections of some of the $E(U)$. Consequently, declaring the complements of compact open sets to be open (in forming $(\Pi \mathcal{S}_i)^B$) is equivalent to making the complements of the sets $E(U)$ open. But the complement of $E(U)$ is $E(\mathcal{S}_i - U)$, and $\mathcal{S}_i - U$ is a typical open set of $\mathcal{S}_i^B$. Thus $\Pi \mathcal{S}_i^B$ and $(\Pi \mathcal{S}_i)^B$ have the same topology. ●

Lemma 17. *The product of compact Stone spaces is again a compact Stone space.*

Proof. Let $\mathscr{S}_i$, $i \in I$, be Stone spaces. Then $\mathscr{S} = \Pi \mathscr{S}_i$ is T_0 and compact (Exercises 16 and 21). Since $\mathscr{S}_i^B$ is Boolean (Lemma 13), so is $\Pi \mathscr{S}_i^B$ (Exercises 20, 21, and 22). By Lemma 16, $\mathscr{S}^B = \Pi \mathscr{S}_i^B$, and thus $\mathscr{S}^B$ is Boolean. Therefore, by Lemma 13, $\mathscr{S}$ is a Stone space. ●

Proof of Theorem 15. Let e_i be the ith projection ($e_i: \mathscr{S}(L) \to \mathscr{S}(L_i)$, $fe_i = f(i)$). By Lemma 12, there is a unique $\{0, 1\}$-homomorphism ε_i: $L_i \to L$ satisfying $\mathscr{S}(\varepsilon_i) = e_i$. It is easy to visualize ε_i; think of the elements of L_i as compact open sets of $\mathscr{S}_i$; then $U\varepsilon_i = E(U) = Ue_i^{-1}$. It is obvious from this that ε_i is an embedding.

By applying $\mathscr{S}$ to Figure 4, we obtain Figure 5. Thus the method of defining $\mathscr{S}(\varphi)$ is clear. For $x \in \mathscr{S}(K)$, $x\mathscr{S}(\varphi)$ is a member of $\Pi \mathscr{S}(L_i)$, and $x\mathscr{S}(\varphi)(i) = x\mathscr{S}(\varphi_i)$ for $i \in I$.

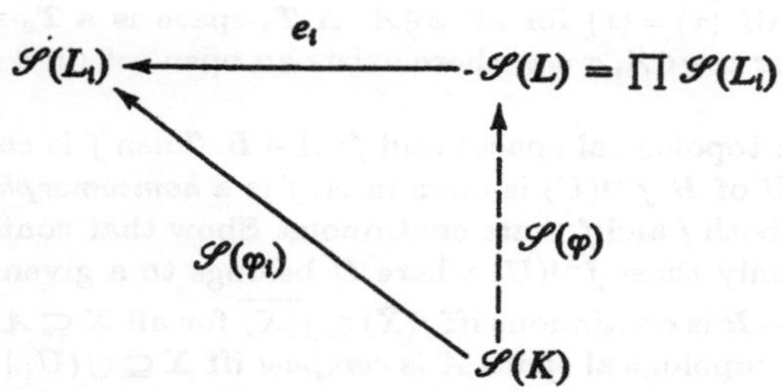

Figure 5

To show that this correspondence is indeed of the form $\mathscr{S}(\varphi)$ for some homomorphism $\varphi: L \to K$, we have to verify that (a) $\mathscr{S}(\varphi)$ is continuous (this statement follows from Exercise 18), and that (b) if V is compact open in $\mathscr{S}(L)$, then $V\mathscr{S}(\varphi)^{-1}$ is compact open in $\mathscr{S}(K)$. It is enough to verify (b) for $V = E(U)$ where U is compact open in some $\mathscr{S}(L_i)$. Then

$$V\mathscr{S}(\varphi)^{-1} = E(U)\mathscr{S}(\varphi)^{-1} = Ue_i^{-1}\mathscr{S}(\varphi)^{-1} = U(\mathscr{S}(\varphi)e_i)^{-1} = U\mathscr{S}(\varphi_i)^{-1},$$

and therefore $V\mathscr{S}(\varphi)^{-1}$ is compact open since $\mathscr{S}(\varphi_i)$ satisfies the condition of Lemma 12. ●

Exercises

The following exercises review the basic concepts and theorems of topology that are utilized in this section.

1. A *topological space* is a set A and a collection T of subsets of A, with $A \in T$, closed under finite intersections and arbitrary unions; a member of T is called an *open set*. Call a set *closed* iff its complement is open. Characterize the family of closed sets.

2. A family of nonvoid sets $B \subseteq T$ is a *base* for open sets iff every open set is a union of members of B. Show that for a set A, $B \subseteq P(A)$ is a base for open sets of some topological space defined on A iff $\bigcup B = A$, and for $X, Y \in B$ and $p \in X \cap Y$, there exists a $Z \in B$ with $p \in Z$, $Z \subseteq X$, and $Z \subseteq Y$.
3. A family of nonvoid sets $C \subseteq P(A)$ is a *subbase* for open sets iff the finite intersections of members of C form a base for open sets. Show that $C \subseteq P(A)$ is a subbase of some topology defined on A iff $\bigcup C = A$.
4. Let A be a topological space and let $X \subseteq A$. Then there exists a smallest closed set $\overline{X}$ containing X; $\overline{X}$ is the *closure* of X. Show that $\overline{\varnothing} = \varnothing$ and that, for $X, Y \subseteq A$, $X \subseteq Y$ implies that $\overline{X} \subseteq \overline{Y}$, $X \subseteq \overline{X}$, $\overline{X \cup Y} = \overline{X} \cup \overline{Y}$, and $\overline{\overline{X}} = \overline{X}$.
5. Prove that the properties of $\overline{X}$ given in Exercise 4 characterize it.
6. Show that $a \in \overline{X}$ iff every open set (in a given base) containing a has a nonvoid intersection with X.
7. A space A is a *T_0-space* iff $\overline{\{x\}} = \overline{\{y\}}$ implies that $x = y$. Show that A is a T_0-space iff, for every $x, y \in A$, $x \neq y$, there exists an open set (in a given base) containing exactly one of x and y.
8. A is a *T_1-space* iff $\overline{\{x\}} = \{x\}$ for all $x \in A$. A T_1-space is a T_0-space. Show that A is a T_1-space iff, for $x, y \in A$, $x \neq y$, there exists an open set (in a given base) containing x but not y.
9. Let A and B be topological spaces and $f\colon A \to B$. Then f is called *continuous* iff for every open set U of B, $f^{-1}(U)$ is open in A. f is a *homeomorphism* iff f is one-to-one and onto and if both f and f^{-1} are continuous. Show that continuity can be checked by considering only those $f^{-1}(U)$ where U belongs to a given subbase.
10. Show that $f\colon A \to B$ is continuous iff $f(\overline{X}) \subseteq \overline{f(X)}$ for all $X \subseteq A$.
11. A subset X of a topological space A is *compact* iff $X \subseteq \bigcup(U_i \mid U_i$ open, $i \in I)$ implies that $X \subseteq \bigcup(U_i \mid i \in I_1)$ for some finite $I_1 \subseteq I$. The space A is *compact* iff $X = A$ is compact. Show that A is compact iff, for every family F of closed sets, if $\bigcap F_1 \neq \varnothing$ for all finite $F_1 \subseteq F$, then $\bigcap F \neq \varnothing$.
12. Let A be a compact topological space and let X be a closed set in A. Show that X is compact.
13. Prove that a space A is compact iff, in the lattice of closed sets of A, every maximal dual ideal is principal.
*14. Show that a T_1-space A is compact iff it has a subbase C of closed sets (that is, $\{A - X \mid X \in C\}$ is a subbase for open sets) with the property: If $\bigcap D = \varnothing$ for some $D \subseteq C$, then $\bigcap D_1 = \varnothing$ for some finite $D_1 \subseteq D$ (J. W. Alexander [1939]).
15. Let A_i, $i \in I$, be topological spaces and set $A = \Pi(A_i \mid i \in I)$. For $U \subseteq A_i$, set $E(U) = \{f \mid f \in A, f(i) \in U\}$. The *product topology* on A is the topology determined by taking all the sets $E(U)$ as a subbase for open sets, where U ranges over all open sets of A_i for all $i \in I$. Show that the projection map $e_i\colon f \to f(i)$ is a continuous map of A onto A_i.
16. Show that if the A_i are T_0-spaces (T_1-spaces), so is $A = \Pi(A_i \mid i \in I)$.
17. A map $f\colon A \to B$ is *open* iff $f(U)$ is open in B for every open $U \subseteq A$. Show that the projection maps (see Exercise 15) are open.
18. Prove that a function $f\colon B \to \Pi A_i$ is continuous iff, for each $i \in I$, $e_i f\colon B \to A_i$ is continuous.
19. A space A is a *Hausdorff space* (*T_2-space*) iff, for $x, y \in A$ with $x \neq y$, there exist open sets U, V such that $x \in U$, $y \in V$, $U \cap V = \varnothing$. Show that:
    (i) A is Hausdorff iff $\Delta = \{\langle x, x\rangle \mid x \in A\}$ is closed in $A \times A$.
    (ii) A compact subset of a T_2-space is closed.

20. Prove that a product of Hausdorff spaces is a Hausdorff space.
21. Show that a product of compact spaces is compact (Tihonov's Theorem). (Hint: use Exercise 14.)
22. A space A is *totally disconnected* iff, for $x, y \in A$, $x \neq y$, there exists a closed open set U with $x \in U$, $y \notin U$. Show that the product of totally disconnected sets is totally disconnected.

* * *

23. Let I and J be ideals of a join-semilattice. Verify that

$$I \vee J = \{t \mid t \leq i \vee j,\ i \in I,\ j \in J\}.$$

24. Show that for a join-semilattice L, $I(L)$ is a lattice iff any two elements of L have a lower bound.
25. Give a detailed proof of Lemma 2.
26. Prove that every join-semilattice can be embedded in a Boolean lattice (considered as a join-semilattice).
27. Show that a finite distributive join-semilattice is a distributive lattice.
28. Let L be a join-semilattice and let Θ be a *join-congruence*, that is, an equivalence relation on L having the Substitution Property for join. Then L/Θ is also a join-semilattice. Show that the distributivity of L does not imply the distributivity of L/Θ.
29. Let F be a free join-semilattice; let F_0 be F with a new 0 added. Show that F_0 is a distributive join-semilattice.
30. Let φ be a join-homomorphism of the join-semilattice F_0 onto the join-semilattice F_1. Show that, for distributive join-semilattices F_0, F_1, the proper homomorphism concept is the one requiring that if P is a prime ideal of F_1, then $P\varphi^{-1}$ is a prime ideal of F_0.
31. Show that there is no "free distributive join-semilattice" with the homomorphism concept of Exercise 30.
32. Does Theorem 1.22 generalize to distributive join-semilattices?
33. Characterize the Stone spaces of finite Boolean lattices and of finite chains.
34. Let $\mathscr{S}_0$ and $\mathscr{S}_1$ be disjoint topological spaces; let $\mathscr{S} = \mathscr{S}_0 \cup \mathscr{S}_1$ and call $U \subseteq \mathscr{S}$ open iff $U \cap \mathscr{S}_0$ and $U \cap \mathscr{S}_1$ are open. Show that if $\mathscr{S}_0$ and $\mathscr{S}_1$ are Stone spaces, then so is $\mathscr{S}$.
35. If, in Exercise 34, $\mathscr{S}_i = \mathscr{S}(L_i)$, $i = 0, 1$, then $\mathscr{S} = \mathscr{S}(L_0 \times L_1)$.
36. To prove Theorem 11, pick an element $a(P, Q) \in P - Q$ for all $P, Q \in \mathscr{P}(B)$ with $P \neq Q$. Show that the $a(P, Q)$ R-generate all of B.
37. In Lemma 12, characterize $\mathscr{S}(\varphi)$ for one-to-one and for onto φ.
38. Determine the connection between the Stone space of a lattice and the Stone space of a sublattice.
39. Call the Stone space of a generalized Boolean lattice a *generalized Boolean space*; characterize it. (Compactness of $\mathscr{S}$ should be replaced by *local compactness*: For every $p \in \mathscr{S}$ there exists an open set U with $p \in U$ and a set V with $U \subseteq V$, such that V is compact.)
40. Show that the product of (generalized) Boolean spaces is (generalized) Boolean.
41. Call the join-semilattice L *modular* iff, for $a, b, c \in L$, $a \leq b$ and $b \leq a \vee c$ imply the existence of $c_1 \in L$ with $c_1 \leq c$ and $b = a \vee c_1$. Show that a distributive join-semilattice is modular.
42. Show that Lemma 1 remains valid if all occurrences of the word "distributive" are replaced by the word "modular."
43. Show that the set of all finitely generated normal subgroups of a group (and also the finitely generated ideals of a ring) form a modular join-semilattice.

44. The lattice of congruence relations of a join-semilattice L is distributive iff any pair of elements with a lower bound are comparable (D. Papert [1964], R. A. Dean and R. H. Oehmke [1964]).

* * *

We have seen in this section how Theorem I.5.24 can be used to show the existence of free distributive products. The same method, however, does not apply to distributive lattices with 0 and 1. Nevertheless, the idea of the proofs of Theorems I.5.5 and I.5.24 can be used to get a much stronger result on the existence of free products; it is easiest to formulate this result (G. Grätzer [1968], Theorem 29.2) for universal algebras. To do so we have to introduce some concepts.

A *type* τ of algebras is a sequence $\langle n_0, n_1, \dots, n_\gamma, \dots\rangle$ of non-negative integers, $\gamma < o(\tau)$, where $o(\tau)$ is an ordinal called the *order of* τ. An *algebra* $\mathfrak{A}$ *of type* τ is an ordered pair $\langle A; F\rangle$, where A is a nonvoid set and F is a sequence $\langle f_0, \dots, f_\gamma, \dots\rangle$, $\gamma < o(\tau)$, where f_γ is an n_γ-ary operation on A. If $o(\tau)$ is finite, $o(\tau) = n$, then we write $\langle A; f_0, \dots, f_{n-1}\rangle$ for $\langle A; F\rangle$.

45. Define the concepts of subalgebra, polynomial, identity, and equational class for algebras of a given type τ. Show that if $\mathbf{K}$ is an equational class, $\mathfrak{A}$ is an algebra in $\mathbf{K}$, and $\mathfrak{B}$ is a subalgebra of $\mathfrak{A}$, then $\mathfrak{B}$ is in $\mathbf{K}$.
46. Define the concepts of homomorphism, homomorphic image, and direct product for algebras of a given type. Show that an equational class is closed under the formation of homomorphic images and direct products.
47. Let $\mathfrak{A} = \langle A; F\rangle$ be an algebra, let $H \subseteq A$, and let $H \neq \varnothing$. Show that there exists a smallest subset $[H]$ of A, $[H] \supseteq H$ such that $\langle [H]; F\rangle$ is a subalgebra of $\mathfrak{A}$. (This subalgebra is said to be *generated by* H.)
48. Show that $|[H]| \le |H| + |F| + \aleph_0$.
49. Modify Definition 14 for algebras. Show that the φ in (ii) is unique.
50. Let $\mathfrak{B}$ and $\mathfrak{C}$ be free $\mathbf{K}$-products of $\mathfrak{A}_i$, $i \in I$, with the embeddings ε_i, $i \in I$, and χ_i, $i \in I$, respectively. Show that there exists an isomorphism $\alpha: B \to C$ such that $\varepsilon_i\alpha = \chi_i$ and $\chi_i\alpha^{-1} = \varepsilon_i$, for all $i \in I$.
51. Let $\mathbf{K}$ be an equational class of algebras and let $\mathfrak{A}_i \in \mathbf{K}$, for $i \in I$. Choose a set S satisfying

$$|S| \ge \sum |A_i| + |F| + \aleph_0.$$

Let Q be the set of all pairs $\langle \mathfrak{B}, (\varphi_i \mid i \in I)\rangle$ such that $B \subseteq S$, φ_i is a homomorphism of $\mathfrak{A}_i$ into $\mathfrak{B}$, and $B = [\bigcup (A_i\varphi \mid i \in I)]$. Form $\mathfrak{A} = \Pi(\mathfrak{B} \mid \langle \mathfrak{B}, (\varphi_i \mid i \in I)\rangle \in Q)$ (direct product), and, for $a \in A_i$, define $f_a \in A$ by $f_a(\langle \mathfrak{B}, (\varphi_i \mid i \in I)\rangle) = a\varphi_i$. Finally, let $\mathfrak{R}$ be the subalgebra generated by the f_a, $a \in A_i$, $i \in I$. Show that $\mathfrak{R} \in \mathbf{K}$, $a \to f_a$ is a homomorphism ε_i of $\mathfrak{A}_i$ into $\mathfrak{R}$, for $i \in I$, and that $\mathfrak{R}$ is generated by $\bigcup (A_i\varepsilon_i \mid i \in I)$.
52. Show that ε_i is one-to-one iff, for $i \in I$, $a, b \in A_i$, $a \neq b$, there exists an algebra $\mathfrak{C} \in \mathbf{K}$ and homomorphisms $\psi_j: \mathfrak{A}_j \to \mathfrak{C}$ such that $a\psi_i \neq b\psi_i$.
53. Combine the previous exercises to prove the following result.
Existence Theorem for Free Products: *Let* $\mathbf{K}$ *be an equational class of algebras,* $\mathfrak{A}_i \in \mathbf{K}$ *for* $i \in I$. *A free* $\mathbf{K}$*-product of* $\mathfrak{A}_i$, $i \in I$, *exists iff, for* $i \in I$, $a, b \in A_i$, $a \neq b$, *there exist a* $\mathfrak{C} \in \mathbf{K}$, *homomorphisms* $\psi_j: \mathfrak{A}_j \to \mathfrak{C}$ *for* $j \in I$ *such that* $a\psi_i \neq b\psi_i$.
54. Show that in proving the existence of free distributive products and free $\{0, 1\}$-distributive products, we can always choose $\mathfrak{C} = \mathfrak{C}_2$, the two-element chain, in applying Exercise 53.
55. Show that the two definitions of free $\mathbf{K}$-product are equivalent for any class $\mathbf{K}$ of lattices.

56. Show that the free Boolean algebra on $\mathfrak{m}$ generators is a free $\{0, 1\}$-distributive product of $\mathfrak{m}$ copies of the free Boolean algebra on one generator.
57. Prove that the free Boolean algebra on $\mathfrak{m}$ generators can be represented by the clopen subsets of $\{0, 1\}^{\mathfrak{m}}$ where $\{0, 1\}$ is the two-element discrete topological space.
58. Find a topological representation for the free distributive lattice on $\mathfrak{m}$ generators (G. Ja. Areškin [1953a]).

6. Distributive Lattices with Pseudocomplementation

In this section we shall deal exclusively with pseudocomplemented distributive lattices. There are two concepts that we should be able to distinguish: a lattice, $\langle L; \wedge, \vee \rangle$, in which every element has a pseudocomplement; and an algebra $\langle L; \wedge, \vee, {}^*, 0, 1 \rangle$, where $\langle L; \wedge, \vee, 0, 1 \rangle$ is a bounded lattice and where, for every $a \in L$, the element a^* is the pseudocomplement of a. We shall call the former a *pseudocomplemented lattice* and the latter a *lattice with pseudocomplementation* (as an operation)—the same kind of distinction that we make between Boolean lattices and Boolean algebras. As defined in the exercises of Section 5, a pseudocomplemented lattice is an algebra of type $\langle 2, 2 \rangle$, whereas a lattice with pseudocomplementation is an algebra of type $\langle 2, 2, 1, 0, 0 \rangle$. To see the difference in viewpoint, consider the lattice of Figure 1. As a distributive lattice it has twenty-five sublattices and eight congruences; as a lattice with pseudocomplementation it has three subalgebras and five congruences.

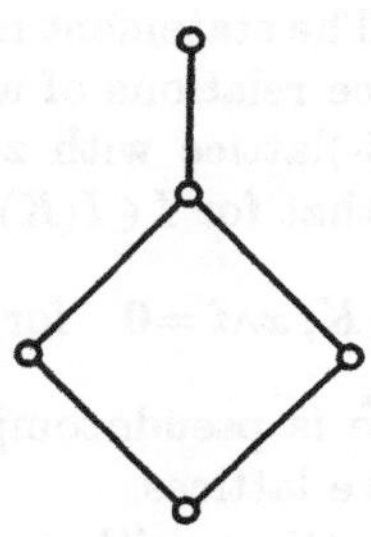

Figure 1

Thus, for a lattice with pseudocomplementation L, a *subalgebra* L_1 is a $\{0, 1\}$-sublattice of L closed under * (that is, $a \in L_1$ implies that $a^* \in L_1$). A *homomorphism* φ is a $\{0, 1\}$-homomorphism that also satisfies $(x\varphi)^* = x^*\varphi$. Similarly, a *congruence relation* Θ shall have the Substitution Property also for *: $a \equiv b\ (\Theta)$ implies that $a^* \equiv b^*\ (\Theta)$.

A wide class of examples is provided by

Theorem 1. *Any complete lattice that satisfies the Join Infinite Distributive Identity* (JID) *is a pseudocomplemented distributive lattice.*

PROOF. Let L be such a lattice. For $a \in L$, set

$$a^* = \bigvee(x \mid x \in L, a \wedge x = 0).$$

Then, by (JID),

$$a \wedge a^* = a \wedge \bigvee(x \mid a \wedge x = 0) = \bigvee(a \wedge x \mid a \wedge x = 0) = \bigvee 0 = 0.$$

Furthermore, if $a \wedge x = 0$, then $x \leq a^*$ by the definition of a^*; thus a^* is indeed the pseudocomplement of a. ●

COROLLARY 2. *Every distributive algebraic lattice is pseudocomplemented.*

PROOF. Let L be a distributive algebraic lattice. By Theorem 3.13 and Lemma 5.1 we can assume that $L = I(S)$, where S is a distributive join-semilattice with 0. Let $I, I_j \in I(S)$, for $j \in J$. Then $I \wedge I_k \subseteq I \wedge \bigvee(I_j \mid j \in J)$ for any $k \in J$, and thus

$$\bigvee(I \wedge I_j \mid j \in J) \subseteq I \wedge \bigvee(I_j \mid j \in J).$$

To prove the reverse inclusion, let $a \in I \wedge \bigvee(I_j \mid j \in J)$, that is, $a \in I$ and $a \in \bigvee(I_j \mid j \in J)$. The latter implies that $a \leq t_1 \vee \cdots \vee t_n$, where $t_1 \in I_{j_1}, \ldots, t_n \in I_{j_n}, j_1, \ldots, j_n \in J$. Thus $a \in I_{j_1} \vee \cdots \vee I_{j_n}$ and so, using the distributivity of $I(L)$, we obtain

$$a \in I \wedge (I_{j_1} \vee \cdots \vee I_{j_n}) = (I \wedge I_{j_1}) \vee \cdots \vee (I \wedge I_{j_n}) \subseteq \bigvee(I \wedge I_j \mid j \in J),$$

completing the proof of the (JID). The statement now follows from Theorem 1. ●

Thus, the lattice of all congruence relations of an arbitrary lattice and the lattice of all ideals of a distributive (semi-)lattice with zero are examples of pseudocomplemented distributive lattices. Note that for $I \in I(K)$,

$$I^* = \{x \mid x \in K, x \wedge i = 0 \quad \text{for all} \quad i \in I\}.$$

Also, any finite distributive lattice is pseudocomplemented. Therefore, our investigations include all finite distributive lattices.

The first class of distributive lattices with pseudocomplementation, other than the class of Boolean algebras, to be examined in detail was the class of Stone algebras. A distributive lattice with pseudocomplementation L is called a *Stone algebra* iff it satisfies the *Stone identity*:

$$a^* \vee a^{**} = 1.$$

The corresponding pseudocomplemented lattice is called a *Stone lattice*. To understand the meaning of this identity, define the *skeleton* of L:

$$S(L) = \{a^* \mid a \in L\}.$$

The elements of $S(L)$ are called *skeletal*. By Theorem I.6.4, $\langle S(L); \wedge, \vee, {}^*, 0, 1\rangle$ is a Boolean algebra. For a Stone algebra L, $S(L)$ is a subalgebra of L:

Lemma 3. *For a distributive lattice with pseudocomplementation L, the following conditions are equivalent:*

(i) *L is a Stone algebra.*

(ii) $(a\wedge b)^* = a^*\vee b^*$ *for* $a, b\in L$.

(iii) $a, b\in S(L)$ *implies that* $a\vee b\in S(L)$.

(iv) *$S(L)$ is a subalgebra of L.*

Proof. The proofs that (ii) implies (iii), that (iii) implies (iv), and that (iv) implies (i) are trivial. Now let L be a Stone algebra; we show that $a^*\vee b^*$ is the pseudocomplement of $a\wedge b$. Indeed, $(a\wedge b)\wedge(a^*\vee b^*) = (a\wedge b\wedge a^*)\vee(a\wedge b\wedge b^*) = 0\vee 0 = 0$. If $(a\wedge b)\wedge x = 0$, then $(b\wedge x)\wedge a = 0$, and so $b\wedge x \leq a^*$. Meeting both sides by a^{**} yields $b\wedge x\wedge a^{**} \leq a^*\wedge a^{**} = 0$; that is, $x\wedge a^{**}\wedge b = 0$, implying that $x\wedge a^{**}\leq b^*$. By the Stone identity, $a^*\vee a^{**} = 1$, and thus $x = x\wedge 1 = x\wedge(a^*\vee a^{**}) = (x\wedge a^*)\vee(x\wedge a^{**}) \leq a^*\vee b^*$. ●

This is already enough to yield the structure theorem for finite Stone algebras (G. Grätzer and E. T. Schmidt [1957a]):

Corollary 4. *A finite distributive lattice is a Stone lattice iff it is the direct product of finite distributive* dense *lattices, that is, finite distributive lattices with only one atom.*

Proof. By Lemma 3, a Stone lattice L has a complemented element $a\notin\{0, 1\}$ iff $S(L)\neq\{0, 1\}$; thus the decomposition of Theorem 1.6 can be repeated until each factor L_i satisfies $S(L_i) = \{0, 1\}$. In a direct product, $*$ is formed componentwise; therefore, all the L_i are Stone lattices. For a finite lattice K with $S(K) = \{0, 1\}$, the condition that K has exactly one atom is equivalent to K being a Stone lattice. ●

Another significant subset of a Stone algebra is the *dense set*:

$$D(L) = \{a \mid a^* = 0\}.$$

The elements of $D(L)$ are called *dense.*

We can easily check that $D(L)$ is a dual ideal of L and that $1\in D(L)$: thus $D(L)$ is a distributive lattice with 1. Since $a\vee a^*\in D(L)$ for every $a\in L$, we can interpret the identity

$$a = a^{**}\wedge(a\vee a^*)$$

to mean that every $a\in L$ can be represented in the form $a = b\wedge c$, where $b\in S(L)$, $c\in D(L)$. Such an interpretation correctly suggests that if we know $S(L)$ and $D(L)$ and the relationships between elements of $S(L)$ and $D(L)$, then we can describe L. The relationship is expressed by the homomorphism $\varphi(L)\colon S(L)\to\mathcal{D}(D(L))$ defined by

$$\varphi(L)\colon a\to\{x \mid x\in D(L),\ x\geq a^*\}.$$

Theorem 5 (C. C. Chen and G. Grätzer [1969a]). *Let L be a Stone algebra. Then $S(L)$ is a Boolean algebra, $D(L)$ is a distributive lattice with 1, and $\varphi(L)$ is a $\{0, 1\}$-homomorphism of $S(L)$ into $\mathcal{D}(D(L))$. The triple $\langle S(L), D(L), \varphi(L)\rangle$ characterizes L up to isomorphism.*

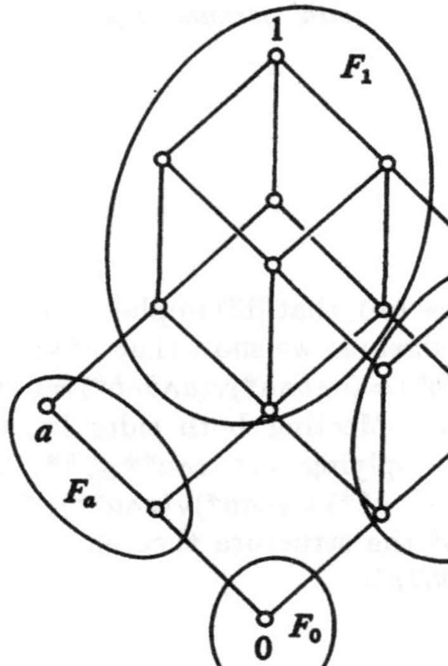

Figure 2

Proof. The first statement is easily verified. For $a \in S(L)$, set

$$F_a = \{x \mid x^{**} = a\}.$$

The sets $\{F_a \mid a \in S(L)\}$ form a partition of L; for a simple example, see Figure 2. Obviously, $F_0 = \{0\}$ and $F_1 = D(L)$. The map $x \to x \vee a^*$ sends F_a into $F_1 = D(L)$; in fact, the map is an isomorphism between F_a and $a\varphi(L) \subseteq D(L)$. Thus $x \in F_a$ is completely determined by a and $x \vee a^* \in a\varphi(L)$—that is, by a pair $\langle a, z\rangle$, where $a \in S(L)$, $z \in a\varphi(L)$—and every such pair determines one and only one element of L. To complete our proof we have to show how the partial ordering on L can be determined by such pairs.

Let $x \in F_a$ and $y \in F_b$. Then $x \leq y$ implies that $x^{**} \leq y^{**}$, that is, $a \leq b$. Since $x \leq y$ iff

$$a \vee x \leq a \vee y \quad \text{and} \quad x \vee a^* \leq y \vee a^*,$$

and since the first of these two conditions is trivial, we obtain:

$$x \leq y \quad \text{iff} \quad a \leq b \quad \text{and} \quad x \vee a^* \leq y \vee a^*.$$

Identifying x with $\langle x \vee a^*, a\rangle$ and y with $\langle y \vee b^*, b\rangle$, we see that the preceding conditions are stated in terms of the components of the ordered pairs, except that $y \vee a^*$ will have to be expressed by the triple.

Because $\varphi(L)$ is a $\{0, 1\}$-homomorphism and a is the complement of a^*, we conclude that $a\varphi(L)$ and $a^*\varphi(L)$ are complementary dual ideals of $D(L)$. Thus, every z can be written in a unique fashion in the form $z = z\varrho_a \wedge z_1$, where $z\varrho_a \in a\varphi(L)$ and $z_1 \in a^*\varphi(L)$. Observe that ϱ_a is expressed in terms of the triple. Finally, $y \vee a^* = y \vee b^* \vee a^* = (y \vee b^*)\varrho_a$. Thus for $u \in a\varphi(L)$ and $v \in b\varphi(L)$, we have $\langle u, a\rangle \leq \langle v, b\rangle$ iff $a \leq b$ and $u \leq v\varrho_a$. ●

This theorem shows that for Stone algebras, the behavior of the skeleton and the dense set is decisive. This conclusion leads us to formulate the goal of research for Stone algebras and *for all* distributive lattices with pseudocomplementation as follows:

A problem for distributive lattices with pseudocomplementation is considered solved if it can be reduced to two problems: one for Boolean algebras and one for distributive lattices with 1.

By applying Zorn's Lemma to prime dual ideals of a lattice with 0, we obtain that every prime dual ideal is contained in a maximal prime dual ideal, or, equivalently, we get that every prime ideal contains a *minimal prime ideal* P, that is, a prime ideal P such that $Q \subset P$ for no prime ideal Q. Minimal prime ideals play an important role in the theory of distributive lattices with pseudocomplementation as illustrated by the following result:

Theorem 6 (G. Grätzer and E. T. Schmidt [1957a]). *Let L be a distributive lattice with pseudocomplementation. Then L is a Stone algebra iff $P \vee Q = L$ for any two distinct minimal prime ideals P and Q.*

Proof. Let L be a Stone algebra and let P and Q be distinct minimal prime ideals. Choose $a \in P - Q$ (note that $P \not\subset Q$, since Q is minimal, and $Q \neq P$, hence $P - Q \neq \varnothing$); since $a \wedge a^* = 0 \in Q$, $a \notin Q$, and Q is prime, we obtain that $a^* \in Q$.

$L - P$ is a maximal dual prime ideal, hence by the dual of Corollary 1.18, it is a maximal dual ideal of L. Thus $(L-P) \vee [a) = L$ and so $0 = a \wedge x$ for some $x \in L - P$. Therefore, $a^* \geq x \in L - P$ and so $a^* \notin P$.

Hence $a^* \in Q - P$. Similarly, $a^{**} \in P - Q$, and so $1 = a^* \vee a^{**} \in P \vee Q$, yielding $P \vee Q = L$.

To prove the converse (for this proof see J. Varlet [1966]), let us assume that L is not a Stone algebra and let $a \in L$ such that $a^* \vee a^{**} \neq 1$. Let R be a prime ideal (see Corollary 1.17) such that $a^* \vee a^{**} \in R$.

We claim that $(L - R) \vee [a^*) \neq L$. Indeed, if $(L - R) \vee [a^*) = L$, then there exists an $x \in L - R$ such that $x \wedge a^* = 0$. Then $a^{**} \geq x \in L - R$, hence $a^{**} \in L - R$, a contradiction.

Let F be a maximal dual prime ideal containing $(L - R) \vee [a^*)$ and, similarly, let G be a maximal dual prime ideal containing $(L - R) \vee [a^{**})$. We set $P = L - F$ and $Q = L - G$.

Then P and Q are minimal prime ideals. Moreover, $P \neq Q$, because $a^* \in F = L - P$ and hence $a^* \notin P$; thus $a^{**} \in P$, while $a^{**} \notin Q$. Finally, $P, Q \subseteq R$, hence $P \vee Q \neq L$. ●

We conclude this section by proving two representation theorems for Stone algebras which correspond to the two representation theorems for distributive lattices given in Section 1. The proofs we present use the Subdirect Product Representation Theorem of G. Birkhoff [1944]. Direct proofs are possible but we shall present a proof that can be generalized to other equational classes of distributive lattices with pseudocomplementation.

In the remainder of this section "algebra" means universal algebra. For the purposes of this book, the reader can substitute "lattice" or "lattice with pseudocomplementation" for "algebra". Just as for posets and lattices, we write A for the algebra $\mathfrak{A} = \langle A; F \rangle$ if there is no danger of confusion.

Definition 7. *An algebra A is called* subdirectly irreducible *iff there exist elements $u, v \in A$ such that $u \neq v$ and $u \equiv v$ (Θ) for all congruences $\Theta > \omega$.*

In other words, A has at least two elements and $C(A) = \{\omega\} \cup [\Theta(u, v))$. An equivalent form is:

Corollary 8. *The algebra A is subdirectly irreducible iff $\omega = \bigwedge(\Theta_i \mid i \in I)$ $(\Theta_i \in C(A)$ for $i \in I)$ implies that $\Theta_i = \omega$ for some $i \in I$.*

Example 9. A distributive lattice L is subdirectly irreducible iff $|L| = 2$.

Proof. If $|L| = 1$, then L is not subdirectly irreducible by definition. If $|L| = 2$, then obviously L is subdirectly irreducible. Let $|L| > 2$. Then there exist $a, b, c \in L$, $a < b < c$. We claim that $\Theta(a, b) \wedge \Theta(b, c) = \omega$, which by Corollary 2 shows that L is not subdirectly irreducible. Let $x \equiv y$ $(\Theta(a, b) \wedge \Theta(b, c))$. By Theorem 3.3 this implies that $x \vee b = y \vee b$ and $x \wedge b = y \wedge b$; thus $x = y$ by Corollary 1.3. ●

Example 10. $\mathfrak{B}_2$ is the only subdirectly irreducible Boolean algebra.

Proof. Let B be Boolean. The statement is obvious for $|B| \leq 2$. If $|B| > 2$, then B has a direct product representation, $B = B_1 \times B_2$, $|B_1|, |B_2| \geq 2$ (use Exercise 1.5); thus B cannot be subdirectly irreducible. ●

We shall need a simple universal algebraic lemma:

Lemma 11 (The Second Isomorphism Theorem). *Let L be an algebra and let Θ be a congruence relation of L. For any congruence Φ of L such that $\Phi \geq \Theta$, define the relation Φ/Θ on L/Θ by*

$$[x]\Theta \equiv [y]\Theta \ (\Phi/\Theta) \quad \text{iff} \quad x \equiv y \ (\Phi).$$

Then Φ/Θ is a congruence of L/Θ. Conversely, every congruence Ψ of L/Θ can be (uniquely) represented in the form $\Psi = \Phi/\Theta$ for some congruence $\Phi \geq \Theta$. In particular, the congruence lattice of L/Θ is isomorphic with the dual ideal $[\Theta)$ of the congruence lattice of L.

Proof. We have to prove that Φ/Θ is well-defined, it is an equivalence relation, and it has the Substitution Property. To represent Ψ, define Φ by

$$x \equiv y \ (\Phi) \quad \text{iff} \quad [x]\Theta \equiv [y]\Theta \ (\Psi).$$

Again, we have to verify that Φ is a congruence. $\Phi/\Theta = \Psi$ follows from the definition of Φ. The details are trivial and left to the reader. ●

Equational classes of universal algebras can be introduced by defining polynomials and identities, just as in the case of lattices. However, in the next theorem the reader can avoid the use of this terminology by substituting for "equational class" the phrase "class closed under the formation of subalgebras, homomorphic images, and direct products". (This does not make the result more general, see Theorem V.1.3.)

Theorem 12 (The Subdirect Product Representation Theorem. G. Birkhoff [1944]). *Let* $\mathbf{K}$ *be an equational class of algebras. Every algebra* A *in* $\mathbf{K}$ *can be embedded in a direct product of subdirectly irreducible algebras in* $\mathbf{K}$.

Proof. For $a, b \in A$, $a \neq b$, let $\mathfrak{X}$ denote the set of all congruences Θ of A satisfying $a \not\equiv b\ (\Theta)$. Since $\omega \in \mathfrak{X}$, $\mathfrak{X}$ is not empty. Let C be a chain in $\mathfrak{X}$. Then $\Theta = \bigcup(\Phi \mid \Phi \in C)$ is a congruence, $a \not\equiv b\ (\Theta)$, and thus every chain in $\mathfrak{X}$ has an upper bound. By Zorn's Lemma, $\mathfrak{X}$ has a maximal element $\Psi(a, b)$. We claim that $A/\Psi(a, b)$ is subdirectly irreducible; in fact, $u = [a]\Psi(a, b)$ and $v = [b]\Psi(a, b)$ satisfy the condition of Definition 7. Indeed, if Θ is any congruence of $A/\Psi(a, b)$, $\Theta \neq \omega$, then by Lemma 11, $\Theta = \Phi/\Psi(a, b)$. Since $\Theta \neq \omega$, we obtain $\Phi > \Psi(a, b)$, and so $a \equiv b\ (\Phi)$. Thus $u \equiv v\ (\Theta)$, as claimed.

Let $B = \Pi(A/\Psi(a, b) \mid a, b \in A, a \neq b)$; then B is a direct product of subdirectly irreducible algebras. We embed A into B by $\varphi\colon x \to f_x$, where f_x takes on the value $[x]\Psi(a, b)$ in the algebra $A/\Psi(a, b)$. Clearly, φ is a homomorphism. To show that φ is one-to-one, assume that $f_x = f_y$. Then $x \equiv y\ (\Psi(a, b))$ for all $a, b \in A$, $a \neq b$. Therefore, $x \equiv y\ (\bigwedge(\Psi(a, b) \mid a, b \in A, a \neq b)$, and so $x = y$. ●

We got a little bit more than claimed. If we pick $x \in A/\Psi(a, b)$, then $x = [y]\Psi(a, b)$ for some $y \in A$. Thus there is an element in the representation of A whose component in $A/\Psi(a, b)$ is x; such a representation is called *subdirect*.

Corollary 13. *In an equational class* $\mathbf{K}$, *every algebra can be represented as a subdirect product of subdirectly irreducible algebras in* $\mathbf{K}$.

Observe how strong Theorem 12 is. If combined with Example 9 it yields Theorem 1.19; when combined with Example 10 we obtain Corollary 1.21.

The reader should note that subdirect representations of an algebra A are in one-to-one correspondence with families $(\Theta_i \mid i \in I)$ of congruence relations of A satisfying

$$\bigwedge(\Theta_i \mid i \in I) = \omega.$$

A subdirect representation by subdirectly irreducible algebras corresponds to such families $(\Theta_i \mid i \in I)$ which satisfy in addition the property that all Θ_i are completely meet-irreducible (an element Θ is *completely meet-irreducible* iff $\Theta = \bigwedge(\Phi_j \mid j \in J)$ implies that $\Theta = \Phi_j$ for some $j \in J$). Thus Lemma 11 and Theorem 12 combine to yield

Corollary 14. *Every congruence relation of an algebra is a meet of completely meet-irreducible elements.*

Let $\mathfrak{S}_1$ denote the three-element chain $\{0, e, 1\}$ $(0 < e < 1)$ as a distributive lattice with pseudocomplementation.

Theorem 15. *Up to isomorphism, $\mathfrak{B}_2$ and $\mathfrak{S}_1$ are the only subdirectly irreducible Stone algebras.*

Proof. $\mathfrak{B}_2$ and $\mathfrak{S}_1$ are obviously subdirectly irreducible (the congruence lattice of $\mathfrak{S}_1$ is a three-element chain).

Now let L be a subdirectly irreducible Stone algebra. By Lemma 3, $S(L)$ is a subalgebra of L. By definition, $|L| > 1$. If $|S(L)| > 2$, then $S(L)$ is directly decomposable and therefore so is L. Thus $|S(L)| = 2$, that is, $S(L) = \{0, 1\}$. If $|D(L)| > 2$, then there exist congruences Θ and Φ on $D(L)$ such that $\Theta \wedge \Phi = \omega$ on $D(L)$ (by Example 9). Extend Θ and Φ to L by defining $\{0\}$ as the only additional congruence class. We conclude that L is subdirectly reducible.

Thus $S(L) = \{0, 1\}$ and so $L = D(L) \cup \{0\}$ and $|D(L)| \leq 2$, yielding that $L \cong \mathfrak{B}_2$ or $L \cong \mathfrak{S}_1$. ●

Corollary 16 (G. Grätzer [1969]). *Every Stone algebra can be embedded in a direct product of two- and three-element chains.*

Proof. Combine Corollary 13 and Theorem 15. ●

Every distributive lattice can be embedded in some $P(X)$. O. Frink [1962] asked whether every Stone algebra can be embedded in some $I(P(X))$?

Theorem 17 (G. Grätzer [1963]). *A distributive lattice with pseudocomplementation L is a Stone algebra iff it can be embedded into some $I(P(X))$.*

Proof. $I(P(X))$, and therefore any of its subalgebras, is a Stone algebra by Corollary 2.

It is obvious that the class of Stone algebras that can be embedded into some $I(P(X))$ is closed under the formation of direct products and subalgebras. Hence by Corollary 16 it is sufficient to prove that $\mathfrak{B}_2$ and $\mathfrak{S}_1$ can be so embedded. For $\mathfrak{B}_2$ this is obvious. To embed $\mathfrak{S}_1$, take an infinite set X and embed $\mathfrak{S}_1$ into $I(P(X))$ as follows:

$$0 \to \{\varnothing\},$$
$$e \to \{A \mid A \subseteq X \text{ and } A \text{ is finite}\},$$
$$1 \to P(X).$$

It is obvious that this is an embedding. ●

Exercises

1. Show that every bounded chain is a pseudocomplemented distributive lattice.
2. Let L be a lattice with 1. Adjoin a new zero 0 to L: $L_1 = L \cup \{0\}$, $0 < x$ for all $x \in L$. Show that L_1 is a pseudocomplemented lattice.
3. Call a lattice with 0 *dense* iff 0 is meet-irreducible. Show that every bounded dense lattice K is pseudocomplemented and that every such lattice can be constructed by the method of Exercise 2 with $L = D(K)$.
4. Find an example of a complete distributive lattice L that is not pseudocomplemented.
5. Prove that if L is a complete Stone lattice, then so is $I(L)$. (Hint: $I^* = (a]$, where $a = \bigwedge(x^* \mid x \in I)$.)
6. Show that a distributive pseudocomplemented lattice is a Stone lattice iff $(a \vee b)^{**} = a^{**} \vee b^{**}$ for $a, b \in L$.
7. Find a small set of identities characterizing Stone algebras.
8. Let L be a Stone algebra. Show that $S(L)$ is a retract of L.
9. Let L be a Stone algebra, $a, b \in S(L)$, and $a \equiv b$. Prove that $x \to (x \vee a^*) \wedge b$ embeds F_a into F_b.
10. Let B be a Boolean algebra. Define $B^{[2]} \subseteq B^2$ by $\langle a, b\rangle \in B^{[2]}$ iff $a \leq b$. Verify that $B^{[2]}$ is a sublattice of B^2 but not a subalgebra of B^2. Show that $B^{[2]}$ is a Stone lattice.
11. Let L be a pseudocomplemented distributive lattice. Show that, for $a, b \in L$,

$$(a \vee b)^* = a^* \wedge b^*$$

and

$$(a \wedge b)^{**} = a^{**} \wedge b^{**}.$$

12. Prove that a prime ideal P of a Stone algebra L is minimal iff P as an ideal of L is generated by $P \cap S(L)$.
13. Show that a distributive lattice with pseudocomplementation is a Stone algebra iff every prime ideal contains exactly one minimal prime ideal (G. Grätzer and E. T. Schmidt [1957a]).

*14. Prove that a poset Q is isomorphic to the poset of all prime ideals of a Stone algebra iff (i) every element of Q contains exactly one minimal element, (ii) for every minimal element m of Q, the poset $\{x \mid x > m, x \in Q\}$ is isomorphic to the poset of all prime ideals of some distributive lattice with 1 (C. C. Chen and G. Grätzer [1969b]).

15. Give a detailed proof of the Second Isomorphism Theorem.
16. Prove Corollary 16 directly.

Exercises 17–29 are from C. C. Chen and G. Grätzer [1969a]. Exercises 30 and 31 are from C. C. Chen and G. Grätzer [1969b].

Let B be a Boolean algebra, let D be a distributive lattice with 1, and let φ be a $\{0, 1\}$-homomorphism of B into $\mathcal{D}(D)$. Set $L = \{\langle x, a\rangle \mid a \in B,\ x \in a\varphi\}$ and define $\langle x, a\rangle \leq \langle y, b\rangle$ iff $a \leq b$ and $x \leq y\varrho_a$, where $[y\varrho_a) = a\varphi \wedge [y)$.

17. Verify the following formulas:
    (i) If $a \in B$ and $d \in D$, then $d\varrho_a = d$ iff $d \in a\varphi$.
    (ii) $d\varrho_a \geq d$ for $a \in B$, $d \in D$.
    (iii) $d\varrho_a \wedge d\varrho_{a'} = d$ for $a \in B$, $d \in D$ (where a' is the complement of a in B).
    (iv) $\varrho_a\varrho_b = \varrho_{a \wedge b}$ for $a, b \in C$.

18. Prove that:
    (i) $d\varrho_a \wedge d\varrho_b = d\varrho_{a \vee b}$ for $a, b \in B$, $d \in D$.
    (ii) $d\varrho_{a \wedge b} = d\varrho_a \vee d\varrho_b$ for $a, b \in B$, $d \in D$.
19. Show that L is a poset under the given partial ordering.
20. For $\langle x, a\rangle, \langle y, b\rangle \in L$, verify that $\langle x, a\rangle \wedge \langle y, b\rangle = \langle x\varrho_b \wedge y\varrho_a, a \wedge b\rangle$.
*21. Show that $\langle x, a\rangle \vee \langle y, b\rangle = \langle (x\varrho_{b'} \wedge y) \vee (x \wedge y\varrho_{a'}), a \vee b\rangle$.
22. For $\langle x, a\rangle, \langle y, b\rangle, \langle z, c\rangle \in L$, let

$$A = (\langle x, a\rangle \wedge \langle y, b\rangle) \vee \langle z, c\rangle$$

and

$$B = (\langle x, a\rangle \vee \langle z, c\rangle) \wedge (\langle y, b\rangle \vee \langle z, c\rangle).$$

Compute A; show that $B = \langle d, (a \vee c) \wedge (b \vee c)\rangle$, where $d = d_0 \vee d_1 \vee d_2 \vee d_3$, and $d_0 = x\varrho_{b \wedge c'} \wedge y\varrho_{a \wedge c'} \wedge z$, $d_1 = x\varrho_{b \wedge c'} \wedge y\varrho_{a \wedge c} \wedge z$, $d_2 = x\varrho_{b \vee c} \wedge y\varrho_{a \wedge c'} \wedge z$, and $d_3 = x\varrho_{b \vee c} \wedge y\varrho_{b \vee c} \wedge z\varrho_{a' \vee b'}$.
23. Show that $d_0 \geq d_1$ and $d_0 \geq d_2$; therefore, $d = d_0 \vee d_3$.
24. Show that L is distributive.
25. Show that L is a Stone lattice.
26. Identify $b \in B$ with $\langle 1, b\rangle$ and $d \in D$ with $\langle d, 1\rangle$. Verify that $S(L) = B$, $D(L) = D$, and $\varphi(L) = \varphi$. In other words, we have proved the Construction Theorem of C. C. Chen and G. Grätzer [1969a]: *Given a Boolean algebra B, a distributive lattice D with 1, and a $\{0, 1\}$-homomorphism $\varphi: B \to \mathfrak{D}(D)$, there exists a Stone algebra L whose triple is $\langle B, D, \varphi\rangle$.*
27. Describe isomorphisms and homomorphisms of Stone algebras in terms of triples.
28. Describe subalgebras of Stone algebras in terms of triples.
29. For a given Boolean algebra B with more than one element and distributive lattice D with 1, construct a Stone algebra with $S(L) \cong B$ and $D(L) \cong D$. ($S(L)$ and $D(L)$ are independent.)
30. Show that a Stone algebra L is complete if $S(L)$ and $D(L)$ are complete.
*31. Characterize the completeness of Stone algebras in terms of triples.
*32. Show that a distributive lattice with pseudocomplementation L has the *Congruence Extension Property*: for a subalgebra K of L and for a congruence relation Θ of K, there exists a congruence relation Φ of L extending Θ, that is, for $a, b \in K$, $a \equiv b\ (\Theta)$ iff $a \equiv b\ (\Phi)$ (G. Grätzer and H. Lakser [1971]).
33. Let L be a lattice or a lattice with additional operations. If $a, b \in L$ and $[b, a]$ is simple, then the $\Psi(a, b)$ (defined in the proof of Theorem 12) is unique.

Further Topics and References

A number of papers are devoted to the subject of guaranteeing the distributivity of a lattice by imposing equations on a fixed generating set. For instance, if L is generated by a, b, and c, then L is distributive iff $x \wedge (y \vee z) = (x \wedge y) \vee (x \wedge z)$ and its dual holds for any permutation x, y, z of a, b, c and $(a \wedge b) \vee (b \wedge c) \vee (c \wedge a) = (a \vee b) \wedge (b \vee c) \wedge (c \vee a)$, by a result of O. Ore [1940]. These seven conditions (there are only seven by commutativity) are independent but if L is assumed to be modular, all seven are equivalent, as it is evident from Figure I.5.7. Generalizations can be found in M. Kolibiar [1972] and S. Tamura [1971], [1971a], and for the modular case in B. Jónsson [1955], R. Musti and E. Buttafuoco [1956], and R. Balbes [1969].

For some special classes of lattices, Theorem 1.1 and 1.2 have various stronger forms that claim the existence of very large or very small pentagons and diamonds. For instance, a bounded relatively complemented nonmodular lattice always contains a pentagon as a $\{0, 1\}$-sublattice. The same is true of the diamond in certain complemented modular lattices; such results are implicit in J. von Neumann [1936/37]. If the lattice is finite modular nondistributive, then it contains a diamond in which a, b, c cover o and i covers a, b, c. If L is finite and nonmodular, then the pentagon it contains can be required to satisfy $a \succ b$.

It seems hard to generalize the uniqueness of an irredundant join-representation of an element of a finite distributive lattice. The best generalization is that of R. P. Dilworth and P. Crawley [1960] to distributive algebraic lattices in which every proper interval has an atom. See the survey article by R. P. Dilworth [1961] and S. Kinugawa and J. Hashimoto [1966]. Some results on and references to the modular and semimodular cases can be found in Chapter IV.

By Theorem 1.19, every distributive lattice is a subdirect product of copies of $\mathfrak{C}_2$. Distributive lattices that are subdirect products of copies of $\mathfrak{C}_n$ are characterized in F. W. Anderson and R. L. Blair [1961].

For a finite distributive lattice L, what is the smallest k such that L is a subdirect product of k chains? For $a \in L$, let n_a be the number of elements of L covering a. Then $k = \max\{n_a \mid a \in L\}$. This is an easy application of a well-known combinatorial result of R. P. Dilworth [1950] (first discovered by T. Gallai in 1936) according to which a poset P is a union of at most k chains iff P is of width $\leqq k$. See also H. Tverberg [1967].

The size of maximal sublattices of a finite distributive lattice is investigated in I. Rival [1973] and [1974]; maximal $\{0, 1\}$-sublattices of finite Boolean lattices are considered in H. Sharp [1968] and D. Steven [1968]. $\{0, 1\}$-sublattices of finite Boolean lattices and finite topologies are closely related. See also R. P. Stanley [1973].

Theorems 1.9 and 1.19 can be generalized by using ideals other than prime ideals In this connection see Theorem IV.5.17 and G. Birkhoff and O. Frink [1948].

The problem of determining $|F_D(n)|$ goes back to R. Dedekind [1900]. For some older and some recent contributions (including the tabulation of $|F_D(n)|$ for $n \leq 7$) see R. Church [1940], E. N. Gilbert [1954], G. Hansel [1967], D. Kleitman [1969], R. Church [1965] (this gives the correct value for $|F_D(7)|$), and F. Lunnon [1971].

Infinitary Boolean polynomials are considered in H. Gaifman [1964] and A. W. Hales [1964]; they prove that free complete Boolean algebras do not exist.

Free distributive lattices have many interesting properties. All chains are finite or countable (the proof of this is similar to that of Theorem VI. 2.10). If a and H are such that $x \wedge y = a$ for all $x, y \in H$, $x \neq y$, call H *a-disjoint*. In a free distributive lattice all a-disjoint sets are finite, see R. Balbes [1967].

All chains of a free Boolean algebra are also finite or countable, see I. Reznikoff [1963] and A. Horn [1968].

A more general form of Theorem 2.5 can be found in R. Sikorski [1964]; for the universal algebraic background see Theorem 12.2 in G. Grätzer [1968].

Some properties of $\Theta[I]$ can be generalized to certain ideals of a general lattice, see Chapter III.

Let L be a lattice and let f be an n-ary function on L, that is, $f: L^n \to L$. We say that

f has the *Congruence Substitution Property* iff, for every congruence relation Θ of L and $a_i, b_i \in L, 1 \leq i \leq n, a_i \equiv b_i(\Theta), 1 \leq i \leq n$, imply that $f(a_1, \ldots, a_n) \equiv f(b_1, \ldots, b_n)$ (Θ). On a Boolean algebra B, a function has the Congruence Substitution Property iff it is a Boolean algebraic function, that is, a Boolean polynomial in which elements of B are substituted for some variables. Functions satisfying the Congruence Substitution Property on a bounded distributive lattice were described in G. Grätzer [1964].

The method given in Theorem 3.9 is not the only one used to introduce ring operations in a generalized Boolean lattice. G. Grätzer and E. T. Schmidt [1958d] prove that ring operations $+, \cdot$ can be introduced on a distributive lattice L such that $+$ and $\cdot$ satisfy the Congruence Substitution Property iff L is relatively complemented. Furthermore, $+$ and $\cdot$ are uniquely determined by the zero of the ring, which can be an arbitrary element of L.

Whether every distributive algebraic lattice is isomorphic to the congruence lattice of some lattice is one of the longest-standing problems of lattice theory. The method used in Section 3 for the finite case can be easily extended to infinite algebraic lattices in which every element is a finite join of join-irreducible elements. A further extension of this result is in E. T. Schmidt [1962] and [1968]. (In reading the two papers, the reader should disregard the Theorem and Lemmas 9 and 10 of the first paper.) See also E. T. Schmidt [1969], [1974], [1975] and J. Berman [1972].

Congruence lattices of distributive lattices are not considered in this text because their characterization problem is trivial by Lemma 4.5: A lattice is isomorphic to the congruence lattice of a distributive lattice iff it is isomorphic to $I(B)$, where B is a generalized Boolean lattice; $I(B)$, by Theorem 3.13, is characterized as a distributive algebraic lattice in which the compact elements form a relatively complemented sublattice.

Algebraic lattices originated in A. Komatu [1943], G. Birkhoff and O. Frink [1948], L. Nachbin [1949], and J. R. Büchi [1952]. The original definition in G. Birkhoff and O. Frink [1948] is as follows: (i) L is complete; (ii) in L every element is the join of join-inaccessible elements; (iii) L is join-continuous. An element a of L is *join-accessible* iff there is a nonvoid subset H of L that is *directed* (for $x, y \in H$ there exists an upper bound $z \in H$), $\bigvee H = a$, and $a \notin H$; otherwise, a is *join-inaccessible*. L is *join-continuous* iff for any $a \in L$ and directed $H \subseteq L$, we have $a \wedge \bigvee H = \bigvee (a \wedge h \mid h \in H)$. The significance of algebraic lattices for universal algebras is discussed in the Concluding Remarks.

Interestingly, it is sufficient to formulate conditions (i) and (iii) of the previous paragraph for chains only. In other words, a lattice L is complete iff $\bigwedge C$ and $\bigvee C$ exist for any chain C of L; and a (complete) lattice L is join-continuous iff $a \wedge \bigvee C = \bigvee (a \wedge c \mid c \in C)$ for any chain C of L. These statements are immediate consequences of the following result of T. Iwamura [1944]. Let H be an infinite directed set. Then H has a decomposition $H = \bigcup (H_\gamma \mid \gamma < \alpha)$, where each H_γ is directed; for $\gamma < \delta < \alpha$ we have $H_\gamma \subseteq H_\delta$, and for $\gamma < \alpha$ we have $|H_\gamma| < H$. Extensions of lattices preserving (iii) are considered in D. A. Kappos and F. Papangelou [1966].

The results of Section 4 were presented in FC without the assumption that L has a 0. The necessity of this assumption is pointed out in R. D. Byrd and R. A. Mena [1975] and R. D. Byrd, R. A. Mena, and L. A. Troy [1975]. These papers give a detailed analysis of what happens if this assumption is dropped.

W. Hanf [1976] proves that there is no algorithmic way to find a generating chain in all countable Boolean algebras. There are, however, always generating chains of a rather special order type; see R. S. Pierce [1973].

The problem of completions of a lattice has been extensively studied. The standard method is the MacNeille completion (see Exercises 4.9–4.14, which are from H. M. MacNeille [1937]). This method is described in detail in G. Bruns [1962]; see also Y. Sampei [1953]. One of the important shortcomings of this method is that it does not preserve identities, not even distributivity; see M. Cotlar [1944] and N. Funayama [1944]. See also P. Crawley [1962], R. P. Dilworth and J. E. McLaughlin [1952], and W. H. Cornish [1974]. What is preserved for complemented modular lattices is examined in J. E. McLaughlin [1961].

If we define a *completion* $\hat{L}$ of a lattice L as any complete lattice $\hat{L}$ containing L as a sublattice such that all infinite meets and joins that exist in L are preserved in $\hat{L}$, we can ask whether there is *any* distributive completion of a distributive lattice. The answer, in general, is in the negative; see P. Crawley [1962].

For Boolean algebras, set representations preserving all joins and meets, or joins and meets of certain types, play an important role, especially in relationship with infinite distributive identities. The larger part of the book by R. Sikorski [1964] is devoted to this problem. For distributive lattices these questions become even more complicated—see, for instance, G. Bruns [1959], [1961], and [1962a]; C. C. Chang and A. Horn [1962]; R. P. Dilworth and J. E. McLaughlin [1952]; A. Horn [1962]; J. Jakubik [1972]; and G. N. Raney [1952], [1953]; see also R. Balbes and Ph. Dwinger [1974].

Cardinalities of maximal chains in Boolean algebras are considered by J. Jakubik [1957], [1958] and G. W. Day [1970]. See also Exercise 4.28.

R. S. Pierce [1972] gave a deep analysis of the structure of countable Boolean algebras. A generalization of Boolean algebras generated by chains can be found in R. W. Quackenbush and H. C. Reichel [1975].

The method used in establishing Lemma 4.22 was employed by B. Jónsson [1956] to construct "universal" lattices, that is, lattices containing a large class of lattices as sublattices (see also B. Jónsson [1960] and M. D. Morley and R. L. Vaught [1962]).

For a distributive lattice L, $\mathscr{P}(L)$ is a poset and, by Section 5, a topological space. These two approaches are married in H. A. Priestley [1972] where $\mathscr{P}(L)$ is considered, and characterized, as an ordered topological space. This approach proved very convenient in applications; see H. A. Priestley [1972], [1974], and [1975]; see also M. E. Adams [1974], [1975], [1976], B. A. Davey [1973], [1974], T. P. Speed [1972a]. T. P. Speed [1972] characterizes $\mathscr{P}(L)$ for bounded L by the property that $\mathscr{P}(L) \cup \{0, 1\}$ is a profinite poset. See also M. E. Adams [1975], R. Balbes [1971], B. A. Davey [1973], T. P. Speed [1972a] and [1974].

A solution to the word problem of free $\{0, 1\}$-distributive products is given in G. Grätzer and H. Lakser [1969]. In the same paper it is proved that if $\mathfrak{m}$ is a regular cardinal and L_i, $i \in I$, are bounded distributive lattices satisfying $|C| < \mathfrak{m}$ for any chain C in any L_i, then the same holds in the free $\{0, 1\}$-distributive product. In particular, if all L_i satisfy the Countable Chain Condition, so does the free $\{0, 1\}$-distributive product; see also Section 12 of FC. M. E. Adams and D. Kelly [a] has just extended this result to free products of lattices. See also B. Jónsson [1971].

The analogous problem for α-disjoint sets is considered in H. Lakser [1973a] and M. E. Adams and D. Kelly [b]. An interesting description of free {0, 1}-distributive products can be found in R. W. Quackenbush [1972a].

A weaker form of Corollary 6.16 can be found in T. P. Speed [1969].

Subdirect products of lattices have an interesting property due to G. M. Bergman (see K. A. Baker and A. F. Pixley [a]): a subdirect product L of $L_1, \ldots, L_n$ is determined up to isomorphism by its projections onto $L_i \times L_j$, $1 \leq i < j \leq n$. Subdirect products of ideal lattices are investigated in R. Freese [1975].

The interdependence of the various "finiteness conditions" for distributive lattices is investigated in S. P. Avann [1964] and S. Rudeanu [1964].

Boolean algebras with only the trivial automorphism are constructed in M. Katětov [1951], B. Jónsson [1951], and L. Rieger [1951]; Katětov's example has $2^{\aleph_0}$ elements, the others are very large.

Some categorical concepts have been studied for distributive lattices. For a class **K** of algebras, $P \in \mathbf{K}$ is called *projective in* **K** iff, for any $A, B \in \mathbf{K}$, homomorphisms $\gamma: A \to B$ and $\beta: P \to B$, if γ is onto, then there is a homomorphism $\alpha: P \to A$ satisfying $\alpha\gamma = \beta$. Free algebras are projective.

R. Balbes [1967] proves that a finite distributive lattice L is projective in **D** iff the join of any two meet-irreducible elements is meet-irreducible. See also R. Balbes and A. Horn [1970b], G. Grätzer and B. Wolk [1970]. R. Engelking [1965] has shown that a subalgebra of a free Boolean algebra need not be projective; but every countable Boolean algebra is projective, see P. R. Halmos [1961].

I is *injective in* **K** iff, for any $A, B \in \mathbf{K}$, embedding $\gamma: B \to A$ and homomorphism $\beta: B \to I$, there is a homomorphism $\alpha: A \to I$ satisfying $\gamma\alpha = \beta$. Injective Boolean algebras are the complete ones, see P. R. Halmos [1961]; and injective distributive lattices are the complete Boolean lattices. For more details on categorical concepts for distributive lattices, see Section 13 of FC.

R. Dedekind found the distributive identity by investigating ideals of number fields. Rings with a distributive lattice of ideals have been investigated by E. Noether [1927], L. Fuchs [1949] (who names such rings *arithmetical rings*), I. S. Cohen [1950], and Chr. U. Jensen [1963]. Equational classes of rings with distributive ideal lattices were considered in G. Michler and R. Wille [1970] and in H. Werner and R. Wille [1970]. E. A. Behrens [1960] and [1961] considered rings in which one-sided ideals form a distributive lattice. Rings with a distributive lattice of subrings were classified in P. A. Freĭdman [1967]. In this context G. M. Bergman's work on the distributive-divisor-lattice of free algebras should be mentioned, see Chapter 4 of P. M. Cohn [1971].

For groups the corresponding problem is much simpler: The subgroup lattice of a group G is distributive iff G is locally cyclic (see O. Ore [1937] and [1938]). H. L. Silcock [1977] has proved that every finite distributive lattice is isomorphic to the lattice of normal subgroups of a suitable group.

The distributivity of congruence lattices of lattices has a number of important consequences. B. Jónsson [1967] discovered that many of these results hold for arbitrary universal algebras with distributive congruence lattices. His results have found applications that go far beyond lattice theory—they have already been applied to lattice-ordered algebras, closure algebras, nonassociative lattices, cylindric algebras, monadic algebras, lattices with pseudocomplementation, primal algebras, and multi-valued logics.

The foregoing examples show the central role played by distributive lattices in applications of the lattice concept.

Except for V. Glivenko's early work [1929], the study of pseudocomplemented distributive lattices started only in 1956 with a solution of Problem 70 of G. Birkhoff [1948] giving a characterization of Stone lattices by minimal prime ideals (G. Grätzer and E. T. Schmidt [1957a]; for a simplified proof see J. Varlet [1966]), see Theorem 6.6. The idea of a triple was conceived by the author as a tool to prove Frink's conjecture (see O. Frink [1962] and Theorem 6.17). This attempt failed and as a result triples were not utilized until C. C. Chen and G. Grätzer [1969a] and [1969b]. Frink's conjecture was solved using the Compactness Theorem in G. Grätzer [1963] (see Theorem 6.17) and the proof was simplified in G. Bruns [1965] and G. Grätzer [1969]. An interesting generalization can be found in H. Lakser [1971]. Projective and injective Stone algebras are investigated in R. Balbes and G. Grätzer [1971]; see also R. Beazer [1975]. See also Sections 16 and 17 of FC.

Let $\mathbf{B}_n$ denote the equational class of distributive lattices with pseudocomplementation satisfying the identity

$$(\mathrm{L}_n) \qquad (x_1 \wedge \cdots \wedge x_n)^* \vee (x_1^* \wedge \cdots \wedge x_n)^* \vee \cdots \vee (x_1 \wedge \cdots \wedge x_n^*)^* = 1$$

for $n \geq 1$. Then $\mathbf{B}_1$ is the class of Stone algebras. K. B. Lee [1970] has proved that $\mathbf{B}_n$, $-1 \leq n \leq \omega$, is a complete list of equational classes of distributive lattices with pseudocomplementation, where $\mathbf{B}_{-1}$ is the trivial class, $\mathbf{B}_0$ is the class of Boolean algebras, and $\mathbf{B}_\omega$ is the class of all distributive lattices with pseudocomplementation. Moreover,

$$\mathbf{B}_{-1} \subset \mathbf{B}_0 \subset \mathbf{B}_1 \subset \cdots \subset \mathbf{B}_n \subset \cdots \subset \mathbf{B}_\omega.$$

In H. Lakser [1971] and G. Grätzer and H. Lakser [1971] and [1972] most of the structure theorems known for Stone algebras have been generalized to the classes $\mathbf{B}_n$. In these papers the amalgamation class of $\mathbf{B}_n$ (in the sense of Section V.4) is also considered. See also Chapter 3 of FC where many of these results are presented in detail.

In more than 30 papers T. Katriňák investigates various generalizations and specializations of distributive lattices with pseudocomplementation and establishes triple constructions for many classes. These classes include relative Stone algebras, generalized Stone algebras, pseudocomplemented posets, generalized Stone join-semilattices, distributive pseudocomplemented meet-semilattices, modular pseudocomplemented lattices, modular pseudocomplemented lattices satisfying $x^* \vee x^{**} = 1$, double Stone algebras, and so on. Free algebras, equational classes, injective hulls and many other questions are considered. The reader is referred to the papers of T. Katriňák listed in the Bibliography and to R. Balbes and A. Horn [1970], R. Balbes and Ph. Dwinger [1974], G. Bruns and G. Kalmbach [1971] and [1972], A. Day [1970a], H. Lakser [1970] and [1973], J. Varlet [1963], [1966], and [1968], P. V. R. Murty and V. V. R. Rao [1974], U. M. Swamy and V. V. R. Rao [1975].

Relatively pseudocomplemented distributive lattices (often called *Heyting algebras*) arise from nonclassical logic and were first investigated by T. Skolem about 1920.

For a detailed development, see H. B. Curry [1963], which contains all the important rules of computation with $a * b$. Injectives and projectives were investigated in the work of R. Balbes and A. Horn [1970a] and A. Day [1970a]; see also A. Horn [1969] and [1969a], W. C. Nemitz [1965].

In connection with nonclassical logic, many algebras emerged that are distributive lattices endowed with some additional structure. The best known of these, the Post algebras, happen to be Stone lattices. In fact, a *Post algebra* is a free $\{0, 1\}$-distributive product of a finite chain and a Boolean algebra, in which the elements of the chain are regarded as nullary operations; see E. L. Post [1921], P. C. Rosenbloom [1942], G. Epstein [1960], T. Traczyk [1963] and [1967], C. C. Chang and A. Horn [1961], G. Rousseau [1970], R. Balbes and Ph. Dwinger [1971] and [1971a], and M. Mandelker [1970].

For some other types of distributive lattices with additional operations, see N. D. Belnap and J. H. Spencer [1966], J. M. Dunn and N. D. Belnap [1968], H. Rasiowa and R. Sikorski [1963], and J. Varlet [1968].

Two books have appeared recently dealing with distributive lattices with additional operations. R. Balbes and Ph. Dwinger [1974] deals with them from a lattice theoretical viewpoint while H. Rasiowa [1974] investigates their usefulness to logic.

Some of the results for pseudocomplemented distributive lattices referred to above have been extended to ideals of general distributive lattices; see R. Beazer [1975a], W. H. Cornish [1972], [1974a], B. A. Davey [1974a], M. Mandelker [1970].

Some results do not require the distributivity of the whole lattice; for instance, to prove that $I(L)$ is pseudocomplemented it is sufficient that L be *0-distributive*, that is, $a \wedge b = a \wedge c = 0$ implies that $a \wedge (b \vee c) = 0$, for $a, b, c \in L$. This and some related concepts have been in the literature for 20–30 years. Here are some recent references: W. H. Cornish [1974b], M. F. Janowitz [1975], J. Varlet [1968a].

Problems

II.1 (FC 28). Find short one-identity axioms characterizing Boolean algebras.

II.2 (FC 30). Characterize the automorphism groups of Boolean algebras. (B. Jónsson has proved (unpublished result) that the finite automorphism groups of Boolean algebras are exactly the finite symmetric groups. See also R. N. McKenzie and J. D. Monk [1973].)

II.3 (FC 31). Let B and C be infinite Boolean algebras and let $\mathfrak{m}$ be a cardinal number satisfying $\mathfrak{m} \geq |B|$ and $|C|$. Does there exist a Boolean algebra D containing B as a subalgebra and such that the automorphism groups of C and D are isomorphic? (This is a variant of FC 31. Call a Boolean algebra *rigid* iff it has a one-element automorphism group. F. W. Lozier [1969] constructs a rigid Boolean algebra of cardinality $2^{\mathfrak{m}}$ for every infinite cardinal $\mathfrak{m}$. V. Kannan and M. Rajagopalan [1971] announce that there exist rigid Boolean algebras of any infinite cardinality. In September, 1975, J. D. Monk communicated to me the following partial solution using some unpublished results of R. N.

McKenzie and S. Shelah, based on R. N. McKenzie and J. D. Monk [1973]: (GCH) there is a cardinal number $\mathfrak{p} = \mathfrak{p}(B, C)$, such that there is a Boolean algebra D as required for all $\mathfrak{m} \geq \mathfrak{p}$.)

II.4 (FC 33). Investigate further the poset $\mathcal{P}(L)$ of prime ideals of a distributive lattice.

II.5 (FC 34). Characterize $\mathcal{P}(L)$ under the additional assumption that L has a unit and/or a zero. (If L has a unit, then every chain has a supremum. T. Tan pointed out that the converse does not hold.)

II.6 (FC 35). For which classes $\mathbf{K}$ of finite lattices can we claim that if $L \in \mathbf{K}$ and L is nonmodular, then L contains a "minimal" pentagon?

II.7 (FC 36). Characterize the congruence lattices of lattices. (Conjecture: distributive algebraic lattices).

II.8 (FC 37). Characterize the congruence lattices of sectionally complemented lattices. (Conjecture: distributive algebraic lattices).

II.9 (FC 38). Determine $f_{\mathbf{D}}(k, n)$.

II.10 (FC 40). Investigate the cardinalities of maximal chains in complete Boolean algebras.

II.11 (FC 41). What can be proved about the cardinalities of maximal chains in a complete and atomic Boolean algebra without the Generalized Continuum Hypothesis? For instance, do they form a convex set?

II.12 (FC 42). In a free $\{0, 1\}$-distributive product there are two normal forms for an element: normal $\bigvee$-representation and normal $\bigwedge$-representation. (See G. Grätzer and H. Lakser [1969] and also Section 12 of FC.) Compare the number and form of the terms in a normal $\bigvee$-representation and normal $\bigwedge$-representation of an element in a free $\{0, 1\}$-distributive product.

II.13 (FC 43). For a bounded distributive lattice L, let $F(L)$ and $F_n(L)$ denote the lattice of all functions and all n-ary functions on L, respectively, with the Congruence Substitution Property. To what extent do $F(L)$ and $F_n(L)$ determine the structure of L?

II.14 (FC 44). Characterize $F(L)$ and $F_n(L)$.

II.15 (FC 45). Describe the functions with the Congruence Substitution Property on unbounded distributive lattices.

II.16 (FC 46). Study problems II.13 and II.14 in the unbounded case.

II.17. For a natural number n, let $g(n)$ denote the smallest integer such that for every distributive lattice D of length n there is a lattice L of length at most $g(n)$ satisfying $C(L) \cong D$. Determine $g(n)$. ($g(n) \leq 2n - 1$ by Exercise 3.35. See also J. Berman [1972] and E. T. Schmidt [1975].)

II.18.[1] Let K be a lattice and let G be a group. Does there exist a lattice L satisfying $C(L) \cong C(K)$ and $\mathrm{Aut}(L) \cong G$? ($\mathrm{Aut}(L)$ is the automorphism group of L.)

II.19.[1] In II.18, if K is finite, can L be chosen to be finite?

II.20. Is there a fixed type of algebras such that every algebraic lattice is

---

[1] These two problems have been solved by A. Urquhart, A topological representation theory for lattices. Algebra Universalis 8 (1978).

isomorphic to the congruence lattice of an algebra of that type? Can this type be $\langle 2\rangle$? (See Concluding Remarks.)

II.21. Is the statement "There exist two bounded distributive lattices in which every a-disjoint set is countable but there exists an uncountable a-disjoint set in their free $\{0, 1\}$-distributive product" equivalent to the Souslin Hypothesis? (See M. E. Adams and D. Kelly [b].)

II.22.[1] Prove that for every infinite distributive lattice A there exist distributive lattices B and C such that the free distributive products of A, B and A, C are elementarily equivalent but B and C are not elementarily equivalent. (Two lattices are *elementarily equivalent* iff a first-order sentence holding for one, also holds for the other. For A finite such B and C do not exist, see B. Jónsson and Ph. Olin [a].)

II.23. Prove that a distributive lattice L is equationally compact iff L is complete and satisfies (JID) and (MID).

II.24. Let $\mathbf{K}$ be an equational class of lattices such that the word problem of $F_{\mathbf{K}}(\aleph_0)$ can be solved. Give a solution to the word problem of (bounded) free $\mathbf{K}$-products.

II.25 (FC 53). Describe the projective Stone algebras.

II.26 (FC 58). Characterize free $\mathbf{B}_n$-products ($n > 1$).

II.27 (FC 59). Describe $\mathbf{Amal}(\mathbf{B}_n)$ for $n > 2$.

II.28 (FC 60). Determine the projectives in $\mathbf{B}_n$, $n > 1$.

II.29 (FC 61). Let $n \geq 0$. Determine all *injective structures* in the sense of J.-M. Maranda [1964] in the category $\mathbf{B}_n$.

II.30 (FC 62). For every identity I for distributive lattices with pseudocomplementation, there exists a first-order sentence $\Phi(I)$ such that I holds for L iff $\Phi(I)$ holds for $\mathcal{P}(L)$. (In K. B. Lee [1970], the sentence for (L_n) is given as follows: "Every element contains at most n minimal elements." For $n = 0$ use Theorem II.1.22.) Is there a natural class of first-order sentences properly containing all identities for which the same statement can be made?

II.31 (FC 64). Investigate the lattice of implicational classes of distributive lattices with pseudocomplementation.

II.32 (FC 66). For a distributive lattice L, define the number $n(L)$ as follows: Let $n(L)$ be the smallest integer n such that $I(L) \in \mathbf{B}_n$ if L has a zero and $I_0(L) \in \mathbf{B}_n$ if L does not have a zero. Classify and investigate distributive lattices according to the value of $n(L)$. (The case $n = 1$ was considered in T. Katriňák [1966] and [1968].)

II.33 (FC 67). Let $\mathbf{K}$ be an equational class of algebras satisfying $\mathbf{HS}(\mathbf{K}_1) = \mathbf{ISH}(\mathbf{K}_1)$ for all $\mathbf{K}_1 \subseteq \mathbf{K}$. Does $\mathbf{K}$ satisfy the Congruence Extension Property? Does the assumption that the congruence lattices of algebras in $\mathbf{K}$ are distributive make any difference?

II.34. Find maximal classes $\mathbf{K} \subset \mathbf{D}$ such that for $L_1, L_2 \in \mathbf{K}$, $\mathcal{P}(L_1) \cong \mathcal{P}(L_2)$ implies that $L_1 \cong L_2$. (See M. E. Adams [1975].)

---

[1] This was solved by Ph. Olin, Elementary properties of free products. Abstract. Notices Amer. Math. Soc. 23 (1976), 735–E1.

CHAPTER III

CONGRUENCES AND IDEALS

1. Weak Projectivity and Congruences

Let a, b, c, and d be elements of a lattice L; if, for any congruence relation Θ of L, $a \equiv b\ (\Theta)$ implies that $c \equiv d\ (\Theta)$, then we can say that "$a \equiv b$ forces $c \equiv d$". It is necessary to understand "forcing" in order to study the structure of congruence relations of lattices; this will be accomplished in the present section.

In Section I.3 we proved that $a \equiv b\ (\Theta)$ iff $a \wedge b \equiv a \vee b\ (\Theta)$ and so it is enough to deal with pairs of comparable elements. To simplify our notation, let a/b denote an ordered pair of elements a, b of a lattice L satisfying $b \leq a$; a/b is called a *quotient* of L. (This notation obviously imitates quotient groups: G/H.) c/d is a *subquotient* of a/b iff $b \leq d \leq c \leq a$; we call a/b a *proper quotient* iff $b < a$. If $b \prec a$ then a/b is called a *prime quotient*. We write $a/b = c/d$ iff $a = c$ and $b = d$.

Now take a look at Figures 1 and 2 permitting the degenerate cases: $a = c$ and $b = d$. In both cases $a \equiv b\ (\Theta)$ iff $c \equiv d\ (\Theta)$ because $x \wedge y \equiv x\ (\Theta)$ iff $y \equiv x \vee y\ (\Theta)$. In

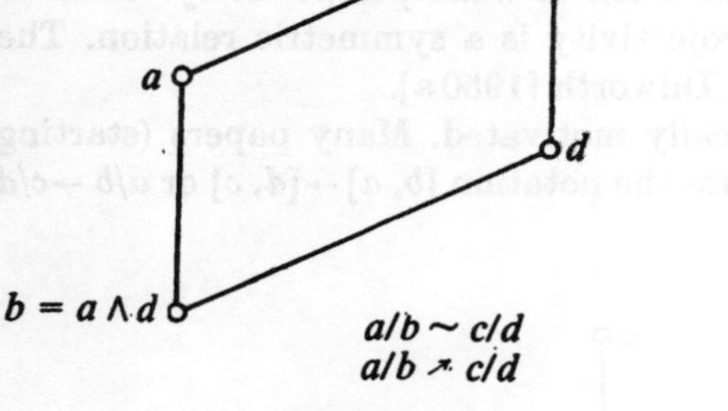

Figure 1

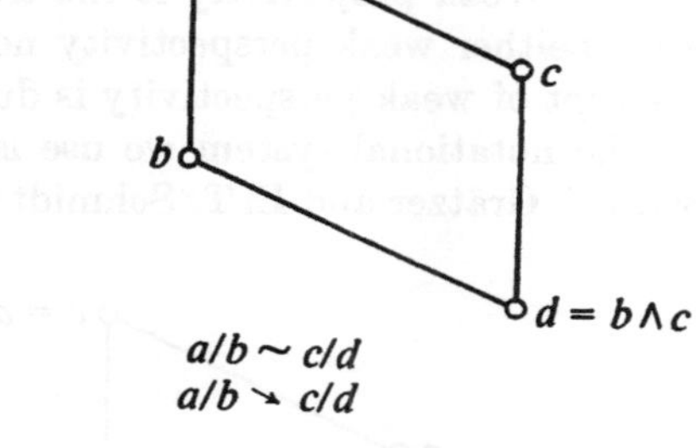

Figure 2

either case we shall write $a/b \sim c/d$, and say that *a/b is perspective to c/d*. If we want to show whether the perspectivity is "up" or "down", we shall write $a/b \nearrow c/d$ in the first case and $a/b \searrow c/d$ in the second case.

If for some natural number n there exist $a/b = e_0/f_0, e_1/f_1, \ldots, e_n/f_n = c/d$ such that $e_i/f_i \sim e_{i+1}/f_{i+1}$, $i = 0, \ldots, n-1$, then we shall say that *a/b is projective to c/d*, and write $a/b \approx c/d$. (Note that $a/b \approx a/b$ with $n = 1$ and $a/b \sim c/d$ implies that $a/b \approx c/d$.) Projectivity is the transitive extension of perspectivity. Observe that $a/b \approx c/d$ and $a = b$ imply that $c = d$.

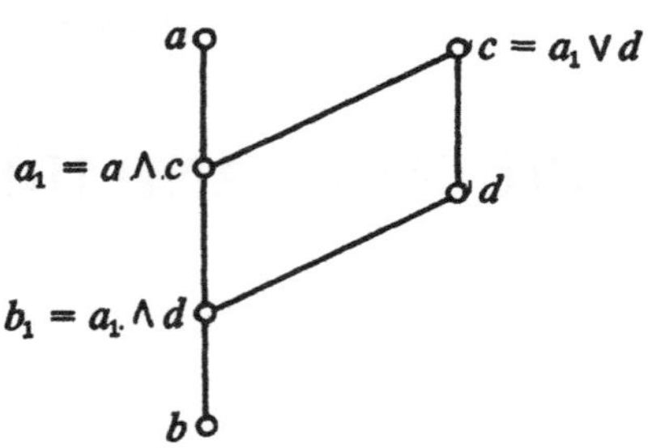

Figure 3

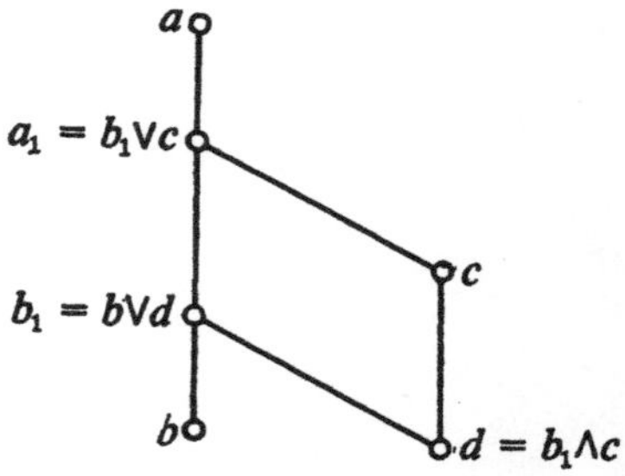

Figure 4

The concept of projectivity is sufficient for the study of "forcing" in many large classes of lattices (for instance, in the class of modular lattices, see Section IV.1). In general, however, we have to introduce somewhat more cumbersome notions: weak perspectivity and weak projectivity.

Consider Figures 3 and 4 permitting the degenerate cases $a_1 = c$ and $b_1 = d$. If $a \equiv b\ (\Theta)$, then $a_1 \equiv b_1\ (\Theta)$ by Lemma I.3.7; since $a_1/b_1 \sim c/d$, $a \equiv b\ (\Theta)$ implies that $c \equiv d\ (\Theta)$. Let us say that c/d is weakly projective into a/b iff we can get from c/d into a/b in finitely many steps as described in Figures 3 and 4. Figures 3 and 4 then should describe the concept c/d weakly perspective into a/b. It turns out however that a special case of Figures 3 and 4 suffices to describe the same.

We write $c/d \searrow_w a/b$ iff $b \leq d$ and $c = a \vee d$; similarly, $c/d \nearrow_w a/b$ iff $c \leq a$ and $d = b \wedge c$ (see Figures 5 and 6). If $c/d \nearrow_w a/b$ or $c/d \searrow_w a/b$, then c/d is *weakly perspective into* a/b, in symbols, $c/d \sim_w a/b$. If for some natural number n and $c/d = e_0/f_0, e_1/f_1, \ldots, e_n/f_n = a/b$ we have $e_i/f_i \sim_w e_{i+1}/f_{i+1}$, $i = 0, \ldots, n-1$, then c/d is *weakly projective into* a/b, in notation, $c/d \approx_w a/b$. (Note that c/d is weakly projective into c/d with $n = 1$.) Weak projectivity is the transitive extension of weak perspectivity. Observe that neither weak perspectivity nor weak projectivity is a symmetric relation. The concept of weak perspectivity is due to R. P. Dilworth [1950 a].

The notational system we use is geometrically motivated. Many papers (starting with G. Grätzer and E. T. Schmidt [1958 d]) use the notation $[b, a] \rightarrow [d, c]$ or $a/b \rightarrow c/d$

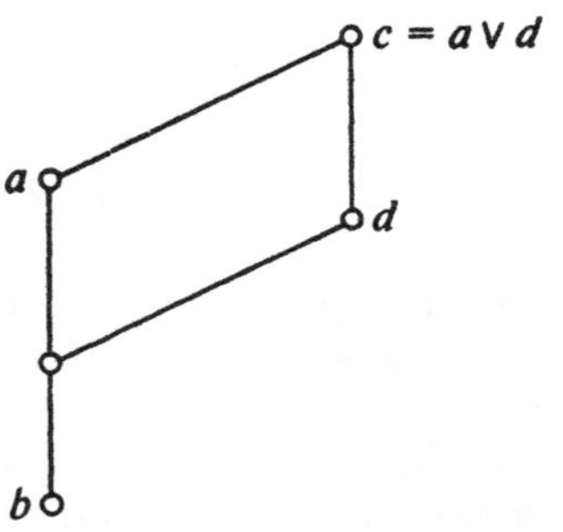

$c/d \sim_w a/b$
$c/d \searrow_w a/b$

Figure 5

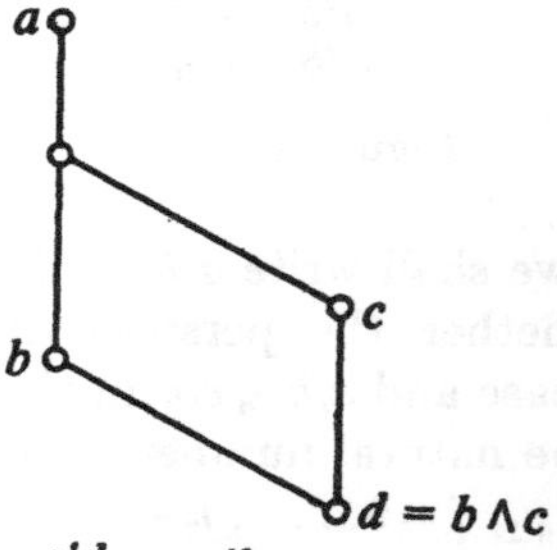

$c/d \sim_w a/b$
$c/d \nearrow_w a/b$

Figure 6

for $c/d \approx_w a/b$; these notations appear to be more natural from a universal algebraic point of view and have the advantage of emphasizing the nonsymmetric nature of weak projectivity.

The following elementary observation shows the equivalence of the various forms of the definition of weak projectivity.

Lemma 1. *Let L be a lattice, $a, b, c, d \in L$, $b \leq a$, and $d \leq c$. Then the following conditions are equivalent*:

(i) *c/d is weakly projective into a/b.*

(ii) *There is an integer m and there are elements $e_0, \ldots, e_{m-1} \in L$ such that*

$$p_m(a, e_0, \ldots, e_{m-1}) = c \quad \text{and} \quad p_m(b, e_0, \ldots, e_{m-1}) = d,$$

where the polynomial p_m is defined by

$$p_m(x, y_0, \ldots, y_{m-1}) = \cdots (((x \wedge y_0) \vee y_1) \wedge y_2) \vee \cdots.$$

(iii) *There is an integer n and there are quotients $c/d = e_0/f_0$, e_0'/f_0', e_1/f_1, $e_1'/f_1' \ldots$, $e_n/f_n = a/b$ such that e_i'/f_i' is a subquotient of e_{i+1}/f_{i+1} and $e_i/f_i \sim e_i/f_i$, for $i = 0, 1, \ldots, n-1$.*

Remark. Condition (iii) is the definition of weak projectivity using Figures 3 and 4.

Proof. Figures 5 and 6 are special cases of Figures 3 and 4 respectively; hence (i) implies (iii). Now let (iii) hold. Then for each $i = 0, 1, \ldots, n-1$, either $e_i/f_i \nearrow e_i'/f_i'$ or $e_i/f_i \searrow e_i'/f_i'$. In both cases,

$$e_i = p_4(e_{i+1}, e_i', f_i', e_i, f_i),$$

$$f_i = p_4(f_{i+1}, e_i', f_i', e_i, f_i).$$

Repeating these steps n times we get (ii) with $m \leq 4n$. Thus (iii) implies (ii). Finally, let (ii) hold. Then

$$a \wedge e_0/b \wedge e_0 \nearrow_w a/b,$$

$$(a \wedge e_0) \vee e_1/(b \wedge e_0) \vee e_1 \searrow_w a \wedge e_0/b \wedge e_0,$$

and so on; in m steps this yields that $c/d \approx_w a/b$. ●

We shall write $c/d \overset{k}{\approx}_w a/b$ iff Lemma 1(ii) holds with $m = k$. This is slightly artificial. It corresponds to requiring that the series of k weak perspectivities end with $\nearrow_w$ and that $\nearrow_w$ and $\searrow_w$ alternate throughout.

Intuitively, "$a \equiv b$ forces $c \equiv d$" iff c/d is put together from pieces each weakly projective into a/b. To state this more precisely, we describe $\Theta(a, b)$, the smallest congruence relation under which $a \equiv b$ (see Section II.3).

Theorem 2 (R. P. Dilworth [1950a]). *Let L be a lattice, $a, b, c, d \in L$, $b \leq a$, $d \leq c$. Then $c \equiv d$ $(\Theta(a, b))$ iff for some sequence $c = e_0 \geq e_1 \geq \cdots \geq e_m = d$ we have*

$$e_j/e_{j+1} \approx_w a/b, \quad \text{for} \quad j = 0, \ldots, m-1.$$

PROOF. Let Φ denote the following relation on L: $x \equiv y\ (\Phi)$ iff $x \vee y = c$ and $x \wedge y = d$ satisfy the condition of the theorem. We first prove that Φ is a congruence relation by verifying the conditions of Lemma I.3.8. Φ is reflexive since for any $c \in L$ we get

$$c/c \nearrow_w a \vee c/b \vee c \searrow_w a/b.$$

It is also obvious that if $a_1 \geq a_2 \geq a_3$ and $a_1 \equiv a_2\ (\Phi)$, $a_2 \equiv a_3\ (\Phi)$, then $a_1 \equiv a_3\ (\Phi)$. Indeed, take the sequences establishing $a_1 \equiv a_2\ (\Phi)$ and $a_2 \equiv a_3\ (\Phi)$; putting the two sequences together we get a sequence establishing $a_1 \equiv a_3\ (\Phi)$.

Now let $c \equiv d\ (\Phi)$, $c \geq d$, and $f \in L$. Let $c = e_0 \geq e_1 \geq \cdots \geq e_m = d$ be the sequence establishing $c \equiv d\ (\Phi)$, that is, $e_i/e_{i+1} \approx_w a/b$ for $i = 0, \ldots, m-1$. Then $c \wedge f = e_0 \wedge f \geq e_1 \wedge f \geq \cdots \geq e_m \wedge f = d \wedge f$ and $e_i \wedge f/e_{i+1} \wedge f \nearrow_w e_i/e_{i+1} \approx_w a/b$, hence $e_i \wedge f/e_{i+1} \wedge f \approx_w a/b$ for $i = 0, \ldots, n-1$; this proves that $c \wedge f \equiv d \wedge f\ (\Phi)$. Similarly, $c \vee f \equiv d \vee f\ (\Phi)$.

Thus by Lemma I.3.8, Φ is a congruence relation. $a \equiv b\ (\Phi)$ and so Φ is a congruence relation under which $a \equiv b$. Now let Θ be any congruence relation satisfying $a \equiv b\ (\Theta)$. It is easy to see that for $x \geq y$ and $u \geq v$, $x \equiv y\ (\Theta)$ and $u/v \sim_w x/y$ imply that $u \equiv v\ (\Theta)$. By a trivial induction, $x \equiv y\ (\Theta)$ and $u/v \approx_w x/y$ imply that $u \equiv v\ (\Theta)$. So finally, let $c \equiv d\ (\Phi)$, established by $c \vee d = e_0 \geq e_1 \geq \cdots \geq e_m = c \wedge d$. Since $e_i/e_{i+1} \approx_w a/b$ we conclude that $e_i \equiv e_{i+1}\ (\Theta)$, for $i = 0, \ldots, m-1$. Therefore, by the transitivity of Θ, we obtain that $c \equiv d\ (\Theta)$. This proves that Φ is the smallest congruence relation under which $a \equiv b$, and so $\Phi = \Theta(a, b)$. ●

Let L be a lattice and $H \subseteq L^2$. To compute $\Theta(H)$, the smallest congruence relation Θ under which $a \equiv b\ (\Theta)$ for all $\langle a, b \rangle \in H$, we use the formula (Lemma II.3.2)

$$\Theta(H) = \bigvee(\Theta(a, b) \mid \langle a, b \rangle \in H),$$

and we need a formula for joins:

LEMMA 3. *Let L be a lattice and let Θ_i, $i \in I$, be congruence relations of L. Then $a \equiv b\ (\bigvee(\Theta_i \mid i \in I))$ iff there is a sequence $z_0 = a \wedge b \leq z_1 \leq \cdots \leq z_n = a \vee b$ such that for each j with $0 \leq j < n$ there is an $i_j \in I$ satisfying $z_j \equiv z_{j+1}\ (\Theta_{i_j})$.*

The proof of Lemma 3 is the same as that of Theorem I.3.9, namely, a direct application of Lemma I.3.8.

By combining Theorem 2 and Lemma 3 we get:

COROLLARY 4. *Let L be a lattice, let $H \subseteq L^2$, and let $a, b \in L$ with $b \leq a$. Then $a \equiv b\ (\Theta(H))$ iff for some integer n there exists a sequence $a = c_0 \geq c_1 \geq \cdots \geq c_n = b$ such that for each i, with $0 \leq i < n$, there exists a $\langle d, e \rangle \in H$ satisfying*

$$c_i/c_{i+1} \approx_w d \vee e/d \wedge e.$$

COROLLARY 5. *Let L be a lattice, let I be an ideal of L, and let $a, b \in L$, $b \leq a$. Then $a \equiv b\ (\Theta[I])$ iff for some integer n there exists a sequence $a = c_0 \geq c_1 \geq \cdots \geq c_n = b$ such that for each i, with $0 \leq i < n$, there exist $d, e \in I$ with $e \leq d$ satisfying $c_i/c_{i+1} \approx_w d/e$.*

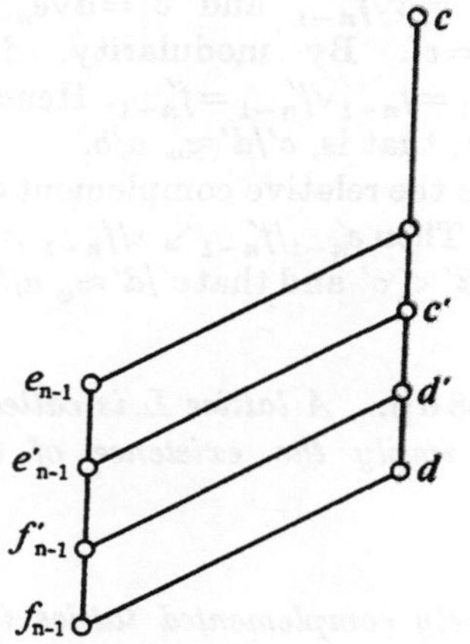

Figure 7

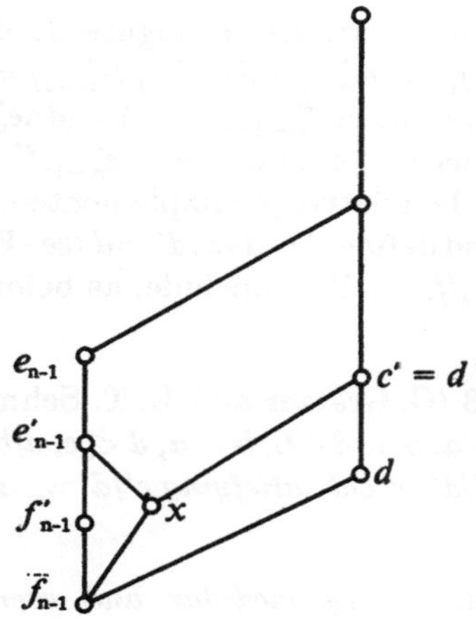

Figure 8

Recall that an ideal I is called the (ideal) kernel of a congruence relation Θ iff I is a congruence class modulo Θ (Section I.3).

Corollary 6. *Let L be a lattice and let I be an ideal of L. Then I is a kernel of a congruence relation iff*

$$a/b \approx_w c/d, \quad a \in L, \quad \text{and} \quad b, c, d \in I \quad \text{imply that} \quad a \in I.$$

Proof. Combine Corollary 5 with the observation that I is a congruence kernel of some congruence relation iff I is the kernel of $\Theta[I]$. ●

In distributive lattices every ideal is the kernel of some congruence relation, in fact this property characterizes distributivity. In general lattices we shall introduce various classes of ideals that are congruence kernels for which $\Theta[I]$ can be nicely described. This will be done and applied in Sections 2–4.

In a sense, weak projectivity describes the structure of congruence relations of a lattice. It is not surprising, therefore, that many important classes of lattices can be described by weak projectivities. We give two examples.

To introduce the first class, the class of weakly modular lattices, we need a lemma for motivation.

Lemma 7. *Let L be a lattice, $a, b, c, d \in L, b < a, d < c$, and let $a/b \approx_w c/d$. If L is modular or if L is relatively complemented, then there exists a proper subquotient c'/d' of c/d such that $c'/d' \approx_w a/b$.*

Proof. Let $a/b = e_0/f_0 \sim_w e_1/f_1 \sim_w \cdots \sim_w e_n/f_n = c/d$; we prove the statement by induction on n. By duality, we can assume that $e_{n-1}/f_{n-1} \nearrow_w c/d$ and by the induction hypothesis there exist $e'_{n-1} > f'_{n-1}$, such that $e'_{n-1}/f'_{n-1} \approx_w a/b$ and $f_{n-1} \leq f'_{n-1} < e'_{n-1} \leq e_{n-1}$.

Let L be modular. As in Figure 7, define $d' = d \vee f'_{n-1}$ and $c' = d \vee e'_{n-1}$. Then $d' \vee e'_{n-1} = (d \vee f'_{n-1}) \vee e'_{n-1} = d \vee (f'_{n-1} \vee e'_{n-1}) = d \vee e'_{n-1} = c'$. By modularity, $d' \wedge e'_{n-1} = (d \vee f'_{n-1}) \wedge e'_{n-1} = (\text{since } f'_{n-1} \leq e'_{n-1}) = (d \wedge e'_{n-1}) \vee f'_{n-1} = f_{n-1} \vee f'_{n-1} = f'_{n-1}$. Hence $c'/d' \sim e'_{n-1}/f'_{n-1}$. Thus $d' < c'$ and $c'/d' \sim e'_{n-1}/f'_{n-1} \approx_w a/b$; that is, $c'/d' \approx_w a/b$.

Next, let L be relatively complemented. Let x be the relative complement of f'_{n-1} in $[f_{n-1}, e'_{n-1}]$ and define $c' = d \vee x$, $d' = d$ (see Figure 8). Then $e'_{n-1}/f'_{n-1} \searrow x/f_{n-1} \nearrow c'/d'$ and so $c'/d' \approx e'_{n-1}/f'_{n-1}$. We conclude, as before, that $d' < c'$ and that $c'/d' \approx_w a/b$. ●

Definition 8 (G. Grätzer and E. T. Schmidt [1958 d]). *A lattice L is called* weakly modular *iff $a, b, c, d \in L$, $b < a$, $d < c$, $a/b \approx_w c/d$ imply the existence of a proper subquotient c'/d' of c/d satisfying $c'/d' \approx_w a/b$.*

Corollary 9. *Every modular and every relatively complemented lattice is weakly modular.*

The importance of the class of weakly modular lattices will be illustrated in Sections 2 and 4.

To introduce the second class of lattices we again need a lemma.

Lemma 10. *Let L be a lattice and let $a_i, b_i \in L$, $b_i < a_i$, for $i = 1, \ldots, n$. Then*

$$\bigwedge(\Theta(a_i, b_i) \mid i = 1, 2, \ldots, n) \neq \omega$$

iff there exists a proper quotient a/b of L such that

$$a/b \approx_w a_i/b_i, \quad \text{for} \quad i = 1, 2, \ldots, n.$$

Proof. If such a, b exist, then $a \equiv b\ (\Theta(a_i, b_i))$, for $i = 1, 2, \ldots, n$. Hence $a \equiv b\ (\bigwedge(\Theta(a_i, b_i) \mid i = 1, 2, \ldots, n))$ and so $\bigwedge(\Theta(a_i, b_i) \mid i = 1, 2, \ldots, n) \neq \omega$. We prove the converse by induction on n in a somewhat stronger form: if $v < u$ and $u \equiv v\ (\bigwedge(\Theta(a_i, b_i) \mid i = 1, 2, \ldots, n))$, then there exists a proper subquotient a/b of u/v such that $a/b \approx_w a_i/b_i$, for $i = 1, 2, \ldots, n$. For $n = 1$, apply Theorem 2. Assuming the statement proved for $n - 1$ we get a'/b' a proper subquotient of u/v satisfying $a'/b' \approx_w a_i/b_i$, for $i = 1, \ldots, n - 1$. Since $u \equiv v\ (\Theta(a_n, b_n))$ we get $a' \equiv b'\ (\Theta(a_n, b_n))$. Hence by Theorem 2 we obtain a proper subquotient a/b of a'/b' with $a/b \approx_w a_n/b_n$. Since $a/b \approx_w a'/b'$ we also have $a/b \approx_w a_i/b_i$, for $i = 1, 2, \ldots, n - 1$. ●

Since in a subdirectly irreducible lattice ω is meet-irreducible we conclude:

Corollary 11. *Let L be a subdirectly irreducible lattice and let a_i/b_i be proper quotients of L, for $i = 1, \ldots, n$. Then there exists a proper quotient a/b of L satisfying $a/b \approx_w a_i/b_i$, $i = 1, \ldots, n$.*

In fact, Corollary 11 holds for any lattice in which ω is meet-irreducible.

Definition 12 (R. Wille [1972]). *Let L be a lattice and let P be a finite nonvoid subset of L. Then P is called* primitive *iff*

$$\bigwedge(\Theta(x, x\vee y) \mid x, y \in P, x \neq x\vee y) \neq \omega.$$

Combining Lemma 10 and Definition 12 we obtain

Corollary 13. *P is primitive iff there exists a proper quotient a_P/b_P of L, such that $a_P/b_P \approx_w c\vee d/c$ for all $c, d \in P$, $c \neq c\vee d$.*

Observe also that by Corollary 11 every finite nonvoid subset of a subdirectly irreducible lattice is primitive.

It will be pointed out in Chapter V that important classes of lattices can be defined by the property of not containing a primitive subset (poset) isomorphic to a member of a given collection of posets.

We study weak projectivities to enable us to obtain descriptions of quotient lattices such as $L/\Theta(a, b)$. Sometimes however L/Θ can be identified very simply within L: if there is a sublattice L_1 of L having one and only one element in every congruence class, then $L/\Theta \cong L_1$. Such congruence relations are called *representable.* It happens much more often that we can get a meet-subsemilattice or a join-subsemilattice L_1 of L having one and only one element in every congruence class; in such cases we call Θ *meet-representable* or *join-representable*, respectively.

Lemma 14. *Let L be a lattice and let Θ be a congruence relation of L. If every congruence class of Θ has a minimal element, then Θ is join-representable.*

Proof. Let L_1 be the set of minimal elements of congruence classes of Θ; then L_1 contains exactly one element of each congruence class. Now let $a, b \in L_1$. We show that $a\vee b$ is the smallest element of $[a\vee b]\Theta = [a]\Theta \vee [b]\Theta$. Indeed, if $c < a\vee b$ and $c \equiv a\vee b\ (\Theta)$, then $a\wedge c \equiv a\ (\Theta)$ and $b\wedge c \equiv b\ (\Theta)$. Since $c < a\vee b$ we get that $a\wedge c < a$ or $b\wedge c < b$, say $a\wedge c < a$. Then $a\wedge c < a$ and $a\wedge c \in [a]\Theta$ contradict that $a \in L_1$. ●

Thus if L satisfies the Descending Chain Condition, then every congruence relation is join-representable, and dually.

Exercises

1. For all pairs of quotients x/y, u/v of $\mathfrak{N}_5$ investigate when "$x \equiv y$ forces $u \equiv v$". Repeat the investigation for $\mathfrak{M}_3$.
2. Show that if L is *sectionally complemented*, that is, L has a zero and for every $a \in L$ $[0, a]$ is complemented, then in order to learn the congruence structure of L it is sufficient to consider "$a \equiv b$ forces $c \equiv d$" in the special case $b = d = 0$.
3. Prove that if a lattice is modular or relatively complemented, then $a/b \approx_w c/d$ iff $a/b \approx c'/d'$ for some subquotient c'/d' of c/d.

4. Show, by an example, that the conclusion of Exercise 3 is false for lattices in general.
5. Let L be a distributive lattice and let $a/b \approx_w c/d$. Prove that there exists a quotient e/f and a subquotient c'/d' of c/d such that

$$a/b \nearrow e/f \searrow c'/d'.$$

6. Let $p = p(x_0, \ldots, x_{n-1})$ be a lattice polynomial, and let L be a lattice and $a_1, \ldots, a_{n-1} \in L$. Consider the algebraic function $q(x) = p(x, a_1, \ldots, a_{n-1})$. Under what conditions on p is it true that for all quotients a/b of L,

$$q(a)/q(b) \approx_w a/b.$$

(This is true for $p = \cdots((x_0 \wedge x_1) \vee x_2) \wedge \cdots$ by Lemma 1.)
7. Referring to the proof of Lemma 1, axiomatize those properties of p_2 which make it possible to define p_m from p_2 to obtain generalizations of Lemma 1 to some equational classes of algebras.
8. Show that $p_m(x, y_0, \ldots, y_i, y_i, y_{i+2}, \ldots, y_{m-1})$ does not depend on $x, y_0, \ldots, y_{i-1}$ (that is, the value of $p_m(a, e_0, \ldots, e_i, e_i, e_{i+2}, \ldots, e_{m-1})$ is independent of $a, e_0, \ldots, e_{i-1}$).
9. Prove that $p_m(x, c, d, c, d, \ldots) = x$ for any $d \leq x \leq c$.
10. Rephrase and prove Lemma 1 using $q_m = \cdots((x \vee y_0) \wedge y_1) \vee \cdots$. Show that $\overset{m}{\sim}_w$, defined in terms of q_m, may differ from $\overset{m}{\sim}_w$ defined by p_m.
11. Show that in any sequence of weak perspectivities, any number of steps $\nearrow_w$ and $\searrow_w$ can be put in. Observe that in any nonredundant sequence of weak perspectivities, $\nearrow_w$ and $\searrow_w$ have to alternate.
12. Find examples to show that m cannot be bounded in Theorem 2. (Make all your examples planar modular lattices.)
13. Find examples to show that in Theorem 2, $\approx_w$ cannot be replaced by $\overset{n}{\approx}_w$ for any natural number n. (Make all your examples planar modular lattices.)
14. Prove that an ideal I is a kernel iff I is the kernel of $\Theta[I]$.
15. Let L be a lattice, and let I and J be ideals of L with $J \subseteq I$. Assume that J is a kernel in the lattice I and that I is a kernel in the lattice L. Is J a kernel in L?
16. Show that the ideal kernels of the lattice L form a sublattice $I_C(L)$ of $I(L)$, the lattice of all ideals of L. In fact, $I_C(L)$ is closed under arbitrary joins and under all meets existing in $I(L)$.
17. Let L be a distributive lattice, let I be an ideal of L, and let a/b be a quotient of L. Then $a \equiv b\ (\Theta[I])$ iff $a/b \sim a'/b'$ for some quotient a'/b' of I.
18. Show that Exercise 5 characterizes distributivity.
19. Show that Exercise 17 characterizes distributivity.
In Exercises 20–23 (which are due to R. Wille [1972]) let L and L' be lattices and let φ be a homomorphism of L onto L'.
20. Let a'/b' be a quotient of L' and let c/d be a quotient of L with $a'/b' \approx_w c\varphi/d\varphi$. Prove that there is a quotient a/b of L with $a/b \approx_w c/d$, $a\varphi = a'$, and $b\varphi = b'$.
21. Let $c_1/d_1, \ldots, c_n/d_n$ be quotients of L such that $c_1\varphi = c_2\varphi = \cdots = c_n\varphi$ and $d_1\varphi = d_2\varphi = \cdots = d_n\varphi$. Show that there is a quotient a/b of L satisfying $a/b \approx_w c_i/d_i$ and $a\varphi/b\varphi = c_i\varphi/d_i\varphi$, for $i = 1, 2, \ldots, n$.
22. Let P' be a primitive subset of L', and let P be a subset of L such that φ restricted to P is a (poset) isomorphism of P with P'. Show that P is primitive in L.
23. Let P be a primitive subset of L, let a_P, b_P be given as in Corollary 13, and let $a_P\varphi \neq b_P\varphi$. Show that $P\varphi$ is primitive in L' and that φ restricted to P is a poset isomorphism of P with $P\varphi$.

24. Let L be a lattice and let Ψ be a congruence relation of L. Verify the formula:

$$\Theta([a]\Psi, [b]\Psi) = (\Theta(a, b) \vee \Psi)/\Psi.$$

25. Let $a_i, b_i \in L$, for $i = 1, \ldots, n$. Show that

$$\wedge(\Theta([a_i]\Psi, [b_i]\Psi) \mid i = 1, 2, \ldots, n) = (\wedge(\Theta(a_i, b_i) \mid i = 1, 2, \ldots, n) \vee \Psi)/\Psi.$$

26. Use Exercises 24 and 25 to verify Exercises 22 and 23.
27. Show that for ideals I and J of a lattice L

$$\Theta[I] \vee \Theta[J] = \Theta[I \vee J].$$

28. Let L be a lattice and let J_i, $i \in I$, be ideals of L. Prove that

$$\vee(\Theta[J_i] \mid i \in I) = \Theta[\vee(J_i \mid i \in I)].$$

29. Show that $\Theta[I] \wedge \Theta[J] = \Theta[I \wedge J]$ does not hold in general.
30. Verify

$$\Theta[I] \wedge \Theta[J] = \Theta[I \wedge J]$$

for ideals I and J of a distributive lattice.
31. Show that the formula of Exercise 30 holds also under the condition that every ideal is the kernel of at most one congruence relation.
32. Show that Corollary 11 does not hold for infinitely many a_i/b_i.
33. Show that in a finite lattice (or in a lattice in which all chains are finite) every congruence relation is meet- and join-representable.
34. Find a congruence relation Θ of the lattice of Figure 9 which is not representable. (Observe that Θ is both meet- and join-representable.)
35. Let L be a finite lattice and let $L/\Theta \cong \mathfrak{M}_3$. Show that Θ is representable.
36. Give a formal proof of the statement that if L_1 represents the congruence relation Θ of L, then $L/\Theta \cong L_1$.

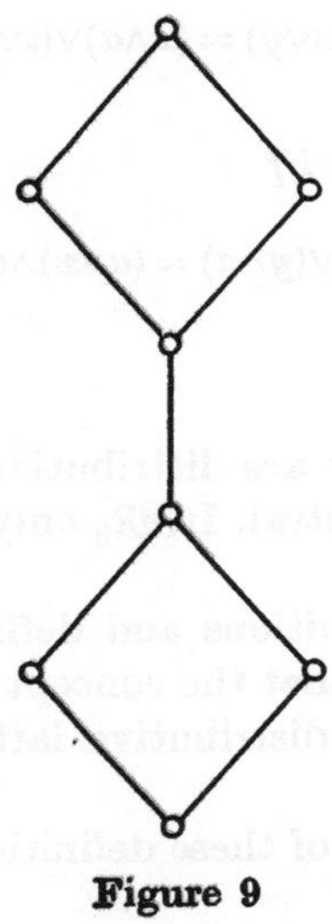

Figure 9

37. A congruence relation Θ of a lattice L is *order-representable* iff there exists a subset $H \subseteq L$ satisfying $[H]\Theta = L$ and

$$a \leq b \text{ in } L \text{ iff } [a]\Theta \leq [b]\Theta \text{ in } L/\Theta, \text{ for } a, b \in H.$$

Show that H as a subposet of L is a lattice and $H \cong L/\Theta$.

38. Prove that every meet-representable congruence relation is order-representable.
39. Let L be a lattice and let Θ be a congruence relation of L. Verify that if L/Θ is finite or countable, then Θ is order-representable.
40. Find a chain C, a lattice L, and a congruence relation Θ of L, such that $L/\Theta \cong C$ but Θ is not order-representable.

2. Distributive, Standard, and Neutral Elements

The three types of elements of a lattice mentioned in the title of this section were discovered by O. Ore [1935], G. Grätzer [1959], and G. Birkhoff [1940a], respectively. It turned out later that all three can be defined using distributive equations only. This may be a coincidence, but it could be considered as a confirmation of the principle that distributive lattice theory provides the foundation of general lattice theory.

Definition 1. *Let L be a lattice and let a be an element of L.*

(i) *The element a is called* distributive *iff*

$$a \vee (x \wedge y) = (a \vee x) \wedge (a \vee y),$$

for all $x, y \in L$.

(ii) *The element a is called* standard *iff*

$$x \wedge (a \vee y) = (x \wedge a) \vee (x \wedge y),$$

for all $x, y \in L$.

(iii) *The element a is called* neutral *iff*

$$(a \wedge x) \vee (x \wedge y) \vee (y \wedge a) = (a \vee x) \wedge (x \vee y) \wedge (y \vee a),$$

for all $x, y \in L$.

For instance, in $\mathfrak{N}_5$ o, i, a, and c are distributive; o, i, and a are standard but c is not standard; only o and i are neutral. In $\mathfrak{M}_3$ only o and i are distributive; they are also standard and neutral.

We can also dualize these definitions and define *dually distributive elements* and *dually standard elements*. Observe that the concept of neutrality is selfdual.

Of course, every element of a distributive lattice is distributive, standard, and neutral.

Various useful equivalent forms of these definitions are given in the three theorems that follow.

Theorem 2 (O. Ore [1935]). *Let L be a lattice and let a be an element of L. The following conditions on a are equivalent*:

(i) *a is distributive.*

(ii) *The map*

$$\varphi: x \to a \vee x \quad (x \in L)$$

is a homomorphism of L onto $[a)$.

(iii) *The binary relation Θ_a on L defined by*

$$x \equiv y\ (\Theta_a) \quad \textit{iff} \quad a \vee x = a \vee y$$

is a congruence relation.

Remark. (i) and (ii) are equivalent since φ is a homomorphism iff it preserves meets, while (ii) and (iii) are equivalent because $\Theta_a = \mathrm{Ker}(\varphi)$. A more formal proof follows.

Proof. (i) *implies* (ii). φ maps L into $[a)$ for any element a of L; in fact φ is always onto since $b\varphi = b$ for all $b \geq a$. The map φ is always a join-homomorphism since

$$x\varphi \vee y\varphi = (a \vee x) \vee (a \vee y) = a \vee (x \vee y) = (x \vee y)\varphi.$$

In view of Definition 1(i), if a is distributive, we also have

$$x\varphi \wedge y\varphi = (a \vee x) \wedge (a \vee y) = a \vee (x \wedge y) = (x \wedge y)\varphi,$$

and so φ is a homomorphism. ◑

(ii) *implies* (iii). Θ_a is the kernel of the homomorphism φ and therefore Θ_a is a congruence relation. ◑

(iii) *implies* (i). Since $x \vee a = (a \vee x) \vee a$, therefore $x \equiv a \vee x\ (\Theta_a)$. Similarly, $y \equiv a \vee y\ (\Theta_a)$. Therefore,

$$x \wedge y \equiv (a \vee x) \wedge (a \vee y)\ (\Theta_a).$$

By the definition of Θ_a we get

$$a \vee (x \wedge y) = a \vee ((a \vee x) \wedge (a \vee y)) = (a \vee x) \wedge (a \vee y),$$

proving (i). ●

Theorem 3 (G. Grätzer and E. T. Schmidt [1961]). *Let L be a lattice and let a be an element of L. The following conditions are equivalent*:

(i) *a is standard.*

(ii) *The binary relation Θ_a on L defined by*

$$x \equiv y\ (\Theta_a) \quad \textit{iff} \quad (x \wedge y) \vee a_1 = x \vee y \quad \textit{for some} \quad a_1 \leq a$$

is a congruence relation.

(iii) *a is distributive and, for $x, y \in L$,*

$$a \wedge x = a \wedge y \quad \textit{and} \quad a \vee x = a \vee y \quad \textit{imply that} \quad x = y.$$

Proof. (i) *implies* (ii). Let a be standard and let Θ_a be defined as in (ii). We use Lemma I.3.8 to verify that Θ_a is a congruence relation. By definition, Θ_a is reflexive and $x \equiv y\ (\Theta_a)$ iff $x \wedge y \equiv x \vee y\ (\Theta_a)$. If $x \leq y \leq z$ $x \equiv y\ (\Theta_a)$, and $y \equiv z\ (\Theta_a)$, then $x \vee a_1 = y$ and $y \vee a_2 = z$ for some $a_1, a_2 \leq a$. Hence $x \vee (a_1 \vee a_2) = z$, showing that $x \equiv z\ (\Theta_a)$ since $a_1 \vee a_2 \leq a$. Finally, let $x \leq y$ and $x \equiv y\ (\Theta_a)$, that is, $x \vee a_1 = y$ with $a_1 \leq a$. For any $t \in L$, $(x \vee t) \vee a_1 = y \vee t$, hence $x \vee t \equiv y \vee t\ (\Theta_a)$. To show the Substitution Property for meet, observe that

$$y \wedge t \leq y = x \vee a_1 \leq x \vee a,$$

and so, using the fact that a is standard we get

$$y \wedge t = (y \wedge t) \wedge (x \vee a) = ((y \wedge t) \wedge x) \vee ((y \wedge t) \wedge a) = (x \wedge t) \vee a_2,$$

where $a_2 = y \wedge t \wedge a \leq a$. Thus the conditions of Lemma I.3.8 are verified and Θ_a is a congruence relation. ◑

(ii) *implies* (iii). Let us assume that the Θ_a defined in (ii) is a congruence relation. We can show that a is distributive just as in Theorem 2. Now let $a \wedge x = a \wedge y$ and $a \vee x = a \vee y$. Since $y \equiv a \vee y\ (\Theta_a)$, meeting both sides with x and using $a \vee y = a \vee x$ we obtain

$$x \wedge y \equiv x \wedge (a \vee y) = x \wedge (a \vee x) = x\ (\Theta_a).$$

Thus $x = (x \wedge y) \vee a_1$ for some $a_1 \leq a$. Also, $a_1 \leq x$, hence $a_1 \leq a \wedge x = a \wedge y$, and so $a_1 \leq x \wedge y$. We conclude that $x = x \wedge y$. Similarly, $y = x \wedge y$, and so $x = y$. ◑

(iii) *implies* (i). Let us assume (iii), let $x, y \in L$, and define $b = x \wedge (a \vee y)$ and $c = (x \wedge a) \vee (x \wedge y)$. In order to show $b = c$, by (iii), it will be sufficient to prove that $a \wedge b = a \wedge c$ and $a \vee b = a \vee c$. To prove the first,

$$a \wedge x \leq a \wedge c \leq (\text{using } c \leq b) \leq a \wedge b = a \wedge x \wedge (a \vee y) = a \wedge x,$$

hence $a \wedge c = a \wedge b$. To prove the second, we compute using the fact that a is distributive:

$$a \vee b = a \vee ((x \wedge (a \vee y))) = (a \vee x) \wedge (a \vee y) = a \vee (x \wedge y) = a \vee (x \wedge a) \vee (x \wedge y) = a \vee c. \quad ●$$

Theorem 4. *Let L be a lattice and let a be an element of L. The following conditions are equivalent:*

(i) *a is neutral.*

(ii) *a is distributive, a is dually distributive, and $a \wedge x = a \wedge y$ and $a \vee x = a \vee y$ imply $x = y$ for any $x, y \in L$.*

(iii) *There is an embedding φ of L into a direct product $A \times B$ where A has a 1 and B has a 0 and $a\varphi = \langle 1, 0 \rangle$.*

(iv) *For any $x, y \in L$, the sublattice generated by a, x, and y is distributive.*

Remark. The equivalence of (ii)–(iv) is due to G. Birkhoff [1940a]; in fact, G. Birkhoff used (iv) as a definition of neutrality. It was conjectured in G. Grätzer and E. T. Schmidt [1961] that (i) is equivalent to (ii)–(iv). This was proved in G. Grätzer [1962], J. Hashimoto and S. Kinugawa [1963], and Iqbalunissa [1964].

Proof.

(i) *implies* (ii). Let a be neutral. Then

$$(*)\quad a\vee(x\wedge y)=x\wedge(a\vee y)\quad \text{for}\quad x\geq a.$$

Indeed,

$$a\vee(x\wedge y)=(a\wedge x)\vee(x\wedge y)\vee(y\wedge a)=(\text{by Definition 1(iii)})$$
$$=(a\vee x)\wedge(x\vee y)\wedge(y\vee a)=x\wedge(a\vee y).$$

To show that a is distributive, compute:

$$a\vee(x\wedge y)=a\vee(a\wedge x)\vee(x\wedge y)\vee(y\wedge a)=(\text{by Definition 1(iii)})$$
$$=a\vee((a\vee x)\wedge(x\vee y)\wedge(y\vee a))$$
$$(\text{apply } (*) \text{ to } a,\ a\vee x, \text{ and } (x\vee y)\wedge(y\vee a))$$
$$=(a\vee x)\wedge(a\vee((x\vee y)\wedge(y\vee a)))$$
$$(\text{apply } (*) \text{ to } a,\ y\vee a, \text{ and } x\vee y)$$
$$=(a\vee x)\wedge(y\vee a)\wedge(a\vee x\vee y)=(a\vee x)\wedge(a\vee y),$$

as claimed. By duality we get that a is dually distributive.

Finally, let $a\wedge x=a\wedge y$ and $a\vee x=a\vee y$. Then

$$x=x\wedge(a\vee x)\wedge(a\vee y)\wedge(x\vee y)=x\wedge((a\wedge x)\vee(x\wedge y)\vee(a\wedge y))$$
$$=x\wedge((a\wedge x)\vee(x\wedge y))=(a\wedge x)\vee(x\wedge y)=(a\wedge x)\vee(a\wedge y)\vee(x\wedge y).$$

Since the right-hand side is symmetric in x and y we conclude that $x=y$. ◐

(ii) *implies* (iii). Let (ii) hold for a and define $A=(a]$ and $B=[a)$. Let

$$\varphi: x\to\langle x\wedge a, x\vee a\rangle.$$

Since a is distributive and dually distributive, φ is a homomorphism of L into $A\times B$ (by Theorem 1(ii) and its dual). The map φ is one-to-one, since if $x\varphi=y\varphi$ for $x, y\in L$, then

$$\langle x\wedge a, x\vee a\rangle=\langle y\wedge a, y\vee a\rangle,$$

that is, $x\wedge a=y\wedge a$ and $x\vee a=y\vee a$; thus $x=y$ by (ii). So φ is an embedding, $a\varphi=\langle a, a\rangle$, and a is the unit element of A and the zero of B. ◐

(iii) *implies* (iv). The following three statements are obvious: (iv) holds for the zero and the unit in any lattice; (iv) holds for $\langle a_0, a_1\rangle$ in $A_0\times A_1$ iff it holds for a_i in A_i, for $i=0, 1$; if a is an element of L_0, L_0 is a sublattice of L_1, and (iv) holds for a in L_1, then (iv) holds for a in L_0. (iii) along with these three statements proves (iv). ◐

(iv) *implies* (i). Obvious by Exercise I.4.7. ●

The results stated in Theorems 2–4 make it possible to verify the most important properties of distributive, standard, and neutral elements.

Theorem 5.

(i) *Every neutral element is standard.*

(ii) *Every standard element is distributive.*

(iii) *Every standard and dually standard (dually distributive) element is neutral.*

(iv) *If a is distributive or standard, then the Θ_a in Theorem 2(iii) and Theorem 3(ii), respectively, agrees with $\Theta[(a]]$.*

Proof.

(i) and (iii). By Theorem 4(ii) and Theorem 3(iii). ◑

(ii) By Theorem 3(iii). ◑

(iv) If $u \equiv a\ (\Theta)$ for any $u \leq a$ and a congruence relation Θ, then $a \vee x = a \vee y$ implies that $x = x \vee (a \wedge x) \equiv x \vee a = y \vee a \equiv y \vee (a \wedge y) = y\ (\Theta)$, hence for the relation Θ_a given by Theorem 2(iii) we have $\Theta_a \leq \Theta$. Similarly, if $(x \wedge y) \vee a_1 = x \vee y$ for some $a_1 \leq a$, then $x \wedge y = (x \wedge y) \vee (a \wedge x \wedge y) \equiv (x \wedge y) \vee a_1 = x \vee y\ (\Theta)$, and so $x \equiv y\ (\Theta)$; hence for the relation Θ_a of Theorem 3(ii) we obtain $\Theta_a \leq \Theta$. ●

For a principal ideal $(a] = I$ let $\Theta[I]$ be denoted by Θ_a. Then by Theorem 5(iv), the Θ_a of Theorem 2(iii) and the Θ_a of Theorem 3(ii) are indeed the Θ_a just defined in the case a is distributive or standard. Hence 2(iii) and 3(ii) are definitions of distributive and standard elements, respectively, in terms of the properties of Θ_a.

As we have already seen, the converse of Theorem 5(i), as well as that of Theorem 5(ii), is false. Two wide classes of lattices in which the converse holds are the class of modular lattices and the class of relatively complemented lattices.

Theorem 6 (G. Grätzer and E. T. Schmidt [1961]). *Let L be a weakly modular lattice. Then an element of L is distributive iff it is neutral.*

Before proving the theorem we verify a lemma connecting the distributivity of an element with weak projectivity.

Lemma 7. *Let L be a lattice and let a be an element of L. The element a is distributive iff $u \leq z \leq a \leq y \leq x$ and $x/y \approx_w z/u$ imply $x = y$.*

Proof. Let a be a distributive element, $u \leq z \leq a \leq y \leq x$, and $x/y \approx_w z/u$. Since $u \leq z \leq a$, we get $u \equiv z\ (\Theta_a)$, hence $x \equiv y\ (\Theta_a)$. By Theorem 2(iii), $x = x \vee a = y \vee a = y$. Let us now assume that a is not distributive. Then there exist $b, c \in L$ such that $a \vee (b \wedge c) \neq (a \vee b) \wedge (a \vee c)$. Since $a \equiv a \wedge b \wedge c\ (\Theta_a)$ we obtain

$$(a \vee b) \wedge (a \vee c) = a \vee ((a \vee b) \wedge (a \vee c)) \equiv a \vee (((a \wedge b \wedge c) \vee b) \wedge ((a \wedge b \wedge c) \vee c))$$
$$= a \vee (b \wedge c)\ (\Theta_a)$$

so by Corollary 1.5 there exist $x, y, u \in L$ satisfying, with $z = a$,

$$u < z \leq a \leq (a \vee b) \wedge (a \vee c) \leq y < x \leq a \vee (b \wedge c), \quad x/y \approx_w z/u. \quad ●$$

Proof of Theorem 6. Let L be a weakly modular lattice, and let a be a distributive element. If a is not dually distributive, then by Lemma 7 there exist $u < z \leq a \leq y \leq x$ satisfying $z/u \approx_w x/y$. By weak modularity, there exists a proper subquotient x_1/y_1 of x/y satisfying $x_1/y_1 \approx_w z/u$, contradicting by Lemma 7 that a is distributive.

Now let $a \wedge x = a \wedge y$, $a \vee x = a \vee y$, and $x \neq y$ for some $x, y \in L$. Proceeding just as in the step "(ii) implies (iii)" in the proof of Theorem 3, we conclude that

$$x \equiv y \ (\Theta(a, a \wedge x))$$

and dually

$$x \equiv y \ (\Theta(a, a \vee x)).$$

Applying Theorem 1.2 to $x \equiv y \ (\Theta(a, a \wedge x))$ we get a proper subquotient x_1/y_1 of $x \vee y/x \wedge y$ satisfying $x_1/y_1 \approx_w a/a \wedge x$. Since $x \equiv y \ (\Theta(a, a \vee x))$ we also have $x_1 \equiv y_1$ $(\Theta(a, a \vee x))$, and again by Theorem 1.2 we get a proper subquotient x_2/y_2 of x_1/y_1 such that $x_2/y_2 \approx_w a \vee x/a$. Using the weak modularity of L, we obtain a proper subquotient x_3/y_3 of $a \vee x/a$ satisfying $x_3/y_3 \approx_w x_2/y_2$. Hence

$$x_3/y_3 \approx_w x_2/y_2 \approx_w x_1/y_1 \approx_w a/a \wedge x,$$

contradicting Lemma 7. By Theorem 4(ii), a is neutral. ●

Corollary 8. *In a weakly modular lattice every standard element is neutral.*

Theorem 9. *Let L be a lattice.*

(i) *Let D denote the set of all distributive elements of L. Then a and $b \in D$ imply that $a \vee b \in D$.*

(ii) *Let S denote the set of all standard elements of L. Then a and $b \in S$ imply that $a \wedge b$ and $a \vee b \in S$.*

(iii) *Let N denote the set of all neutral elements of L. Then a and $b \in N$ imply that $a \wedge b$ and $a \vee b \in N$.*

Proof.

(i) Let $a, b \in D$ and compute:
$(a \vee b) \vee (x \wedge y) = a \vee b \vee (x \wedge y) = a \vee ((b \vee x) \wedge (b \vee y)) = (a \vee b \vee x) \wedge (a \vee b \vee y)$,
so $a \vee b \in D$. ◑

(ii) Let $a, b \in S$. First we do the join:
$x \wedge (a \vee b \vee y) = (x \wedge a) \vee (x \wedge (b \vee y)) = (x \wedge a) \vee (x \wedge b) \vee (x \wedge y) = (x \wedge (a \vee b)) \vee (x \wedge y)$,
proving that $a \vee b \in S$. Now we verify the formula

$$\Theta_a \wedge \Theta_b = \Theta_{a \wedge b},$$

where Θ_a, Θ_b, and $\Theta_{a \wedge b}$ are the relations described by Theorem 3(ii). Since $\Theta_a \wedge \Theta_b \geq \Theta_{a \wedge b}$ is trivial, let $x \equiv y \ (\Theta_a \wedge \Theta_b)$. Then $x \equiv y \ (\Theta_a)$, and so $(x \wedge y) \vee a_1 = x \vee y$ for some $a_1 \leq a$. We also have $x \equiv y \ (\Theta_b)$ and therefore $a_1 = a_1 \wedge (x \vee y) \equiv a_1 \wedge x \wedge y$ (Θ_b). Thus for some $b_1 \leq b$ we have $a_1 = (a_1 \wedge x \wedge y) \vee b_1$. Now $(x \wedge y) \vee b_1 = (x \wedge y) \vee (a_1 \wedge x \wedge y) \vee b_1 = (x \wedge y) \vee a_1 = x \vee y$; since $b_1 \leq b$ and $b_1 \leq a_1 \leq a$, we obtain $b_1 \leq a \wedge b$; these verify that $x \equiv y \ (\Theta_{a \wedge b})$.

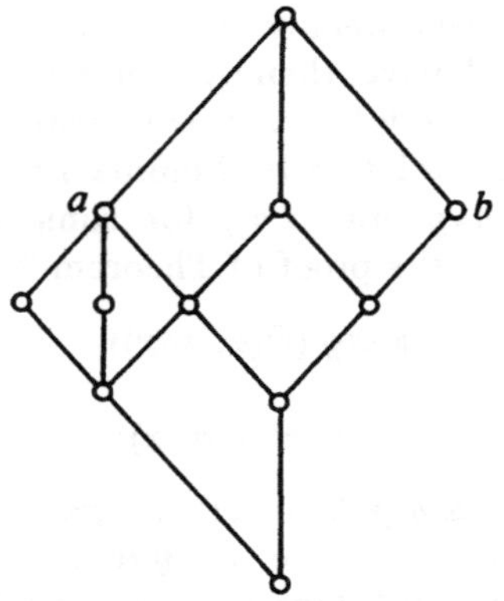

Figure 1

This formula shows that if $a, b \in S$, then the relation $\Theta_{a \wedge b}$ of Theorem 3(ii) is a congruence relation, hence $a \wedge b \in S$ by Theorem 3. ◐

(iii) Let $a, b \in N$. By Theorem 5, $a, b \in S$. Hence by (ii), $a \wedge b \in S$. By Theorems 3 and 4, to show that $a \wedge b \in N$ we have to prove only that $a \wedge b$ is dually distributive. Since a and b are dually distributive, they are distributive elements of the dual of L, hence by (i), $a \vee b$ in the dual of L is distributive and so $a \wedge b$ in L is dually distributive. ●

Figure 1 shows that $a, b \in D$ does not imply that $a \wedge b \in D$.

Exercises

1. Let L be a bounded lattice. Show that 0 and 1 are distributive, standard, and neutral.
2. Let L be the lattice of Figure 2. Then $(a]$ is the kernel of some congruence relation but a is not distributive.
3. If a is a distributive element of L, then $L/\Theta_a \cong (a]$. To what extent does $L/\Theta_a \cong (a]$ characterize the distributivity of a?
4. Find distributive elements which are not dually distributive.

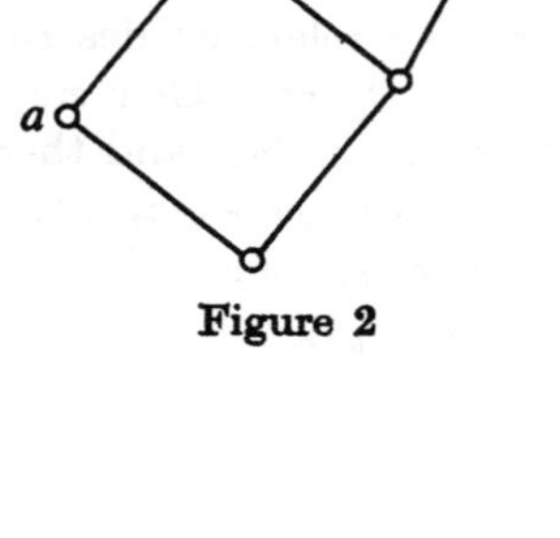

Figure 2

5. Investigate the relation Θ_a of Theorem 2(iii) as a congruence relation of the join-semilattice.
6. Let L be a lattice, let φ be a homomorphism of L onto L', and let L_1 be a sublattice of L'. Let a be an element of L satisfying $a\varphi \in L_1$. Show that if a is distributive in L, then $a\varphi$ is distributive in L_1.
7. Prove the analogue of Exercise 6 for standard and neutral elements.
8. Let a be a distributive element of L. Show that $(a]$ is a distributive element of $I(L)$.
9. Prove the analogue of Exercise 8 for standard and neutral elements.
10. Show that in Theorem 3(iii) (and in Theorem 4(ii)) the condition can be weakened by assuming that $x \geq y$.
11. Verify that the element a of a lattice L is standard iff $x \leq a \vee y$ implies that $x = (x \wedge a) \vee (x \wedge y)$ (G. Grätzer and E. T. Schmidt [1961]).
12. Prove that the join of two distributive elements is again distributive by verifying the formula $\Theta_a \vee \Theta_b = \Theta_{a \vee b}$.
13. Find classes of lattices in which the meet of two distributive elements is distributive.
14. Show that the map $a \to \Theta_a$ for standard elements is an embedding of the sublattice of standard elements into the congruence lattice.
15. Let L be a lattice and let $a, b, c \in L$. Show that a, b, and c generate a distributive sublattice iff for any permutation x, y, z of a, b, c we have

$$x \wedge (y \vee z) = (x \wedge y) \vee (x \wedge z),$$
$$x \vee (y \wedge z) = (x \vee y) \wedge (x \vee z),$$
$$(x \wedge y) \vee (y \wedge z) \vee (z \wedge x) = (x \vee y) \wedge (y \vee z) \wedge (z \vee x)$$

(O. Ore [1940]).
16. Let L be a lattice and let a, b be standard elements of L. Show that for any $c \in L$, a, b, and c generate a distributive sublattice.
17. Do three distributive elements generate a distributive sublattice?
18. Verify directly that in a modular lattice distributive element = standard element = neutral element.
19. Verify the conclusion of Exercise 18 in a relatively complemented lattice.
20. An element of a modular lattice is neutral iff it has at most one relative complement in any interval containing it (G. Grätzer and E. T. Schmidt [1961]).
21. Let L be a modular lattice and let $a \in L$. Then a is neutral iff a is neutral in every interval of the form $[a \wedge x, a \vee x]$, for $x \in L$.
22. Show that the conclusion of Exercise 21 fails in weakly modular lattices. (See Figure 3.)

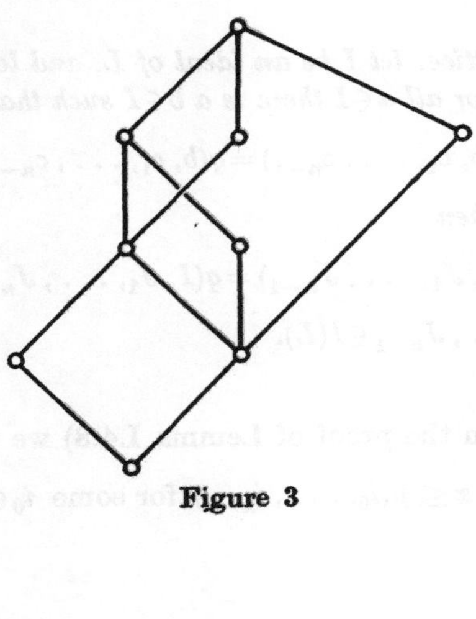

Figure 3

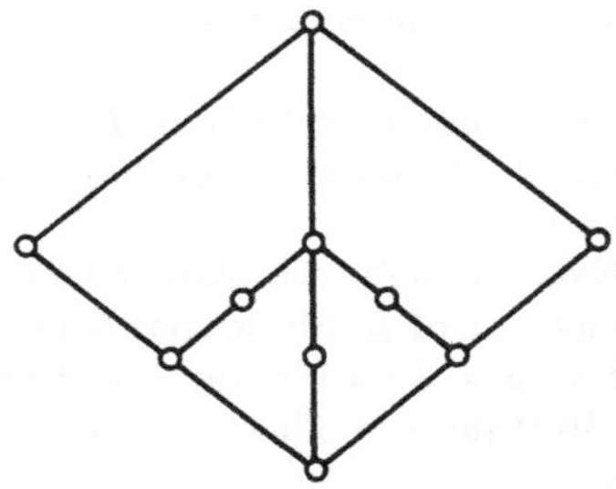

Figure 4

23. Show that Exercise 21 also fails for weakly modular lattices and standard elements. (See Figure 4.)
24. Let L be a bounded relatively complemented lattice. Then an element a of L is neutral iff it has a unique complement (J. Hashimoto and S. Kinugawa [1963]).
25. Show that the N of Theorem 9 is the intersection of the maximal distributive sublattices of L (G. Birkhoff [1940a]).
26. Prove Theorem 9(iii) using Theorem 4(iii).

3. Distributive, Standard, and Neutral Ideals

The three types of ideals mentioned in the title of this section derive naturally from the concepts introduced in Section 2.

DEFINITION 1. *An ideal I of a lattice L is called* distributive, standard, *or* neutral, *respectively, iff I is distributive, standard, or neutral, respectively, as an element of $I(L)$, the lattice of all ideals of L.*

To establish a connection between the type of elements and the type of ideals they generate we need a lemma:

LEMMA 2. *Let L be a lattice, let I be an ideal of L, and let p and q be n-ary polynomials. Let us assume that for all $a \in I$ there is a $b \in I$ such that $a \leq b$ and*

$$p(b, c_1, \ldots, c_{n-1}) = q(b, c_1, \ldots, c_{n-1}),$$

for all $c_1, \ldots, c_{n-1} \in L$. Then

$$p(I, J_1, \ldots, J_{n-1}) = q(I, J_1, \ldots, J_{n-1})$$

holds in $I(L)$ for all $J_1, \ldots, J_{n-1} \in I(L)$.

PROOF. In Section I.4 (in the proof of Lemma I.4.8) we proved the formula:

$$p(I_0, \ldots, I_{n-1}) = \{x \mid x \leq p(i_0, \ldots, i_{n-1}) \text{ for some } i_0 \in I_0, \ldots, i_{n-1} \in I_{n-1}\}.$$

Thus we obtain:

$$\begin{aligned}
&p(I, J_1, \ldots, J_{n-1})\\
&=\{x \mid x \leq p(a, j_1, \ldots, j_{n-1}) \text{ for some } a \in I, j_1 \in J_1, \ldots, j_{n-1} \in J_{n-1}\}\\
&=\{x \mid x \leq p(b, j_1, \ldots, j_{n-1}) \text{ where } b \in I \text{ satisfies } p(b, c_1, \ldots, c_{n-1})\\
&\qquad = q(b, c_1, \ldots, c_{n-1}) \text{ for all } c_1, \ldots, c_{n-1} \in L \text{ and } j_1 \in J_1, \ldots, j_{n-1} \in J_{n-1}\}\\
&=\{x \mid x \leq q(b, j_1, \ldots, j_{n-1})\} = q(I, J_1, \ldots, J_{n-1}).
\end{aligned}$$ ●

Corollary 3. *Let L be a lattice and let $a \in L$. Then a is distributive, standard, or neutral, respectively, iff $(a]$ as an ideal is distributive, standard, or neutral, respectively.*

The main characterization theorems can be proved similarly to the proofs of Theorems 2.2–2.4.

Theorem 4. *Let L be a lattice and let I be an ideal of L. The following conditions on I are equivalent:*

(i) *I is distributive.*

(ii) *$\Theta[I]$ can be described as follows:*

$$x \equiv y\ (\Theta[I]) \quad \textit{iff} \quad x \vee i = y \vee i \quad \textit{for some} \quad i \in I.$$

Proof. If in (ii) we put $(x] \vee I = (y] \vee I$ in place of $x \vee i = y \vee i$, then the equivalence can be proved as in Section 2 since $x \equiv y\ (\Theta[I])$ in L iff $(x] \equiv (y]\ (\Theta_I)$ in $I(L)$. Now, if $x \vee i = y \vee i$ for some $i \in I$, then $(x] \vee I = (y] \vee I$ is obvious. Conversely, if $(x] \vee I = (y] \vee I$, then $x \leq y \vee i_0$ and $y \leq x \vee i_1$ for some $i_0, i_1 \in I$. Therefore, $x \vee i = y \vee i$ with $i_0 \vee i_1 = i \in I$. ●

Theorem 5.[1] *Let L be a lattice and let I be an ideal of L. The following conditions on I are equivalent:*

(i) *I is standard.*

(ii) *The equality*

$$(a] \wedge (I \vee (b]) = ((a] \wedge I) \vee (a \wedge b]$$

holds for all $a, b \in L$.

(iii) *For any ideal J of L,*

$$I \vee J = \{i \vee j \mid i \in I \textit{ and } j \in J\}.$$

(iv) *$\Theta[I]$ can be described by*

$$x \equiv y\ (\Theta[I]) \quad \textit{iff} \quad (x \wedge y) \vee i = x \vee y \quad \textit{for some} \quad i \in I.$$

---

[1] This result and all the other unreferenced results in this section are based on G. Grätzer [1959] and G. Grätzer and E. T. Schmidt [1961].

(v) *I is a distributive ideal and, for all* $J, K \in I(L)$,

$$I \wedge J = I \wedge K \quad \textit{and} \quad I \vee J = I \vee K \quad \textit{imply that} \quad J = K.$$

Proof. The equivalence of the five conditions can be verified as in Section 2. Only (iii) is new. But (iii) follows from (ii): if I satisfies (ii) and $a \in I \vee J$, then $a \leq i \vee j$ for some $i \in I$ and $j \in J$; hence $(a] = (a] \wedge (I \vee (j]) = (\text{by (ii)}) = ((a] \wedge I) \vee (a \wedge j]$. Therefore, $a \leq i_1 \vee j_1$, where $i_1 \in (a] \wedge I$ and $j_1 \leq a \wedge j$. Consequently, $a = i_1 \vee j_1$. Finally, observe that when proving the analogue of "(i) implies (iv)" it is sufficient to use (iii). ●

Theorem 6. *Let L be a lattice and let I be an ideal of L. The following conditions on I are equivalent:*

(i) *I is neutral.*

(ii) *For all* $j, k \in L$,

$$(I \wedge (j]) \vee ((j] \wedge (k]) \vee ((k] \wedge I) = (I \vee (j]) \wedge ((j] \vee (k]) \wedge ((k] \vee I).$$

(iii) *For all* $J, K \in I(L)$, *I, J, and K generate a distributive sublattice of* $I(L)$.

(iv) *I is distributive and dually distributive, and, for all* $J, K \in I(L)$,

$$I \wedge J = I \wedge K \quad \textit{and} \quad I \vee J = I \vee K \quad \textit{imply that} \quad J = K.$$

Proof. We can verify that (i) is equivalent to (ii) by using the argument of Lemma 2. The rest of the proof is the same as in Section 2. ●

Observe that every distributive ideal I of a lattice L is the kernel of $\Theta[I]$. Indeed, if $i \in I$, $a \in L$, and $i \equiv a\ (\Theta[I])$, then $i \vee j = a \vee j$ for some $j \in I$, thus $a \leq i \vee j \in I$, and so $a \in I$. This explains why $L/\Theta[I]$ can always be described. This description is especially simple if I is principal, $I = (a]$. Then $\Theta[(a]] = \Theta_a$ is a representable congruence relation and $[a)$ represents Θ_a, hence

$$L/\Theta_a \cong [a).$$

If I is not principal, then we describe $I(L/\Theta[I])$ as follows:

Theorem 7. *Let I be a distributive ideal of a lattice L. Then* $I(L/\Theta[I])$ *is isomorphic with the lattice of all ideals of L containing I, that is, with* $[I, L]$ *in* $I(L)$.

Proof. Let φ be the homomorphism $x \to [x]\Theta$ of L onto L/Θ. Then the map ψ: $K \to K\varphi^{-1}$ maps $I(L/\Theta)$ into $[I, L]$. To show that this map is onto it is sufficient to see that $[J]\Theta = J$ for all $J \supseteq I$. Indeed, if $j \in J$, $a \in L$ and $j \equiv a\ (\Theta[I])$, then $j \vee i = a \vee i$ for some $i \in I$ and so $a \leq i \vee j \in J$ and $a \in J$, as claimed. Since ψ is obviously isotone and one-to-one, we conclude that it is an isomorphism. ●

Corollary 8. $L/\Theta[I]$ *is determined by the interval* $[I, L]$ *of* $I(L)$, *provided that I is a distributive ideal of L.*

Proof. We know (Section II. 3) that $I(L/\Theta[I])$ determines $L/\Theta[I]$ up to isomorphism. ●

Call a congruence relation Θ *distributive*, *standard*, or *neutral*, respectively, iff $\Theta=\Theta[I]$ where I is an ideal which is distributive, standard, or neutral, respectively. The following result was first established by J. Hashimoto [1952] for neutral congruences and by G. Grätzer and E. T. Schmidt [1961] in its present form.

The congruences Φ and Ψ of a lattice L are said to *permute* (or are *permutable*) iff $\Phi\vee\Psi=\Phi\cdot\Psi$ where $\Phi\cdot\Psi$ is the binary relation defined by

$$a\equiv b\ (\Phi\cdot\Psi)\quad \text{iff there exists a } c\in L \text{ with } a\equiv c\ (\Phi) \text{ and } c\equiv b\ (\Psi).$$

An equivalent definition is that Φ and Ψ permute iff $\Phi\cdot\Psi=\Psi\cdot\Phi$, which is, in turn, equivalent to $\Phi\cdot\Psi$ being a congruence relation.

Theorem 9. *Any two standard congruences of a lattice permute.*

Proof. Let Φ and Ψ be standard congruences of the lattice L, that is, $\Phi=\Theta[I]$ and $\Psi=\Theta[J]$, where I and J are standard ideals of L. Let $a\equiv b\ (\Phi)$ and $b\equiv c\ (\Psi)$, $a, b, c\in L$. We want to show that there exists a $d\in L$ satisfying $a\equiv d\ (\Psi)$ and $d\equiv c\ (\Phi)$.

If $x\leq y\leq z$, $x\equiv y\ (\Phi)$, and $y\equiv z\ (\Psi)$, then $y=x\vee i$ for some $i\in I$ and $z=y\vee j$ for some $j\in J$. With $u=x\vee j$ we have $z=u\vee i$, hence $x\equiv u\ (\Psi)$ and $u\equiv z\ (\Phi)$.

Applying this observation to $a\equiv a\vee b\ (\Phi)$, $a\vee b\equiv a\vee b\vee c\ (\Psi)$, and to $c\equiv b\vee c\ (\Psi)$, $b\vee c\equiv a\vee b\vee c\ (\Phi)$, we obtain elements $e, f\in L$ satisfying $a\equiv e\ (\Psi)$, $e\equiv a\vee b\vee c\ (\Phi)$, $a\leq e\leq a\vee b\vee c$, and $c\equiv f\ (\Phi)$, $f\equiv a\vee b\vee c\ (\Psi)$, $c\leq f\leq a\vee b\vee c$. Set $d=e\wedge f$. Then

$$a\equiv e=e\wedge(a\vee b\vee c)\equiv e\wedge f=d\ (\Psi),$$
$$c\equiv f=f\wedge(a\vee b\vee c)\equiv f\wedge e=d\ (\Phi).\quad ●$$

Theorem 9 is significant because in a wide class of lattices all congruences are standard. Recall, that a lattice L is *sectionally complemented* iff L has a zero, 0, and all intervals $[0, a]$ of L are complemented.

Theorem 10. *In a sectionally complemented lattice L all congruence relations are standard.*

Proof. Let Θ be a congruence relation of L and let $I=[0]\Theta$ be the ideal kernel of Θ. Let $a, b\in L$ and $a\equiv b\ (\Theta)$. Let c be the complement of $a\wedge b$ in $[0, a\vee b]$. Then

$$c=c\wedge(a\vee b)\equiv c\wedge(a\wedge b)=0\ (\Theta),$$

hence $c\in I$. Thus $a\equiv b\ (\Theta)$ iff $(a\wedge b)\vee i=a\vee b$ for some $i\in I$. By Theorem 5(iv), this shows that I is standard and $\Theta=\Theta[I]$. ●

Corollary 11. *Let L be a weakly modular sectionally complemented lattice. If L satisfies the Ascending Chain Condition, then all congruences of L are neutral, in fact, of the form Θ_a where a is a neutral element.*

Proof. Let Θ be a congruence relation of L. By Theorem 10, $\Theta = \Theta[I]$ where I is a standard ideal of L. By the Ascending Chain Condition, $I = (a]$. By Corollary 3, a is standard. Since L is weakly modular, by Theorem 2.6, a is neutral. Hence $\Theta = \Theta_a$ with a neutral. ●

Exercises

1. Prove that an ideal generated by a set of distributive elements is distributive.
2. Show that an ideal generated by a set of standard elements is standard.
3. Do the analogues of Exercises 1 and 2 hold for neutral ideals?
4. Verify that the converse of Exercise 2 does not hold. (Hint: Consider the lattice of Figure 1.)

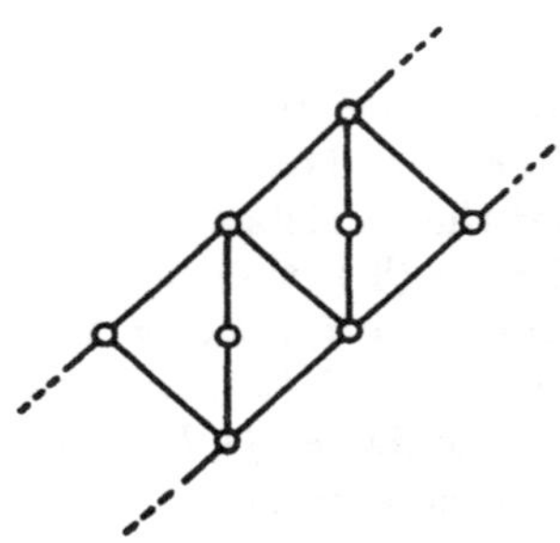

Figure 1

5. Prove Corollary 3 directly.
6. Consider the lattice L as a sublattice of $I(L)$ under the natural embedding $x \to (x]$. Show that every congruence relation of L can be extended to $I(L)$.
7. Characterize those congruence relations of $I(L)$ which are extensions of congruences of L.
8. For any ideal I of a lattice L relate the congruence relation $\Theta[I]$ of L with the congruence relation Θ_I of $I(L)$.
9. Characterize a standard ideal I of L in terms of the congruence relation Θ_I on $I(L)$.
10. Show that in Theorem 5 it is sufficient to assume that condition (iii) holds for principal ideals.

*11. Construct a lattice L and an ideal I of L such that Theorem 5(v) holds for all principal ideals J and K yet I is not standard (Iqbalunissa [1965]).

12. Show that in Theorem 5(v) and Theorem 6(iv) we can assume that $J \supseteq K$.
13. Show that the congruence relation Θ and Φ of the lattice L permute iff, for all $a, b, c \in L$ with $a \leq b \leq c$, $a \equiv b\ (\Theta)$, and $b \equiv c\ (\Phi)$, there exists a $d \in L$ satisfying $a \leq d \leq c$, $a \equiv d\ (\Phi)$, and $d \equiv c\ (\Theta)$.
14. Use Exercise 13 to reprove Theorem 9.
15. Show that $I \to \Theta[I]$ is an isomorphism between the lattice of standard ideals and the lattice of standard congruence relations.
16. Construct a lattice L and standard ideals I_j, $j \in J$, of L such that $I = \wedge(I_j \mid j \in J) \neq \varnothing$ and I is not a standard ideal. (Hint: Consider the lattice of Figure 2.)

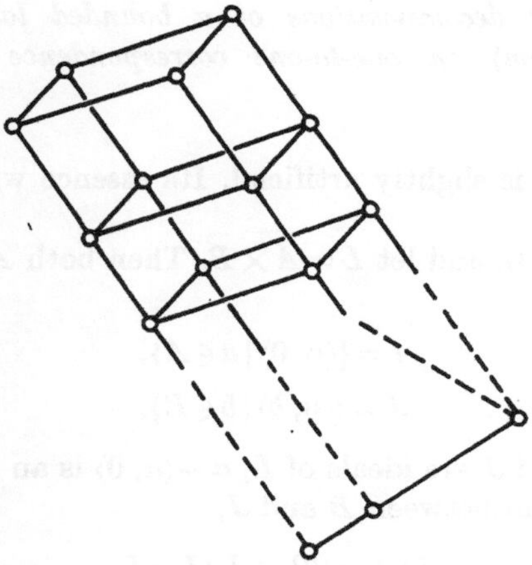

Figure 2

17. Let L be a lattice and let I and J be ideals of L. If I is standard and $I \wedge J$, $I \vee J$ are principal, then J is principal.
18. For a lattice L and standard ideal I of L, let L/I denote the quotient lattice $L/\Theta[I]$. Verify the First Isomorphism Theorem for Standard Ideals:
For any ideal J of L, $I \wedge J$ is a standard ideal of I and

$$I \vee J/I \cong J/I \wedge J.$$

19. Prove the Second Isomorphism Theorem for Standard Ideals:
Let L be a lattice, let I and J be ideals of L, $J \subseteq I$, and let J be standard. Then I is standard iff I/J is standard in L/J, and in this case

$$L/I \cong (L/J)/(I/J).$$

20. State and verify the Second Isomorphism Theorem for Neutral Ideals (J. Hashimoto [1952]).

4. Structure Theorems

Let A and B be bounded lattices and $L = A \times B$. Then $a = \langle 1, 0 \rangle$ and $b = \langle 0, 1 \rangle$ are neutral elements, and they are complementary. Conversely, if L is a bounded lattice, $a, b \in L$, a and b are complementary elements, and a is neutral, then, by Theorem 2.4, the map

$$\varphi: x \to \langle x \wedge a, x \vee a \rangle$$

embeds L into $(a] \times [a)$. Observe that φ is also onto: if $\langle u, v \rangle \in (a] \times [a)$, then $x\varphi = \langle u, v \rangle$ for $x = u \vee (v \wedge b)$. Indeed, the first component of $x\varphi$ is $(u \vee (v \wedge b)) \wedge a = (u \wedge a) \vee (v \wedge a \wedge b) = u$, since $u \leq a$, $a \wedge b = 0$, and a is dually distributive. Similarly, $x \vee a = u \vee (v \wedge b) \vee a = u \vee ((v \vee a) \wedge (a \vee b)) = v$, since a is distributive, $u \leq a \leq v$, and $a \vee b = 1$.

Theorem 1. *The direct decompositions of a bounded lattice L into two factors are (up to isomorphism) in one-to-one correspondence with the complemented neutral elements of L.*

The proof given above is slightly artificial. Its essence will come out better if we consider lattices with 0.

Let L be a lattice with 0, and let $L = A \times B$. Then both A and B have 0 so we can set

$$I = \{\langle a, 0\rangle \mid a \in A\},$$
$$J = \{\langle 0, b\rangle \mid b \in B\}.$$

It is easily seen that I and J are ideals of L, $a \to \langle a, 0\rangle$ is an isomorphism of A and I, $b \to \langle 0, b\rangle$ is an isomorphism between B and J,

$$I \wedge J = \{0\}, \quad I \vee J = L.$$

The last equality holds because $\langle a, b\rangle = \langle a, 0\rangle \vee \langle 0, b\rangle$. For the same reason, every ideal K of L has a unique representation of the form

$$K = I_1 \vee J_1, \quad I_1 \subseteq I, \quad J_1 \subseteq J, \quad I_1, J_1 \in I(L)$$

(with $I_1 = I \cap K$ and $J_1 = J \cap K$) and every such pair $\langle I_1, J_1\rangle$ occurs (indeed for $K = I_1 \vee J_1$). Therefore, the map

$$K \to \langle K \cap I, K \cap J\rangle$$

is a direct product representation of $I(L)$ and so I and J are neutral ideals by Theorem 2.4.

Conversely, let I and J be complementary neutral ideals of L. Then, for any $x \in L$,

$$x \in (x] = (x] \wedge (I \vee J) = ((x] \cap I) \vee ((x] \cap J),$$

and so $x \leq i \vee j$, $i \in (x] \cap I$, $j \in (x] \cap J$. But $i \leq x$, $j \leq x$, hence $x = i \vee j$. So every element x of L has a representation of the form $x = i \vee j$, $i \in I$, $j \in J$. This representation is unique since if $x = i \vee j$, $i \in I$, $j \in J$, then

$$(x] \cap I = (i \vee j] \cap I = ((i] \vee (j]) \cap I = (i] \vee ((j] \cap I) = (i],$$

and similarly

$$(x] \cap J = (j].$$

Since every pair $\langle i, j\rangle$ occurs in some representation we obtain that

$$x \to \langle i, j\rangle$$

is an isomorphism between L and $I \times J$. Generalizing slightly we get the following:

Theorem 2. *Let L be a lattice with 0. There is (up to isomorphism) a one-to-one correspondence between direct decomposition*

$$L = A_0 \times \cdots \times A_{n-1}$$

and n-tuples of neutral ideals $\langle I_0, \ldots, I_{n-1}\rangle$ *which satisfy*

$$I_i \wedge I_j = 0 \quad \textit{for} \quad i \neq j \quad \textit{and} \quad I_0 \vee \cdots \vee I_{n-1} = L.$$

Corollary 3. *Let L be a lattice with 0. Let L have two direct decompositions:*

$$L = A_0 \times \cdots \times A_{n-1},$$
$$L = B_0 \times \cdots \times B_{m-1}.$$

Then L has a direct decomposition

$$L = C_{0,0} \times C_{0,1} \times \cdots \times C_{0,n-1} \times C_{1,0} \times \cdots \times C_{1,n-1} \times \cdots \times C_{m-1,0} \times \cdots \times C_{m-1,n-1}$$

such that

$$A_i \cong C_{0,i} \times \cdots \times C_{m-1,i}, \quad \textit{for} \quad 0 \leq i < n,$$

and

$$B_i \cong C_{i,0} \times \cdots \times C_{i,n-1}, \quad \textit{for} \quad 0 \leq i < m.$$

Proof. Let $\langle I_0, \ldots, I_{n-1}\rangle$ and $\langle J_0, \ldots, J_{m-1}\rangle$ be the neutral ideals associated with the two decompositions as in Theorem 2. Then

$$\langle I_0 \wedge J_0, I_1 \wedge J_0, \ldots, I_{n-1} \wedge J_0, \ldots, I_0 \wedge J_{m-1}, \ldots, I_{n-1} \wedge J_{m-1}\rangle$$

is again a sequence of neutral ideals satisfying the conditions of Theorem 2. (We only need Theorem 2.9 to help verify this.) The direct decomposition associated with this sequence will yield the decomposition required. ●

A lattice L is called *directly indecomposable* iff L has no representation in the form $L = A \times B$ where both A and B have more than one element.

Corollary 4. *Let L be a lattice with 0. If L has a representation*

$$L = A_0 \times \cdots \times A_{n-1},$$

where each A_i, $0 \leq i < n$, *is directly indecomposable, then for any other decomposition*

$$L = B_0 \times \cdots \times B_{m-1}$$

of L into directly indecomposable factors we have that $n = m$ *and that there exists a permutation* α *of* $\{0, \ldots, n-1\}$ *such that* $A_i \cong B_{i\alpha}$, *for all* $0 \leq i < n$.

Results analogous to Corollaries 3 and 4 also hold for general lattices, see the exercises.

In trying to sharpen Corollary 4 to a structure theorem there are two difficulties we have to overcome. A lattice need not have a decomposition into directly indecomposable factors. Directly indecomposable lattices are hard to accept as "building blocks"; one would rather have *simple lattices*, that is, lattices with only the two trivial congruences, ω and ι.

The first difficulty is easy to overcome by chain conditions. Observe that if A and B are lattices with 0, $C=\{0, a_1, a_2, \ldots, a_n\}$ is a chain in A, and $D=\{0, b_1, \ldots, b_m\}$ is a chain in B, then

$$E=\{\langle 0,0\rangle, \langle a_1, 0\rangle, \ldots, \langle a_n, 0\rangle, \langle a_n, b_1\rangle, \ldots, \langle a_n, b_m\rangle\}$$

is a chain in $A \times B$ of length $n+m$. This easily implies that $l(A\times B)=l(A)+l(B)$ for lattices A and B of finite length.

Lemma 5. *Let L be a bounded lattice. If L is of finite length, then L is isomorphic to a direct product of directly indecomposable lattices.*

The passage from " directly indecomposable " to " simple " requires much stronger hypotheses.

Theorem 6. *Let L be a sectionally complemented weakly modular lattice. If L is of finite length, then L can be represented as a direct product of simple lattices.*

Proof. First observe that if $L \cong A\times B$ and L satisfies the hypotheses of this theorem then so do A and B; only the statement about weak modularity has to be verified and it follows from the observation that

$$\langle a_1, a_2\rangle/\langle b_1, b_2\rangle \approx_w \langle c_1, c_2\rangle/\langle d_1, d_2\rangle \quad \text{implies that} \quad a_1/b_1 \approx_w c_1/d_1.$$

By Lemma 5, $L\cong L_1\times\cdots\times L_k$ where all L_i are directly indecomposable. By the observation above, all L_i satisfy the hypotheses of this theorem. So it is sufficient to prove that if L satisfies the hypotheses of the theorem and L is directly indecomposable, then L is simple. Indeed, if Θ is a congruence relation of L, then $\Theta=\Theta[I]$ where I is a standard ideal by Theorem 3.10. By the chain condition, $I=(a]$, and a is standard by Corollary 3.3. By Theorem 2.6 a is neutral. Since L is complemented, by Theorem 1 $(a]$ is a direct factor of L. But L is directly indecomposable, hence $a=0$ or $a=1$, yielding $\Theta=\omega$ or $\Theta=\iota$, verifying that L is simple. ●

Since modular lattices and relatively complemented lattices are special cases of weakly modular lattices we obtain two famous structure theorems as special cases:

Corollary 7 (The Birkhoff-Menger Theorem, G. Birkhoff [1935b] and K. Menger [1936]). *Let L be a complemented modular lattice. If L is of finite length, then L is isomorphic to a direct product of simple lattices.*

We shall see in Chapter IV that these simple lattices are exactly $\mathfrak{C}_2$ and the nondegenerate projective geometries of finite dimension.

Corollary 8 (R. P. Dilworth [1950]). *Let L be a relatively complemented lattice. If L is of finite length, then L is isomorphic to a direct product of simple lattices.*

If L is a direct product of finitely many simple lattices $L_1, \ldots, L_k$, then by Theorem I.3.13

$$C(L) \cong C(L_1) \times \cdots \times C(L_k) \cong (\mathfrak{C}_2)^k,$$

that is, $C(L)$ is a Boolean lattice. This led G. Birkhoff to suggest (in G. Birkhoff [1948]) the study of lattices L for which $C(L)$ is Boolean. We are going to present a solution to this problem.

Let us call a congruence relation Θ of a lattice L *separable* iff, for all $a, b \in L$, $a < b$, there exists a sequence $a = x_0 < x_1 < \cdots < x_n = b$ such that for each $0 \leq i < n$, $x_i \equiv x_{i+1}\ (\Theta)$ or $u \equiv v(\Theta)$ for no proper subquotient u/v of x_{i+1}/x_i.

Theorem 9 (G. Grätzer and E. T. Schmidt [1958 d]). *For a lattice L, $C(L)$ is Boolean iff L is weakly modular and all congruences of L are separable.*

Proof. Let $C(L)$ be Boolean. Let

$$a/b \approx_w c/d,\ a, b, c, d \in L,\ a \neq b;$$

we wish to find a proper subquotient c'/d' of c/d satisfying $c'/d' \approx_w a/b$. Let $\Theta = \Theta(a, b)$ and let Θ' be the complement of Θ. Then $\Theta \vee \Theta' = \iota$ and so $c \equiv d\ (\Theta \vee \Theta')$. By Theorem I.3.9, there is a sequence $d = x_0 < x_1 < \cdots < x_n = c$ such that, for each $0 \leq i < n$,

$$x_i \equiv x_{i+1}\ (\Theta) \quad \text{or} \quad x_i \equiv x_{i+1}\ (\Theta').$$

If, for all $0 \leq i < n$, $x_i \equiv x_{i+1}\ (\Theta')$, then $c \equiv d\ (\Theta')$; since $a/b \approx_w c/d$, $a \equiv b\ (\Theta')$. But $\Theta = \Theta(a, b)$ and so $a \equiv b\ (\Theta)$. We conclude that $a \equiv b\ (\Theta \wedge \Theta')$, that is, $a \equiv b\ (\omega)$, contradicting $a \neq b$. Therefore, $x_i \equiv x_{i+1}\ (\Theta)$ for some $0 \leq i < n$. Applying Theorem 1.2 to $x_i \equiv x_{i+1}\ (\Theta(a, b))$, we obtain a proper subquotient c'/d' of x_{i+1}/x_i satisfying $c'/d' \approx_w a/b$, proving that L is weakly modular.

Now let Θ be a congruence relation of L and let $a, b \in L$, $a < b$. Then $a \equiv b\ (\Theta \vee \Theta')$, where Θ' is the complement of Θ, and so by Theorem I.3.9 there exists a sequence $a = x_0 < x_1 < \cdots < x_n = b$ such that, for each $0 \leq i < n$,

$$x_i \equiv x_{i+1}\ (\Theta) \quad \text{or} \quad x_i \equiv x_{i+1}\ (\Theta').$$

We claim that the same sequence establishes the separability of Θ. Indeed, if for some i, $x_i \not\equiv x_{i+1}\ (\Theta)$, and $u, v \in [x_i, x_{i+1}]$, $u \equiv v\ (\Theta)$, then $x_i \equiv x_{i+1}\ (\Theta')$ and so $u \equiv v\ (\Theta')$, implying that $u \equiv v\ (\Theta \wedge \Theta')$, that is $u = v$.

To prove the converse, let us start out with a weakly modular lattice L and a congruence relation Θ of L. We claim:

The binary relation Φ defined on L by

$$a \equiv b\ (\Phi) \quad \text{iff} \quad u \equiv v\ (\Theta) \quad \text{for no proper subquotient } u/v \text{ of } a \vee b/a \wedge b$$

is a congruence relation; in fact, Φ is the pseudocomplement of Θ.

We prove the first part of this claim by verifying that Φ satisfies conditions (i)–(iii) of Lemma I.3.8. Condition (i) is clear by the definition of Φ. To verify (ii), let a, b,

$c \in L$, $a \le b \le c$, $a \equiv b$ (Φ), and $b \equiv c$ (Φ); we wish to show that $a \equiv c$ (Φ). Assume to the contrary that $a \not\equiv c(\Phi)$; then $u \equiv v$ (Θ) for some proper subquotient u/v of c/a. Since $u \equiv v$ $(\Theta(a, c))$ and $\Theta(a, c) = \Theta(a, b) \vee \Theta(b, c)$, by Theorem 1.2 and Lemma 1.3 there exists a proper subquotient x/y of u/v satisfying $x/y \approx_w c/b$ or $x/y \approx_w b/a$, say, $x/y \approx_w c/b$. Therefore, $b < c$. By weak modularity, there exists a proper subquotient c_1/b_1 of c/b satisfying $c_1/b_1 \approx_w x/y$. But $x, y \in [v, u]$ and $u \equiv v$ (Θ), so $x \equiv y$ (Θ). Hence $c_1 \equiv b_1(\Theta)$, contradicting $c_1, b_1 \in [b, c]$ and $b \equiv c$ (Φ). This shows that (ii) holds.

To verify (iii), assume to the contrary that $a, b, c \in L$, $a \le b$, $a \equiv b$ (Φ), but $a \wedge c \not\equiv b \wedge c$ (Φ); then there exists a proper subquotient u/v of $b \wedge c/a \wedge c$ satisfying $v \equiv u$ (Θ). Then $u/v \approx_w b \wedge c/a \wedge c \approx_w b/a$, hence $u/v \approx_w b/a$; so $a < b$ and by weak modularity $b_1/a_1 \approx_w u/v$, where b_1/a_1 is a proper subquotient of b/a. Since $u \equiv v$ (Θ) we obtain $b_1 \equiv a_1$ (Θ), contradicting $a \equiv b$ (Φ). By duality, $a \vee c \equiv b \vee c$ (Φ).

This shows that Φ is a congruence relation. $\Theta \wedge \Phi = \omega$ is trivial. Now if $\Theta \wedge \Psi = \omega$, $a \equiv b$ (Ψ), $u, v \in [a \wedge b, a \vee b]$, and $u \equiv v$ (Θ), then also $u \equiv v$ (Ψ), hence $u = v$, showing that $\Psi \subseteq \Phi$. This completes the proof of the claim.

Now to complete the proof of the theorem, let L be weakly modular and let the congruences of L be separable. Let Θ be a congruence of L. Let Φ be the congruence constructed in the claim. For $a, b \in L$, $a < b$, let $a = x_0 < x_1 < \cdots < x_n = b$ be the chain establishing the separability of Θ. Then for each $0 \le i < n$, either $x_i \equiv x_{i+1}$ (Θ), or, by the definitions of separability and of Φ, $x_i \equiv x_{i+1}$ (Φ). Therefore, $a \equiv b$ $(\Theta \vee \Phi)$, and so $\Theta \vee \Phi = \iota$. This shows that Φ is a complement of Θ. Since $C(L)$ is distributive (Theorem II.3.11) this proves that $C(L)$ is Boolean. ●

We get a large number of corollaries from Theorem 9 yielding that $C(L)$ is Boolean for special classes of lattices. Some of these will be discussed in the exercises.

The *centre* of a bounded lattice L is the sublattice $\mathrm{Cen}(L)$ of complemented neutral elements. It is obvious that $0, 1 \in \mathrm{Cen}(L)$ and $\mathrm{Cen}(L)$ is a Boolean lattice. In some chapters of lattice theory it is important to know when $\mathrm{Cen}(L)$ is a complete lattice. (If this is the case L can often be "coordinatized" over $\mathrm{Cen}(L)$.) This was established for continuous geometries by J. von Neumann [1936/37].

Let L be a complete lattice and $K \subseteq L$. Then K is a *complete sublattice* of L iff $\bigwedge X$ and $\bigvee X \in K$ for all $X \subseteq K$ ($\bigwedge X$ and $\bigvee X$ are formed in L).

THEOREM 10. *Let L be a complete, sectionally complemented, dually sectionally complemented, weakly modular lattice. Then the centre of L is a complete sublattice of L.*

PROOF. Let $X \subseteq \mathrm{Cen}(L)$. Set $a = \bigwedge X$. Then $(a]$ is the kernel of the congruence relation

$$\bigwedge (\Theta_x \mid x \in X).$$

Hence, by Theorem 3.10, $(a]$ is standard; by Corollary 3.3, a is standard. Now Corollary 2.8 tells us that a is neutral. Since a is complemented, $a \in \mathrm{Cen}(L)$.

By duality, we obtain $\bigvee X \in \mathrm{Cen}(L)$. ●

Corollary 11 (M. F. Janowitz [1967]). *The centre of a complete relatively complemented lattice is a complete sublattice.*

This result has some interesting implications concerning direct decompositions of complete relatively complemented lattices; see S. Maeda [1966].

Exercises

1. Show that the representations of a lattice L with 0 as a direct product of two lattices are (up to isomorphism) in one-to-one correspondence with pairs of ideals $\langle I, J\rangle$ satisfying $I \cap J = \{0\}$ and every element a of L has exactly one representation of the form $a = i \vee j$, $i \in I$, $j \in J$.
2. Let $L = A_1 \times A_2$. Define the binary relation Θ_i on L by $\langle a_1, a_2\rangle \equiv \langle b_1, b_2\rangle\ (\Theta_i)$ iff $a_i = b_i$ $(i = 1, 2)$. Show that Θ_1 and Θ_2 are congruence relations of L, $\Theta_1 \wedge \Theta_2 = \omega$, and $\Theta_1 \vee \Theta_2 = \iota$.
3. Show that Θ_1 and Θ_2 of Exercise 2 are permutable.
4. Prove that the representations of a lattice L as a direct product of two lattices are (up to isomorphism) in a one-to-one correspondence with pairs of congruence relations $\langle \Theta_1, \Theta_2\rangle$ which are complementary and permutable.
5. Show that if in Exercise 4 we pass from two to n factors, then we get an n-tuple of congruences $\langle \Theta_1, \ldots, \Theta_n\rangle$ such that $\Theta_1 \wedge \cdots \wedge \Theta_n = \omega$, $(\Theta_1 \wedge \cdots \wedge \Theta_{i-1}) \vee \Theta_i = \iota$, and $\Theta_1 \wedge \cdots \wedge \Theta_{i-1}$ and Θ_i permute, for $i = 2, \ldots, n$.
6. Can we replace in Exercise 5 the condition "$\Theta_1 \wedge \cdots \wedge \Theta_{i-1}$ and Θ_i permute" by "any two distinct Θ_i and Θ_j permute"?
7. Use Exercise 5 to verify Corollary 3 for arbitrary lattices.
8. Verify Corollary 4 for arbitrary lattices.
9. Relate Exercise 5 to Theorem 2.
10. Let B be the Boolean lattice R-generated by the rational interval $[0, 1]$. Show that, for any natural number n, B has a representation as a direct product of n lattices $L_1, \ldots, L_n$, $|L_i| > 1$ for $i = 1, \ldots, n$, but B has no representation as a direct product of directly indecomposable lattices.
11. Construct a lattice which is not of finite length but every chain in the lattice is finite.
12. Statements 5–8 of this section deal with lattices of finite length. Which of these statements remain valid for lattices in which every chain is finite?
13. Prove that every congruence relation of a finite lattice or of a lattice of finite length is separable.[1]
14. Verify that if in a lattice L, for every $a, b \in L$, $a < b$, there is a finite maximal chain in $[a, b]$, then all congruence relations of L are separable. This holds, in particular, if the lattice is *locally finite*, that is, if all intervals are finite.
15. Let L be a distributive lattice, $a, b, a_1, a_2, \ldots \in L$, $a = a_1 < a_2 < a_3 < \cdots < b$. Then $\Theta = \bigvee(\Theta(a_{2i-1}, a_{2i}) \mid i = 1, 2, \ldots)$ is not a separable congruence relation.
16. For a distributive lattice L, $C(L)$ is Boolean iff L is locally finite. (Use Exercises 14 and 15. J. Hashimoto [1952].)
17. The separable congruences of a lattice L form a sublattice of $C(L)$.

[1] Exercises 13–19 are based on G. Grätzer and E. T. Schmidt [1958d].

18. Let L be a lattice with 1. A neutral congruence relation $\Theta[I]$ is separable iff I is principal.
19. Let L be a complemented modular lattice. Then $C(L)$ is Boolean iff all neutral ideals of L are principal (Shih-chiang Wang [1953]).
20. Find a complete lattice L and a sublattice K of L such that K is a complete lattice but not a complete sublattice of L.

Further Topics and References

The properties of weak projectivity and of the congruences $\Theta[I]$ are discussed in greater detail in G. Grätzer and E. T. Schmidt [1958d]. For universal algebras A. I. Mal'cev introduced similar concepts; the difference is that while for lattices it is sufficient to consider " unary algebraic functions " of a special form (namely, ... $(((x \wedge a_0) \vee a_1) \wedge a_2) \cdots$), for universal algebras in general we have to consider arbitrary unary algebraic functions (see, for instance G. Grätzer [1968]). The polynomial p_2 is also of special interest; identities of p_2 that hold for lattices imply that the congruence lattices of lattices are distributive. A general condition for the distributivity of congruence lattices of algebras in a given equational class of algebras can be found in B. Jónsson [1967].

Weak modularity is a rather complicated condition. It can be somewhat simplified for finite lattices: a finite lattice is weakly modular iff $a/b \approx_w c/d$ and $a > b \geq c > d$ imply the existence of a proper subquotient c'/d' of c/d satisfying $c'/d' \approx_w a/b$ (G. Grätzer [1963a]).

In general, if in a lattice L any interval contains a finite maximal chain, then it is sufficient to consider weak projectivities of prime quotients, see for instance N. Funayama [1942], J. Jakubik [1955a] and D. T. Finkbeiner [1960]. Under this condition, L is simple iff any pair of prime quotients are projective. In general, no bound can be put on the number of perspectivities. However, if L is a relatively complemented lattice of length n, $a, b \succ 0$ and $a/0 \approx b/0$, then $a/0 \overset{k}{\approx} b/0$, where $k \leq 2\left[\frac{1}{2}(n+1)\right]$, see J. E. McLaughlin [1951] and [1953].

The investigation of the number of projectivities becomes important in the study of equational classes of lattices, see Section V.3 and C. Herrmann [1973].

Local conditions implying the projectivity of any pair of prime quotients can be found in F. A. Smith [1974].

Modular lattices and relatively complemented lattices share a property stronger than weak modularity. In P. Crawley and R. P. Dilworth [1973] a lattice is said to have the *Projectivity Property* iff whenever a/b is weakly projective into c/d, then a/b is projective to some subquotient c'/d' of c/d.

Weak modularity seems to be a very natural concept. It appears quite surprisingly in some results. The following result of Iqbalunissa [1966] is a good illustration of this point: if in the lattice L every congruence relation is neutral, then L is weakly modular.

A lot more information on standard elements and ideals can be found in G. Grätzer [1959] and G. Grätzer and E. T. Schmidt [1961]. Among the topics discussed in these papers are the Isomorphism Theorems, the Zassenhaus Lemma, the Jordan-Hölder Theorem for standard ideals, and the Schreier extension problem. It is also shown that in a finite modular lattice exactly the neutral ideals satisfy the First Isomorphism Theorem. Some of these results were inspired by J. Hashimoto [1952].

In the proof of Theorem 4.9 we describe the pseudocomplement Φ of a congruence relation Θ of a weakly modular lattice. Iqbalunissa [1966] observes that the converse also holds: if in a lattice L the relation Φ given by any congruence relation Θ (as described in the proof of Theorem 4.9) is always a congruence relation, then L is weakly modular.

For some additional results on standard ideals see M. F. Janowitz [1964], [1964a], and [1965], in which some types of ideals, more general than standard ideals, are discussed. Congruences of relatively complemented lattices, in particular a description of $\Theta(a, b)$, is given in M. F. Janowitz [1968]. Some counterexamples are given in Iqbalunissa [1965] and [1965a].

Lattices whose congruences form a Boolean lattice are discussed also in T. Tanaka [1952], J. Hashimoto [1957], P. Crawley [1960]. See also D. T. Finkbeiner [1960] and J. Jakubik [1955]. Lattices whose congruences form a Stone lattice are studied in M. F. Janowitz [1968] and [1968a], and Iqbalunissa [1971].

Standard elements in the lattice of all subsemigroups of a group are studied in S. G. Ivanov [1966].

It is pointed out in S. Maeda [1974], that if L is the dual of the lattice of all T_1-topologies on an infinite set, then L has infinitely many standard elements, but only the elements 0 and 1 are neutral.

The concept of a standard ideal has recently been extended to convex sublattices in E. Fried and E. T. Schmidt [1975].

Distributivity of a pair of lattice elements is investigated in P. G. Kontorovič, S. G. Ivanov, and G. P. Kondrašov [1965]; see also F. and S. Maeda [1970]. We shall consider modular pairs of elements in Section IV.2.

A great deal of useful information on congruence relations of lattices, in particular about regularity and permutability, can be found in J. Hashimoto [1963].

There is a connection between projectivity and representability. If L/Θ is projective (in a class containing L), then Θ is representable. A. Day [1973] considers a weaker concept implying the representability of Θ for finite L.

Problems

III. 1. Generalize the concepts of distributive and neutral ideals to convex sublattices. (For standard convex sublattices, see E. Fried and E. T. Schmidt [1975].)

III. 2. Let $L = I_0 \supseteq I_1 \supseteq \cdots \supseteq I_{n+1} = I$ be a descending sequence of ideals. If I_{j+1} is a standard ideal of I_j for $j = 0, \ldots, n$, then I is a *standard ideal of order n* of L. Investigate standard ideals of order 2 (order n). Under what conditions do they form a sublattice?

III. 3. An ideal I of a lattice L is said to *satisfy the First Isomorphism Theorem* (G. Grätzer and E. T. Schmidt [1958d]) iff $(I \vee J)/\Theta[I] \cong J/\Theta[I \cap J]$ for any ideal J of L (under the natural isomorphism). Investigate this concept and relate it to standard ideals.

III. 4. Under what conditions do the ideals of a weakly modular lattice form a weakly modular lattice?

III. 5. Investigate lattices whose congruences form a Stone lattice.

III. 6. For $n > 1$, investigate lattices whose congruence lattices, as distributive lattices with pseudocomplementation, belong to $\mathbf{B}_n$.

III. 7. Develop structure theorems for lattices all of whose congruences are standard (distributive, neutral).

CHAPTER IV

MODULAR AND SEMIMODULAR LATTICES

1. Modular Lattices

The modular identity is unquestionably the most important identity apart from the distributive identity. In this section we examine the most important consequences of modularity.

THEOREM 1. *For a lattice L, the following conditions are equivalent:*

(i) *L is modular, that is,*
$x \geq z$ *implies that* $x \wedge (y \vee z) = (x \wedge y) \vee z$.

(ii) *L satisfies the* shearing identity[1]:
$x \wedge (y \vee z) = x \wedge ((y \wedge (x \vee z)) \vee z)$.

(iii) *L does not contain a pentagon.*

REMARK. In Section II.1 we have already proved the equivalence of (i) and (iii). The importance, or convenience, of the shearing identity is that it can be applied to any expressions of the form $x \wedge (y \vee z)$ without any assumptions. Observe also the dual of the shearing identity:

$$x \vee (y \wedge z) = x \vee ((y \vee (x \wedge z)) \wedge z).$$

PROOF. (i) *implies* (ii). Since $x \vee z \geq z$, by modularity, $(y \wedge (x \vee z)) \vee z = (y \vee z) \wedge (x \vee z)$, and so

$$x \wedge ((y \wedge (x \vee z)) \vee z) = x \wedge ((y \vee z) \wedge (x \vee z)) = x \wedge (y \vee z). \quad ◐$$

(ii) *implies* (iii). In $\mathfrak{N}_5$ (ii) fails; indeed,

$$a \wedge (c \vee b) = a \wedge i = a \quad \text{and} \quad a \wedge ((c \wedge (a \vee b)) \vee b) = a \wedge ((c \wedge a) \vee b) = a \wedge b = b.$$

Thus (ii) implies (iii). ◑

(iii) *implies* (i) as in Section II.1. ●

The most important form of modularity is the following:

---

[1] This identity was named by I. Halperin.

Theorem 2 (The Isomorphism Theorem). *Let L be a modular lattice and let $a, b \in L$. Then*

$$\varphi_b\colon x \to x \wedge b$$

is an isomorphism of $[a, a \vee b]$ and $[a \wedge b, b]$. The inverse isomorphism is

$$\psi_a\colon x \to x \vee a.$$

(See Figure 1.)

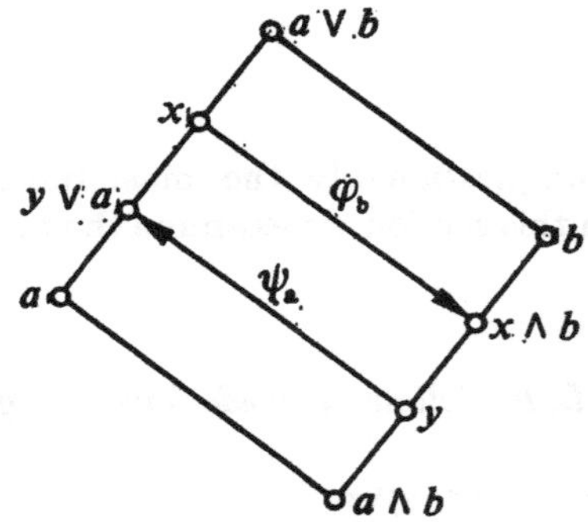

Figure 1

Proof. It is sufficient to show that, for $x \in [a, a \vee b]$, $x\varphi_b\psi_a = x$. Indeed, if this is true, then by duality, $x\psi_a\varphi_b = x$ for all $x \in [a \wedge b, b]$. The isotone maps φ_b and ψ_a thus compose into the identity maps, hence they are isomorphisms, as claimed.

So let $x \in [a, a \vee b]$. Thus $x\varphi_b\psi_a = (x \wedge b) \vee a$. Since $x \in [a, a \vee b]$, we have $a \leq x$ and so modularity applies:

$$x\varphi_b\psi_a = (x \wedge b) \vee a = x \wedge (b \vee a) = x,$$

because $x \leq a \vee b$. ●

Corollary 3. *Let L be a modular lattice and let $a, b \in L$. Then for $x, y \in [a \wedge b, b]$ we have*

$$a \vee (x \wedge y) = (a \vee x) \wedge (a \vee y).$$

Most applications of the Isomorphism Theorem are like Corollary 3: the special form of the isomorphism is used. There are two important applications where the form of the isomorphism plays no role.

Let us write $a \preceq b$ ($b \succeq a$) for $a \prec b$ or $a = b$. A lattice L is said to satisfy the *Upper Covering Condition* iff $a \preceq b$ implies that $a \vee c \preceq b \vee c$, for $a, b, c \in L$. The *Lower Covering Condition* is the dual.

Theorem 4. *A modular lattice satisfies both the Upper and the Lower Covering Condition.*

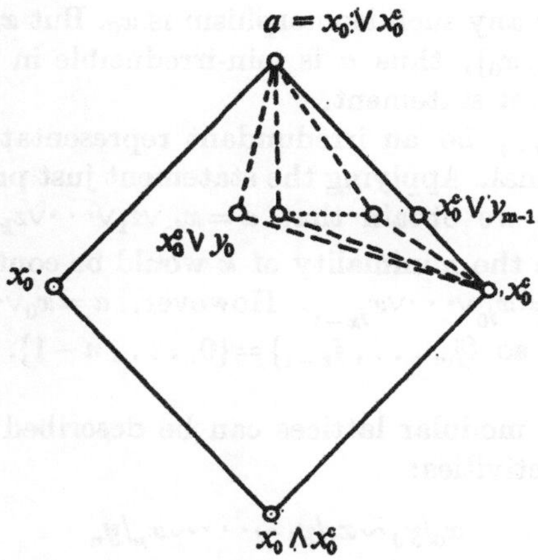

Figure 2

Proof. Let L be a modular lattice. Let $a, b, c \in L$ and $a \prec b$. If $a \vee c = b \vee c$ we have nothing to prove. If $a \vee c \neq b \vee c$, then $a \nleq b \vee c$, and so $a \wedge (b \vee c) = b$. Applying the Isomorphism Theorem to a and $b \vee c$ we obtain

$$[b, a] \cong [b \vee c, a \vee c].$$

Since $[b, a]$ is a prime interval, so is $[b \vee c, a \vee c]$, that is, $b \vee c \prec a \vee c$. By duality, we get the Lower Covering Condition. ●

The second application deals with representations of elements. In Corollary II.1.13 we proved that in a finite distributive lattice the irredundant representation of an element as a join of join-irreducible elements is unique. This is obviously false in $\mathfrak{M}_3$. But we have the following result:

Theorem 5 (The Kuroš-Ore Theorem, A. G. Kuroš [1935], O. Ore [1936]). *Let L be a modular lattice and let $a \in L$. If $a = x_0 \vee \cdots \vee x_{n-1}$ and $a = y_0 \vee \cdots \vee y_{m-1}$ are irredundant representations of a as joins of join-irreducible elements, then for every x_i there is a y_j such that*

$$a = x_0 \vee \cdots \vee x_{i-1} \vee y_j \vee x_{i+1} \vee \cdots \vee x_{n-1}$$

and $n = m$.

Proof. Let us prove the first statement say for x_0. Let $x_0^c = x_1 \vee \cdots \vee x_{n-1}$. (See Figure 2.) Since $y_0 \vee \cdots \vee y_{m-1} = a$ we obtain that

$$(x_0^c \vee y_0) \vee (x_0^c \vee y_1) \vee \cdots \vee (x_0^c \vee y_{m-1}) = a,$$

where $x_0^c \vee y_0, \ldots, x_0^c \vee y_{m-1} \in [x_0^c, a]$. By the Isomorphism Theorem,

$$[x_0^c, a] \cong [x_0 \wedge x_0^c, x_0],$$

and the image of a under any such isomorphism is x_0. But x_0 is join-irreducible in L, and, therefore, in $[x_0 \wedge x_0^c, x_0]$; thus a is join-irreducible in $[x_0^c, a]$. Hence $x_0^c \vee y_j^c = a$ for some j, proving the first statement.

Now let $a = z_0 \vee \cdots \vee z_{k-1}$ be an irredundant representation of a as join of join-irreducibles with k minimal. Applying the statement just proved to $a = z_0 \vee \cdots \vee z_{k-1}$, z_0, and $a = x_0 \vee \cdots \vee x_{n-1}$, we obtain that $a = x_{j_0} \vee z_1 \vee \cdots \vee z_{k-1}$. This representation is irredundant, otherwise the minimality of k would be contradicted. Repeating this we eventually obtain $a = x_{j_0} \vee \cdots \vee x_{j_{k-1}}$. However, $a = x_0 \vee \cdots \vee x_{n-1}$ is an irredundant representation and so $\{j_0, \ldots, j_{k-1}\} = \{0, \ldots, n-1\}$. This shows that $n \leq k$. Thus $k = n$ $(= m)$. ●

Weak projectivities in modular lattices can be described rather precisely. Let us call a sequence of perspectivities:

$$x_0/y_0 \sim x_1/y_1 \sim \cdots \sim x_n/y_n$$

alternating iff, for each i with $0 < i < n$, either $x_{i-1}/y_{i-1} \nearrow x_i/y_i \searrow x_{i+1}/y_{i+1}$ or $x_{i-1}/y_{i-1} \searrow x_i/y_i \nearrow x_{i+1}/y_{i+1}$. If, in addition, $x_i = x_{i-1} \vee x_{i+1}$ in the first case and $y_i = y_{i-1} \wedge y_{i+1}$ in the second, we call the sequence *normal*. (See Figure 3.)

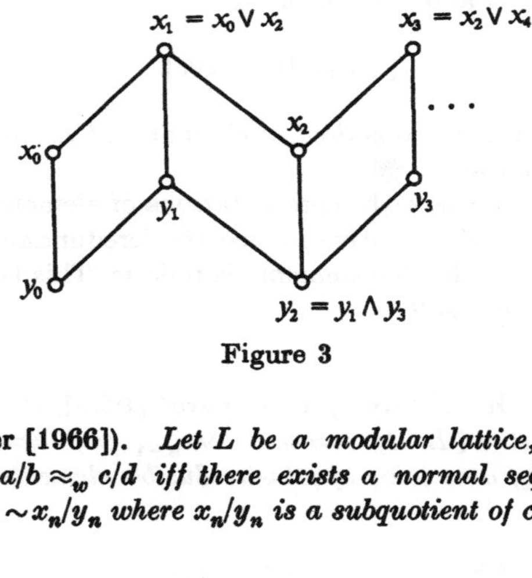

Figure 3

Theorem 6 (G. Grätzer [1966]). *Let L be a modular lattice, and let a/b and c/d be quotients of L. Then $a/b \approx_w c/d$ iff there exists a normal sequence of perspectivities $a/b = x_0/y_0 \sim x_1/y_1 \sim \cdots \sim x_n/y_n$ where x_n/y_n is a subquotient of c/d.*

Proof. It follows from the Isomorphism Theorem that if $x/y \nearrow u/v$ and x_1/y_1 is a subquotient of x/y, then $x_1/y_1 \nearrow u_1/v_1$ where $u_1 = y_1 \vee v$ and $v_1 = x_1 \vee v$. This, the dual statement, and a simple induction show that if $a/b \approx_w c/d$, then there is an alternating sequence of perspectivities

$$a/b = x_0/y_0 \sim \cdots \sim x_n/y_n$$

where x_n/y_n is a subquotient of c/d. Starting with this sequence we show by induction on n that there is a normal sequence of perspectivities

$$a/b = u_0/v_0 \sim u_1/v_1 \sim \cdots \sim u_n/v_n = x_n/y_n.$$

This statement is obvious for $n \leq 1$. By duality, we can assume that $u_{n-1}/v_{n-1} \searrow u_n/v_n$. By the induction hypothesis, there is a normal sequence

$$a/b = u_0/v_0 \sim \cdots \sim u_{n-1}/v_{n-1} = x_{n-1}/y_{n-1}.$$

We define a new sequence as follows:

$$a/b = u_0/v_0 \sim u_1/v_1 \sim \cdots \searrow u_{n-2}/v_{n-2} \nearrow u_{n-2} \vee x_n/v_{n-1} \wedge (u_{n-2} \vee x_n) \searrow x_n/y_n.$$

We claim that this is a normal sequence of perspectivities. The following formulas have to be checked; (i) and (ii) are the perspectivities, and (iii) the normality:

(i) $u_{n-2}/v_{n-2} \nearrow u_{n-2} \vee x_n/v_{n-1} \wedge (u_{n-2} \vee x_n)$;
(ii) $u_{n-2} \vee x_n/v_{n-1} \wedge (u_{n-2} \vee x_n) \searrow x_n/y_n$;
(iii) $v_{n-2} = v_{n-3} \wedge (v_{n-1} \wedge (u_{n-2} \vee x_n))$.

Since $v_{n-2} \leq v_{n-1} \wedge (u_{n-2} \vee x_n) \leq v_{n-1}$,

$$u_{n-2} \wedge (v_{n-1} \wedge (u_{n-2} \vee x_n)) = v_{n-2}$$

is obvious. Now compute:

$$\begin{aligned} &u_{n-2} \vee (v_{n-1} \wedge (u_{n-2} \vee x_n)) \quad \text{(by modularity and } u_{n-2} \leq u_{n-2} \vee x_n) \\ &= (u_{n-2} \vee v_{n-1}) \wedge (u_{n-2} \vee x_n) \\ &= u_{n-2} \vee x_n, \end{aligned}$$

verifying (i). The proof of (ii) is similar. For (iii), observe that $v_{n-2} \leq u_{n-2} \vee x_n$ and $v_{n-3} \wedge v_{n-1} = v_{n-2}$, hence

$$v_{n-3} \wedge (v_{n-1} \wedge (u_{n-2} \vee x_n)) = (v_{n-3} \wedge v_{n-1}) \wedge (u_{n-2} \vee x_n) = v_{n-2}.$$

This completes the induction step. ●

We can do even more than this; we can determine the sublattices generated by three consecutive quotients.

Theorem 7 (G. Grätzer [1966]). *Let L be a modular lattice and let $x_0/y_0 \nearrow x_1/y_1 \searrow x_2/y_2$ be a normal sequence of quotients. Let A be the sublattice of L generated by x_0, x_1, x_2, y_0, y_1, and y_2. Then either A is distributive, in which case A is the lattice of Figure 4 or some quotient thereof and we also have*

$$x_0/y_0 \searrow (x_0 \wedge x_2)/(y_0 \wedge y_2) \nearrow x_2/y_2,$$

or A is not distributive and A is the lattice of Figure 5 or some quotient not collapsin the diamond in A.

Proof. A is generated by the elements x_0, y_1, and x_2, hence A is isomorphic to a quotient lattice of $F_M(3)$, see Figure I.5.7. The generators satisfy the relation $x_0 \vee y_1 = y_1 \vee x_2 = x_0 \vee x_2$ (the last one comes in because of normality). The corresponding quotient lattice is given in Figure 5. A quotient of this is distributive iff the diamond is collapsed, yielding Figure 4. ●

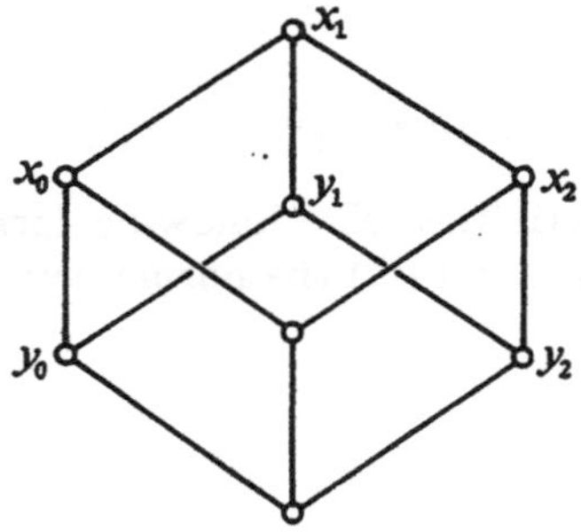

Figure 4

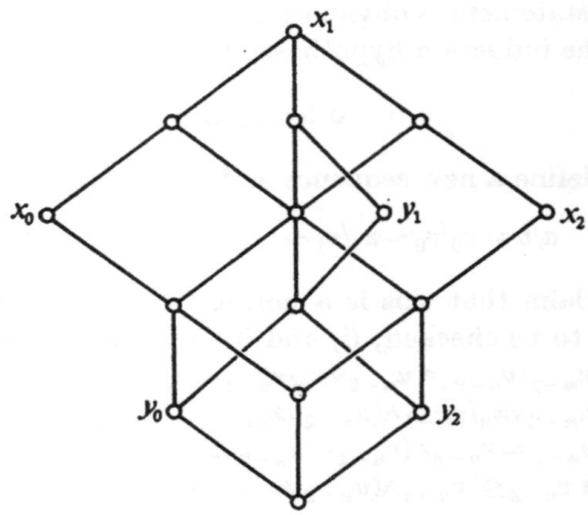

Figure 5

Corollary 8. *Let L be a modular lattice and let $a/b \approx c/d$ in L. There is a shortest normal sequence of quotients $a/b = x_0/y_0 \sim x_1/y_1 \sim \cdots \sim x_n/y_n = c/d$. Then either $n \leq 2$, or for each i, $0 < i < n$, the sublattice generated by $x_{i-1}, x_i, x_{i+1}, y_{i-1}, y_i, y_{i+1}$ is isomorphic to one of the lattices of Figure 5, Figure 6, or their duals.*

Proof. If $n > 2$, then the sublattice generated by $x_{i-1}, x_i, x_{i+1}, y_{i-1}, y_i, y_{i+1}$ cannot be distributive, otherwise we could turn the down arrow up and the up arrow down and get a sequence of length $n-1$. Thus these elements generate the lattice of Figure 5, or its dual, or a quotient lattice of Figure 5, or its dual. The only nontrivial quotient lattice of the lattice of Figure 5 is given by Figure 6. ●

As a typical example of the advantages of using the shearing identity we consider independence in the sense of J. von Neumann [1936/37]. This plays a very important role in the applications of lattice theory to direct decompositions of groups and rings and also in continuous geometries.

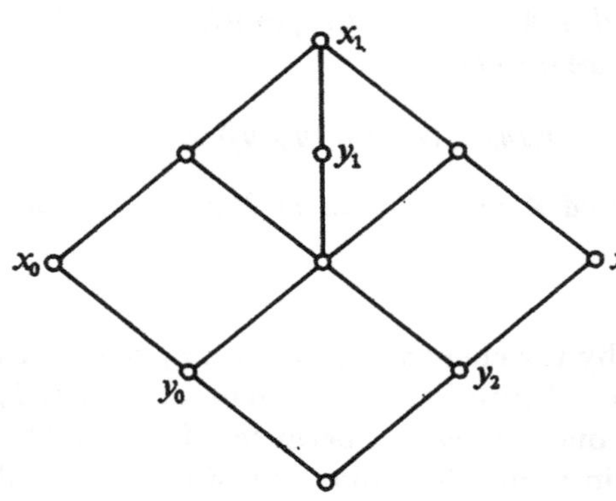

Figure 6

Definition 9. *Let L be a lattice with 0. A subset I of $L-\{0\}$ is called* independent *iff for any two finite subsets X, Y of I we have*

$$\bigvee X \wedge \bigvee Y = \bigvee (X \cap Y).$$

Corollary 10. *A subset I of a lattice L is independent iff*

$$\varphi: X \rightarrow \bigvee X$$

is an isomorphism between $[I]$ and the generalized Boolean lattice of all finite subsets of I.

Proof. $\bigvee X \vee \bigvee Y = \bigvee (X \cup Y)$ and so if I is independent then φ is a homomorphism. If φ is not one-to-one, then $\bigvee X = \bigvee Y$ for some $X \neq Y$, $X, Y \subseteq I$, and X, Y finite. Let, say, $X \nsubseteq Y$ and $a \in X - Y$. Then $a \leq \bigvee Y$, $a \notin Y$. Therefore,

$$a = a \wedge \bigvee Y = \bigvee \{a\} \wedge \bigvee Y = \bigvee (\{a\} \cap Y) = \bigvee \varnothing = 0,$$

a contradiction. Thus φ is an isomorphism. The converse is obvious. ●

A singleton $\{a\}$ ($a \in L-\{0\}$) is always independent. $\{a, b\}$ is independent iff $a \wedge b = 0$ ($a, b \in L-\{0\}$ and $a \neq b$). For $\{a, b, c\}$ to be independent ($a, b, c \in L-\{0\}$ and a, b, c all distinct) we have to require that

$$a \wedge (b \vee c) = 0, \; b \wedge (a \vee c) = 0, \; c \wedge (a \vee b) = 0$$
$$(a \vee b) \wedge (a \vee c) = a, \; (b \vee c) \wedge (b \vee a) = b, \; (c \vee a) \wedge (c \vee b) = c.$$

For modular lattices, fewer relations will do:

Theorem 11. *Let L be a modular lattice with 0. Then an n element set $\{a_1, \ldots, a_n\} \subseteq L-\{0\}$ is independent iff*

$$(a_1 \vee \cdots \vee a_i) \wedge a_{i+1} = 0, \quad \textit{for} \quad i = 1, 2, \ldots, n-1.$$

Proof. We obtain the necessity of the condition by setting $X = \{a_1, \ldots, a_i\}$, $Y = \{a_{i+1}\}$. Now assume that $\{a_1, \ldots, a_n\}$ satisfies the condition. Let $X, Y \subseteq \{a_1, \ldots, a_n\}$, $X \cap Y = \varnothing$. We claim that

$$\bigvee X \wedge \bigvee Y = 0.$$

Indeed, let $a_k \in X \cup Y$ with k maximal. Let, say, $a_k \in Y$. Apply the shearing identity to $\bigvee X \wedge \bigvee Y$ with $x = \bigvee X$, $y = a_k$, and $z = \bigvee (Y - \{a_k\})$:

$$\begin{aligned} \bigvee X \wedge \bigvee Y &= \bigvee X \wedge (a_k \vee \bigvee (Y - \{a_k\})) \\ &= \bigvee X \wedge ((a_k \wedge (\bigvee X \vee \bigvee (Y - \{a_k\}))) \vee \bigvee (Y - \{a_k\})) \\ &= \bigvee X \wedge \bigvee (Y - \{a_k\}), \end{aligned}$$

since $a_k \wedge (\bigvee X \vee \bigvee (Y - \{a_k\})) \leq (a_1 \vee \cdots \vee a_{k-1}) \wedge a_k = 0$.

Proceeding thus we can eliminate all the a_i belonging to $X \cup Y$, getting $\bigvee X \wedge \bigvee Y = \bigvee \varnothing = 0$. Now in the general case,

$$\begin{aligned}\bigvee X \wedge \bigvee Y &= \bigvee X \wedge (\bigvee (X \cap Y) \vee \bigvee (Y - X)) \quad \text{(by modularity)}\\ &= \bigvee (X \cap Y) \vee (\bigvee X \wedge \bigvee (Y - X)) \quad \text{(by } X \cap (Y-X) = \varnothing)\\ &= \bigvee (X \cap Y). \quad \bullet\end{aligned}$$

We conclude this section with three important "sublattice theorems".

Theorem 12 (J. von Neumann [1936/37]). *Let L be a modular lattice and let $a, b, c \in L$. The sublattice of L generated by a, b, and c is distributive iff $a \wedge (b \vee c) = (a \wedge b) \vee (a \wedge c)$.*

Proof. By inspection of the diagram of $F_M(3)$ (see Figure I.5.7). If a, b, and c are the generators and $a \wedge (b \vee c) \equiv (a \wedge b) \vee (a \wedge c)$ (Θ) where Θ is a congruence relation, then Θ collapses the only diamond and so $F_M(3)/\Theta$ is distributive. $\bullet$

One can view the definition of modularity as requiring that any sublattice generated by three elements, two of which are comparable, has to be distributive. This is true in general for any two chains:

Theorem 13 (G. Birkhoff [1940]). *Let L be a modular lattice and let C_0 and C_1 be chains in L. The sublattice of L generated by $C_0 \cup C_1$ is distributive.*

Proof. Since a lattice is distributive iff every finitely generated sublattice is distributive, it is sufficient to verify this result for finite C_0 and C_1. Let $C_0 = \{a_0, \ldots, a_{m-1}\}$ and $C_1 = \{b_0, \ldots, b_{n-1}\}$, $a_0 < \cdots < a_{m-1}$, $b_0 < \cdots < b_{n-1}$. To simplify the notation we shall write $a_m = b_n = a_{m-1} \vee b_{n-1}$. Let us define

$$x(r, \alpha, \beta) = (a_{\alpha(1)} \wedge b_{\beta(1)}) \vee \cdots \vee (a_{\alpha(r)} \wedge b_{\beta(r)}),$$

where $\alpha = \langle \alpha(1), \ldots, \alpha(r) \rangle$, $m \geq \alpha(1) > \alpha(2) > \cdots > \alpha(r) \geq 1$ and $\beta = \langle \beta(1), \ldots, \beta(r) \rangle$, $1 \leq \beta(1) < \beta(2) < \cdots < \beta(r) \leq n$. Let A denote the set of all elements of L of the form $x(r, \alpha, \beta)$. Then $C_0, C_1 \subseteq A$, since $a_i = a_i \wedge b_n$ and $b_i = a_m \wedge b_i$. It is also easy to see that A is closed under join: if we form $x(r, \alpha, \beta) \vee x(s, \alpha', \beta')$, then if we have both $a_i \wedge b_j$ and $a_i \wedge b_k$, we can eliminate one by absorption; if $a_i < a_j$, $b_k < b_t$ and we have $a_i \wedge b_k$ and $a_j \wedge b_t$, the former is eliminated by absorption; what is left can be written in the form $x(r', \alpha'', \beta'')$.

Now we prove by induction the formula

$$x(r, \alpha, \beta) = a_{\alpha(1)} \wedge (b_{\beta(1)} \vee a_{\alpha(2)}) \wedge \cdots \wedge (b_{\beta(r-1)} \vee a_{\alpha(r)}) \wedge b_{\beta(r)}$$

and its dual. The case $r = 1$ is obvious. Let us assume that this formula and its dual have been verified for $r - 1$. Now compute:

$$x(r, \alpha, \beta) = (a_{\alpha(1)} \wedge b_{\beta(1)}) \vee \cdots \vee (a_{\alpha(r)} \wedge b_{\beta(r)}) =$$

(observe $a_{\alpha(1)} \geq (a_{\alpha(2)} \wedge b_{\beta(2)}) \vee \cdots \vee (a_{\alpha(r)} \wedge b_{\beta(r)})$ and apply modularity)

$$= a_{\alpha(1)} \wedge (b_{\beta(1)} \vee (a_{\alpha(2)} \wedge b_{\beta(2)}) \vee \cdots \vee (a_{\alpha(r)} \wedge b_{\beta(r)}))$$

(observe $b_{\beta(r)} \geq b_{\beta(1)} \vee \cdots \vee (a_{\alpha(r-1)} \wedge b_{\beta(r-1)})$ and apply modularity)

$$= a_{\alpha(1)} \wedge (b_{\beta(1)} \vee (a_{\alpha(2)} \wedge b_{\beta(2)}) \vee \cdots \vee (a_{\alpha(r-1)} \wedge b_{\beta(r-1)}) \vee a_{\alpha(r)}) \wedge b_{\beta(r)}$$

(apply to the expression in parentheses the dual of the formula for $r-1$)

$$= a_{\alpha(1)} \wedge (b_{\beta(1)} \vee a_{\alpha(2)}) \wedge (b_{\beta(2)} \vee a_{\alpha(3)}) \wedge \cdots \wedge (b_{\beta(r-1)} \vee a_{\alpha(r)}) \wedge b_{\beta(r)},$$

completing the proof.

Now we easily see that A is a sublattice. Indeed, $x(r, \alpha, \beta) \wedge x(s, \alpha', \beta')$ can be rewritten by the above formula as a meet of $b_i \vee a_j$; but by the dual of the above formula the result can again be put in the form $x(r, \alpha, \beta)$.

So we conclude that A is the sublattice generated by $C_0 \cup C_1$ and therefore $|A| \leq$ the number of expressions of the form $x(r, \alpha, \beta)$.

Now consider the set $X = [0, n+1] \times [0, m+1]$, let F be the sublattice of $P(X)$ generated by

$$a_i = \{\langle x, y\rangle \mid y \leq i+1\}, \qquad \text{for} \quad i = 0, 1, \ldots, m-1,$$

$$b_i = \{\langle x, y\rangle \mid x \leq i+1\}, \qquad \text{for} \quad i = 0, 1, \ldots, n-1,$$

and let $C_0 = \{a_0, \ldots, a_{m-1}\}$ and $C_1 = \{b_0, \ldots, b_{n-1}\}$. In this lattice

$$x(r, \alpha, \beta) = \bigcup(\{\langle x, y\rangle \mid x \leq \alpha(i)+1, y \leq \beta(i)+1\} \mid i = 1, 2, \ldots, r)$$

and so all these represent different elements. Consequently, F is the free modular lattice generated by $C_0 \cup C_1$. Since F is distributive, any modular lattice generated by $C_0 \cup C_1$ is distributive. ●

Finally, we generalize Theorem II.1.6 to modular lattices:

Theorem 14. *Let L be a modular lattice and let $a, b \in L$. Then the sublattice of L generated by $[a \wedge b, a] \cup [a \wedge b, b]$ is isomorphic to $[a \wedge b, a] \times [a \wedge b, b]$.*

Remark. In the distributive case $[[a \wedge b, a] \cup [a \wedge b, b]] = [a \wedge b, a \vee b]$; this does not hold for modular lattices as exemplified by $\mathfrak{M}_3$.

Proof. The isomorphism we set up is

$$\varphi: \langle x, y\rangle \to x \vee y, \text{ for } x \in [a \wedge b, a] \text{ and } y \in [a \wedge b, b].$$

Using the formula for the dual of $x(r, \alpha, \beta)$ in the proof of Theorem 13 we obtain:

$$x \vee y = (a \vee y) \wedge (b \vee x).$$

Hence

$$a \wedge (x \vee y) = a \wedge (a \vee y) \wedge (b \vee x) = a \wedge (b \vee x) = (a \wedge b) \vee x = x$$

and $b\wedge(x\vee y)=y$, proving that φ is one-to-one. φ is obviously onto and preserves join. Let $x, x_1 \in [a\wedge b, a]$ and $y, y_1 \in [a\wedge b, b]$. Then

$$\begin{aligned}(x\vee y)\wedge(x_1\vee y_1) &= (a\vee y)\wedge(b\vee x)\wedge(a\vee y_1)\wedge(b\vee x_1)\\ &= (a\vee y)\wedge(a\vee y_1)\wedge(b\vee x)\wedge(b\vee x_1) \qquad \text{(use Corollary 3)}\\ &= (a\vee(y\wedge y_1))\wedge(b\vee(x\wedge x_1)) = (x\wedge x_1)\vee(y\wedge y_1),\end{aligned}$$

proving that φ is an isomorphism. ●

Exercises

1. Show that a lattice L is modular iff it satisfies the identity
$$(x\vee(y\wedge z))\wedge(y\vee z) = (x\wedge(y\vee z))\vee(y\wedge z).$$
2. Let $p=(x\vee y)\wedge(y\vee z)\wedge(z\vee x)$, $d_0=a\vee b\vee c$, and define, for $n\geq 0$, $d_{n+1}=p(a\vee d_n, b\wedge d_n, c\wedge d_n)$. Show that the identity $d_1=d_2$ is equivalent to modularity but $d_2=d_3$ is not (B. Wolk).
3. A finite lattice L is modular iff it does not contain a pentagon $\{0, a, b, c, i\}$ satisfying $a \succ b$.
4. Can the numbers of covering pairs in Exercise 3 be increased?
5. Let L be a lattice and $a, b\in L$. Then φ_b: $x\to x\wedge b$ maps $[a, a\vee b]$ into $[a\wedge b, b]$, and ψ_a: $x\to x\vee a$ maps $[a\wedge b, b]$ into $[a, a\vee b]$. Show that $\varphi_b\psi_a\varphi_b=\varphi_b$ and $\psi_a\varphi_b\psi_a=\psi_a$.
6. Using the notation of Exercise 5, prove that $\{x \mid x\in[a\wedge b, b], x\psi_a\varphi_b=x\}$ is isomorphic to $\{x \mid x\in[a, a\vee b], x\varphi_b\psi_a=x\}$. (Exercises 5 and 6 are due to W. Schwan [1948], [1948a], [1949], and [1949a].)
7. Does Corollary 3 characterize the modularity of a lattice?
8. Find a nonmodular lattice satisfying the Upper and Lower Covering Conditions.
9. Show that the Kuroš-Ore Theorem fails in the first lattice of Figure 2.1.
10. Prove that in the Kuroš-Ore Theorem the representation $a=x_0\vee\cdots\vee x_{i-1}\vee y_j\vee x_{i+1}\vee\cdots\vee x_{n-1}$ is always irredundant.
11. Using the notation of the Kuroš-Ore Theorem, show that there is a permutation π of $\{0,\ldots, m-1\}$ such that $a=x_0\vee\cdots\vee x_{i-1}\vee y_{i\pi}\vee x_{i+1}\vee\cdots\vee x_{n-1}$ holds for all i, $0\leq i<n$ (R. P. Dilworth [1946]).
12. Can the Kuroš-Ore Theorem be sharpened so as to require that all x_i can be simultaneously replaced by y_i? (No, inspect Figure 3.4b.)
13. Does Theorem 6 characterize modularity?
14. Prove Theorem 7 directly, that is, without reference to $F_M(3)$.
15. Show that for finite modular lattices it is sufficient to consider projectivities of prime intervals. Simplify Theorems 6 and 7 for prime intervals.
16. Extend Exercise 15 to locally finite lattices.
17. Show that Exercise 15 does not generalize to nonmodular lattices.
18. Show that to define independence in modular lattices it is sufficient to assume that $\bigvee X\wedge\bigvee Y=\varnothing$ for $X\cap Y=\varnothing$, $X, Y\subseteq I$, X and Y finite.
19. Let L be a complete lattice and let I be a set of compact elements of L. Then I is independent iff $X\to\bigvee X$ is an isomorphism between the Boolean lattice of all subsets of I and the complete sublattice generated by I.

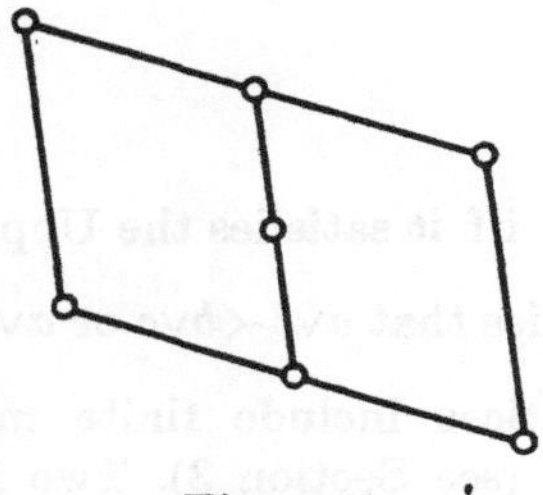

Figure 7

20. Let L be a complemented modular lattice and let $a_0 = 0 < a_1 < \cdots < a_n = 1$ be elements of L. Let $b_1 = a_1$, let b_2 be a relative complement of a_1 in $[0, a_2]$, . . . , let b_n be a relative complement of a_{n-1} in $[0, a_n]$. Then $B = \{b_1, \ldots, b_n\}$ is an independent set.
21. Prove the converse of Exercise 20.
*22. Let L be a modular lattice and let C_0, C_1, and C_2 be chains in L. The sublattice generated by $C_0 \cup C_1 \cup C_2$ is distributive iff $a \wedge (b \vee c) = (a \wedge b) \vee (a \wedge c)$ for all $a \in C_0$, $b \in C_1$, and $c \in C_2$ (B. Jónsson [1955]).
23. Let $\mathbf{K}$ be an equational class of lattices and let P be a poset such that $F_{\mathbf{K}}(P)$ exists in the sense of Definition I.5.2. Let A be the class of all *finite* lattices L such that L is generated by some homomorphic image of P (in the sense defined in the proof of Theorem I.5.5). Let us assume that there is an $L \in A$ such that $|L| \geq |L'|$ for any $L' \in A$. Is it true that $F_{\mathbf{K}}(P)$ exists and is finite, in fact, is it true we can take $F_{\mathbf{K}}(P) = L$?
24. Prove that in a modular lattice L, the sublattice generated by $[a \wedge b, a] \cup [a \wedge b, b]$ equals $[a \wedge b, a \vee b]$ for any $a, b \in L$ iff L is distributive.
25. Find equations to replace " $a \wedge (b \vee c) = (a \wedge b) \vee (a \wedge c)$ " in Theorem 12.
26. Find a direct proof of the statement that in a modular lattice every distributive element is neutral.
27. Show that in a distributive lattice $a/b \approx c/d$ and c/d a subquotient of a/b imply that $a/b = c/d$.
28. Let C be a bounded chain. Define a subset M of C^3 by $\langle x, y, z \rangle \in M$ iff $x \wedge y = y \wedge z = z \wedge x$. Then M as a subposet of C^3 is a lattice. For $\alpha, \beta \in M$, $\alpha \wedge \beta$ in M is the same as $\alpha \wedge \beta$ in C^3. Compute $\alpha \vee \beta$.
29. Show that the lattice M of Exercise 28 is a modular lattice.
30. Show that in M of Exercise 28, $\langle 1, 0, 0\rangle/\langle 0, 0, 0\rangle \approx \langle x, 0, 0\rangle/\langle 0, 0, 0\rangle$ for any $x > 0$ provided that C is infinite. (Exercises 28–30 are due to E. T. Schmidt [1962].)
31. Show that the identity $d_2 = d_3$ of Exercise 2 fails in the lattice of Figure 7 (B. Wolk).
32. Show that a finite modular lattice is dismantlable iff it has breadth two or less (D. Kelly and I. Rival [1974]).
33. Verify that if L is a finite modular lattice satisfying

$$|L| \leq \frac{1}{3}(5l(L) + 7),$$

then L has an m element sublattice for all $m \leq |L|$. (I. Rival [b]. Hint: Use Exercise I.6.32.)
34. Does the fact that φ_b of Theorem 2 is one-to-one for all $a, b \in L$ characterize the modularity of L?

2. Semimodular Lattices

A lattice is called *semimodular* iff it satisfies the Upper Covering Condition, that is,

$$a \prec b \text{ implies that } a \vee c \prec b \vee c \text{ or } a \vee c = b \vee c.$$

Examples of semimodular lattices include finite modular lattices and some important lattices from geometry (see Section 3). Two further examples are given in Figure 1.

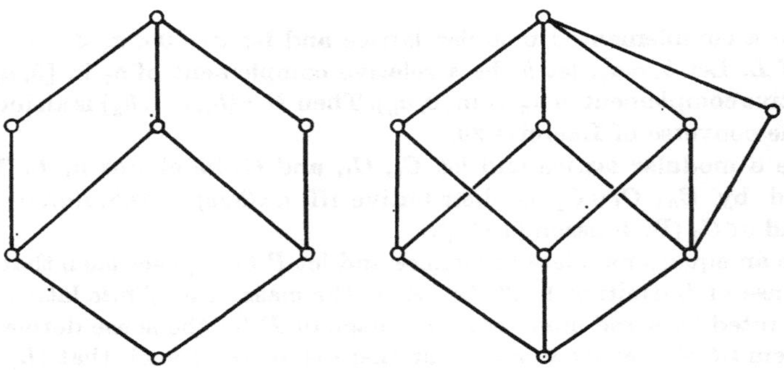

Figure 1

Let L be a lattice with 0. We define the *height* function: for $a \in L$, let $h(a)$ denote the length of a longest maximal chain in $[0, a]$ if there is a finite longest maximal chain; otherwise put $h(a) = \infty$. If L is of finite length, $0 \leq h(a) < \omega$, for all $a \in L$. We first show that for semimodular lattices of finite length, $h(a)$ is the length of *any* maximal chain in $[0, a]$.

THEOREM 1 (The Jordan-Hölder Chain Condition. O. Ore [1943]). *Let L be a lattice of finite length. If L is semimodular, then any two maximal chains of L are of the same length.*

PROOF. Let $C = \{a_0, \ldots, a_n\}$, $0 = a_0 < \cdots < a_n = 1$, be a maximal chain of length n. We prove that any other maximal chain is of length n by induction on n. If $n \leq 1$, the statement is trivial. Let us assume the statement to hold for length $< n$. Let $C' = \{b_0, b_1, \ldots, b_m\}$, $0 = b_0 < b_1 < \cdots < b_m = 1$ be another maximal chain in L. If $a_1 = b_1$, then in the semimodular lattice $[a_1)$ the maximal chain $C - \{a_0\}$ is of length $n-1$, so the maximal chain $C' - \{b_0\}$ has to be of length $n-1$, therefore $n = m$. If $a_1 \neq b_1$ (see Figure 2), then let C'' be a maximal chain in $[a_1 \vee b_1)$ and let k be the length of C''. Because of semimodularity, $a_1 \vee b_1 \succ a_1$ and $a_1 \vee b_1 \succ b_1$. Hence $C'' \cup \{a_1\}$ is a maximal chain of length $k+1$ and $C - \{a_0\}$ is a maximal chain of length $n-1$ in $[a_1)$; thus $k+1 = n-1$. Similarly, $k+1 = m-1$, hence $n = m$. ●

Theorem 1 suggests that semimodularity can be characterized in terms of the height function (the equivalence of (i)–(iii), and that they imply (iv) can be found in G. Birkhoff [1933], see also G. Birkhoff [1967]):

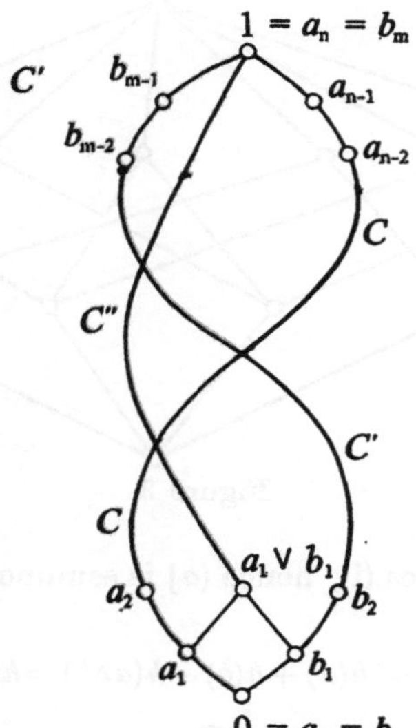

Figure 2

Theorem 2. ***Let L be a lattice of finite length. The following conditions on L are equivalent for all*** $a, b, c \in L$:

(i) ***L is semimodular.***

(ii) ***If*** $a \neq b$, ***a and b cover*** $a \wedge b$, ***then*** $a \vee b$ ***covers a and b.***

(iii) ***If*** $a \leq b$ ***and C is a maximal chain in*** $[a, b]$, ***then*** $\{x \vee c \mid x \in C\}$ ***is a maximal chain in*** $[a \vee c, b \vee c]$.

(iv) $h(a) + h(b) \geq h(a \wedge b) + h(a \vee b)$.

Proof. (ii) *implies* (i). Let $a \prec b$. If $c \leq a$ or $a \vee c \geq b$, then $b \vee c \succeq a \vee c$. If $c \nleq a$ and $a \vee c \ngeq b$, then $b \wedge (a \vee c) = a$. Let $a = a_0 < a_1 < \cdots < a_n = a \vee c$ be a maximal chain in $[a, a \vee c]$. Since b and a_1 cover a and $b \neq a_1$, $b \vee a_1$ covers a_1. An easy induction shows that $b \vee a_i$ covers a_i for $i = 1, 2, \ldots, n$. Thus $b \vee a_n$ covers a_n, that is, $b \vee c$ covers $a \vee c$. ◐

(i) *implies* (iii). Obvious, since if $x \prec y$ in C, then $c \vee x \preceq c \vee y$. ◐

(iii) *implies* (iv). (iii) obviously implies semimodularity and so, by Theorem 1, we can assume that the Jordan-Hölder Chain Condition holds. Let C be a maximal chain in $[a \wedge b, b]$. By the Jordan-Hölder Chain Condition, the length of C is $h(b) - h(a \wedge b)$. By (iii), $D = \{a \vee x \mid x \in C\}$ is a maximal chain in $[a, a \vee b]$. The length of D is at most the length of C, that is, $h(b) - h(a \wedge b)$; on the other hand, the length of D is $h(a \vee b) - h(a)$ by the Jordan-Hölder Chain Condition, and so

$$h(b) - h(a \wedge b) \geq h(a \vee b) - h(a),$$

which was to be proved. ●

(iv) *implies* (ii). Certainly, (ii) holds in (0]. By induction on $h(x)$, let $x \in L$ and let (ii) hold in $(y]$ for all $y < x$. Let $a, b \in (x]$, $a, b \succ a \wedge b$, and $a \neq b$. Then $a < x$, so (ii)

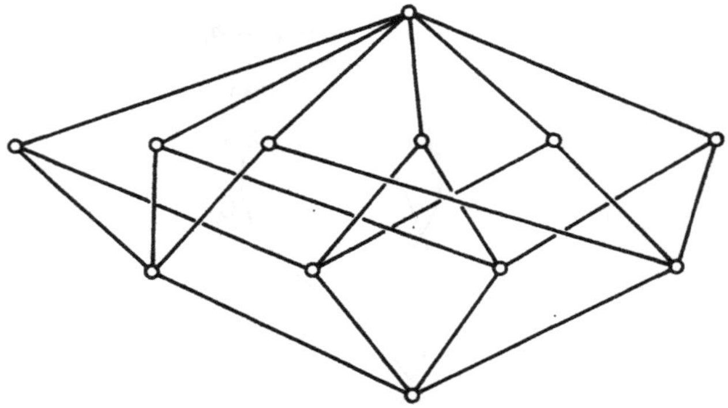

Figure 3

holds in $(a]$ and (ii) implies (i), hence $(a]$ is semimodular. Thus $h(a) - h(a \wedge b) = 1$. Hence by (iv),

$$h(a \vee b) \leq h(a) + h(b) - h(a \wedge b) = h(b) + 1,$$

and so $a \vee b \succ b$. Similarly, $a \vee b \succ a$. ●

Using Theorem 2 it is easy to check that some constructions yield semimodular lattices. We give one example. Take a semimodular lattice L of length n and pick a $k < n$. Let L_k be the set of all $x \in L$ with $h(x) \leq k$ along with 1. It is easy to check that L_k is a semimodular lattice of length $k+1$. The result of this construction with $L = (\mathfrak{C}_2)^4$ and $k = 2$ is shown in Figure 3.

Another application of Theorem 2 is

Corollary 3. *Let L be a lattice of finite length. The following conditions on L are equivalent:*

(i) *L is modular.*

(ii) *L satisfies the Upper and the Lower Covering Conditions.*

(iii) $h(a) + h(b) = h(a \wedge b) + h(a \vee b)$.

Proof. We know from Section 1, that (i) implies (ii). Theorem 2 and its dual (to be more precise, the dual of 2(iii)) yield (iii). Now assume (iii). If L is not modular, then L contains a pentagon $\{o, a, b, c, i\}$. Thus

$$h(i) = h(b) + h(c) - h(o)$$

and

$$h(i) = h(a) + h(c) - h(o),$$

implying $h(a) = h(b)$, a contradiction. ●

Independence in semimodular lattices can be easily handled only for atoms.

Theorem 4. *Let $I = \{a_1, \ldots, a_n\}$ be a set of n atoms of a semimodular lattice. Then the following conditions on I are equivalent:*

(i) *I is independent.*

(ii) $(a_1 \vee \cdots \vee a_i) \wedge a_{i+1} = 0$, *for* $i = 1, 2, \ldots, n-1$.
(iii) $h(a_1 \vee \cdots \vee a_n) = n$.

Proof. (i) *implies* (ii) is obvious. ◑

(ii) *implies* (iii). We prove by induction on i that $h(a_1 \vee \cdots \vee a_i) = i$. This is true for $i = 1$. If $h(a_1 \vee \cdots \vee a_i) = i$, then by (ii) and semimodularity,

$$a_1 \vee \cdots \vee a_i \prec a_1 \vee \cdots \vee a_i \vee a_{i+1}$$

and so $h(a_1 \vee \cdots \vee a_i \vee a_{i+1}) = h(a_1 \vee \cdots \vee a_i) + 1 = i + 1$. ◑

(iii) *implies* (i). We show that the condition of Corollary 1.10 holds for I. It follows from (iii), that $h(\bigvee X) = |X|$ for any $X \subseteq I$ so the map $\varphi: X \to \bigvee X$ is one-to-one. Obviously, φ is join preserving. Now take $X, Y \subseteq I$, and let $a = \bigvee X \wedge \bigvee Y$ and $b = \bigvee (X \cap Y)$. Then $a \geq b$ and by Theorem 2(iv)

$$h(\bigvee X) + h(\bigvee Y) \geq h(a) + h(\bigvee (X \cup Y)),$$

so by (iii),

$$|X| + |Y| \geq h(a) + |X \cup Y|.$$

We conclude that

$$h(a) \leq |X| + |Y| - |X \cup Y| = |X \cap Y|.$$

On the other hand,

$$h(a) \geq h(b) = |X \cap Y|,$$

hence $h(a) = h(b)$ and $a = b$. ●

Let A be a set of atoms. Then $G(\subseteq A)$ *spans* A iff for every $a \in A$ there is a finite $G_1 \subseteq G$ such that $a \leq \bigvee G_1$. The following generalizes a result familiar from any first course on vector spaces.

Theorem 5. *Let A be a set of atoms of a semimodular lattice, let I be an independent subset of A, and let $G \supseteq I$ span A. Then there is an independent subset J of A such that $G \supseteq J \supseteq I$ and J spans A.*

Proof. Let $\mathfrak{X}$ be the set of all independent subsets X of A with $I \subseteq X \subseteq G$. If $C \subseteq \mathfrak{X}$ and C is a chain, then $\bigcup C$ is again independent, since independence is tested with the finite subsets of $\bigcup C$ and every finite subset of $\bigcup C$ is also a finite subset of some $X \in C$. Hence by Zorn's Lemma (Section II.1) there is a maximal independent subset J of A with $I \subseteq J \subseteq G$. We wish to show that J spans A. It is sufficient to show that J spans G. Indeed, let $g \in G$. If $g \in J$ there is nothing to prove. If $g \in G - J$, then $J \cup \{g\}$ is not independent, and so there is a finite $J_1 \subseteq J$, such that $J_1 \cup \{g\}$ is not independent, that is, by Theorem 4(iii) and by semimodularity,

$$h(\bigvee (J_1 \cup \{g\})) < |J_1| + 1.$$

Since $h(\bigvee J_1) = |J_1|$, we obtain that $\bigvee (J_1 \cup \{g\}) = \bigvee J_1$, that is, $g \leq \bigvee J_1$, proving that $\{g\}$ is spanned by J. ●

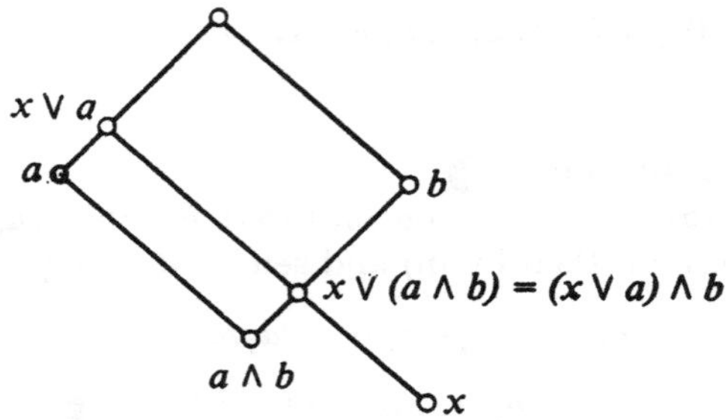

Figure 4

The definition of semimodularity was given for arbitrary lattices but it is obvious that it is not very useful for lattices without many prime intervals: lattices without prime intervals are always semimodular. Various attempts have been made to rectify this situation, that is, to come up with a definition agreeing with semimodularity for lattices of finite length, that also selects an interesting class of lattices which are not of finite length.

Definition 6 (L. R. Wilcox [1939] and S. Maeda [1965]). *Let L be a lattice. A pair of elements $\langle a, b\rangle$ of L is called* modular (see Figure 4), *in notation, aMb, iff*

$$x \leq b \text{ implies that } x\vee(a\wedge b) = (x\vee a)\wedge b.$$

*The lattice L is called M-*symmetric *iff aMb implies that bMa for any $a, b \in L$.*

Now it is obvious that L is modular iff aMb for all $a, b \in M$. In the proof of $[a\wedge b, b] \cong [a, a\vee b]$ in the Isomorphism Theorem we only used that aMb in L and bMa in the dual of L, hence

Corollary 7. *Let L be a lattice and $a, b \in L$. If aMb in L and bMa in the dual of L, then $[a\wedge b, b] \cong [a, a\vee b]$. This isomorphism is given by ψ_a whose inverse is φ_b.*

The notation φ_b and ψ_a are from Section 1 (see Figure 1.1).
Half of this conclusion still holds if only aMb is assumed.

Lemma 8. *Let L be a lattice and let $a, b \in L$. The following conditions are equivalent:*
(i) *aMb.*
(ii) *φ_b is onto.*
(iii) *ψ_a is one-to-one.*

Proof. (i) *implies* (ii). If $x \in [a\wedge b, b]$, then $x \leq b$, and so

$$(a\vee x)\varphi_b = (a\vee x)\wedge b = x\vee(a\wedge b) = x. \quad ◑$$

(ii) *implies* (iii). Let $x, y \in [a\wedge b, b]$, $x\psi_a = y\psi_a$, and $x \neq y$. Then $(x\vee y)\psi_a = x\psi_a = y\psi_a$ and x or $y < x\vee y$, say $x < x\vee y$. There exists a $z \in [a, a\vee b]$ such that $z\varphi_b = x$. We have $z \geq x$ and $z \geq a$, hence $z \geq x \vee a = x\psi_a = (x \vee y)\psi_a$. Thus $z \geq x \vee y$ and so $z\varphi_b \geq x \vee y > x$, a contradiction. ◐

(iii) *implies* (i). Let $x \leq b$. Then

$$(x \vee (a \wedge b))\psi_a = x \vee (a \wedge b) \vee a = x \vee a,$$

$$((x \vee a) \wedge b)\psi_a = ((x \vee a) \wedge b) \vee a = x \vee a,$$

hence, using that ψ_a is one-to-one, we obtain aMb. To verify that $((x \vee a) \wedge b) \vee a = x \vee a$, observe that $x \vee a \geq a$ and $x \vee a \geq (x \vee a) \wedge b$, hence $x \vee a \geq ((x \vee a) \wedge b) \vee a$. Moreover, if $t \geq (x \vee a) \wedge b$ and $t \geq a$, then $t \geq (x \vee a) \wedge b \geq x$ (since $b \geq x$), and so $t \geq a \vee x$. ●

The following result is implicit in L. R. Wilcox [1939].

Theorem 9. *Let L be a lattice of finite length. Then L is semimodular iff L is M-symmetric.*

Proof. Let L be a semimodular lattice of finite length. We shall prove that

$$aMb \quad \text{iff} \quad h(a) + h(b) = h(a \wedge b) + h(a \vee b),$$

from which M-symmetry trivially follows.

Using the notation of "(iii) implies (iv)" in the proof of Theorem 2, the length of C is $h(b) - h(a \wedge b)$ and the length of D is $h(a \vee b) - h(a)$. So if aMb, then ψ_a is one-to-one, $|C| = |D|$, and $h(b) - h(a \wedge b) = h(a \vee b) - h(a)$. Conversely, if aMb fails, then ψ_a is not one-to-one, and C can be chosen so as to include $x, y \in [a \wedge b, b]$, $x\psi_a = y\psi_a$. Then $|D| < |C|$ and we obtain $h(b) - h(a \wedge b) > h(a \vee b) - h(a)$.

To prove the converse we do not have to assume that L is of finite length. So let L be an M-symmetric lattice, let $a, b, c \in L$, and let $b \succ a$. If $b \vee c = a \vee c$ we have nothing to prove. If $b \vee c > a \vee c$, then put $d = a \vee c$ and we have $b \wedge d = a$, $b \vee d = b \vee c$. We have to prove that $b \vee d \succ d$. Indeed, let $b \vee d > x \geq d$. Then $x \ngeq b$ and so $b \wedge x = a$ and $b \vee x = b \vee d$. Since $b \succ b \wedge x$, φ_b as a map from $[x, x \vee b]$ into $[x \wedge b, b]$ is an onto map and so, by Lemma 8, we obtain xMb. By M-symmetry, bMx, which means by definition that for any $y \leq x$,

$$y \vee (b \wedge x) = (y \vee b) \wedge x.$$

Let $y = d$, then we obtain

$$d = d \vee (b \wedge x) = (d \vee b) \wedge x = x,$$

that is, $b \vee c \succ a \vee c$. ●

Examples of M-symmetric lattices not of finite length include the lattice of all closed subspaces of a Banach space and the projection lattice of a von Neumann algebra.

Exercises

1. Show that a lattice L is semimodular iff $x \succ x \wedge y$ implies that $x \vee y \succ y$.
2. Modify the proof of the Jordan-Hölder Theorem. Assume only that C' is a chain and $n < m$, and derive a contradiction. What conclusion can be drawn from this proof?

3. Let A, B, and C be sets of atoms of a semimodular lattice. Show that if A spans B and B spans C, then A spans C.
4. Show that (i)–(iii) of Lemma 8 are equivalent to (iv) $\psi_a\varphi_b$ is the identity map.
5. Let L be a semimodular lattice. Prove that if p and q are atoms of L, $a \in L$, and $a < a \vee q \leq a \vee p$, then $a \vee p = a \vee q$ (*Steinitz-MacLane Exchange Axiom*).
6. Let L be a lattice. Show that all sublattices of L are semimodular iff L is modular.
7. Show that direct products and convex sublattices of semimodular lattices are again semimodular.
8. Prove that a homomorphic image of a semimodular lattice of finite length is again semimodular.
9. Investigate the statements of Exercises 6–8 for M-symmetric lattices.
10. Show that Part(A), the lattice of all partitions on a set A, is a semimodular lattice.
11. Show that the lattice of all congruence relations of a semilattice is a semimodular lattice (R. Freese and J. B. Nation [1973]).
12. Let L be a modular lattice and let I be an ideal of L. Then $L' = (L - I) \cup \{0\}$ is a lattice. Under what conditions is L' semimodular?
13. Is the lattice L' of Exercise 12 always M-symmetric?
14. Let L be a lattice. Prove that Sub(L) is semimodular iff L is a chain (K. M. Koh [1973]).
15. A poset P is called *graded* iff an integer-valued function h can be defined on P satisfying, for $x, y \in P$:

$$x \leq y \quad \text{and} \quad h(x) + 1 = h(y) \quad \text{iff} \quad x \prec y.$$

Show that P is graded iff every interval of P is of finite length and satisfies the Jordan-Hölder Chain Condition.
16. Let L be a lattice. Show that Sub(L) is graded iff the dual of Sub(L) is semimodular, which in turn is equivalent to the condition that L has no sublattice isomorphic to $\mathfrak{C}_2 \times \mathfrak{C}_3$ (H. Lakser [1973b]).
17. Prove that a lattice of finite length is semimodular iff it is graded and satisfies $h(x) + h(y) \geq h(x \wedge y) + h(x \vee y)$ (G. Birkhoff [1933]).
18. The *Infinite Jordan-Hölder Chain Condition* holds for a lattice L iff for any two maximal chains C_1 and C_2 of L we have $|C_1| = |C_2|$. Show that this fails for the lattice $L = A \times B$, where A is the [0, 1] real interval and B is the [0, 1] rational interval.
19. Show that the Infinite Jordan-Hölder Chain Condition holds for the Boolean lattice of all subsets of a set A iff A is finite.
20. A lattice L is said to satisfy the *MacLane Condition* iff for $a, b, c \in L$ satisfying $a \parallel b$ and $a \wedge b < c \leq a$ there exists a $d \in L$ satisfying $a \wedge (b \vee d) = c$. Show that the MacLane Condition implies the Upper Covering Condition.
21. Prove that a lattice of finite length is semimodular iff it satisfies the MacLane Condition. (Exercises 20 and 21 are from S. MacLane [1938].)
22. Find M-symmetric lattices that fail to satisfy the MacLane Condition.
23. Find lattices satisfying the MacLane Condition that are not M-symmetric.

3. Geometric Lattices

Just as most lattices arising out of algebraic examples are algebraic (Section II.3, see also Concluding Remarks), most lattices arising out of geometry are geometric in the following sense:

Definition 1. *A lattice L is called* geometric *iff L is semimodular, L is algebraic, and the compact elements of L are exactly the finite joins of atoms of L.*

Equivalently, L is complete, L is *atomistic* (that is, every element of L is a join of atoms), all atoms are compact, and L is semimodular.

This concept was introduced by G. Birkhoff [1935a], influenced by H. Whitney [1935], for lattices of finite length. Geometric lattices with no restriction on length were introduced and investigated by S. MacLane [1938] under the name *exchange lattices*. The name, geometric lattice, is due to M. L. Dubreil-Jacotin, L. Lesieur, and R. Croisot [1953]. G. Birkhoff [1967] and some others retain in their definition the requirement that the lattice be of finite length. Many authors call these lattices *matroid lattices*. (For combinatorial definitions of matroids, see the Further Topics and References for this chapter.)

Later in this section we shall investigate the connection between geometric lattices and geometries. We start out, however, by investigating the lattice theoretic consequences of this definition.

Corollary 2.

(i) *Let L be a geometric lattice. Then the set F of elements of finite height is an ideal of L. The lattice F is semimodular and every element of F is a finite join of atoms. L is isomorphic to $I(F)$, the lattice of all ideals of F.*

(ii) *An interval of a geometric lattice is again a geometric lattice.*

Proof.

(i) Let $a, b \in F$, $c \in L$, and $c \leq a$. Then $h(c) \leq h(a)$, hence $c \in F$. By Theorem 2.2, $h(a \vee b) \leq h(a) + h(b) - h(a \wedge b)$ and so $a \vee b \in F$. Thus F is an ideal. The second statement of (i) is obvious. The third statement follows from the proof of Theorem II.3.13. ◐

(ii) An interval $[a, b]$ of an algebraic lattice L is again algebraic. Moreover, $\{a \vee p \mid p$ an atom of L, $p \nleq a$ and $p \leq b\}$ is by semimodularity the set of all atoms of $[a, b]$. The rest is easy. ●

Now we utilize the theory of independence.

Lemma 3. *Let L be a geometric lattice. Every element a of L is a join of an independent set of atoms, in fact, of any maximal independent set of atoms of L contained in $(a]$.*

Proof. Let A be a maximal independent set of atoms of L contained in $(a]$ (A exists by Theorem 2.5 with $I = \varnothing$ and G the set of all atoms in $(a]$). By Theorem 2.5 and the compactness of atoms, A spans all the atoms in $(a]$. Since L is atomistic $a = \bigvee A$. ●

Theorem 4 (G. Birkhoff [1935a] and S. MacLane [1938]). *A geometric lattice L is complemented. In fact, it is relatively complemented.*

Proof. Let $a \in L$, let A be a maximal independent set of atoms in $(a]$, and let K be the set of all atoms not contained in $(a]$. Then by Theorem 2.5 there is a maximal independent set I of atoms of L satisfying $A \subseteq I \subseteq A \cup K$. Set $b = \bigvee(I - A)$. Since $\bigvee I = 1$ by Theorem 2.5, we obtain $a \vee b = 1$. Let $c = a \wedge b$. If $c \neq 0$, then there is an atom $p \leq c$. Since $p \leq \bigvee I$ and $p \leq \bigvee(I - A)$, by the compactness of p there exist finite $I_1 \subseteq I$ and $I_2 \subseteq I - A$ such that $p \leq \bigvee I_1$ and $p \leq \bigvee I_2$. By the definition of independence,

$$p \leq \bigvee I_1 \wedge \bigvee I_2 = \bigvee(I_1 \cap I_2) = \bigvee \varnothing = 0,$$

a contradiction. Hence $a \wedge b = 0$, proving that L is complemented. The second statement follows from the first and from Corollary 2. ●

Our next task is to prove the structure theorem of geometric lattices due to F. Maeda [1951] and in its sharper form due to U. Sasaki and S. Fujiwara [1952].

Theorem 5. *Every geometric lattice is isomorphic to a direct product of directly indecomposable geometric lattices.*

This theorem is augmented by a characterization theorem of direct indecomposability. Let us call the elements a and b *perspective*, in symbol $a \sim b$, iff they have a common complement, that is, $a \vee x = b \vee x = 1$ and $a \wedge x = b \wedge x = 0$ for some element x. The reader should not confuse the perspectivity of elements with perspectivity of quotients. (See Exercise 13.)

Theorem 6. *A geometric lattice is directly indecomposable iff any two atoms are perspective.*

For the proof of Theorem 5 we need a lemma.

Lemma 7. *Let L be a geometric lattice. If $a, b, x \in L$, $a \wedge x = b \wedge x = 0$, and $a \vee x = b \vee x$, then $a \sim b$. If $a, b \in L$, $a \sim b$, and $h(a)$, $h(b) < \infty$, then there is an $x \in L$ satisfying $a \wedge x = b \wedge x = 0$, $a \vee x = b \vee x$, and $h(x) < \infty$.*

Proof. Let $a \wedge x = b \wedge x = 0$ and $a \vee x = b \vee x$. Let y be the relative complement (which exists by Theorem 4) of $a \vee x = b \vee x$ in $[x, 1]$. Then $a \wedge y = a \wedge (a \vee x) \wedge y = a \wedge x = 0$ and $a \vee y = a \vee x \vee y = 1$; similarly, $b \wedge y = 0$ and $b \vee y = 1$, that is, $a \sim b$. To prove the second statement, let $a \wedge y = b \wedge y = 0$, $a \vee y = b \vee y = 1$, and observe that $a \leq b \vee y$ and $h(a)$, $h(b) < \infty$, hence, by the compactness of a there is an $x_1 \leq y$, $h(x_1) < \infty$ satisfying $a \leq b \vee x_1$. Similarly choose $x_2 \leq y$, $h(x_2) < \infty$ satisfying $b \leq a \vee x_2$. Then $x = x_1 \vee x_2$ will satisfy the lemma. ●

Proof of Theorem 5. For $a, b \in L$, let us write $a \approx b$, and call a and b *projective*, iff $a = x_0 \sim x_1 \sim \cdots \sim x_n = b$ for some $x_1, \ldots, x_{n-1} \in L$. For an atom p of L we define

$$R(p) = \{q \mid q \text{ is an atom of } L \text{ and } p \approx q\}$$

and

$$Z=\{R(p) \mid p \text{ is an atom of } L\}.$$

For $x\in Z$, $x=R(p)$, we set $c(x)=c(p)=\bigvee R(p)$.

We are going to show that L is isomorphic to $\Pi([0, c(x)] \mid x\in Z)$.

Every element a of L is a join of atoms, hence if we set $a_{R(p)}=\bigvee((a]\wedge R(p))$, then $a=\bigvee(a_{R(p)} \mid R(p)\in Z)$, where $a_{R(p)}\leq c(p)$.

To establish that

$$a\to\langle a_{R(p)} \mid R(p)\in Z\rangle$$

is the required isomorphism it is sufficient by Exercise 21 to show that every $a\in L$ has a unique representation in the form

$$a=\bigvee(a_{R(p)} \mid R(p)\in Z) \text{ with } a_{R(p)}\leq c(p).$$

Indeed, let a have two such distinct representations:

$$a=\bigvee(x_{R(p)} \mid R(p)\in Z)=\bigvee(y_{R(p)} \mid R(p)\in Z),\quad x_{R(p)}, y_{R(p)}\leq c(p)$$

and for at least one atom, say q, $x_{R(q)}\neq y_{R(q)}$. By taking the join of the two representations we can assume that $x_{R(p)}\leq y_{R(p)}$ and so $x_{R(q)}<y_{R(q)}$.

We now need the following simple statement:

If $A\cup\{p\}$ is a set of atoms of L and $p\leq\bigvee A$, then $p\sim r$ for some $r\in A$.

Indeed, by the compactness of p, $p\leq\bigvee A_1$ for some finite $A_1\subseteq A$. Choose A_1 minimal. Then $p\sim r$ for any $r\in A_1$, because if $x=\bigvee(A_1-\{r\})$ then $r\vee x=\bigvee A_1$; $p\nleq x$ (since A_1 is minimal), hence $h(p\vee x)=h(x)+1=h(\bigvee A_1)$ and so $p\vee x=\bigvee A_1$. By minimality, A_1 is independent, hence $r\wedge x=0$ and $p\nleq x$, hence $p\wedge x=0$. Thus $p\sim r$ by Lemma 7.

It follows from this statement that there is an atom $t\in R(q)$ such that $t\leq y_{R(q)}$ and $t\nleq x_{R(q)}$. It also follows that if we set

$$a_1=x_{R(q)} \quad\text{and}\quad a_2=\bigvee(x_{R(p)} \mid p\nsim q),$$

then $a=a_1\vee a_2$ and $a_1\wedge a_2=y_{R(q)}\wedge a_2=0$. So we obtain elements t, a_1, and a_2 satisfying

$$t\wedge a_1=0,\quad a_1\wedge a_2=0,\quad t\leq a_1\vee a_2$$

(see Figure 1). Let x_1 be the relative complement of $t\vee a_1$ in $[a_1, a_1\vee a_2]$ and x_2 the relative complement of $a_2\wedge x_1$ in $[0, a_2]$. We claim that

$$t\wedge x_1=x_2\wedge x_1=0 \quad\text{and}\quad t\vee x_1=x_2\vee x_1$$

and therefore $t\sim x_2$ by Lemma 7. Indeed,

$$t\wedge x_1=t\wedge(t\vee a_1)\wedge x_1=t\wedge a_1=0,$$
$$x_2\wedge x_1=a_2\wedge x_2\wedge x_1=x_2\wedge(a_2\wedge x_1)=0,$$
$$t\vee x_1=t\vee a_1\vee x_1=a_1\vee a_2,$$
$$x_2\vee x_1=x_2\vee(a_2\wedge x_1)\vee x_1=a_2\vee x_1=a_1\vee a_2.$$

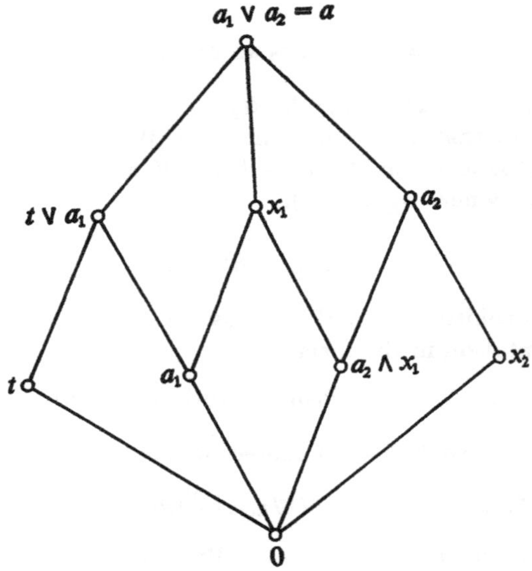

Figure 1

Now observe that $x_1 \prec t \vee x_1 = x_1 \vee x_2$. Therefore, for any atom $u \leq x_2$, we have $x_1 \vee u = t \vee x_1$ and, of course, $x_1 \wedge u = 0$. Hence $t \sim u$ by Lemma 7. Now recall that $t \in R(q)$, so $t \sim u$ implies that $u \in R(q)$. But $u \leq a_2 = \bigvee(x_{R(p)} \mid p \not\approx q)$, hence $u \in R(p)$ for some $p \neq q$, a contradiction.

Finally, we have to show that $[0, c(p)]$ is directly indecomposable. If it is not, then by Theorem III.4.1, it has a pair a, b of complemented neutral elements, $a, b \notin \{0, c(p)\}$. So we can choose atoms u, v of $[0, c(p)]$ such that $u \leq a$ and $v \leq b$. Since $u \approx v$ in L, $u \approx v$ in $[0, c(p)]$ (since $[0, c(p)]$ is a direct factor). Taking the homomorphism $\varphi: x \to x \wedge a$ of $[0, c(p)]$ onto $[0, a]$, we obtain that $u\varphi \approx v\varphi$. But $u = u\varphi$, $v\varphi = 0$, and $u \approx 0$ imply that $u = 0$, a contradiction. ●

It is clear from the proof of Theorem 5, that indecomposability is equivalent with the projectivity of atoms. Therefore to prove Theorem 6 it suffices to prove the following statement:

Theorem 6′. *In a geometric lattice, perspectivity of atoms is transitive.*

Proof. Let p, q, and r be atoms of L, $p \sim q$, $q \sim r$; by Lemma 7, we can choose x and y such that

$$p \wedge x = q \wedge x = 0, \quad p \vee x = q \vee x, \; h(x) = m,$$

$$q \wedge y = r \wedge y = 0, \quad q \vee y = r \vee y, \; h(y) = n.$$

Choose x so as to minimize $h(x)$. Let $\{s_1, \ldots, s_m\}$ be an independent set of atoms in $(x]$, and $\{t_1, \ldots, t_n\}$ in $(y]$, and let $x_i = s_1 \vee \cdots \vee s_{i-1} \vee s_{i+1} \vee \cdots \vee s_m$, for $1 \leq i \leq m$. We claim that $p \nleq x_i \vee q$. Otherwise, $p \leq x_i \vee q$ and so $x_i \vee p \leq x_i \vee q$. However, $h(x_i) =$

$m-1$ and therefore $h(x_i \vee p) = h(x_i \vee q) = m$, implying that $x_i \vee p = x_i \vee q$ and then x_i could replace x, contradicting the minimality of $h(x)$. So $h(p \vee q \vee x_i) = 2 + m - 1 = m+1$, and

$$p \vee q \vee x_i = p \vee x = q \vee x, \quad \text{for} \quad 1 \leq i \leq m.$$

Now by Theorem 2.4 $\{s_1, \ldots, s_m, p\}$ is an independent set of atoms so, by Theorem 2.5, we can choose a subset $\{t_1', \ldots, t_k'\}$ of $\{t_1, \ldots, t_n\}$ such that

$$\{p, s_1, \ldots, s_m, t_1', \ldots, t_k'\}$$

is independent. Set $y' = t_1' \vee \cdots \vee t_k'$. Then $\{p, s_1, \ldots, s_m, y'\}$ and also $\{q, s_1, \ldots, s_m, y'\}$ are independent. Therefore, using the definition of independence, we obtain

$$y' = (x \vee y') \wedge (q \vee x_1 \vee y') \wedge \cdots \wedge (q \vee x_m \vee y').$$

Since $r \wedge y = 0$, we have $r \nleq y$; hence $r \nleq y'$ and, therefore, either $r \nleq x \vee y'$ or $r \nleq q \vee x_i \vee y'$, for some $1 \leq i \leq m$. Set $a = x \vee y'$ if $r \nleq x \vee y'$ or $a = q \vee x_i \vee y'$ if $r \nleq q \vee x_i \vee y'$. Then

$$r \wedge a = 0 \quad \text{and} \quad r \leq p \vee a$$

(the latter because $r \leq q \vee y \leq p \vee x \vee y = p \vee x \vee y' = p \vee q \vee x_i \vee y'$).
Finally, since $r \nleq a$, we obtain $p \nleq a$, that is,

$$p \wedge a = 0,$$

thus $a \leq a \vee r \leq a \vee p$ and $a \prec a \vee p$, and so

$$a \vee r = a \vee p,$$

proving that $p \sim r$. ●

If the geometric lattice L is of finite length, then L is said to be *finite dimensional*. In this case, Theorem 5 follows from Lemma III.4.5. With some effort, the general case could be reduced to the finite dimensional case. However, without the present proof of Theorem 5 it would be more difficult to prove Theorem 6.

Theorem III.4.6 could be used to prove the result that every finite dimensional geometric lattice is isomorphic to a direct product of simple geometric lattices. The factors we obtain from Theorems 5 and 6 are such that any two atoms are perspective. Now if L is a geometric lattice of finite length in which any two atoms are perspective, then L is simple. Indeed, if Θ is a congruence relation of L and $\Theta \neq \omega$, then there exist $a, b \in L$, $a < b$ such that $a \equiv b\ (\Theta)$. Choose an atom p such that $p \leq b$ and $p \nleq a$. Then $p \equiv 0\ (\Theta)$, since $p = p \wedge b \equiv p \wedge a = 0\ (\Theta)$. Now let q be any atom of L. Since $p \sim q$, p and q have a common complement x. Then

$$p/0 \nearrow 1/x \searrow q/0,$$

and so $q \equiv 0\ (\Theta)$. Since $h(1) < \infty$, $1 = q_1 \vee \cdots \vee q_n$ for some finite set of atoms $\{q_1, \ldots, q_n\}$ and $q_i \equiv 0\ (\Theta)$, $i = 1, \ldots, n$, hence $1 \equiv 0\ (\Theta)$. Thus $\Theta = \iota$ and L is simple.

Exactly the same proof yields that if L is a geometric lattice in which any two atoms are perspective, then L is subdirectly irreducible; the only atom Φ of $C(L)$ contained in every $\Theta \neq \omega$ is the congruence $\Phi = \Theta(p, 0)$, where p is any atom. If $a \in L$ and $h(a) < \infty$, then $a \equiv 0\ (\Phi)$. In general, however, L is not simple.

By a *geometry* we mean a set A certain subsets of which are called *subspaces* (or *flats*) such that certain basic properties hold; we shall list these properties in the next definition.

DEFINITION 8. *A* geometry $\langle A, ^-\rangle$ *is a set A and a function $X \to \bar{X}$ of $P(A)$ into itself satisfying the following conditions:*

(i) $X \subseteq \bar{X}$; *if* $X \subseteq Y$, *then* $\bar{X} \subseteq \bar{Y}$; $\bar{\bar{X}} = \bar{X}$.

(ii) $\bar{\varnothing} = \varnothing$, *and* $\overline{\{x\}} = \{x\}$ *for* $x \in A$.

(iii) *If* $x \in \overline{X \cup \{y\}}$, *but* $x \notin \bar{X}$, *then* $y \in \overline{X \cup \{x\}}$.

(iv) *If* $x \in \bar{X}$, *then* $x \in \bar{X}_1$ *for some finite* $X_1 \subseteq X$.

(i) means that $\langle A, ^-\rangle$ is a closure space. A subset X of a closure space is called *closed* iff $\bar{X} = X$. It follows simply form (i), that $\bar{X}$ is the smallest closed set containing X.

LEMMA 9. *Let $\langle A, ^-\rangle$ be a closure space. Then $L\langle A, ^-\rangle = \{\bar{X} \mid X \subseteq A\}$ is a complete lattice with respect to set inclusion, and $\bigwedge Y = \bigcap Y$ for $Y \subseteq L\langle A, ^-\rangle$. Conversely, if $L \subseteq P(A)$ is closed under arbitrary intersection, then*

$$\bar{X} = \bigcap (Y \mid Y \supseteq X \text{ and } Y \in L\}$$

defines a closure space $\langle A, ^-\rangle$ such that a set is closed iff it belongs to L.

PROOF. Let $Y \subseteq L\langle A, ^-\rangle$ and $Z = \bigcap Y$. Then $Z \subseteq X$ for all $X \in Y$, so $\bar{Z} \subseteq \bar{X} = X$, and therefore $\bar{Z} \subseteq \bigcap Y = Z$. Since $Z \subseteq \bar{Z}$ always holds, $Z = \bar{Z}$. The remaining steps in the proof of Lemma 9 are very easy and will be left to the reader. ●

Thus a closure space is completely determined by the lattice L of closed sets. In case of a geometry we have $\overline{\{x\}} = \{x\}$ meaning that the elements of A can be identified as atoms of L, hence L determines the geometry even if L is known only up to isomorphism.

For geometries, a closed set is called a *subspace* and $\bar{X}$ is called the *subspace spanned by* X. Thus (iv) means that if x belongs to the subspace spanned by X, then x belongs to the subspace spanned by some finite $X_1 \subseteq X$. A closure space satisfying (iv) is usually called *algebraic*.

LEMMA 10. *A lattice L is algebraic iff L is isomorphic to the lattice of closed sets of an algebraic closure space.*

PROOF. Let L be an algebraic lattice. By Theorem II.3.13, $L \cong I(F)$ where F is a join-semilattice with 0. Take $A = F$ and for $X \subseteq A$ define $\bar{X} = (X]$, the ideal generated by X. The verification of (i) and (iv) for $\langle A, ^-\rangle$ is contained in the proof of Theorem II.3.13. Then $I(F)$ is the set of closed sets and so $L \cong I(F)$ by assumption.

Conversely, if $\langle A, ^-\rangle$ is an algebraic closure space and L is the lattice of closed sets, then it is easily seen that $X \in L$ is compact iff $X = \overline{X}_1$ for some finite $X_1 \subseteq X$, from which it follows immediately that L is algebraic. ●

Condition (iii) makes it possible to define the *rank* or *dimension* of a subspace. Let $r(\varnothing) = 0$ and $r(\{a\}) = 1$. If X is a subspace, $r(X) = n$, and $y \notin X$, then $r(\overline{X \cup \{y\}}) = n + 1$. (iii) implies that this defines r uniquely.

Finally, (ii) states that $\varnothing$ is a subspace and so is every singleton $\{x\}$. If we drop this last condition we get what is known as a *pregeometry* from which a geometry can be obtained by identifying any two elements, x, y of A, for which $\overline{\{x\}} = \overline{\{y\}}$.

THEOREM 11. *Let $\langle A, ^{-}\rangle$ be a geometry. Then $L = L\langle A, ^{-}\rangle$ is a geometric lattice. Conversely, if L is a geometric lattice, A is the set of atoms of L, and for $X \subseteq A$, $\overline{X}$ is the set of atoms spanned by X (in the sense defined in Section 2), then $\langle A, ^{-}\rangle$ is a geometry and $L \cong L\langle A, ^{-}\rangle$.*

PROOF. Let $\langle A, ^{-}\rangle$ be a geometry. Then by Lemma 9 $L = L\langle A, ^{-}\rangle$ is algebraic. Since X is compact iff $X = \overline{X}_1$ for some finite $X_1 \subseteq X$, we see that X is compact iff X is a finite join of atoms (for $x \in A$, $\overline{\{x\}}$ is an atom by 8(ii)). It remains to show that L is semimodular. Let $X, Y \in L$, $Y = \overline{X \cup \{x\}}$, and $x \notin X$. We claim that $X \prec Y$. Indeed, if $Z \in L$ and $X \subset Z \subseteq Y$, then there is a $z \in Z - X$ and $z \in Z \subseteq Y = \overline{X \cup \{x\}}$, and so by 8(iii), $x \in \overline{X \cup \{z\}} \subseteq Z$. Thus, $Y = \overline{X \cup \{x\}} \subseteq Z$; we conclude that $Y = Z$, proving $X \prec Y$. Now let $U \in L$. Then $Y \vee U = \overline{Y \cup U} = \overline{X \cup U \cup \{x\}}$ and $X \vee U = \overline{X \cup U}$, hence either $x \in \overline{X \cup U}$ and so $X \vee U = Y \vee U$ or $x \notin \overline{X \cup U}$ and in this case $X \vee U \prec Y \vee U$.

Conversely, let L be a geometric lattice, let A be the set of atoms of L, and let $\overline{X}$ be the set of atoms spanned by X for $X \subseteq A$. It is immediate that $\langle A, ^{-}\rangle$ is a closure space. 8(ii) and 8(iv) are also clear by definition. To verify 8(iii), let $x \in \overline{X \cup \{y\}}$ and $x \notin \overline{X}$. Since $\{y\}$ is an atom, by semimodularity, $\overline{X \cup \{y\}} = \overline{X} \vee \{y\} \succ X$; thus $X \subset \overline{X \cup \{x\}} \subseteq \overline{X \cup \{y\}}$ implies that $\overline{X \cup \{x\}} = \overline{X \cup \{y\}}$; and so $y \in \overline{X \cup \{x\}}$.

Now let φ be the map $X \to \bigvee X$ for $X \subseteq A$, $X \in L\langle A, ^{-}\rangle$. Then φ maps $L\langle A, ^{-}\rangle$ into L. Since every element of L is a join of atoms, $X \subseteq Y$ iff $\bigvee X \leq \bigvee Y$. Therefore, φ is onto, one-to-one and both φ and φ^{-1} are isotone. Hence φ is an isomorphism. ●

The lattice of subspaces thus completely determines the geometry and vice versa. The diagrams of geometric lattices are as a rule too complicated to be of any use. Many times, however, we can draw a picture of the geometry associated with the lattice.

Let us call an element of height 2 a *line*. A line is the join of any two distinct points in the line. Some geometries (for instance, projective geometries, see Section 5) are completely determined by their lines, some are not. When drawing the picture of a geometry we try to represent lines by straight lines or by curves. Lines having exactly two points are not drawn as a rule. Figures 2a, 3a, and 4a are pictures of geometries and Figures 2b, 3b, and 4b are the corresponding lattice diagrams. Observe the simplicity of geometries compared to the intricate diagrams of the lattices.

Closure spaces satisfying the additional requirements 8(ii)–8(iv) can be found in

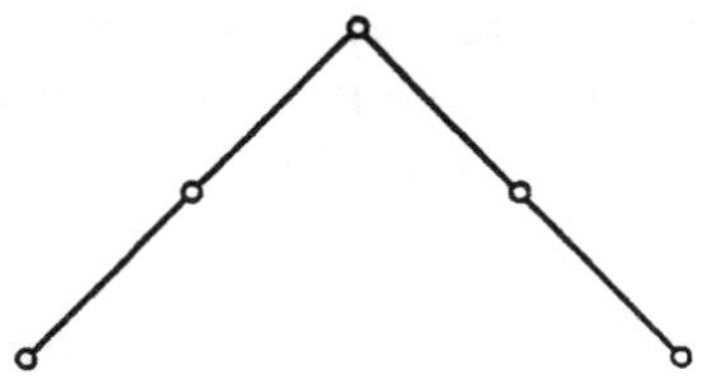
Figure 2a

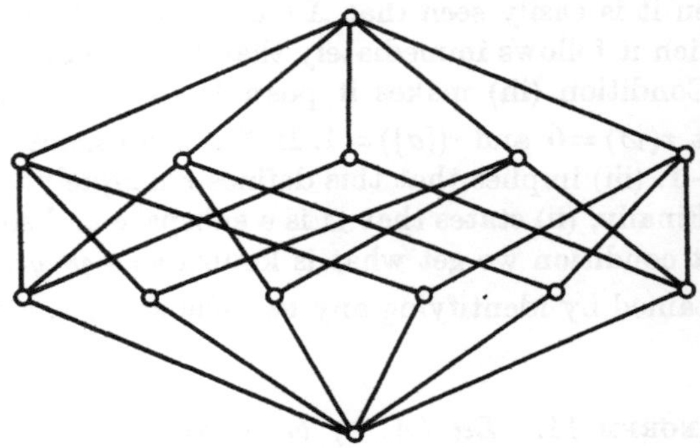
Figure 2b

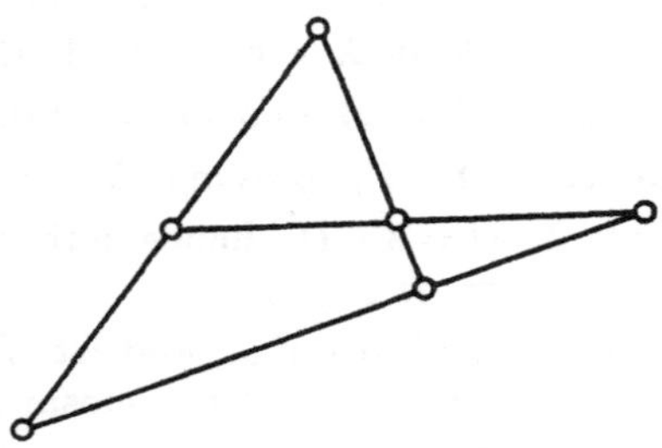
Figure 3a

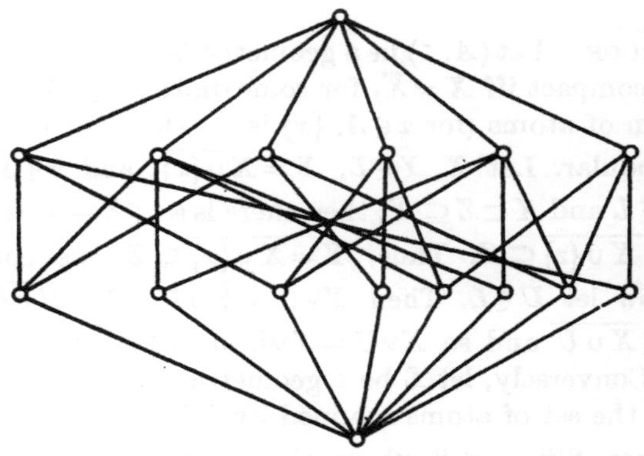
Figure 3b

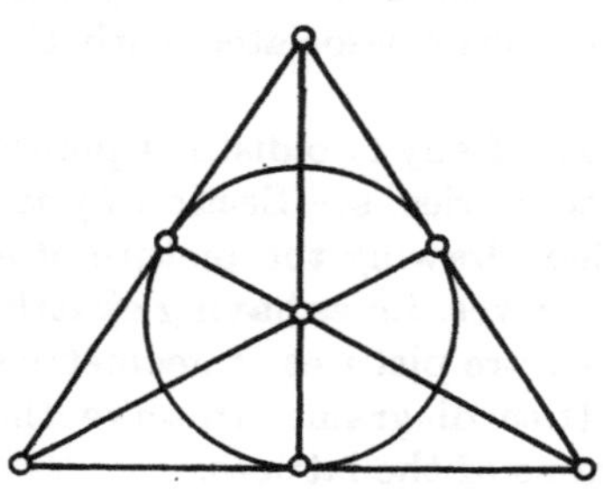
Figure 4a

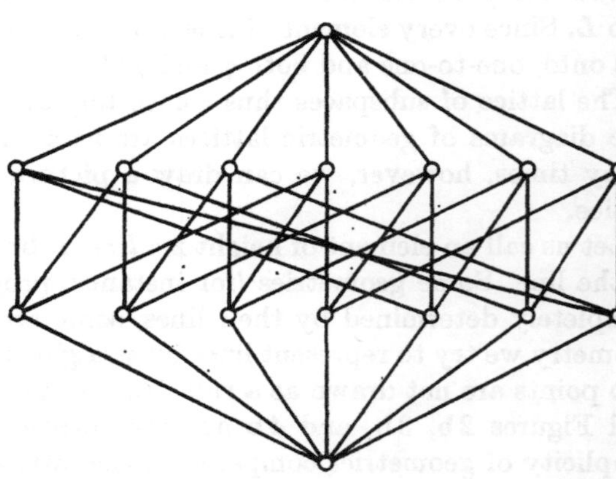
Figure 4b

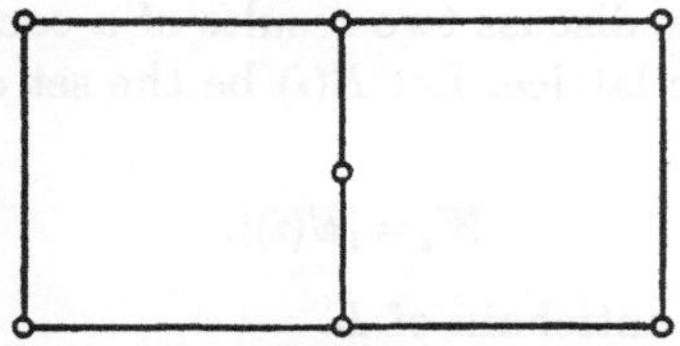

Figure 5

large numbers, especially in algebra, geometry, and combinatorics. Here is an example from combinatorics.

A *graph* is a set G with a fixed set E (the edges) of two- element subsets of G (unoriented graph without loops). Figure 5 shows a diagram of a graph. Two points a and b (elements of G) are connected with a straight line segment iff $\{a, b\} \in E$.

Let $a, b \in G$ and $A \subseteq E$. We shall call a and b *A-connected* iff there is a sequence of edges $\{x_0, x_1\}, \{x_1, x_2\}, \ldots, \{x_{n-1}, x_n\} \in A$ such that $a = x_0$ and $x_n = b$.

Now we define a geometry, called the *edge geometry*, on E: for $\{a, b\} \in E$ and $A \subseteq E$, let $\{a, b\} \in \bar{A}$ iff a and b are A-connected.

THEOREM 12. *$\langle E, ^- \rangle$ is a geometry.*

PROOF. All the properties of a geometry obviously hold except perhaps 8(iii). Let a and b be A-connected, then there is a connecting sequence $\{x_0, x_1\}, \ldots, \{x_{n-1}, x_n\}$ for which n is minimal. We claim that no edge can occur twice in this sequence. Indeed, let $0 \leq i < j \leq n-1$ and $\{x_i, x_{i+1}\} = \{x_j, x_{j+1}\}$. Now if $x_i = x_j$ and $x_{i+1} = x_{j+1}$, then by dropping $\{x_{i+1}, x_{i+2}\}, \ldots, \{x_j, x_{j+1}\}$ from the sequence we get a shorter connecting sequence. If $x_i = x_{j+1}$ and $x_{i+1} = x_j$, then we can drop all of $\{x_i, x_{i+1}\}, \ldots, \{x_j, x_{j+1}\}$.

Now we prove 8(iii). Let $x \in \overline{X \cup \{y\}}$ but $x \notin \bar{X}$, where $x = \{a, b\}$ and $y = \{c, d\}$. Let $e_0, \ldots, e_{n-1} \in X \cup \{y\}$ be a shortest sequence connecting a and b. Since $x \notin \bar{X}$, one of the $e_0, \ldots, e_{n-1}$ must be y. By the observation above, exactly one of $e_0, \ldots, e_{n-1}$, say e_i, equals y. But then $e_{i+1}, \ldots, e_{n-1}, x, e_0, \ldots, e_{i-1}$ will connect c and d, hence $y \in \overline{X \cup \{x\}}$. ●

For instance, if we take the graph of Figure 6, then Figure 2a shows the edge geometry and Figure 2b the lattice, called the *edge lattice*, associated with it.

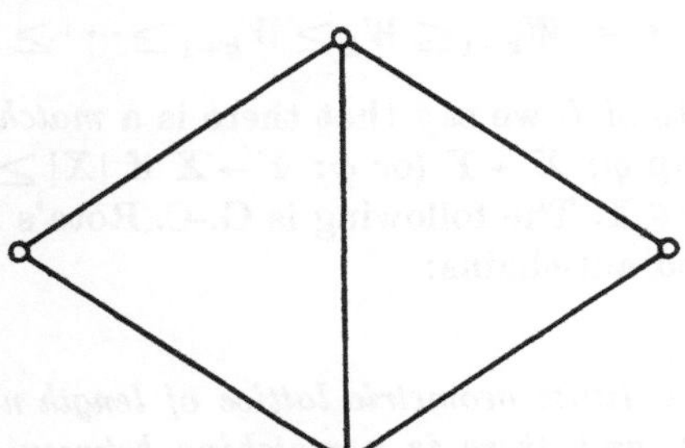

Figure 6

To conclude this section we discuss two results of a combinatorial nature.

Let L be a finite geometric lattice. Let $E(i)$ be the set of elements of L of height i and

$$W_i = |E(i)|.$$

Observe that each $E(i)$ is an antichain of L.

An antichain E of L is a *maximum sized antichain* iff $|X| \leq |E|$ for any antichain X of L. The following two theorems will establish the size of a maximum sized antichain for some classes of geometric lattices. The first is credited to R. P. Dilworth and L. H. Harper in K. Baker [1969].

Theorem 13. *Let the finite geometric lattice L have the property that every element x of height i is covered by a_i and covers b_i elements (a_i and b_i depend only on i). Then a maximum sized antichain of L has* $\max(W_i \mid 0 \leq i \leq l(L))$ *elements.*

Proof. Let $n = l(L)$ and let $\max(W_i \mid 0 \leq i \leq n) = W_k$; then $E(k)$ is an antichain with W_k elements. We have to prove that if S is an antichain, then $|S| \leq W_k$.

For $x \in L$, let $s(x)$ denote the number of maximal chains going through x. It is obvious, that if $h(x) = i$, then $s(x) = b_1 \cdots b_i \cdot a_i \cdots a_{n-1}$, hence $s(x)$ depends only on i. Since every maximal chain has exactly one element of order i, we obtain that $s(x) \cdot W_i = s$ where s is the number of maximal chains in L. Using $W_i \leq W_k$ we obtain

$$s(x) = \frac{s}{W_i} \geq \frac{s}{W_k}.$$

Finally, a maximal chain goes through at most one element of the antichain S. Therefore,

$$s \geq \sum(s(x) \mid x \in S) \geq |S| \cdot \frac{s}{W_k},$$

that is, $|S| \leq W_k$. ●

Theorem 13 as well as the next result was motivated by *Sperner's Lemma* (E. Sperner [1928]) stating the conclusions of Theorem 13 for $P(X)$.

A finite geometric lattice L of length n is *unimodal* iff for some integer k we have that

$$W_1 \leq \cdots \leq W_{k-1} \leq W_k \geq W_{k+1} \geq \cdots \geq W_{n-1}.$$

If X and Y are subsets of L we say that there is a *matching* between X and Y iff there is a one-to-one map $\varphi\colon X \to Y$ (or $\varphi\colon Y \to X$ if $|X| \geq |Y|$) such that x is comparable with $x\varphi$ for all $x \in X$. The following is G.-C. Rota's approach to the problem of finding maximum sized antichains:

Theorem 14. *Let L be a finite geometric lattice of length n. If L is unimodal ($W_1 \leq \cdots \leq W_k \geq \cdots \geq W_{n-1}$) and there is a matching between $E(i)$ and $E(i+1)$ for all $i < n$, then a maximum sized antichain of L has W_k elements.*

Proof. For a subset S of L let

$$t(S) = \max(h(x) - h(y) \mid x, y \in S)$$

be the *thickness* of S. Let S be an antichain of L. We shall prove $|S| \leq W_k$ by induction on $t(S)$.

If $t(S) = 0$, then $S \subseteq E(i)$ for some i, hence $|S| \leq W_i \leq W_k$. Now let $t(S) > 0$ and assume the inequality for antichains with a smaller thickness. Since $t(S) > 0$, there is an $x \in S$ with $h(x) \neq k$, say $h(x) < k$. Let $i = \min(h(x) \mid x \in S)$, and set $S = S_0 \cup S_1$ where $S_1 = S \cap E(i)$ and $S_0 = S - S_1$. By assumption, $i < k$, and so $|E(i)| \leq |E(i+1)|$, hence there is a matching $\varphi\colon E(i) \to E(i+1)$. Define

$$S' = S_0 \cup S_1\varphi .$$

S_0 and $S_1\varphi$ are disjoint, because if $x \in S_0$ and $x \in S_1\varphi$, that is, $x = y\varphi$ with $y \in S_1$, then $x, y \in S$, $x \neq y$, and x and y are comparable, contradicting that S is an antichain. Thus $|S'| = |S|$. Moreover, S' is an antichain; indeed, if $x, y \in S'$, $x \neq y$, x and y are comparable, then $x \in S_0$ and $y \in S_1\varphi$, that is, $y = z\varphi$ for some $z \in S_1$. Since $h(y) = i+1$, we have $h(y) \leq h(x)$, therefore $x < y$ is impossible. Thus $y < x$, contradicting $z \leq y$ and $z \parallel x$.

Thus S' is an antichain, $t(S') = t(S) - 1$, and so

$$|S| = |S'| \leq W_k. \quad \bullet$$

Exercises

1. Let F be a semimodular lattice with 0 in which every element is a finite join of atoms. Show that $I(F)$ is a geometric lattice.
2. Is the converse of Exercise 1 true?
3. Call a lattice L *atomic* iff L has a 0 and for every $a \in L$, $a \neq 0$, there is an atom $p \leq a$. A lattice L is (*upper*) *continuous* iff

$$a \wedge \bigvee D = \bigvee(a \wedge x \mid x \in D)$$

for any (upper) directed subset D of L (D is *directed* iff for all $x, y \in D$ there is a $z \in D$ satisfying $x \leq z$ and $y \leq z$). Prove that a lattice L is geometric iff L is atomic, relatively complemented, continuous, and semimodular (G. Birkhoff [1948] and F. Maeda [1951]).
4. Show that in Exercise 3 " semimodular " can be replaced by the condition that if a and b cover $a \wedge b$, then $a \vee b$ covers a and b.
5. Prove that in Exercise 3, " semimodularity " can be replaced by "M-symmetry " (F. and S. Maeda [1970]).
6. Let F be a semilattice and let $C(F)$ be the lattice of all congruence relations of F. Prove that $C(F)$ is geometric iff $C(F)$ is atomistic.
7. Let K be a subfield of the field K'. Let L be the lattice of all relatively algebraically closed subfields $M \supseteq K$. (M is *relatively algebraically closed* iff a polynomial with coefficients in K which factors in K' also factors in M.) Prove that L is a geometric lattice.
8. Prove that the lattice L of Exercise 7 is modular iff the transcendence degree of K' over K is at most two (S. MacLane [1938]).

9. Let G be an abelian group. Let L denote the lattice of all subgroups H with the property that G/H (the quotient group) has no element of finite order excepting zero. Prove that L is a geometric lattice.
10. Consider the concepts of p-independence and p-basis in fields of characteristic p (see, for instance, B. L. van der Waerden [1931]). Can these be viewed as independence in a suitable geometric lattice?
11. Let L be a geometric lattice. Show that $a \sim b$ in L iff $a \wedge x = b \wedge x$ and $a \vee x = b \vee x$ for some $x \in L$.
12. Let a and b be atoms in the geometric lattice L. Show that $a \sim b$ iff there exists a finite independent set of atoms I such that $b \in I$ and for $I_1 \subseteq I$, $a \leq \bigvee I_1$ iff $I_1 = I$.
13. Show that if $a \sim b$, then $a/0 \nearrow 1/x \searrow b/0$ for some x.
14. Prove that in a geometric lattice $a \sim b$ iff $a/0 \sim x/y \sim b/0$ for some quotient x/y.
15. Relate $a \approx b$ and $a/0 \approx b/0$ in a general lattice, in a geometric lattice, and in a modular geometric lattice.
16. Let L be a modular geometric lattice. Show that L is simple iff L is a directly indecomposable lattice of finite length.
17. Let I and J be maximal independent sets of atoms of a geometric lattice. Prove that $|I| = |J|$.
18. Show that in a geometric lattice L, for every element a there is a smallest element $c(a)$ of the centre of L satisfying $a \leq c(a)$.
19. Prove that in a geometric lattice $a \sim b$ iff $c(a) = c(b)$ (with $c(x)$ as in Exercise 18).
20. A lattice L is called *left-complemented* (L. R. Wilcox [1942]) iff L is bounded and for $a, b \in L$ there exists $b_1 \leq b$ such that

$$a \vee b_1 = a \vee b, \quad a \wedge b_1 = 0, \quad \text{and} \quad aMb_1.$$

Is every geometric lattice left-complemented?
21. Let L be a complete lattice and let $X \subseteq L$. Then $L \cong \Pi((x] \mid x \in X)$ iff every element $a \in L$ has one and only one expression of the form

$$a = \bigvee(a_x \mid x \in X),$$

where $a_x \leq x$ for all $x \in X$.
22. Define the isomorphism of two geometries. Prove that two geometries are isomorphic iff the associated geometric lattices are isomorphic.
23. Define a *circuit* of a graph as a set of edges $\{x_0, x_1\}, \ldots, \{x_{n-1}, x_n\}, \{x_n, x_0\}$. Define the subspaces of the geometry on the edges of the graph in terms of the circuits. What role is played by the minimal circuits?
24. Draw the lattice associated with the graph of Figure 5.
25. Does the lattice associated with a graph determine the graph (up to isomorphism)?
26. Generalize Theorem 13 to posets.
27. Prove Sperner's Lemma using Theorem 13.
28. Prove Sperner's Lemma using Theorem 14.
29. Does Theorem 13 or Theorem 14 apply to finite projective geometries?
30. Consider the graph G_n of Figure 7. Let L_n be the edge lattice of G_n. If $A \in L_n$ and $\{a, b\} \in A$, then A is determined by $\{i \mid \{a, i\} \text{ and } \{b, i\} \in A\}$. If $\{a, b\} \notin A$, then A is determined by $\{i \mid \{a, i\} \text{ or } \{b, i\} \in A\}$ and a choice function selecting one of $\{a, i\}$ and $\{b, i\}$. Conclude from this, that

$$W_k = 2^k \binom{n}{k} + \binom{n}{k-1}.$$

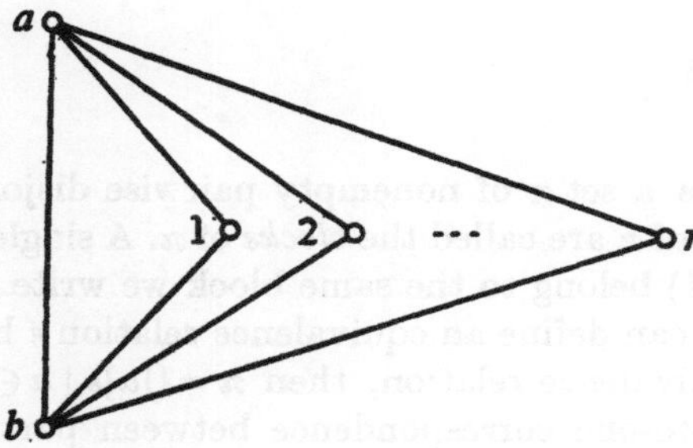

Figure 7

31. Show that in L_{10} there is an antichain contained in $E(6) \cup E(7)$ that has more elements than any $E(k)$. (Exercises 30 and 31 are due to R. P. Dilworth and C. Greene [1971].)

Let P be a finite poset. The *Möbius function* μ is an integer valued function defined on $P \times P$ by the formulas:

$$\mu(x, x) = 1 \quad \text{for} \quad x \in P;$$

$$\mu(x, y) = \begin{cases} 0, & \text{if} \quad x \nleq y, \\ -\Sigma(\mu(x, z) \mid x \leq z < y), & \text{if} \quad x < y. \end{cases}$$

(See A. F. Möbius [1832] and L. Weisner [1935].)

32. Establish the Möbius inversion formula: if f and g are real valued functions on a finite partially ordered set P and $g(x) = \Sigma(f(y) \mid y \leq x)$ for every $x \in P$, then

$$f(x) = \Sigma(g(y)\mu(y, x) \mid y \leq x).$$

33. Let P and Q be finite posets with Möbius functions μ_P and μ_Q, respectively. Then the Möbius function $\mu_{P \times Q}$ of $P \times Q$ satisfies

$$\mu_{P \times Q}(\langle x, y \rangle, \langle u, v \rangle) = \mu_P(x, u)\, \mu_Q(y, v)$$

for $x, u \in P$ and $y, v \in Q$.

34. Show that the Möbius function μ of the Boolean lattice B is given by

$$\mu(x, y) = (-1)^{h(y) - h(x)}$$

for $x \leq y$, where $h(x)$ is the height of x in B.

35. Let μ be the Möbius function of a finite lattice L and let $x, y, z \in L$.

(i) If $x \leq y$ and y is not the join of elements covering x then $\mu(x, y) = 0$ (Ph. Hall [1934]).

(ii) If $x \leq z \leq y$ then

$$\Sigma(\mu(x, t) \mid z \vee t = y) = \begin{cases} \mu(x, y), & \text{if} \quad z = x, \\ 0, & \text{if} \quad z \neq x \end{cases}$$

(L. Weisner [1935]).

36. Show that the Möbius function μ of a finite distributive lattice L is given by: $\mu(x, y) = 0$, if y is not the join of elements covering x; $\mu(x, y) = (-1)^n$, if y is the join of n distinct elements covering x. (Hint: Use Exercises 34 and 35(i).)

37. Let μ be the Möbius function of a finite geometric lattice L and $x, y \in L$, $x \leq y$. Then $\mu(x, y) \neq 0$; $\mu(x, y)$ is positive if $h(y) - h(x)$ is even; $\mu(x, y)$ is negative if $h(y) - h(x)$ is odd (G.-C. Rota [1964]). (Hint: Apply Exercise 35(ii).)

4. Partition Lattices

A *partition* of a set A is a set π of nonempty pairwise disjoint subsets of A whose union is A. The members of π are called the *blocks* of π. A singleton as a block is called *trivial*. If a and b ($a, b \in A$) belong to the same block we write $a \equiv b\ (\pi)$ or $a\pi b$.

Given a partition π we can define an equivalence relation ε by $\langle x, y\rangle \in \varepsilon$ iff $x \equiv y\ (\pi)$. Conversely, if ε is an equivalence relation, then $\pi = \{[a]\varepsilon \mid a \in A\}$ is a partition of A. Thus we set up a one-to-one correspondence between partitions and equivalence relations. Partitions are easier to visualize so we shall use them; if the reader prefers equivalence relations, he can use them throughout.

Part(A) will denote the set of all partitions of A partially ordered by

$$\pi_0 \leq \pi_1 \quad \text{iff} \quad x \equiv y\ (\pi_0) \quad \text{implies that} \quad x \equiv y\ (\pi_1).$$

Part(A) with this partial ordering (which corresponds to set inclusion for the corresponding equivalence relations) forms a complete lattice, called the *partition lattice* (or *equivalence lattice*) of A.

We draw a picture of a partition by drawing the boundary lines of the blocks. Then $\pi_0 \leq \pi_1$ iff the boundary lines of π_1 are also boundary lines of π_0 (but π_0 may have some more boundary lines). Equivalently, the blocks of π_1 are unions of blocks of π_0. (See Figure 1.)

Lemma 1.

(i) Part(A) *is a complete lattice and* $x \equiv y\ (\bigwedge(\pi_i \mid i \in I))$ *iff* $x \equiv y\ (\pi_i)$, *for all* $i \in I$; $x \equiv y\ (\bigvee(\pi_i \mid i \in I))$ *iff there is a natural number* n, $i_0, \ldots, i_{n-1} \in I$, *and* $x_0, \ldots, x_{n+1} \in A$ *such that* $x = x_0$, $y = x_{n+1}$, *and* $x_j \equiv x_{j+1}\ (\pi_{i_j})$, *for* $0 \leq j \leq n$.
The zero of Part(A) *has only trivial blocks and the unit has only one block, namely* A.

(ii) *The atoms of* Part(A) *are the partitions with exactly one nontrivial block, and this block has exactly two elements. The dual atoms of* Part(A) *are the partitions with exactly two blocks.*

(iii) $\pi_0 \prec \pi_1$ *in* Part(A) *iff* π_1 *is the result of replacing two blocks of* π_0 *by their union.*

(iv) *In* Part(A), $[\pi)$ *is isomorphic with the partition lattice of the set* π.

(v) *In* Part(A), $(\pi]$ *is isomorphic with the direct product of all* Part(X) *where* X *ranges over the nontrivial blocks of* π.

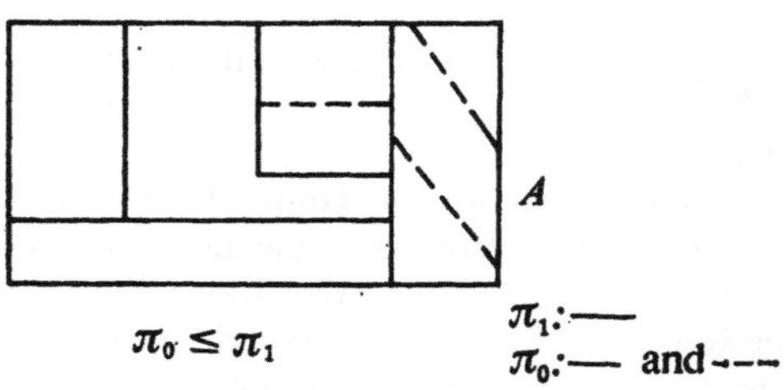

Figure 1

Proof.

(i) This is implicit in the proof of Theorem I.3.9. ◑

(ii) and (iii) are trivial. ◑

(iv) If $\pi \leq \xi$ then each block of ξ is a union of blocks of π. Thus ξ defines a partition of the blocks of π, that is, of π. This sets up the required isomorphism. ◑

(v) If $\xi \leq \pi$, then for each nontrivial block X of π, ξ defines a partition ξ_X of X. Since a block of ξ which intersects X is completely within X, the map $\xi \to \langle \xi_X \mid X$ is a nontrivial block of $\pi\rangle$ sets up the required isomorphism. ●

From Lemma 1 we easily conclude:

Theorem 2 (O. Ore [1942]). Part(A) *is a simple geometric lattice.*

Proof. By Lemma 1(i), Part(A) is complete. It follows from 1(ii) that every element is a join of atoms and, by the formula for join, the atoms are compact. If π_0 and π_1 satisfy the condition of 1(iii), then so do $\pi_0 \vee \xi$ and $\pi_1 \vee \xi$ unless $\pi_0 \vee \xi = \pi_1 \vee \xi$. Therefore, Part$(A)$ is semimodular. This shows that Part(A) is geometric.

Let π and ξ be atoms. By 1(ii), π and ξ have only one nontrivial block each, say $\{a, b\}$ and $\{c, d\}$, respectively. Let τ be a partition with two blocks as follows: if $\{a, b\} \cap \{c, d\} = \varnothing$, the blocks of τ are $A - \{a, c\}$ and $\{a, c\}$; if $\{a, b\} \cap \{c, d\} = \{e\}$, then the blocks of τ are $A - \{e\}$ and $\{e\}$. In either case, τ is a complement of π and ξ and so $\pi \sim \xi$. By the discussion in Section 3, Part(A) is simple if A is finite. Let A be infinite and let Θ be a congruence relation of Part(A), $\Theta \neq \omega$. Then all atoms are congruent modulo Θ to the zero of Part(A). Since A is infinite we can split A up into two disjoint sets A_0 and A_1 of the same cardinality. Let φ be a one-to-one map of A_0 onto A_1. Now we define some partitions of A. Let π have the two blocks, A_0 and A_1. Let ξ_i have exactly one nontrivial block, A_i $(i = 0, 1)$. The blocks of τ are all the sets of the form $\{x, x\varphi\}$ where $x \in A_0$. Then these partitions form a sublattice of Part(A); a diagram of this sublattice is given in Figure 2, where π_0 and π_1 denote the zero and unit of Part(A), respectively.

Let α be any atom $\leq \tau$. Then $\alpha \equiv \pi_0\ (\Theta)$, hence $\pi \equiv \pi \vee \alpha\ (\Theta)$. Since $\pi \prec \pi_1$, we obtain that $\pi \vee \alpha = \pi_1$ and so $\pi_1 \equiv \pi\ (\Theta)$. Taking the meets of both sides with τ and joining with ξ_i we get $\xi_i \equiv \pi_1\ (\Theta)$, $i = 0, 1$, hence $\pi_0 = \xi_0 \wedge \xi_1 \equiv \pi_1\ (\Theta)$. Thus $\Theta = \iota$ and Part(A) is simple. ●

Call a partition π *finite* iff all blocks of π are finite and only finitely many blocks

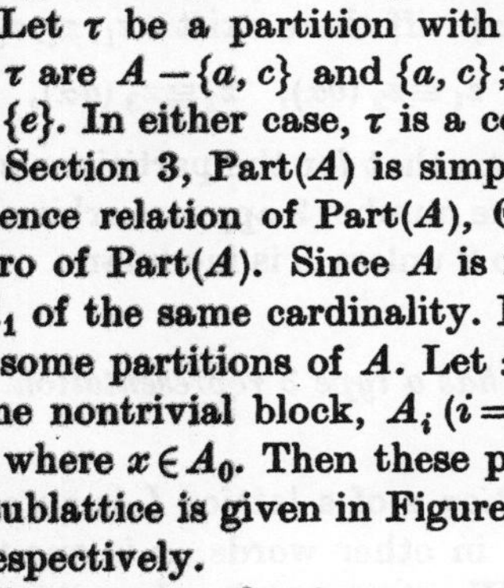

Figure 2

are nontrivial. Then it is clear that π is finite iff it is a finite join of atoms (recall 1(ii)). Since Part(A) is geometric, this is furthermore equivalent to π being compact.

Let $\text{Part}_{\text{fin}}(A)$ be the set of all finite partitions. This is obviously an ideal of Part(A). By Corollary 3.2 we obtain:

COROLLARY 3. $\text{Part}_{\text{fin}}(A)$ *is an ideal of* Part(A) *and the following isomorphism holds*:

$$I(\text{Part}_{\text{fin}}(A)) \cong \text{Part}(A).$$

The importance of partition lattices lies in the fact that they are as universal for lattices as permutation groups are for groups. Ph. M. Whitman [1946] proved that every lattice L can be embedded in some partition lattice. One drawback of this result is that the joins of the partitions representing elements of L have to be computed by 1(i) which can be rather complicated. This was corrected by B. Jónsson [1953]. To state Whitman's theorem as improved by Jónsson, we need some concepts.

A *representation* of a lattice L is an embedding α of L into some Part(A); α is called *of type* 3 iff for all $a, b \in L$ and $x, y \in A$

$$x \equiv y \ ((a \vee b)\alpha) \quad \text{iff there exist} \quad z_1, z_2, z_3 \in A \quad \text{such that}$$
$$x \equiv z_1 \ (a\alpha), \quad z_1 \equiv z_2 \ (b\alpha), \quad z_2 \equiv z_3 \ (a\alpha), \quad \text{and} \quad z_3 \equiv y \ (b\alpha).$$

In other words, type 3 means that for the partitions $a\alpha$, $a \in L$, the join formula in 1(i) always holds with $n \leq 3$. The number 3 appears arbitrary but we shall see in Theorem 5 that 3 cannot be reduced to 2 unless L is modular.

THEOREM 4. *Every lattice has a type* 3 *representation.*

PROOF. *A meet-representation* α of a lattice L is an embedding of L as a meet-semilattice into some Part(A); in other words, α is one-to-one and preserves meets but not necessarily the joins. We start out by describing a simple way of constructing meet-representations of a lattice.

Let A be a set and let L be a lattice with 0. A *distance function* δ maps A^2 *onto* L and it satisfies the following rules:

(i) δ is symmetric, that is, $\delta(x, y) = \delta(y, x)$, for all $x, y \in A$;

(ii) $\delta(x, x) = 0$, for all $x \in A$;

(iii) δ satisfies the *triangle inequality*, that is, $\delta(x, y) \vee \delta(y, z) \geq \delta(x, z)$, for all $x, y, z \in A$.

For instance, if L is a lattice with 0 and $A = L$, then putting $\delta(x, y) = x \vee y$ for $x \neq y$ and $\delta(x, x) = 0$ yields a distance function.

With a distance function δ we can associate a meet-representation $\alpha = \alpha(\delta)$ as follows: for $a \in L$ and $x, y \in A$, let $x \equiv y \ (a\alpha)$ iff $\delta(x, y) \leq a$. Indeed, $a\alpha$ is a partition. $x \equiv y \ (a\alpha \wedge b\alpha)$ iff $x \equiv y \ (a\alpha)$ and $x \equiv y \ (b\alpha)$, which is equivalent to $\delta(x, y) \leq a$ and $\delta(x, y) \leq b$; we obtained $\delta(x, y) \leq a \wedge b$, which is the same as $x \equiv y \ ((a \wedge b)\alpha)$. Thus $a\alpha \wedge b\alpha = (a \wedge b)\alpha$ and α is a meet-homomorphism. To show that α is one-to-one, take $a, b \in L$ and $a \neq b$; we can assume that $a \nleq b$. Since δ is onto, there are $x, y \in A$ such that $\delta(x, y) = a$. Then $x \equiv y \ (a\alpha)$. Since $a \nleq b$, $x \equiv y \ (b\alpha)$ fails to hold, and so $a\alpha \neq b\alpha$.

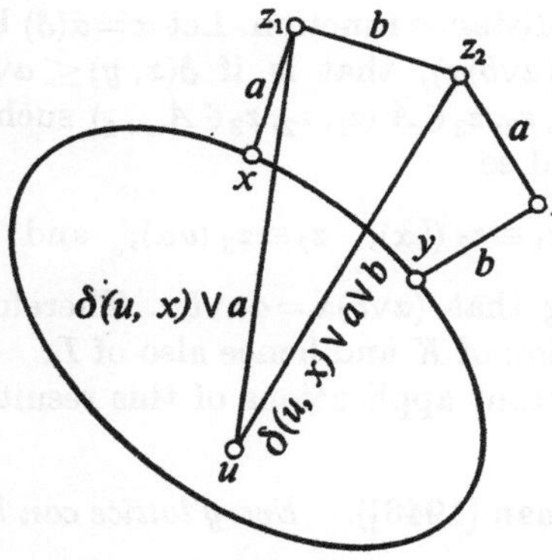

Figure 3

With a distance function $\delta\colon A^2 \to L$ and a quadruple $\langle x, y, a, b\rangle$ ($x, y \in A$ and $a, b \in L$) satisfying $\delta(x, y) \leq a \vee b$ we associate a new set

$$A^* = A \cup \{z_1, z_2, z_3\}$$

with three distinct elements z_1, z_2, z_3 not in A and a new distance function $\delta^*\colon (A^*)^2 \to L$ defined as follows (see Figure 3): δ^* is symmetric, satisfies $\delta^*(x, x) = 0$, for all $x \in A^*$, and for $u, v \in A$, $\delta^*(u, v) = \delta(u, v)$;

$$\delta^*(z_2, z_3) = a, \quad \delta^*(z_1, z_2) = b, \quad \text{and} \quad \delta^*(z_1, z_3) = a \vee b;$$

for $u \in A$,

$$\delta^*(u, z_1) = \delta(u, x) \vee a, \ \delta^*(u, z_2) = \delta(u, x) \vee a \vee b, \ \text{and} \ \delta^*(u, z_3) = \delta(u, y) \vee b.$$

It is easy to verify that δ^* is a distance function by virtue of $\delta(x, y) \leq a \vee b$.

Now let $\{\langle x_\gamma, y_\gamma, a_\gamma, b_\gamma\rangle \mid \gamma < \chi\}$ be a well-ordered sequence of all quadruples satisfying $x_\gamma, y_\gamma \in A$, $a_\gamma, b_\gamma \in L$, and $\delta(x_\gamma, y_\gamma) \leq a_\gamma \vee b_\gamma$. For all $\gamma < \chi$ we define:

$A_0 = A^*$, $\delta_0 = \delta^*$; this construction being performed with the quadruple $\langle x_0, y_0, a_0, b_0\rangle$; $A_\gamma = (\bigcup(A_\xi \mid \xi < \gamma))^*$, $\delta_\gamma = (\bigcup(\delta_\xi \mid \xi < \gamma))^*$, this construction being performed with the quadruple $\langle x_\gamma, y_\gamma, a_\gamma, b_\gamma\rangle$. Finally, $A^+ = \bigcup(A_\gamma \mid \gamma < \chi)$ and $\delta^+ = \bigcup(\delta_\gamma \mid \gamma < \chi)$.

Summarizing, we obtain:

For every set A and distance function δ there exist a set $A^+ \supseteq A$ and a distance function δ^+ on A^+ extending δ such that, for all $x, y \in A$ and $a, b \in A$, if $\delta(x, y) \leq a \vee b$ then there exist $z_1, z_2, z_3 \in A^+$ such that

$$\delta^+(x, z_1) = \delta^+(z_2, z_3) = a \quad \text{and} \quad \delta^+(z_1, z_2) = \delta^+(z_3, y) = b.$$

Now to prove Theorem 4 embed the lattice L into a lattice K having a zero. Let $A_0 = K$ and define δ_0 as above: $\delta_0(x, y) = x \vee y$ if $x \neq y$ and $\delta_0(x, x) = 0$. Again we define inductively sets and distance functions:

$$A_{n+1} = (A_n)^+ \quad \text{and} \quad \delta_{n+1} = (\delta_n)^+, \quad \text{for } n < \omega,$$
$$A = \bigcup(A_n \mid n < \omega), \quad \delta = \bigcup(\delta_n \mid n < \omega).$$

Obviously, $\delta: A^2 \to K$ is a distance function. Let $\alpha = \alpha(\delta)$ be the meet-representation associated with δ. If $x \equiv y\ ((a \vee b)\alpha)$, that is, if $\delta(x, y) \leq a \vee b$, then $x, y \in A_n$ for some $n < \omega$, hence there exists $z_1, z_2, z_3 \in A$ ($z_1, z_2, z_3 \in A_{n+1}$) such that $\delta(x, z_1) = \delta(z_2, z_3) = a$ and $\delta(z_1, z_2) = \delta(z_3, y) = b$, and so

$$x \equiv z_1\ (a\alpha), \quad z_1 \equiv z_2\ (b\alpha), \quad z_2 \equiv z_3\ (a\alpha), \quad \text{and} \quad z_3 \equiv y\ (b\alpha).$$

Thus $x \equiv y\ (a\alpha \vee b\alpha)$, proving that $(a \vee b)\alpha = a\alpha \vee b\alpha$. Therefore, α is a representation, in fact, a type 3 representation of K and hence also of L. ●

We shall give three important applications of this result.

Corollary 5 (Ph. M. Whitman [1946]). *Every lattice can be embedded in the lattice of all subgroups of some group.*

Proof. Let $\alpha: L \to \mathrm{Part}(A)$ be a representation of L. Let G be the group of all permutations of A. To an element $a \in L$, we assign the subgroup H_a generated by all transpositions (xy) such that $x \equiv y\ (a\alpha)$. (The *transposition* (xy) is the permutation interchanging x and y and leaving all other elements of A fixed.) It is easy to see that $(xy) \in H_a$ iff $x \equiv y\ (a\alpha)$, hence H_a determines $a\alpha$. Thus $a \to H_a$ is one-to-one. $H_a \wedge H_b = H_{a \wedge b}$ is immediate. A transposition (xy) belongs to $H_a \vee H_b$ iff $(xy) = (x_1y_1)(x_2y_2) \cdots (x_ny_n)$, where $x_i \equiv y_i\ (a\alpha)$ or $x_i \equiv y_i\ (b\alpha)$, which is equivalent to $x \equiv y\ (a\alpha \vee b\alpha)$. Thus $H_a \vee H_b = H_{a \vee b}$. ●

A well-known problem asks for the characterization of finite lattices that can be embedded in *finite* partition lattices. The next result shows that identities cannot be utilized in the solution of this problem.

Corollary 6 (D. Sachs [1961]). *If an identity holds for all finite partition lattices, then it holds for all lattices.*

Proof. If an identity holds for all finite partition lattices, then it also holds for all $\mathrm{Part}_{\mathrm{fin}}(A)$ (because every principal ideal of $\mathrm{Part}_{\mathrm{fin}}(A)$ is isomorphic to a principal ideal of a finite partition lattice), and so by Lemma I.4.8 and by Corollary 3 it also holds for $I(\mathrm{Part}_{\mathrm{fin}}(A)) \cong \mathrm{Part}(A)$. Since every lattice can be embedded into some $\mathrm{Part}(A)$, this identity has to hold for all lattices. ●

All finite distributive lattices can easily be embedded in finite partition lattices. There is only one result in the literature providing such embeddings for a more comprehensive class of lattices.

Corollary 7 (A. W. Hales [1970]). *Any finite sublattice of a free lattice can be embedded in a finite partition lattice.*

Proof. Let n be a positive integer and let L be a finite sublattice of $F(n)$, the free lattice with free generators $x_1, \ldots, x_n$. Since a finite sublattice of a free lattice is isomorphic to a sublattice of some $F(n)$, there is no restriction of generality. Let

$a, b \in F(n)$, $a \neq b$. Let $a = p(x_1, \ldots, x_n)$ and $b = q(x_1, \ldots, x_n)$ for some n-ary polynomials p and q; we shall consider the identity $\varepsilon: p = q$. Since $a \neq b$, the identity ε fails in $F(n)$. By Corollary 6, it fails in some finite Part(A), that is, there exist $a_1, \ldots, a_n \in$ Part(A) such that $p(a_1, \ldots, a_n) \neq q(a_1, \ldots, a_n)$. Extending the map $x_i \to a_i$ $(i = 1, \ldots, n)$ to a homomorphism we obtain the following statement:

For every $a, b \in F(n)$, $a \neq b$, there exist a finite set $A(a, b)$ and a homomorphism $\varphi(a, b)$ of $F(n)$ into Part$(A(a, b))$ such that

$$a\varphi(a, b) \neq b\varphi(a, b).$$

Now let A be a disjoint union of the sets $A(a, b)$, $a, b \in L$ and for $x \in L$ define the partition $x\varphi$ on A by

$$u \equiv v\ (x\varphi) \quad \text{iff} \quad u, v \in A(a, b) \quad \text{for some} \quad a, b \in L \quad \text{and} \quad u \equiv v\ (x\varphi(a, b)).$$

It is easily seen that φ is an isomorphism of L into Part(A). The proof concludes by the observation that A is finite since L is finite and all $A(a, b)$ are finite. ●

Let us call a representation $\alpha: L \to$ Part(A) *of type* 2 iff for all $a, b \in L$ and $x, y \in A$

$$x \equiv y\ ((a \vee b)\alpha) \quad \text{iff there exist} \quad z_1, z_2 \in A \quad \text{such that}$$

$$x \equiv z_1\ (a\alpha), \quad z_1 \equiv z_2\ (b\alpha), \quad \text{and} \quad z_2 \equiv y\ (a\alpha).$$

Theorem 8 (B. Jónsson [1953]). *A lattice L has a representation of type 2 iff L is modular.*

Proof. Let L have a representation $\alpha: L \to$ Part(A) of type 2. Let $a, b, c \in L$ and $a \geq c$. Since $(a \wedge b) \vee c \leq a \wedge (b \vee c)$ in any lattice, we wish to show that $a \wedge (b \vee c) \leq (a \wedge b) \vee c$. So let $x, y \in A$ and let $x \equiv y\ ((a \wedge (b \vee c))\alpha)$, that is,

$$x \equiv y\ (a\alpha) \quad \text{and} \quad x \equiv y\ ((b \vee c)\alpha).$$

Since α is a type 2 representation, there exist elements z_1 and z_2 such that

$$x \equiv z_1\ (c\alpha), \quad z_1 \equiv z_2\ (b\alpha), \quad \text{and} \quad z_2 \equiv y\ (c\alpha).$$

Using $c \leq a$, we obtain that $z_1 \equiv x\ (a\alpha)$, $x \equiv y\ (a\alpha)$, and $y \equiv z_2\ (a\alpha)$; thus $z_1 \equiv z_2\ (a\alpha)$. Also, $z_1 \equiv z_2\ (b\alpha)$, hence

$$z_1 \equiv z_2\ ((a \wedge b)\alpha).$$

Thus $x \equiv y\ (((a \wedge b) \vee c)\ \alpha)$, implying that $(a \wedge (b \vee c))\alpha \leq ((a \wedge b) \vee c)\alpha$. Therefore, $a \wedge (b \vee c) \leq (a \wedge b) \vee c$, as claimed.

Now to prove the converse, let L be a modular lattice and let K be L with a zero adjoined. Then K is again modular.

For every distance function $\delta: A^2 \to K$ and $\langle x, y, a, b \rangle$ satisfying $\delta(x, y) \leq a \vee b$ we construct a set $A^* = A \cup \{z_1, z_2\}$ having two more elements than A and a distance function δ^* defined as follows (see Figure 4):

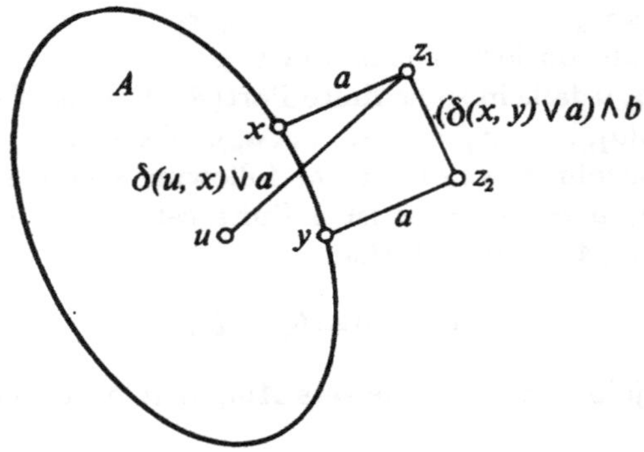

Figure 4

$$\delta^* \text{ is symmetric, satisfies } \delta(x, x) = 0 \text{ for all } x \in A^*, \text{ and } \delta^*(x, y) = \delta(x, y) \text{ for } x, y \in A;$$

$$\delta^*(z_1, z_2) = (\delta(x, y)\vee a)\wedge b;$$

$$\text{for } u \in A, \quad \delta^*(u, z_1) = \delta(u, x)\vee a \quad \text{and} \quad \delta^*(u, z_2) = \delta(u, y)\vee a.$$

The triangle inequality is again trivial excepting the "triangle" $\{u, z_1, z_2\}$. We have to verify the inequality

$$\delta^*(u, z_1)\vee\delta^*(z_1, z_2) \geq \delta^*(u, z_2).$$

Indeed,

$$\begin{aligned} \delta^*(u, z_1)\vee\delta^*(z_1, z_2) &= \delta(u, x)\vee a\vee((\delta(x, y)\vee a)\wedge b) \\ \text{(by modularity)} \qquad &= \delta(u, x)\vee(((\delta(x, y)\vee a)\wedge(a\vee b)) \\ \text{(by absorption, using } \delta(x, y)\leq a\vee b) &= \delta(u, x)\vee\delta(x, y)\vee a \\ \text{(by triangle inequality)} \qquad &\geq \delta(u, y)\vee a = \delta^*(u, z_2). \end{aligned}$$

Now we proceed the same way as the proof of Theorem 4. ●

Under what conditions can we improve the "2" of "type 2" to 1? Call a representation $\alpha: L \to \text{Part}(A)$ *of type* 1 iff, for $x, y \in A$ and $a, b \in L$,

$$x \equiv y \ ((a\vee b)\alpha) \quad \text{iff there exists a} \quad z \in A \quad \text{such that}$$

$$x \equiv z \ (a\alpha) \quad \text{and} \quad z \equiv y \ (b\alpha).$$

This is equivalent to the statement that $a\alpha$ and $b\alpha$ are *permutable*. Lattices having a type 1 representation satisfy a special identity which is not a consequence of the modular identity.

DEFINITION 9 (B. Jónsson [1953])[1]. *Let $x_0, x_1, x_2, y_0, y_1, y_2$ be variables. We define some polynomials:*

$$z_{ij} = (x_i\vee x_j)\wedge(y_i\vee y_j), \qquad 0\leq i<j<3,$$

$$z = z_{01}\wedge(z_{02}\vee z_{12}).$$

[1] See also M. Schützenberger [1945].

The **arguesian identity** *is*

$$(x_0 \vee y_0) \wedge (x_1 \vee y_1) \wedge (x_2 \vee y_2) \leq ((z \vee x_1) \wedge x_0) \vee ((z \vee y_1) \wedge y_0).$$

A lattice satisfying this identity is called **arguesian.**

Remark. This identity is a lattice theoretic form of Desargues' Theorem. To understand the meaning of the identity, the reader should read the discussion of Desargues' Theorem and its connection with the arguesian identity in Section 5.

Theorem 10 (B. Jónsson [1953]). *Any lattice having a type* 1 *representation is arguesian.*

Proof. Let $\alpha: L \to \mathrm{Part}(A)$ be a type 1 representation of the lattice L. We shall show that $L\alpha$ is arguesian (this is sufficient since $L \cong L\alpha$). Let $a_0, a_1, a_2, b_0, b_1, b_2 \in L\alpha$; thus $a_0, \ldots, b_2$ are partitions of the set A. If $x, y \in A$ satisfy

$$x \equiv y\ ((a_0 \vee b_0) \wedge (a_1 \vee b_1) \wedge (a_2 \vee b_2)),$$

then (see Figure 5) there exist $u_0, u_1, u_2 \in A$ such that

$$x \equiv u_i\ (a_i) \quad \text{and} \quad u_i \equiv y\ (b_i), \quad i = 1, 2, 3.$$

Let $c_{ij} = (a_i \vee a_j) \wedge (b_i \vee b_j)$ for $0 \leq i < j < 3$. Then for all such i and j, $u_i \equiv u_j\ (c_{ij})$. Set $c = c_{01} \wedge (c_{02} \vee c_{12})$; then $u_0 \equiv u_1\ (c)$. So $x \equiv u_0\ ((c \vee a_1) \wedge a_0)$ and $u_0 \equiv y\ ((c \vee b_1) \wedge b_0)$, therefore

$$x \equiv y\ (((c \vee a_1) \wedge a_0) \vee ((c \vee b_1) \wedge b_0)),$$

proving that $L\alpha$ is arguesian. ●

Every arguesian lattice is modular. Indeed, to see this it is sufficient to show that $\mathfrak{N}_5$ is not arguesian. If we take $\mathfrak{N}_5$, and substitute $x_0 = y_1 = c$, $x_1 = x_2 = y_0 = b$, $y_2 = a$, then $z = a$, however, the identity demands $a \leq b$, which is false.

It follows from the results of Section 5 that there are modular lattices that are not arguesian.

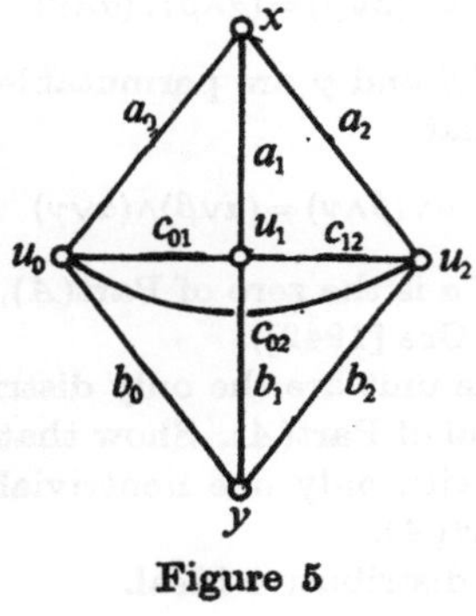

Figure 5

Exercises

1. Compute $|\mathrm{Part}(A)|$ for $|A| \leq 5$.
2. Show that $\mathrm{Part}(A)$ is modular iff $|A| \leq 3$.
3. Let $A \subset B$. Find embeddings of $\mathrm{Part}(A)$ into $\mathrm{Part}(B)$.
4. Let A_i, $i \in I$, be pairwise disjoint sets and let α_i be a partition of A_i, $i \in I$. Define a partition α on $A = \bigcup(A_i \mid i \in I)$ by

$$u \equiv v\,(\alpha) \text{ iff } u, v \in A_i \text{ for some } i \in I \text{ and } u \equiv v\,(\alpha_i).$$

Show that this defines an embedding of $\Pi(\mathrm{Part}(A_i) \mid i \in I)$ into $\mathrm{Part}(A)$.
5. With A_i, $i \in I$, and A as in Exercise 4, is there an $\{0, 1\}$-embedding of $\Pi(\mathrm{Part}(A_i) \mid i \in I)$ into $\mathrm{Part}(A)$?
6. Is there an embedding of $\Pi(\mathrm{Part}(A_i) \mid i \in I)$ into $\mathrm{Part}(\Pi(A_i \mid i \in I))$?
7. Let A be a set and let $B = A \cup \{z\}$ $(z \notin A)$. For $X \subseteq A$ define a partition $\alpha(X)$ on B by

$$x \equiv y\ (\alpha(X)) \quad \text{iff} \quad x, y \in X \cup \{z\}.$$

Show that $X \to \alpha(X)$ is an embedding of $P(A)$ into $\mathrm{Part}(B)$.
8. Using Exercise 7, show that every distributive lattice has an embedding into a partition lattice.
9. Show that every finite distributive lattice can be embedded in a finite partition lattice.
10. Let $\mathfrak{A}$ be a finite (universal) algebra. Define the concept of a congruence relation of an algebra. The congruence relations of $\mathfrak{A}$ form a lattice, $C(\mathfrak{A})$. Verify that the congruence lattice of a finite algebra can be embedded in a finite partition lattice.
11. Following O. Ore [1942], for $\alpha, \beta \in \mathrm{Part}(A)$ we define a graph $G(\alpha, \beta)$ on the set $\alpha \cup \beta$; an edge $\{X, Y\}$ of $G(\alpha, \beta)$ is a block X of α and a block Y of β such that $X \cap Y \neq \varnothing$. Show that the blocks of $\alpha \vee \beta$ are maximal connected subgraphs of $G(\alpha, \beta)$.
12. Let $\alpha, \beta \in \mathrm{Part}(A)$. Prove that the modular equation

$$\alpha \wedge (\beta \vee \gamma) = (\alpha \wedge \beta) \vee \gamma$$

holds for all $\gamma \leq \alpha$ $(\gamma \in \mathrm{Part}(A))$ iff $G(\alpha, \beta)$ is a *tree* (that is, a graph without cycles) (O. Ore [1942]).
13. Let $\beta, \gamma \in \mathrm{Part}(A)$. Show that

$$\alpha \wedge (\beta \vee \gamma) = (\alpha \wedge \beta) \vee (\alpha \wedge \gamma)$$

holds for all $\alpha \in \mathrm{Part}(A)$ iff β and γ are permutable (O. Ore [1942]).
14. Let $\alpha, \gamma \in \mathrm{Part}(A)$. Prove that

$$\alpha \vee (\beta \wedge \gamma) = (\alpha \vee \beta) \wedge (\alpha \vee \gamma)$$

holds for all $\beta \in \mathrm{Part}(A)$ iff α is the zero of $\mathrm{Part}(A)$, or $\alpha \geq \gamma$, or γ has only one nontrivial block and $\gamma \geq \alpha$ (O. Ore [1942]).
15. Show that the zero and the unit are the only distributive elements of $\mathrm{Part}(A)$.
16. Let D be a distributive ideal of $\mathrm{Part}(A)$. Show that every element of D is contained in another element of D with only one nontrivial block. Also show that $\bigvee D$ is a distributive element of $\mathrm{Part}(A)$.
17. Show that $\mathrm{Part}(A)$ has no distributive ideal.

18. Use Theorem III.3.10 and Exercise 17 to show that Part(A) is simple (O. Ore [1942]).
19. Define type n representations. Find a representation which is not of type n for any $n < \omega$.
20. Find a representation which is not associated with a distance function.
21. Disprove the converse of Corollary 7.
*22. Disprove the converse of Theorem 10 (B. Jónsson).

5. Complemented Modular Lattices

The key to the investigation of complemented modular lattices is provided by modular geometric lattices and projective spaces. We begin with some properties of modular geometric lattices.

Lemma 1. *Let L be a modular geometric lattice and let $\langle A, ^{-}\rangle$ be the associated geometry.*

(i) *For $p, q \in A$ and $p \neq q$, $p \sim q$ iff there exists an $r \in A$ such that $r \neq p, q$ and $r \leq p \vee q$. Thus the elements $\{0, p, q, r, p\vee q\}$ form a diamond.*

(ii) *Let $X \subseteq A$. Then X is a subspace iff $p, q \in X$ implies that $r \in X$ for all $r \leq p \vee q$.*

Proof.

(i) Let x be a complement of p and q and define $r = (p\vee q)\wedge x$. Then

$$\begin{aligned} p\vee r &= p\vee((p\vee q)\wedge x) \qquad \text{(by modularity and } p \leq p\vee q)\\ &= (p\vee x)\wedge(p\vee q) = 1\wedge(p\vee q) = p\vee q. \end{aligned}$$

Similarly, $q\vee r = p\vee q$. Obviously, $p\wedge r = q\wedge r = 0$. Hence

$$h(r) = h(p\vee q) + h(0) - h(p) = 1.$$

Thus $r \in A$. Since $r \neq 0$ and $r\wedge p = r\wedge q = 0$, we have $r \neq p, q$. ◐

(ii) Let $X (\subseteq A)$ satisfy the condition that $r \leq p\vee q$ and $p, q \in X$ imply that $r \in X$. We prove by induction on n that

$$r \in A, \quad r \leq p_1\vee\cdots\vee p_n, \quad \text{and} \quad p_1, \ldots, p_n \in X \text{ imply that } r \in X.$$

This is obvious for $n = 1$ and by assumption for $n = 2$. Let us assume this statement for $n-1$ and let $r, p_1, \ldots, p_n$ be given satisfying $r \leq p_1\vee\cdots\vee p_n$. We can assume that $\{p_1, \ldots, p_n\}$ is independent, since otherwise $r \leq p_{i_1}\vee\cdots\vee p_{i_k}$ for some $\{i_1, \ldots, i_k\} \subset \{1, \ldots, n\}$ and so $r \in X$ by induction. With $a = p_2\vee\cdots\vee p_n$, we have then $r\vee a \leq p_1\vee a$ and $h(r\vee a) = h(p_1\vee a) = n$, hence $r\vee a = p_1\vee a$. If $r \leq a$, then $r \in X$ by the induction hypothesis, so we can assume that $r \nleq a$. Thus $r\wedge a = p_1\wedge a = 0$. By the proof of (i), $t = (p_1\vee r)\wedge a$ is an atom and, by the induction hypothesis, $t \in X$. Since $t, p_1 \in X$ and $r \leq t\vee p_1$ we conclude that $r \in X$. This completes the proof of the "if" part of (ii); the "only if" part is trivial. ●

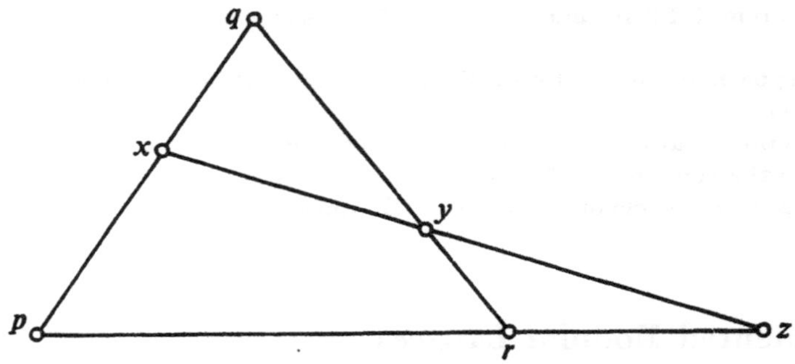

Figure 1

Lemma 2. *Under the conditions of Lemma 1, the geometry* $\langle A, ^- \rangle$ *satisfies the following property* (see Figure 1):

The Pasch Axiom. *If* p, q, r, x, y *are points,* $x \leq p \vee q$, $y \leq q \vee r$, *and* $x \neq y$, *then there is a point* z *such that* $z \leq (p \vee r) \wedge (x \vee y)$.

Remark. Apart from degenerate cases, the Pasch Axiom requires that if a line $x \vee y$ intersects two sides, $p \vee q$ and $q \vee r$, of the triangle $\{p, q, r\}$, then it intersects the third side, $p \vee r$. If $x \neq p, q$ and $y \neq q, r$, then we shall also have $z \neq p, r$.

Proof. If $|\{p, q, r, x, y\}| < 5$, then we can always choose a $z \in \{p, q, r, x, y\}$ satisfying $z \leq (p \vee r) \wedge (x \vee y)$. If $r \leq p \vee q$, again one can choose $z = x$. So let $|\{p, q, r, x, y\}| = 5$, and let $r \nleq p \vee q$. Then $h(p \vee r) = h(x \vee y) = 2$ and $h(p \vee q \vee r) = 3$.

Thus

$$h((p \vee r) \wedge (x \vee y)) = h(p \vee r) + h(x \vee y) - h(p \vee r \vee x \vee y).$$

Now observe that $p \vee q \vee x \vee y = p \vee q \vee r$, hence

$$h((p \vee r) \wedge (x \vee y)) = 2 + 2 - 3 = 1,$$

and so $z = (p \vee r) \wedge (x \vee y)$ is the required atom. ●

We can see from Lemmas 1 and 2 that the important properties of a geometry associated with a modular geometric lattice are formulated only with points and lines. This pattern is followed in the next definition:

Definition 3. *Let* A *be a set and let* L *be a collection of subsets of* A. $\langle A, L \rangle$ *is called a* projective space *iff the following properties hold:*

(i) *Every* $l \in L$ *has at least two elements.*

(ii) *For any two distinct* $p, q \in A$ *there is exactly one* $l \in L$ *satisfying* $p, q \in l$.

(iii) *For* $p, q, r, x, y \in A$ *and* $l_1, l_2 \in L$ *satisfying* $p, q, x \in l_1$ *and* $q, r, y \in l_2$, *there exist* $z \in A$ *and* $l_3, l_4 \in L$ *satisfying* $p, r, z \in l_3$ *and* $x, y, z \in l_4$.

Let us call the members of A *points,* and those of L, *lines.* For $p, q \in A$, $p \neq q$, let $p + q$ denote the (unique) line containing p and q; if $p = q$ set $p + q = \{p\}$.

A set $X \subseteq A$ is called a *linear subspace* iff p and $q \in X$ imply that $p+q \subseteq X$. If X and Y are linear subspaces define

$$X+Y=\bigcup(x+y \mid x \in X \text{ and } y \in Y).$$

LEMMA 4. ***If X and Y are linear subspaces of a projective space, then so is $X+Y$.***

PROOF. **Take the points p, q, r such that $r \in p+q$ and $p, q \in X+Y$. We wish to show that $r \in X+Y$. Now if $p, q \in X$ or $p, q \in Y$, then $r \in X \cup Y \subseteq X+Y$, since X and Y are linear subspaces. If $p \in X$ and $q \in Y$ (or the other way around), then $r \in X+Y$ by the definition of $X+Y$. Therefore we can assume that $p \notin X \cup Y$ and so there exist $p_x \in X$ and $p_y \in Y$ such that $p \in p_x+p_y$. Now we distinguish two cases. Firstly, if $q \in X \cup Y$, say $q \in X$, then consider $q+p_x, q+p, p+p_x$, and $r+p_y$ (see Figure 2). We can assume that q, p, p_x, p_y, and r are all distinct, hence $r+p_y$ has a point in common with $q+p$ and $p+p_y$, hence by 3(iii) there is a $t \in X$ such that $t \in q+p_x$ and $t \in r+p_y$. If $t=p_y$, then $p \in X$, a contradiction. Hence $t \neq p_y$ and so $r \in t+p_y \in X+Y$. Secondly, let $q \notin X \cup Y$. Again we apply 3(iii), this time to $p+p_x, p+q, p_x+q$, and p_y+r (see Figure 3). Hence there is a point $t \in p_x+q$ such that $r \in t+p_y$. By the first case, $t \in X+Y$. Again by the first case, $r \in X+Y$. ●**

Now we come to the crucial step:

THEOREM 5. ***The linear subspaces of a projective space form a modular geometric lattice.***

PROOF. **Since the intersection of any number of linear subspaces is a linear subspace again, we have a closure space $\langle A, ^-\rangle$. For $X \subseteq A$, the closure $\bar{X}$ can be described as follows: set $X_0=X, X_1=X+X, \ldots, X_n=X_{n-1}+X_{n-1}, \ldots$; then $\bar{X}=\bigcup(X_i \mid i=0,1,2,\ldots)$. It follows immediately, that $\langle A, ^-\rangle$ is an algebraic closure space, and so the linear subspaces form an algebraic lattice and for the linear subspaces X and Y, $X \vee Y=\overline{X \cup Y}$.**

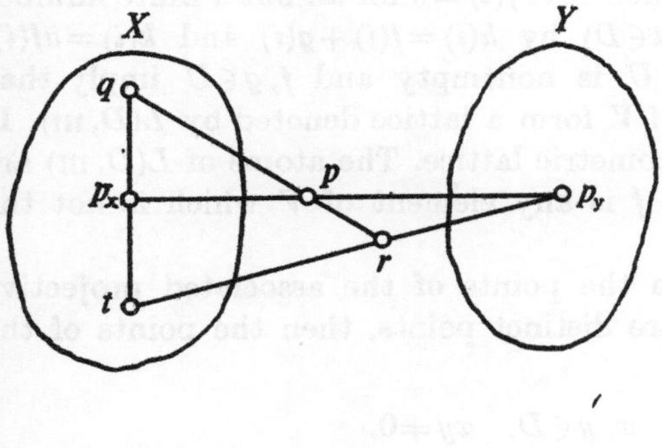

Figure 2

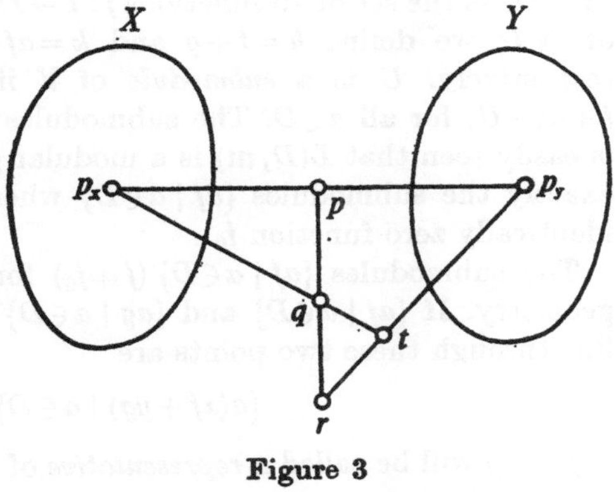

Figure 3

If X, Y, and Z are linear subspaces and $X \supseteq Z$, then $X\wedge(Y\vee Z) \supseteq (X\wedge Y)\vee Z$ obviously holds.

Now let $p \in X\wedge(Y\vee Z)$, that is, $p \in X$ and $p \in Y\vee Z$. Since $p \in Y\vee Z = Y + Z$, there exist $p_y \in Y$ and $p_z \in Z$ such that $p \in p_y + p_z$. From $X \supseteq Z$, it follows that p and $p_z \in X$. If $p = p_z$, then $p \in Z$, and so $p \in (X\wedge Y)\vee Z$. If $p \neq p_z$, then $p_y \in p + p_z \subseteq X$. Thus $p_y \in X\wedge Y$ and $p_z \in Z$; therefore, $p \in (X\wedge Y)\vee Z$. This proves that the lattice is modular.

3.8(i), (ii), (iv) have been verified or are obvious and (iii) follows from modularity. ●

We have proved somewhat more than stated. We also obtained that the linear subspaces define a geometry whose subspaces are exactly the linear subspaces.

Call a geometry *projective* if the associated geometric lattice is modular.

Theorem 6. *There is a one-to-one correspondence between projective spaces (defined by points and lines) and projective geometries (defined as geometries with modular subspace lattices). Under this correspondence, linear subspaces of projective spaces correspond to subspaces of projective geometries.*

Applying the results of Section 3 we obtain that every modular geometric lattice is a direct product of indecomposable modular geometric lattices, which by Lemma 1(i) are characterized by the property that in the associated projective space *every line has at least three points;* such projective spaces (and projective geometries) are called *nondegenerate.*

Corollary 7. *Every modular geometric lattice can be represented as a direct product of modular geometric lattices which are associated with nondegenerate projective geometries.*

The most important example of a modular geometric lattice associated with a nondegenerate projective geometry is the following.

Let D be a *division ring* (that is, an associative ring with unit in which a^{-1} exists for any $a \in D$, $a \neq 0$) and let $\mathfrak{m} > 0$ be a cardinal number (finite or infinite). We take a set I with $|I| = \mathfrak{m}$ and we construct $V = V(D, \mathfrak{m})$, an $\mathfrak{m}$-dimensional vector space over D, as the set of all functions $f: I \to D$ such that $f(i) = 0$ for all but a finite number of $i \in I$; we define $h = f + g$ and $k = af$ $(a \in D)$ by $h(i) = f(i) + g(i)$ and $k(i) = af(i)$, respectively. U is a *submodule* of V iff U is nonempty and $f, g \in U$ imply that $f + ag \in U$, for all $a \in D$. The submodules of V form a lattice denoted by $L(D, \mathfrak{m})$. It is easily seen that $L(D, \mathfrak{m})$ is a modular geometric lattice. The atoms of $L(D, \mathfrak{m})$ are exactly the submodules $\{af \mid a \in D\}$ where f is any element of V which is not the identically zero function f_0.

The submodules $\{af \mid a \in D\}$ $(f \neq f_0)$ form the points of the associated projective geometry. If $\{af \mid a \in D\}$ and $\{ag \mid a \in D\}$ are distinct points, then the points of the line through these two points are

$$\{a(xf + yg) \mid a \in D\}, \quad x, y \in D, \quad xy \neq 0.$$

$f(\neq f_0)$ will be called a *representative* of the point $\{af \mid a \in D\}$.

This geometry is nondegenerate: if f and g are representatives of two distinct points, then $f-g$ represents a third point of the line.

This projective geometry is typical of projective geometries in which Desargues' Theorem holds.

Let us call a set of points X *collinear* iff $X \subseteq l$ for some line l. A triplet $\langle a_0, a_1, a_2\rangle$ of noncollinear points is a *triangle*. Two triangles $\langle a_0, a_1, a_2\rangle$ and $\langle b_0, b_1, b_2\rangle$ are *perspective with respect to the point* p iff $a_i \neq b_i$, $a_i + a_j \neq b_i + b_j$ for $0 \leq i, j < 3$, and the points p, a_i, b_i are collinear for $i = 0, 1, 2$. They are *perspective with respect to a line* l iff $c_{01}, c_{12}, c_{20} \subseteq l$, where c_{ij} is the intersection of $a_i + a_j$ and $b_i + b_j$. The classical *Desargues' Theorem* states that if two triangles are perspective with respect to a point, then they are perspective with respect to a line.

Now we shall prove that the arguesian identity of the last section is a lattice theoretic formulation of Desargues' Theorem. See B. Jónsson [1953], [1954], [1959 b] and M. Schützenberger [1945].

Theorem 8. *Let L be a modular geometric lattice. Then L satisfies the arguesian identity iff Desargues' Theorem holds in the associated projective geometry.*

The proof of Theorem 8 will follow immediately from Lemmas 9–11.

Lemma 9. *Let L be a modular geometric lattice. Then the arguesian identity holds for the atoms of L iff Desargues' Theorem holds in the associated projective geometry.*

Proof. Let us substitute the atoms $a_0, a_1, a_2, b_0, b_1, b_2$ in the arguesian identity. Let us form the c_{ij}, $0 \leq i < j \leq 2$, $c = c_{01} \wedge (c_{02} \vee c_{12})$, and $p = (a_0 \vee b_0) \wedge (a_1 \vee b_1) \wedge (a_2 \vee b_2)$, see Figure 4. We have to verify that

$$p \leq ((c \vee a_2) \wedge a_0) \vee ((c \vee b_2) \wedge b_0).$$

Let us assume first that a_0, a_1, a_2 and b_0, b_1, b_2 are not collinear, $a_i \neq b_i$, $i = 0, 1, 2$, and that the triangles $\langle a_0, a_1, a_2\rangle$ and $\langle b_0, b_1, b_2\rangle$ are perspective with respect to the point p. Then the c_{ij} are all distinct atoms and they are collinear iff $c = c_{01} \wedge (c_{12} \vee c_{02})$ is an atom, in fact, $c = c_{01}$, otherwise $c = 0$. If $c = c_{01}$, then $((c \vee b_0) \wedge a_0) \vee ((c \vee b_1) \wedge a_1)$ is the line $a_0 \vee a_1$, and so $p \leq ((c \vee b_0) \wedge a_0) \vee ((c \vee b_1) \wedge a_1)$, otherwise $c = 0$ and $(c \vee b_0) \wedge a_0 = (c \vee b_1) \wedge a_1 = 0$ and $p \not\leq 0$. Thus, under the conditions stated, the identity holds iff the geometry satisfies Desargues' Theorem.

Now observe that if the above conditions are not satisfied, then $p \leq ((c \vee a_2) \wedge a_0) \vee ((c \vee b_2) \wedge b_0)$ always holds. There are several cases to consider ($p = 0$, p is a line, a_0, a_1, a_2 are collinear, b_0, b_1, b_2 are collinear, $a_i = b_i$ for some i, and so on) but all are trivial. ●

Lemma 10. *Let L be a modular geometric lattice. Let p be an n-ary polynomial in which each variable x_i $(0 \leq i < n)$ occurs at most once. Then for an atom a and elements $b_0, \ldots, b_{n-1} \in L$,*

$$a \leq p(b_0, \ldots, b_{n-1})$$

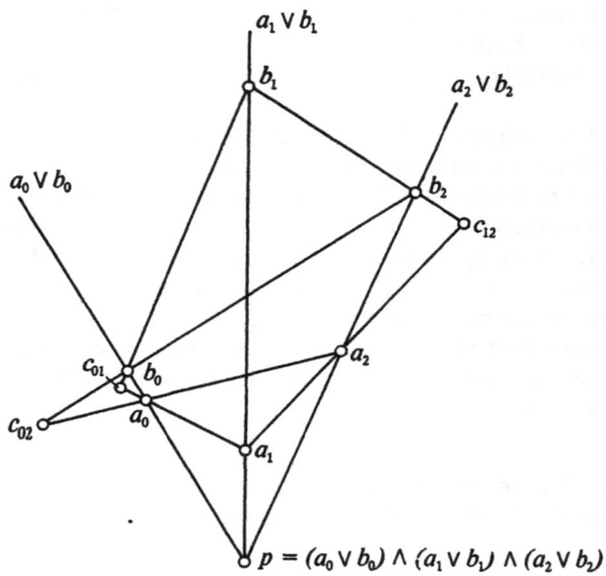

Figure 4

holds iff there exist $a_0, \ldots, a_{n-1} \in L$ *such that*

$$a \leq p(a_0, \ldots, a_{n-1}), \quad a_i \leq b_i \textit{ for all } 0 \leq i < n, \quad \textit{and } a_i \textit{ is an atom if } \quad b_i \neq 0.$$

Proof. We use induction on the rank of p. Let $p = x_i$. Then $a \leq b_i$ and so we can take $a_i = a$. Now let us assume the statement to hold for $p_0 = p_0(x_0, \ldots, x_{n-1})$ and $p_1 = p_1(y_0, \ldots, y_{m-1})$, where $\{x_0, \ldots, x_{n-1}\} \cap \{y_0, \ldots, y_{m-1}\} = \varnothing$. First, let $p = p_0 \wedge p_1$. Now if $a \leq p(b_0, \ldots, b_{n-1}, b_0', \ldots, b_{m-1}') = p_0(b_0, \ldots, b_{n-1}) \wedge p_1(b_0', \ldots, b_{m-1}')$, then $a \leq p_0(b_0, \ldots, b_{n-1})$ and $a \leq p_1(b_0', \ldots, b_{m-1}')$. So by the induction hypothesis, we can choose $a_i \leq b_i$ and $a_j' \leq b_j'$ such that a_i is an atom if $b_i \neq 0$ and a_j' is an atom if $b_j' \neq 0$, and $a \leq p_0(a_0, \ldots, a_{n-1}) \wedge p_1(a_0', \ldots, a_{m-1}') = p(a_0, \ldots, a_{m-1}')$. (Observe that if $x_i = y_j$, then we would not know how to choose $a_i = a_j'$.) Second, let $p = p_0 \vee p_1$. Then $a \leq p_0(b_0, \ldots, b_{n-1}) \vee p_1(b_0', \ldots, b_{m-1}')$. If $p_0(b_0, \ldots, b_{n-1}) = 0$, then $a \leq p_1(b_0', \ldots, b_{m-1}')$, hence we can choose the a_i' by the induction hypothesis. We choose a_i as an arbitrary atom $\leq b_i$ if $b_i \neq 0$ and let $a_i = 0$ if $b_i = 0$. If $p_1(b_0', \ldots, b_{m-1}') = 0$ we proceed similarly. Now if $p_0(b_0, \ldots, b_{n-1}) \neq 0$ and $p_1(b_0', \ldots, b_{m-1}') \neq 0$, then Lemma 4 and Theorem 6 imply that there exist atoms $a' \leq p_0(b_0, \ldots, b_{n-1})$ and $a'' \leq p_1(b_0', \ldots, b_{m-1}')$ such that $a \leq a' \vee a''$. Applying the induction hypothesis twice we obtain $a_0, \ldots, a_{n-1}, a_0', \ldots, a_{m-1}'$ such that $a_i \leq b_i$ $(0 \leq i < n)$, $a_i' \leq b_i'$ $(0 \leq i < m)$, a_i is an atom if $b_i \neq 0$, a_i' is an atom if $b_i' \neq 0$, and $a' \leq p_0(a_0, \ldots, a_{n-1})$, $a'' \leq p_1(a_0', \ldots, a_{m-1})$. Thus

$$a \leq a' \vee a'' \leq p_0(a_0, \ldots, a_{n-1}) \vee p_1(a_0', \ldots, a_{m-1}'),$$

completing the induction. ●

Lemma 11 (G. Grätzer and H. Lakser [1973a]). *Let us assume that we can write the lattice identity ε in the form $p \leq q$, where p and q are lattice polynomials and each variable occurs in p at most once. If ε holds for the atoms and the zero of a modular geometric lattice L, then ε holds for L.*

Proof. Let $b_0, \ldots, b_{n-1} \in L$; we want to verify that

$$p(b_0, \ldots, b_{n-1}) \leq q(b_0, \ldots, b_{n-1}).$$

It is sufficient to show that $a \leq q(b_0, \ldots, b_{n-1})$ for any atom $a \leq p(b_0, \ldots, b_{n-1})$. So let a be an atom and $a \leq p(b_0, \ldots, b_{n-1})$. By Lemma 10 we can choose $a_i \leq b_i$ such that the a_i are atoms or zero, $a_i \leq b_i$ for $0 \leq i < n$, and

$$a \leq p(a_0, \ldots, a_{n-1}).$$

By assumption, ε holds for atoms and zero, hence

$$p(a_0, \ldots, a_{n-1}) \leq q(a_0, \ldots, a_{n-1}).$$

Using these two inequalities and the fact that q is isotone, we obtain

$$a \leq p(a_0, \ldots, a_{n-1}) \leq q(a_0, \ldots, a_{n-1}) \leq q(b_0, \ldots, b_{n-1}). \quad \bullet$$

Proof of Theorem 8. If Desargues' Theorem fails in the projective geometry, then by Lemma 9 the arguesian identity ε fails in L. Conversely, if Desargues' Theorem holds in the projective geometry, then by Lemma 9 ε holds for the atoms of L. A trivial step extends this to the statement that ε holds for the atoms and the zero of L. Observe that ε is of the form required in Lemma 11, hence ε holds in L. $\bullet$

Corollary 12. *Desargues' Theorem holds in the projective space associated with the lattice $L(D, \mathfrak{m})$.*

Proof. With a submodule U we associate an equivalence relation $\alpha(U)$ on $V = V(D, \mathfrak{m})$ defined by $f\alpha(U)g$ iff $f - g \in U$. Then obviously $U \to \alpha(U)$ is a type 1 representation of $L(D, \mathfrak{m})$, hence by Theorem 4.10 $L(D, \mathfrak{m})$ is arguesian. Thus by Theorem 8, Desargues' Theorem holds in the projective space associated with $L(D, \mathfrak{m})$. $\bullet$

The next result is a well-known theorem of geometry; it is the classical proof of Desargues' Theorem in a three-dimensional space.

Theorem 13. *Let L be a modular geometric lattice. Let us assume that the projective geometry associated with L is nondegenerate. If the length of L is at least 4, then L is arguesian.*

Proof. Let the triangles $\langle a_0, a_1, a_2 \rangle$ and $\langle b_0, b_1, b_2 \rangle$ be perspective with respect to the point p. A *plane* α is an element of height 3. $\alpha = a_0 \vee a_1 \vee a_2$ and $\beta = b_0 \vee b_1 \vee b_2$ are planes.

Let us assume that $\alpha \neq \beta$. Since $\alpha \vee \beta = p \vee \alpha = p \vee \beta$, $h(\alpha \vee \beta) = 4$, and so $h(\alpha \wedge \beta) = h(\alpha) + h(\beta) - h(\alpha \vee \beta) = 3 + 3 - 4 = 2$; thus $\alpha \wedge \beta$ is a line. The lines $a_i \vee a_j$ and $b_i \vee b_j$ are in the plane $p \vee a_i \vee a_j$, hence $h(c_{ij}) = h(a_i \vee a_j) + h(b_i \vee b_j) - h(p \vee a_i \vee a_j) = 1$, and so c_{ij} is a point. Since $a_i \vee a_j$ is in the plane α and $b_i \vee b_j$ is in β, c_{ij} is in $\alpha \wedge \beta$. Therefore, c_{01}, c_{02}, and c_{12} are collinear.

Now let us assume that $\alpha = \beta$. Since the length of L is at least 4, α is not the unit element. Take a $\pi \in L, \pi \succ \alpha$. Let l be a relative complement of α in $[p, \pi]$. Then $h(l) = h(\pi) + h(p) - h(\alpha) = 4 + 1 - 3 = 2$, so l is a line. The projective geometry associated with L is nondegenerate, hence we can choose two distinct points p' and p'' on l, $p' \neq p$, $p'' \neq p$.

Now define $d_i = (p' \vee a_i) \wedge (p'' \vee b_i)$, $i = 0, 1, 2$. It is easily seen that $\langle d_0, d_1, d_2 \rangle$ is a triangle and the plane $\delta = d_0 \vee d_1 \vee d_2 \neq \alpha, \beta$. Furthermore, $\langle a_0, a_1, a_2 \rangle$ and $\langle d_0, d_1, d_2 \rangle$ are perspective with respect to p', and $\langle b_0, b_1, b_2 \rangle$ and $\langle d_0, d_1, d_2 \rangle$ are perspective with respect to p''. Thus we can apply the first case to conclude that $(a_0 \vee a_1) \wedge (d_0 \vee d_1)$ and also $(b_0 \vee b_1) \wedge (d_0 \vee d_1)$ are in $l' = \alpha \wedge \delta = \beta \wedge \delta$; but this implies that $(a_0 \vee a_1) \wedge (b_0 \vee b_1) \in l'$, hence $c_{01} \in l'$. Similarly, c_{02} and $c_{12} \in l'$. This shows that the two triangles are perspective with respect to the line l'. ●

Before we come to the classical Coordinatization Theorem we prove one more lemma.

Lemma 14. *A projective space satisfying Desargues' Theorem also satisfies the converse (or "dual") statement:*

If two triangles are perspective with respect to a line, then they are perspective with respect to a point.

Proof. Let us use the notation of Figure 4. We now assume that c_{01}, c_{02}, and c_{12} are on a line l and we do not have the point p. Define p as the intersection of the lines $a_1 + b_1$ and $a_2 + b_2$. To show the two triangles perspective with respect to p we have to verify that a_0, b_0, and p are collinear.

Consider the triangles $\langle a_2, b_2, c_{02} \rangle$ and $\langle a_1, b_1, c_{01} \rangle$; they are perspective with respect to the point c_{12}. By Desargues' Theorem, the intersections of the corresponding sides, a_0, b_0, and p, are collinear. ●

Now we are ready to state the classical Coordinatization Theorem of Projective Geometry.

Theorem 15. *Let L be a directly irreducible arguesian geometric lattice of length at least three. Then there exist a division ring D, unique up to isomorphism, and a unique cardinal number $\mathfrak{m}$ such that $L \cong L(D, \mathfrak{m})$.*

Remark. For lattices of finite length this was already included in O. Veblen and W. H. Young [1910] (of course, in a form appropriate for projective spaces). Some authors claim that this proof is not complete and that no complete proof appeared until J. von Neumann [1936/37]. The condition of finite dimensionality was eliminated in O. Frink [1946].

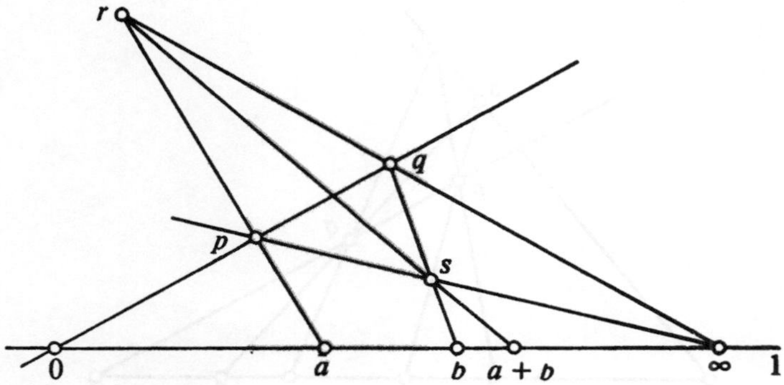

Figure 5

The method described below is " von Staudt's algebra of throws ". A more modern (maybe less intuitive) proof can be found in R. Baer [1952]; see also E. Artin [1940].

Sketch of Proof. A complete proof of this result is too long and of not enough interest for a lattice theory book. However, the mere statement that there is such an isomorphism φ: $L \cong L(D, \mathfrak{m})$ is insufficient for workers in lattice theory. So we choose a middle course: we are going to construct D and φ but will not provide detailed proofs.

Let l be an arbitrary line in L; we fix three distinct points of l and call them 0, 1, and ∞. We set $D = l - \{\infty\}$ and we define addition and multiplication on D.

Let $a, b \in D$. Fix two distinct points p and q not in L such that 0, p, and q are collinear (see Figure 5). Then form the points $r = (a \vee p) \wedge (q \vee \infty)$ and $s = (p \vee \infty) \wedge (b \vee q)$. Then set

$$a + b = (r \vee s) \wedge l.$$

It is easy to see that $a + b$ does not depend on the choice of p and q. Indeed, let p' and q' be another pair of distinct points not in l, such that 0, p', and q' are collinear and let us form r' and s' as before. Then $\langle p, q, r\rangle$ and $\langle p', q', r'\rangle$ and also $\langle p, q, s\rangle$ and $\langle p', q', s'\rangle$ are perspective with respect to the line l; so by Lemma 14, they are perspective with respect to a point, in fact, the same point u, that is

$$u, p, p'; \quad u, q, q'; \quad u, r, r'; \quad u, s, s'$$

are all collinear, respectively. Hence $\langle r, s, q\rangle$ and $\langle r', s', q'\rangle$ are perspective with respect to u, therefore by Desargues' Theorem, they are perspective with respect to a line l'. But $(s \vee q) \wedge (s' \vee q') = b \in l$ and $(r \vee q) \wedge (r' \vee q') = \infty \in l$, hence $l = l'$. Therefore, $(r \vee s) \wedge (r' \vee s') \in l$, that is,

$$(r \vee s) \wedge l = (r' \vee s') \wedge l,$$

showing that $a + b$ is independent of the choice of p and q.

To define ab, choose two distinct points p and q not in l such that p, q, and ∞ are

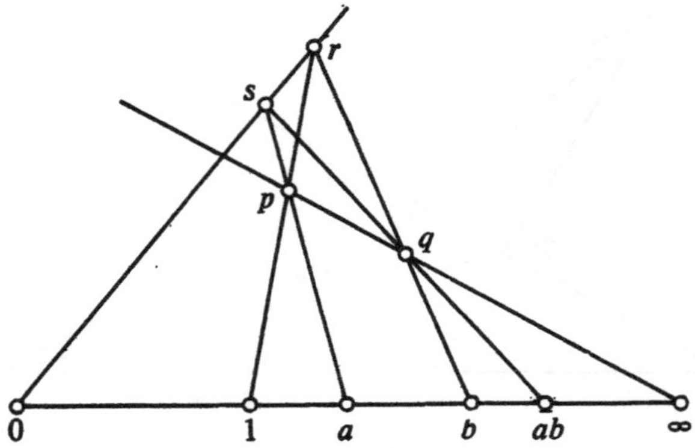

Figure 6

collinear (see Figure 6). Then we form the points $r=(1\vee p)\wedge(q\vee b)$ and $s=(0\vee r)\wedge(p\vee a)$ and we define

$$ab=(q\vee s)\wedge l.$$

Then D is a division ring with 0 as a zero and 1 as the unit element.

Now we choose a maximal independent set I of atoms of L. Let $|I|=\mathfrak{m}$ and we consider the lattice $L(D,\mathfrak{m})$ where each atom is represented by a function $f\colon I\to D$ which is 0 at all but a finite number of elements; we wish to associate with each r of L such a function $d_r\colon I\to D$.

Fix $p, t\in I$, $p\neq t$ and set $l=p\vee t$. We choose a point t_1 on l, $t_1\neq p$ and $t_1\neq t$. We define a division ring D on l with p as a zero, t_1 as the unit, and t as infinity.

Now for any $q\in I$, $q\neq p$ and $q\neq t$, fix a point q_1 of the line $p\vee q$, $q_1\neq p$ and $q_1\neq q$. The map

$$\psi_q\colon x\to(x\vee((t\vee q)\wedge(t_1\vee q_1)))\wedge l$$

is one-to-one and onto between $p\vee q$ and l and it fixes p and takes q_1 into p_1.

We define $d_r\colon I\to D$ for any atom r of L. First, let $r\not\leq\bigvee(I-\{p\})$. Let $r\leq\bigvee I_1$. where I_1 is a minimal (finite) subset of I satisfying this. By assumption, $p\in I_1$. We define $d_r(p)=1$ (the unit of D), $d_r(s)=0$ (the zero of D) for $s\notin I_1$, and for $s\in I_1-\{p\}$,

$$d_r(s)=((r\vee\bigvee(I_1-\{s\}))\wedge(p\vee s))\psi_s\in D.$$

Second, if $r\leq\bigvee(I-\{p\})$, then $r\leq\bigvee I_1$, where I_1 is a minimal (finite) subset of $I-\{p\}$ with this property. By assumption, $p\notin I_1$. Let s be any point on $p\vee r$, $s\neq p$, $s\neq r$. Then $s\not\leq\bigvee(I-\{p\})$, hence d_s has already been defined. Now we define

$$d_r(x)=d_s(x)\quad\text{if}\quad x\neq p\quad\text{and}\quad d_r(p)=0.$$

The claim is then that $r\to d_r$ maps the atoms of L onto the set of functions representing the atoms of $L(D,\mathfrak{m})$ in such a manner that collinearity is preserved. Hence this can be extended to an isomorphism φ of L and $L(D,\mathfrak{m})$. ●

Combining Theorems 13 and 15 we obtain

Corollary 16. *Let L be a directly indecomposable modular geometric lattice of length at least 4. Then $L \cong L(D, \mathfrak{m})$ with a suitable division ring D and cardinal number $\mathfrak{m}$.*

The following embedding theorem is the most important embedding result for complemented modular lattices:

Theorem 17 (O. Frink [1946]). *Every complemented modular lattice L can be embedded in a modular geometric lattice K. This embedding can be chosen to be a $\{0, 1\}$-embedding and K can be chosen to satisfy all the identities of L.*

Remark. The last observation was made in B. Jónsson [1954].

Proof. By Zorn's Lemma, every proper dual ideal of L can be extended to a maximal dual ideal.[1] Let M denote the set of maximal dual ideals of L. For $x \in L$, we set

$$R(x) = \{D \mid x \in D \text{ and } D \in M\}.$$

Take the dual of $\mathfrak{D}(L)$ and let A be the sublattice generated by M. Then $K = I(A)$ is a modular geometric lattice and if we identify $D \in M$ with the principal ideal it generates, M is the set of atoms of K. Thus there is a projective space $\langle M, ^- \rangle$ associated with K. In fact, we can define $\langle M, ^- \rangle$ directly as a projective space:

$$D \in D_1 + D_2 \quad \text{iff} \quad D \supseteq D_1 \cap D_2.$$

We claim that $x \to R(x)$ is a lattice embedding of L into the lattice of subspaces of $\langle M, ^- \rangle$, that is, into K.

This claim proves the theorem, since this is a $\{0, 1\}$-embedding; furthermore, by Lemma I.4.8 and its dual, any identity that holds in L holds in the dual of $\mathfrak{D}(L)$, and therefore in A and $K = I(A)$.

$R(x)$ is a subspace: If $D_0, D_1 \in R(x)$ and $D \supseteq D_0 \cap D_1$, then $x \in D_0 \cap D_1$, hence $x \in D$, and so $D \in R(x)$.

$R(x \wedge y) = R(x) \cap R(y)$: $D \in R(x) \cap R(y)$ iff $x, y \in D$, which is equivalent to $x \wedge y \in D$, that is, $D \in R(x \wedge y)$.

If $x \neq y$, then $R(x) \neq R(y)$: Let $y \ngeq x$, and let x_1 be a relative complement of $x \wedge y$ in $[0, x]$. Let D be a maximal dual ideal containing x_1. (D exists by Zorn's Lemma since L has a 0.) Then $D \in R(x)$ but $D \notin R(y)$, otherwise $y, x_1 \in D$ which contradicts $x \wedge y_1 = 0$.

$R(x \vee y) = R(x) \vee R(y)$: We already know that R is isotone so it is sufficient to prove that if $D \in R(x \vee y)$, then $D \in R(x) \vee R(y)$. This is trivial if x or y is 0. So let $x, y \neq 0$. By Lemma 1(ii) we have to prove the following:

If $x, y \in L$, $x, y \neq 0$, and D is a maximal dual ideal satisfying $x \vee y \in D$, then there exist maximal dual ideals E and F satisfying

$$x \in E, \quad y \in F, \quad E \cap F \subseteq D.$$

---

[1] A *maximal dual ideal* of L is proper ($\neq L$) by definition.

This is obvious if x or $y \in D$; if $x \in D$, choose $E = D$ and let F be any maximal dual ideal F containing y. So let $x, y \notin D$.

First, let us assume that $x \wedge y = 0$.

Let $D_1 = D \cap [y)$. Since $y \notin D_1$ we have $[y) \supset D_1$. Computing in $\mathcal{D}(L)$, we obtain

$$[x) \wedge [y) = [x) \wedge D_1 (= [x \vee y))$$

since in $\mathcal{D}(L)$ the $\wedge$ is set intersection; and so by modularity

$$[x) \vee [y) \supset [x) \vee D_1.$$

Therefore, the dual ideal generated by x and D_1 is proper. Thus there exists a maximal dual ideal $E \supseteq [x) \vee D_1$; in particular, $x \in E$. We repeat this trick. Let $E_1 = E \cap D$. Then $D \supset E_1$ (since E and D are both maximal) and

$$\begin{aligned} [y) \wedge E_1 &= [y) \cap E \cap D \quad (\text{use } E \supseteq [y) \cap D = D_1) \\ &= [y) \cap D. \end{aligned}$$

Hence by modularity,

$$[y) \vee D \supset [y) \vee E_1.$$

Therefore, there exists a maximal dual ideal $F \supseteq [y) \vee E_1$; in particular, $y \in F$.

Now since $F \supseteq E_1 = D \cap E$, we conclude that F is on the line spanned by D and E and since D, E, F are all distinct ($E \neq F$ since $x \in E$, $y \in F$, and $x \wedge y = 0$), D is on the line spanned by E and F, that is, $D \supseteq E \cap F$, which was to be proved.

Second, if $x \wedge y \neq 0$, then let y_1 be a relative complement of $x \wedge y$ in $[0, y]$. Then $x \wedge y_1 = 0$ and $x \vee y_1 = x \vee y$, hence there exist maximal dual ideals E and F satisfying $x \in E$, $y_1 \in F$, and $D \supseteq E \cap F$. Obviously, $y \in F$ so E and F will do for x and y. ●

We can combine the previous results to obtain the following embedding theorems:

Corollary 18. *Every complemented modular lattice can be embedded in a direct product of lattices $L(D_i, \mathfrak{m}_i)$ and subspace lattices of projective planes that do not satisfy Desargues' Theorem.*

Proof. This is immediate from Theorem 17, Corollary 7, and Theorems 13 and 15. ●

Corollary 19. *A complemented modular lattice L can be embedded in a direct product of lattices $L(D_i, \mathfrak{m}_i)$ iff L is arguesian.*

Proof. If L has such an embedding it is arguesian by Corollary 12. If L is arguesian, then by Theorem 17 L can be embedded in an arguesian geometric lattice. The direct factorization of Corollary 7 gives us arguesian directly indecomposable geometric lattices, which, by Theorem 15, are of the form $L(D_i, \mathfrak{m}_i)$. ●

Summarizing these we obtain a result of B. Jónsson [1960a]:

Theorem 20. *Let L be a complemented modular lattice. The following conditions on L are equivalent:*

(i) *L can be embedded in a direct product of $L(D_i, \mathfrak{m}_i)$.*
(ii) *L can be embedded in an arguesian geometric lattice.*
(iii) *L can be embedded in the lattice of all subgroups of an abelian group.*
(iv) *L has a representation of type 1.*
(v) *L is arguesian.*

Proof. (i) is equivalent to (v) by Corollary 19. The equivalence of (i) and (ii) was proved in the proof of Corollary 19. (i) implies (iii) since elements of L are represented by subgroups of the additive group of ΠD_i. (iii) implies (iv) was observed in the proof of Corollary 12. (iv) implies (v) by Theorem 4.10. ●

Theorem 20 explains the significance of the arguesian identity for complemented modular lattices. It is interesting that a weaker version of the arguesian identity holds for all complemented modular lattices.

Let

$$\varepsilon: p(x_1, \ldots, x_6) \leq q(x_1, \ldots, x_6)$$

be the arguesian identity. Choose three new variables x, y, z and define

$$u = (x \vee y) \wedge (y \vee z) \wedge (z \vee x),$$
$$v = (x \wedge y) \vee (y \wedge z) \vee (z \wedge x),$$
$$\bar{x}_i = x_i \wedge ((u \wedge x) \vee v), \quad i = 1, 2, \ldots, 6.$$

Theorem 21 (G. Grätzer and H. Lakser [1973a]). *The identity*

$$\bar{\varepsilon}: p(\bar{x}_1, \ldots, \bar{x}_6) \vee v \leq q(\bar{x}_1, \ldots, \bar{x}_6) \vee v$$

holds in any complemented modular lattice.

Proof. Consider the following lattice property:

(P) If $\{0, a, b, c, i\}$ is a diamond, then $(a]$ is an arguesian lattice.

We claim that if a lattice has property (P), then it satisfies the identity $\bar{\varepsilon}$.

Let $t, r, s \in L$ and let us consider the substitution $x = t$, $y = r$, $z = s$. There are two cases to consider:

Case 1. $(t \vee r) \wedge (r \vee s) \wedge (s \vee t) = (t \wedge r) \vee (r \wedge s) \vee (s \wedge t) = \alpha$. Then $\bar{x}_i = x_i \wedge \alpha$ and so $\bar{\varepsilon}$ becomes

$$p(x_1 \wedge \alpha, \ldots, x_6 \wedge \alpha) \vee \alpha \leq q(x_1 \wedge \alpha, \ldots, x_6 \wedge \alpha) \vee \alpha.$$

But $p(x_1 \wedge \alpha, \ldots, x_6 \wedge \alpha) \leq p(\alpha, \ldots, \alpha) = \alpha$ and similarly for q, so $\bar{\varepsilon}$ becomes $\alpha \leq \alpha$, a triviality.

Case 2. $i = (t \vee r) \wedge (r \vee s) \wedge (s \vee t) \neq (t \wedge r) \vee (r \wedge s) \vee (s \wedge t) = o$. Then it follows from the diagram of $F_M(3)$, that o and i along with $a = (i \wedge t) \vee o$, $b = (i \wedge r) \vee o$, $c = (i \wedge s) \vee o$ form a diamond. Thus property (P) applies and $(a]$ is arguesian. Since $x_1 \wedge a, \ldots, x_6 \wedge a \in (a]$ we obtain

$$p(x_1 \wedge a, \ldots, x_6 \wedge a) \leq q(x_1 \wedge a, \ldots, x_6 \wedge a);$$

joining both sides by o we get that $\bar{\varepsilon}$ holds in L.

Next we claim that property (P) holds for the two types of lattices listed in Corollary 18. Naturally, (P) holds for an $L(D, \mathfrak{m})$ because $L(D, \mathfrak{m})$ is arguesian. Let L be the subspace lattice of a projective plane. If $\{o, a, b, c, i\}$ is a diamond, then a is either a point or a line. In either case, $(a]$ is arguesian.

By Corollary 18, every complemented modular lattice K can be embedded in a direct product of lattices having property (P) and therefore satisfying $\bar{\varepsilon}$. Thus K also satisfies $\bar{\varepsilon}$. ●

It is easy to construct a modular lattice in which $\bar{\varepsilon}$ fails. Let L be the subspace lattice of a projective plane in which Desargues' Theorem fails. By our discussion in Lemma 9, there are seven atoms of L, $a_0, a_1, a_2, b_0, b_1, b_2$, and d, such that

$$p(a_0, a_1, a_2, b_0, b_1, b_2) = d, \quad q(a_0, a_1, a_2, b_0, b_1, b_2) = 0.$$

Let e be a complement of d; e is a dual atom of L. Let $K = L \cup \{b, c, i\}$ such that with $o = e$ and $a = 1$ (the unit of L), $\{o, a, b, c, i\}$ is a diamond. It is immediate that K is a modular lattice. Substitute $x_1 = a_0$, $x_1 = a_1$, $x_2 = a_2$, $x_3 = b_0$, $x_4 = b_1$, $x_5 = b_2$, $x = a(=1)$, $y = b$, and $z = c$. Then $u = i$, $v = o$, $(u \wedge x) \vee o = a(=1)$, hence $x_i = \bar{x}_i$, for $i = 1, 2, \ldots, 6$. Thus $\bar{\varepsilon}$ yields $d \vee o \leq 0 \vee o$, that is, $1 \leq o$, which is not true. We conclude:

Corollary 22 (R. P. Dilworth and M. Hall [1944]). *There is a modular lattice that cannot be embedded in a complemented modular lattice.*

The modular lattice K we constructed above is of length 4. The next result shows that this is best possible.

Theorem 23. *Every modular lattice B of length at most three can be embedded in a complemented modular lattice C. The embedding can always be constructed to be a $\{0, 1\}$-embedding and C can be chosen to have the length of B.*

We shall prove Theorem 23 using the concept of a projective plane.

Definition 24. *A* projective plane $\langle A, L\rangle$ *is a set A and a collection L of subsets of A satisfying:*

(i) *Every $l \in L$ has at least two elements.*

(ii) *For any two points $p_0, p_1 \in A$, $p_0 \neq p_1$, there is exactly one $l \in L$ satisfying $p_0, p_1 \in l$.*

(iii) *For any two distinct lines l_0 and l_1, there is exactly one $p \in A$ satisfying $p \in l_0$ and $p \in l_1$.*

A *partial plane* satisfies (i)–(iii) with "at most one" replacing "exactly one" in (ii) and (iii). Observe that the first two of the three properties already define a partial plane.

It is easily seen that a projective plane is a projective space.

Lemma 25. ***Every partial plane can be embedded in a projective plane, that is, for every partial plane $\langle A, L\rangle$ there is a projective plane $\langle A', L'\rangle$ such that $A \subseteq A'$ and for every $l \in L$ there is an $l' \in L'$ satisfying $l \subseteq l'$.***

Proof. Let $\langle A, L\rangle$ be a partial plane. Define

$$A^+ = A \cup \{p(l, m) \mid l, m \in L \text{ and } l \cap m = \varnothing\},$$

$$L^+ = \{l \cup \{p(l, m) \mid m \in L, l \cap m = \varnothing\} \mid l \in L\} \cup \{\{p, q\} \mid p, q \in A, p \neq q, p, q \in l \text{ for no } l \in L\},$$

where $p(l, m) = p(m, l)$. The new points defined are distinct from each other and from the points in A.

Now define $A_0 = A$, $L_0 = L$, $A_{n+1} = (A_n)^+$, $L_{n+1} = (L_n)^+$, for $n < \omega$, and

$$A' = \bigcup(A_n \mid n < \omega).$$

Lines are defined as follows: let l_n be a line in $\langle A_n, L_n\rangle$; then we obtain the lines l_{k+1} for $k \geq n$:

$$l_{k+1} = l_k \cup \{p(l_k, m) \mid m \in L_k, l_k \cap m = \varnothing\}.$$

L' is the collection of all sets of the form $\bigcup(l_k \mid k \geq n)$. It is obvious that $\langle A', L'\rangle$ satisfies the requirements. ●

Proof of Theorem 23. If the length of B is less than 3, the statement is obvious. So let B be of length 3. We define a partial plane $\langle A, L\rangle$ as follows: the points of $\langle A, L\rangle$ are of two kinds, namely, the atoms of B and for every $a \in B$, $h(a) = 2$, which contains exactly one atom, a new point $p(a)$. For every $a \in B$, $h(a) = 2$, containing at least two atoms we define a line $\{p \mid p \text{ is an atom of } B, p \leq a\}$; if $a \in B$, $h(a) = 2$, and a contains exactly one atom p, then we define the line $\{p, p(a)\}$. It is clear that $\langle A, L\rangle$ is a partial plane. Embed $\langle A, L\rangle$ into a projective plane $\langle A', L'\rangle$ by Lemma 25 and let C be the subspace lattice of $\langle A', L'\rangle$. ●

Exercises

1. Find a geometry $\langle A, ^-\rangle$ in which X is a subspace iff $r \leq p_0 \vee p_1 \vee p_2$ and $p_0, p_1, p_2 \in X$ imply that $r \in X$, but $\langle A, ^-\rangle$ is not a projective geometry.
2. Why do we need $x \neq y$ in the Pasch Axiom? Phrase the Pasch Axiom so that this assumption can be dropped.
3. For points p and q of a projective geometry, define $p \equiv q$ iff there is a third point $r \leq p \vee q$. Show that this relation is an equivalence relation.
4. Derive Corollary 7 from Exercise 3.
5. Prove by direct computation that the projective geometry associated with $L(D, \mathfrak{m})$ satisfies Desargues' Theorem.

6. Show that the Pasch Axiom is equivalent to the following implication holding for all points:

$$\text{If } a_0 \leq a_1 \vee a_2 \quad \text{and} \quad a_1 \wedge (a_0 \vee a_2) \leq a_3 \vee a_4, \quad \text{then}$$

$$a_0 \leq (a_3 \wedge (a_4 \vee (a_1 \wedge (a_0 \vee a_2)))) \vee ((a_0 \vee a_3) \wedge (((a_1 \vee a_0) \wedge a_2)) \vee ((a_3 \vee (a_1 \wedge (a_0 \vee a_2))) \wedge a_4)).$$

7. Prove that the implication of Exercise 6 holds for all subspaces of a projective space.

*8. Prove that the D constructed in the proof of Theorem 15 is a division ring.

9. Let K and L be modular geometric lattices and let φ be a one-to-one map from the atoms of K onto the atoms of L. Prove that φ can be extended to an isomorphism iff

$$a \leq b \vee c \quad \text{iff} \quad a\varphi \leq b\varphi \vee c\varphi$$

holds for any atoms a, b, c of K.

*10. In the proof of Theorem 15, verify that for atoms r, s, t of L, $r \leq s \vee t$ iff $ad_r + bd_s + cd_t = 0$ holds for some $a, b, c \in D$.

11. Combine Exercises 9 and 10 to prove $L \cong L(D, \mathfrak{m})$ in Theorem 15.

12. Show that the identity $\bar{\varepsilon}$ is not self-dual and that the dual of $\bar{\varepsilon}$ could also be used in Theorem 21.

13. The projective plane constructed in Lemma 25 is the *free projective plane* generated by the partial plane. What does " free " mean for projective planes?

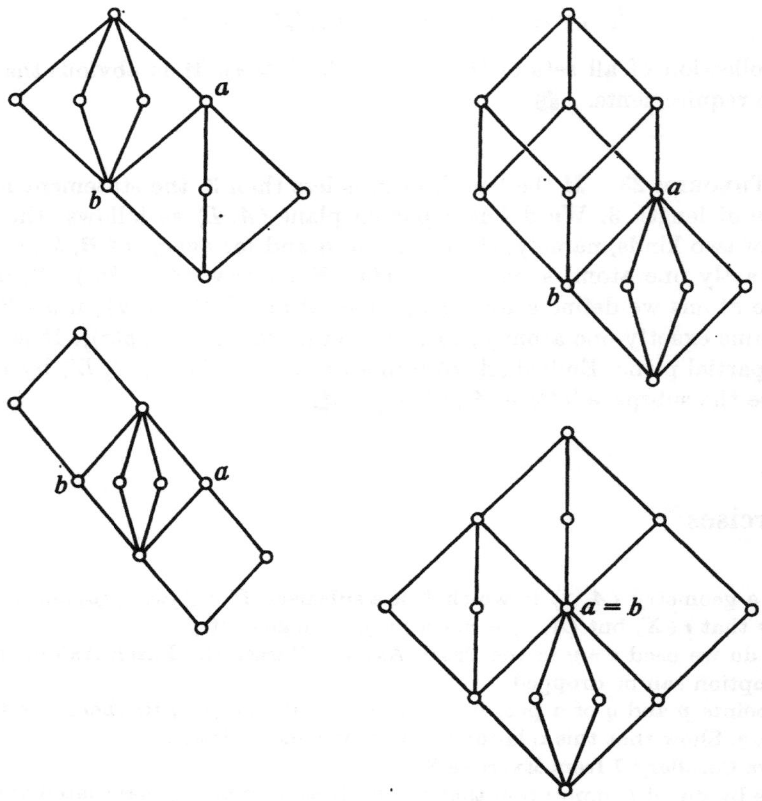

Figure 7

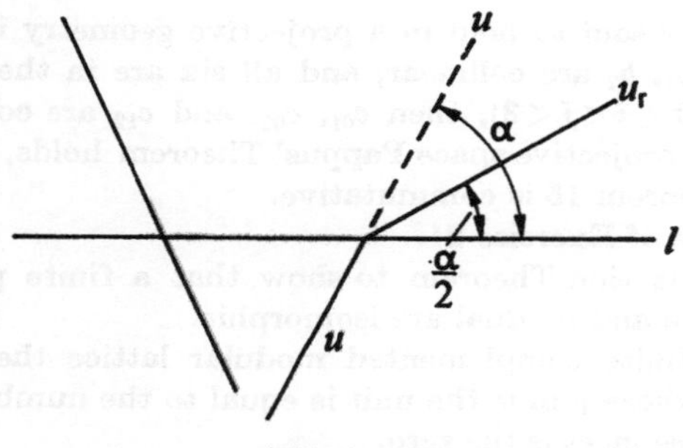

Figure 8

14. **Investigate the lattice theoretic properties of the embedding of the subspace lattice of $\langle A, L\rangle$ into the subspace lattice of $\langle A^+, L^+\rangle$ in the proof of Lemma 25.**

15. **Why could Lemma 25 not be extended to provide embeddings of modular lattices of length 4?**

16. **Let L be a modular lattice of length 3 or 4. Let a be the join of all atoms of L and b the meet of all dual atoms of L. Consider the situations shown in Figure 7. Show that they (along with their duals) suggest a complete classification of all modular lattices of length 3 or 4 with the exception of distributive lattices and of complemented modular lattices (B. Jónsson [1959]).**

17. **Develop a duality theory for projective planes. The dual of $\langle A, L\rangle$ is $\langle L, A \text{ dual}\rangle$, where, for $a \in A$, a dual $= \{l \mid a \in l\}$.**

18. **Show that for a finite projective plane $\langle A, L\rangle$,**

$$|A| = |L|.$$

19. **What is the " dual " of Desargues' Theorem?**

20. **On the Euclidean plane fix a line l (see Figure 8). Let u be a line and let α be the angle determined by u and l. A *refracted line* is defined as follows: if $0 \leq \alpha \leq \pi/2$, then form the line u_r with angle $\alpha/2$; below l the refracted line is u, above l it is u_r. Otherwise, u is the refracted line. Define a projective space as the Euclidean plane with a line at infinity; the lines are the refracted lines with a point at infinity and the line at infinity. Prove that this defines a projective plane in which Desargues' Theorem fails.**

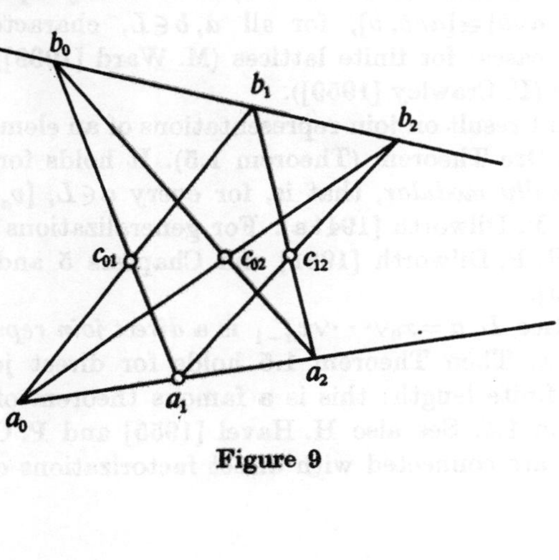

Figure 9

*21. *Pappus' Theorem* is said to hold in a projective geometry iff whenever the points a_0, a_1, a_2 and b_0, b_1, b_2 are collinear, and all six are in the same plane, and $c_{ij} = (a_i \vee b_j) \wedge (a_j \vee b_i)$ $(0 \leq i < j < 3)$, then c_{01}, c_{02}, and c_{12} are collinear. (See Figure 9.) Prove that if in a projective space Pappus' Theorem holds, then the division ring constructed in Theorem 15 is commutative.

*22. Prove the converse of Exercise 21.

*23. Use the Coordinatization Theorem to show that a finite projective geometry of length four or more and its dual are isomorphic.

24. Prove that in a finite complemented modular lattice the number of k-element subsets of atoms whose join is the unit is equal to the number of k-element subsets of dual atoms whose meet is the zero.

25. In a finite modular lattice let L_n (U_n) denote the set of all $(n+1)$-element subsets $\{p, q_1, q_2, \ldots, q_n\}$ such that every q_i covers p (p covers every q_i). Show that $|L_n| = |U_n|$ for every positive integer n. (Hint: use Exercise 24.)

26. Apply Exercise 25 to establish Dilworth's covering theorem for modular lattices (R. P. Dilworth [1954]): in a finite modular lattice the number of elements covered by precisely k elements is equal to the number of elements covering precisely k elements. In particular, in a finite modular lattice the number of join-irreducibles is equal to the number of meet-irreducibles. (This proof is due to B. Ganter and I. Rival [1973].)

Further Topics and References

For finite modular lattices L, the posets $J(L)$ and $M(L)$ do not determine L as they do for finite distributive lattices; but they have many interesting properties. The most celebrated one is a result of R. P. Dilworth [1954]: $|J(L)| = |M(L)|$. (See Exercises 5.23–5.26 and, for a generalization, S. P. Avann [1958a].) By R. Wille [1976], $J(L)$ and $M(L)$ have the same width and, by B. Ganter and I. Rival [a], the same length. In this connection one should mention a result of R. Wille [1974a]: a finitely generated modular lattice of finite length is finite.

The Isomorphism Theorem (Theorem 1.2) is obviously equivalent to modularity. The statement: $[a, a \vee b] \cong [a \wedge b, b]$, for all $a, b \in L$, characterizes the modularity of L only in special cases: for finite lattices (M. Ward [1939]) and, more generally, for algebraic lattices (P. Crawley [1959]).

The most important result on join-representations of an element of a finite modular lattice is the Kuroš-Ore Theorem (Theorem 1.5). It holds for a finite semimodular lattice L iff L is *locally modular*, that is, for every $a \in L$, $[v_a, a]$ is modular, where $v_a = \bigwedge(x \mid a \succ x)$; R. P. Dilworth [1941a]. For generalizations to infinite lattices see the survey article R. P. Dilworth [1961] and Chapters 5 and 6 of P. Crawley and R. P. Dilworth [1973].

In a modular lattice L, $a = x_0 \vee \cdots \vee x_{n-1}$ is a *direct join representation* iff $\{x_0, \ldots, x_{n-1}\}$ is independent. Then Theorem 1.5 holds for direct join representations in modular lattices of finite length; this is a famous theorem of O. Ore [1936], much deeper than Theorem 1.5. See also H. Havel [1955] and P. Crawley [1962a]. Such join representations are connected with direct factorizations of groups, rings, loops, and so on. A number of papers are devoted to this topic started in O. Ore [1937], [1938] and A. G. Kuroš [1943], [1946]; see M. Grayev [1947], A. H. Livšic [1962], E. N. Močulskiĭ [1961], [1962], [1968].

An important new line of research in modular lattices was started in I. M. Gelfand and V. A. Ponomarev [1970]. In this paper all 4-generated (generated by a 4-element set) subdirectly irreducible modular lattices of subspaces of a vector space are determined. This inspired R. Wille, his coworkers, and others to start a systematic search for all 4-generated subdirectly irreducible modular lattices. Partial results of this search can be found in A. Day, C. Herrmann, and R. Wille [1972], R. Freese [1973], C. Herrmann [1973], C. Herrmann, C. M. Ringel, and R. Wille [a], C. Herrmann, M. Kindermann, and R. Wille [a], A. Mitchke and R. Wille [1973], G. Sauer, W. Seibert, and R. Wille [1973], R. Wille [1973], [1973a].

Many of the harder computations in these papers are based on the D_2-lemma and $\mathfrak{M}_3$-lemma of R. Wille [1973]. Two illustrations of these results are given below.

A modular lattice completely freely generated by a finite poset P (that is, $F_{\mathbf{M}}(P^m)$) is finite iff P does not contain a four-element antichain nor does it contain the poset $\{x_0, y_0, y_1, z_0, z_1\}$, where $y_0 < y_1$, $z_0 < z_1$ and $x_0 \| y_i$, $x_0 \| z_j$, $y_i \| z_j$, for $i, j = 0, 1$ (R. Wille [1973a]). This trivially yields an earlier result of R. Wille [1973] determining when a free modular lattice generated by a set of chains is finite.

Let $P = \{o, a, b, c, d, i\}$, $o < a, b, c, d < i$, $a \wedge b = a \wedge c = a \wedge d = b \wedge c = b \wedge d = c \wedge d = o$, $a \vee b = a \vee c = a \vee d = b \vee d = c \vee d = i$; let $\mathfrak{P}$ denote the corresponding partial lattice. Then there are exactly two modular lattices (up to isomorphism) generated by P and containing $\mathfrak{P}$ as a relative sublattice: $\mathfrak{M}_4$ (which is the same as $\mathfrak{M}_3$ but with four atoms) and an infinite (subdirectly irreducible) lattice (A. Day, C. Herrmann, and R. Wille [1972]).

Several attempts have been made to learn something about the structure of free modular lattices (see, for instance, K. Takeuchi [1951a] and [1959]). An interesting result on sublattices of $F_{\mathbf{M}}(4)$ can be found in R. Freese [a].

A proper subclass of the class of finite sublattices of free modular lattices is the class of finite lattices projective in $\mathbf{M}$. The distributive members of this class are described in A. Mitchke and R. Wille [1976].

Although it is still not known whether there is an algorithm deciding when two polynomials agree in a free modular lattice, the word problem for modular lattices has been shown to be recursively unsolvable in G. Hutchinson [1973a] and [a]. His most recent result is that there exists a five-generated finite partial lattice Q defined by a single equation such that there is no algorithm to decide when two polynomials agree in $F_{\mathbf{M}}(Q)$. This may be best possible: if the list of known 4-generated subdirectly irreducible modular lattices is complete, then " five " cannot be reduced to " four " in Hutchinson's results.

G. M. Bergman [1969] and A. P. Huhn [1972], [1972a] defined an important class of modular lattices. A modular lattice L is called ***n-distributive*** iff the identity

$$x \vee \bigwedge (y_i \mid 0 \leq i \leq n) = \bigwedge (x \vee \bigwedge (y_j \mid 0 \leq j \leq n,\ j \neq i) \mid 0 \leq i \leq n)$$

holds in L. An interesting result of A. P. Huhn states that a modular lattice L fails to be n-distributive iff L contains a sublattice $B \cong (\mathfrak{C}_2)^{n+1}$ with bounds o and i and an element a which is a relative complement of all dual atoms of B in $[o, i]$. This result has many applications, see, for instance, A. P. Huhn [1975], R. Freese [a], and Section 2 of C. Herrmann [1973].

Theorem II.3.17 is also true for modular lattices, that is, every finite distributive lattice is isomorphic to the congruence lattice of a suitable modular lattice, by E. T. Schmidt [1974]. The basic construction used to establish this result is further applied in E. T. Schmidt [a] in which a finitely generated simple modular lattice without a prime interval is exhibited.

An equational class **K** of algebras is called *congruence modular* iff, for all $\mathfrak{A} \in \mathbf{K}$, the congruence lattice of $\mathfrak{A}$ is modular. Conditions for this can be found in A. Day [1969]. R. Freese [1975a] exhibits an identity ε, a weaker form of the arguesian identity, which holds in every congruence lattice of a congruence modular equational class, but ε is not implied by the modular identity.

There are two related conjectures. 1 (R. N. McKenzie). If a nontrivial lattice identity ε holds in all congruence lattices of algebras in an equational class **K**, then **K** is congruence modular. 2 (J. B. Nation). If a nontrivial lattice identity ε can be expressed by a *Mal'cev type condition* (in the sense of G. Grätzer [1970a]), then ε implies the modular identity. Some results concerning the first conjecture can be found in J. B. Nation [1974] and A. Day [a]. For some further results and an excellent review of this field see B. Jónsson [1974a].

Necessary and sufficient conditions for the validity of the Jordan-Hölder Chain Condition in a finite poset P is given in O. Ore [1943]; see also S. MacLane [1943]. A *cell* C in P is $C = \{o, a_1, \ldots, a_n, b_1, \ldots, b_m, i\}$, where $\{o, a_1, \ldots, a_n, i\}$ and $\{o, b_1, \ldots, b_m, i\}$ are maximal chains in the interval $[o, i]$ of P and $\inf\{a_k, b_j\} = o$, $\sup\{a_k, b_j\} = i$ for $k = 1, \ldots, n$, $j = 1, \ldots, m$. Ore's condition is: for every cell C, $n = m$ holds (n and m depend on C). This is obviously satisfied in semimodular lattices, where, for every cell, $n = m = 1$. This result has been rediscovered (and published!) half a dozen times since 1943.

The Jordan-Hölder Chain Condition for modular elements is discussed in T. Fujiwara and K. Murata [1953].

Some 23 definitions of semimodularity are discussed in R. Croisot [1951] and [1952/53], especially their interdependence with and without chain conditions.

For further results on M-symmetric lattices, the reader should consult F. and S. Maeda [1970]; see also L. R. Wilcox [1939], [1942], [1944]; S. Maeda [1965], [1969]; and R. Padmanabhan [1974].

For a generalization of semimodularity, see O. Tamaschke [1960].

For more general discussions of independence in semimodular lattices, the reader should consult R. P. Dilworth [1944] and D. T. Finkbeiner [1951].

A sublattice of a semimodular lattice is not necessarily semimodular. In fact, R. P. Dilworth proved that every finite lattice can be embedded in a finite geometric lattice; this result was published in P. Crawley and R. P. Dilworth [1973]. For earlier related results, see R. P. Dilworth [1941a] and D. T. Finkbeiner [1960a].

The treatment of geometry from a lattice theoretic viewpoint started in the twenties by K. Menger; his work was continued by J. von Neumann and G. Birkhoff, and later by O. Frink, L. R. Wilcox, F. Maeda, and others. The reader should consult B. Jónsson [1959b] for a general review of this field; see also the book G. Zappa [1952]. Apart from projective geometries, there are many kinds of geometries that can be classified from a lattice theoretic viewpoint: these include classical concepts such as *affine geometries* and *planar geometries* (X is a subspace iff $\overline{\{p, q, r\}} \subseteq X$ for all $p, q, r \in X$) as well as some new ones, such as the "*n*-stufige Geometrie" of R. Wille [1967], which has a particularly beautiful lattice theoretic characterization. For other lattice theoretic results of a geometric nature, see the books M. L. Dubreil-Jacotin, L. Lesieur, and R. Croisot [1953], F. Maeda [1958], F. and S. Maeda [1970]; the papers of F. Maeda listed in the Bibliography; W. Prenowitz [1948] and U. Sasaki [1952] and [1953].

Geometric lattices, originally called *matroids*, developed from a branch of combinatorics.

The study of *matroids*, in the combinatorial sense, originates with the famous Seven Bridges of Königsberg problem solved over two hundred years ago by L. Euler. Euler demonstrated that a graph is a disjoint union of circuits iff each vertex has an even number of edges incident to it. Thus those subgraphs of a graph which are disjoint unions of circuits are closed under symmetric difference and given two circuits with nonvoid intersection, the symmetric difference will always contain a circuit. This motivates the circuit definition of a matroid:

Let E be a finite set and C, the circuits, a collection of nonvoid subsets of E. Then $\langle E, C\rangle$ is a *matroid* iff

(C1) No circuit properly contains another.

(C2) If C_1 and C_2 are distinct circuits and $e \in C_1 \cap C_2$, then there is a circuit C_3 such that $C_3 \subseteq (C_1 \cup C_2) - \{e\}$.

There exist many matroids that do not correspond to the circuit structure of a graph. As an example, consider a *chain group*, that is, a set of maps from a finite set E to an integral domain R which is closed under pointwise addition and scalar multiplication. For any one of these maps, define the support of the map as the set of those elements of E that do not get mapped onto zero. If the minimal supports (with respect to inclusion) are defined to be the circuits on E, then a matroid results.

An equivalent approach to matroids is by the rank definition: given a finite set E and an integer valued function r on $P(E)$, $\langle E, r\rangle$ is a *matroid* iff, for $A \subseteq E$ and $e, e_1, e_2 \in E$,

(R1) $r(\varnothing) = 0$.

(R2) $r(A) \leq r(A \cup \{e\}) \leq r(A) + 1$.

(R3) $r(A) = r(A \cup \{e_1\}) = r(A \cup \{e_2\})$ implies that $r(A \cup \{e_1, e_2\}) = r(A)$.

Let $\langle E, r\rangle$ be a matroid and $E' = \{e \in E \mid r(\{e\}) = 0\}$. Then for any $A \subseteq E$, $r(A) = r(A - E')$ and hence $\langle E - E', r\rangle$ is a matroid with essentially the same structure as $\langle E, r\rangle$. For $x, y \in E$, let $x \equiv y$ iff $r(\{x, y\}) = 1$. Then $\equiv$ is an equivalence relation and if $[e_1], \ldots, [e_s]$ are equivalence classes, then $r([e_1] \cup \cdots \cup [e_s]) = r(\{e_1, \ldots, e_s\})$ irrespective of the choice of representatives. Thus we may identify related elements with no real loss of generality. Having done this, for $X \subseteq E$ define

$$\overline{X} = \{e \in E \mid r(X \cup \{e\}) = r(X)\}.$$

A geometry has been formed and the rank is completely determined by its definition on the closed sets. Hence the structure of the matroid can be determined from the geometric lattice of subspaces of the geometry.

Matroids have combinatorial application in areas other than the study of the circuit structure of graphs. One such application is in transversal theory. Given sets $A_1, \ldots, A_m$, a set $X \subseteq E = A_1 \cup \cdots \cup A_m$ is called a *partial transversal* iff there is a one-to-one map $\varphi: X \to \{1, \ldots, m\}$ satisfying $x \in A_{x\varphi}$, for $x \in X$. The partial transversals then satisfy the independence axioms for the set E. It immediately follows, for example, that all maximal partial transversals have the same cardinality. Further results concerning the theory of transversals can be found in the definitive book of L. Mirsky [1971].

Another application of matroid theory is to the labeling of edges of graphs. Any set of labelings that are closed under addition and scalar multiplication form a chain group and hence a matroid. If the edges are labeled with elements from the integers modulo two, the set of labelings such that the sum of the labels on edges incident to any given vertex is zero forms a chain group which yields the ordinary circuits of the graph as its matroid. More generally, a chain group can be formed by using labels from an arbitrary integral domain and including in the chain group those labelings such that for every circuit C in the graph the sum, $S(C)$, of the labels on the edges of C is even, i.e., there is an element x in the integral domain such that $2x = S(C)$. An element in this chain group may be formed by assigning a value $g(v)$ to each vertex and then defining $f(e) = g(v_1) + g(v_2)$ where v_1 and v_2 are the two vertices incident to e. In topological terms this is the usual *coboundary operator*. The set of all labelings that can be formed in this manner forms a chain group, and, as a matroid it has the same rank as the original chain group. Hence every chain group element arises in this manner. If the integral domain has characteristic two, and a labeling in this chain group exists whose support is all of E, then in the associated vertex assignment no two joined vertices have the same value. If we call these assignments colours, then we are looking at the classical graph theoretic colouring problem. Thus these chain groups can be used to evaluate the chromatic numbers of graphs. Further considerations of this type are found in the book of H. H. Crapo and G.-C. Rota [1971].

The interested reader should consult the two books referred to above and W. T. Tutte [1959], [1965], [1971] and R. J. Wilson [1973].

Many papers have been published on maximum sized antichains in posets, mostly in geometric lattices. For a finite poset P and maximum sized antichains A and B, set $A \leq B$ iff for every $a \in A$ there is a $b \in B$ such that $a \leq b$ in P. Then the maximum sized antichains form a distributive lattice by R. P. Dilworth [1960]; for a simple proof and an application, see R. Freese [1974].

Many generalizations of Sperner's Lemma are known; see D. A. Drake [1971], M. J. Klass [1974], D. Kleitman, M. Edelberg and D. Lubell [1971], D. Lubell [1966]. It is still not known whether Sperner's Lemma holds for finite partition lattices. Nevertheless, many nice results have been proved for finite geometric lattices about the numbers W_k. Let L be a finite geometric lattice of length n. Then $W_1 \leq W_k$ for $k = 2, 3, \ldots, n-1$ (J. G. Basterfield and L. M. Kelly [1968]) and $W_1 = W_{n-1}$ iff L is modular (C. Greene [1970]); furthermore, for $1 \leq k \leq n-2$,

$$W_1 + \cdots + W_k \leq W_{n-k} + \cdots + W_{n-1},$$

and equality holds iff L is modular (T. A. Dowling and R. M. Wilson [1975]). For finite modular geometric lattices $W_k = W_{n-k}$ for all k (R. P. Dilworth [1954]). For further results see T. A. Dowling and R. M. Wilson [1974].

The proof of these results utilizes the Möbius function, see Exercises 3.32–37 and H. H. Crapo [1966], [1968/69].

Maximum sized antichains in $(\mathfrak{C}_n)^k$ are determined in I. Rosenberg [1967].

The connections of geometric lattices and combinatorics are further explored in H. H. Crapo and G.-C. Rota [1970] and [1971].

Partition lattices were first studied in detail in O. Ore [1942]. Characterizations of partition lattices can be found in R. P. Dilworth [1944], U. Sasaki and S. Fujiwara [1952a], and D. Sachs [1961a].

"Which finite lattices can be embedded in a finite partition lattice?" is one of the oldest open problems of lattice theory. Corollary 4.6 tells us that we shall not obtain an answer by looking at identities. K. A. Baker observed that the same holds for all universal sentences: if a universal sentence Φ holds for all finite lattices, then Φ holds for all lattices. (The proof is easy; it uses that a lattice L can be embedded in an ultra product of its finite partial sublattices.) A. Ehrenfeucht, V. Faber, S. Fajtlowicz, and J. Mycielski [1973] consider the problem of finding, for a finite lattice L, the smallest integer $\mu(L)$ such that L can be embedded in Part(X) with $|X| = \mu(L)$. They obtain results for some special classes of lattices.

Lattices having type 1 representations are characterized in B. Jónsson [1959a] based on some ideas of R. C. Lyndon [1950] and [1956].

An equivalent form of the arguesian identity that formally resembles Desargues' Theorem can be found in G. Grätzer, B. Jónsson, and H. Lakser [1973]. Applying this characterization it is proved that if $\{o, a, b, c, i\}$ is a diamond in a modular lattice, then the arguesian identity holds in $[o, a]$; this is also used in G. Grätzer and J. Sichler [1975]. By B. Jónsson [1959], every arguesian lattice of length ≤ 4 has a type 1 representation.

If a lattice L can be embedded in the lattice of all normal subgroups of a group, then L has a type 1 representation. The converse is false by B. Jónsson [1954]; the counterexample can be chosen to have length 5.

Representations of type 2 and 3 and congruence lattices of universal algebras are considered together in G. Grätzer and E. T. Schmidt [1962] and G. Grätzer and W. A. Lampe [a]; more about this in the Concluding Remarks.

A very simple proof of Whitman's result, in the stronger form due to B. Jónsson (Theorem 4.4), is given in S. K. Thomason [1970] for finite lattices; in fact, the representation found is recursive. This is applied in S. K. Thomason [1970a] to prove that every finite lattice is isomorphic to a sublattice of the lattice of degrees of unsolvability. The reader should not find it difficult to generalize the reasoning of S. K. Thomason [1970] to arbitrary lattices.

An *n-partition* $(n \geq 1)$ of a set A satisfying $|A| \geq n$ is a collection of at least n element subsets of A such that every n-element subset of A is contained in exactly one member of the collection. Let $\text{Part}_n(A)$ denote the set of all n-partitions of A; note that $\text{Part}_1(A) = \text{Part}(A)$. For $P, Q \in \text{Part}_n(A)$, let $P \leq Q$ iff every member of P is contained in a member of Q. This makes $\text{Part}_n(A)$ a lattice, see J. Hartmanis [1956] and [1961]. Hartmanis' main result is that every finite lattice L can be embedded in some finite $\text{Part}_2(A)$. For other related results, see J. Hartmanis [1959] and B. Frontera-Marqués [1964].

The major step in the generalization of the Coordinatization Theorem was made in J. von Neumann [1936/37] replacing the division ring by a regular ring, $L(D, \mathfrak{m})$ by the lattice of principal right ideals of a regular ring, and directly irreducible arguesian geometric lattices by complemented modular lattices with a homogeneous basis with 4 or more elements (a ***homogeneous basis*** of n elements is $a_1, \ldots, a_n$ such that $\{a_1, \ldots, a_n\}$ is independent, $a_1 \vee \cdots \vee a_n = 1$, and $a_i \sim a_j$ for $i \neq j$). Von Neumann's Coordinatization Theorem is considered by many the deepest lattice theoretic result.

Von Neumann's conditions have been relaxed in many ways. The reader should consult I. Amemiya [1957], I. Amemiya and I. Halperin [1959] and [1959a], K. D. Fryer and I. Halperin [1956] and [1958], B. Jónsson [1960a], B. Jónsson and G. S. Monk [1969]; the books F. Maeda [1958], L. A. Skornjakov [1961], the survey article I. Halperin [1961] and the review by I. Halperin of B. Jónsson [1960a] in the Mathematical Reviews.

M. Hall [1954] contains some useful information on projective planes; see also B. Artmann [1969] and R. Wille [b]. For a rather unexpected connection between projective planes and weakly associative lattices, see E. Fried and V. T. Sós [1975].

Complemented modular lattices play an important role as building stones of all modular lattices of finite length; see C. Herrmann [1973c].

Problems

IV. 1. Investigate those finite modular lattices L, for which there is a matching between $J(L) \cup \{0\}$ and $M(L) \cup \{1\}$. (I. Rival.)

IV. 2. Investigate those finite modular lattices L, for which there is an isomorphism φ of the posets $J(L)$ and $M(L)$ that is also a matching. (I. Rival.)

IV. 3. Does every finite (countable) modular lattice L have an embedding in a 4-generated modular lattice K? If L is finite, can K be chosen finite?

IV. 4. Is $\mathbf{M}$ generated by $F_{\mathbf{M}}(4)$ (in the sense of Section V.1)? In other words, if an identity ε fails to hold in a modular lattice, is there a consequence ε' of ε such that ε' has only four variables and ε' also fails in some modular lattice?

IV. 5. Is $\mathbf{M}$ generated by its finite members (members of finite length)?

IV. 6. Which are the finite (simple) sublattices of $F_{\mathbf{M}}(4)$ and $F_{\mathbf{M}}(\aleph_0)$? (For $F(\aleph_0)$ see Concluding Remarks.)

IV. 7. Is $F_{\mathbf{M}}(\aleph_0)$ isomorphic to a sublattice of $F_{\mathbf{M}}(4)$ or of $F_{\mathbf{M}}(n)$ for some $n < \omega$? Are there $n, m < \omega$, $n < m$, such that $F_{\mathbf{M}}(m)$ is isomorphic to a sublattice of $F_{\mathbf{M}}(n)$?

IV. 8. Is the word problem for $F_{\mathbf{M}}(4)$ decidable? Is the word problem for $F_{\mathbf{M}}(\aleph_0)$ decidable?

IV. 9. Is the word problem in four variables decidable for modular lattices?

IV.10. Describe all 4-generated subdirectly irreducible modular lattices.

IV.11. A *modular partial lattice* is a partial lattice obtained from a modular lattice as in Definition I.5.12. Show that there is no finite set of identities characterizing the modularity of a partial lattice. (See Exercise I.5.20 for the concept of validity of an identity in a partial lattice.)

IV.12. Is the modularity of a partial lattice undecidable?

IV.13. Investigate modular lattices L with the following property: if a/b is a subquotient of c/d and $a/b \approx c/d$, then $a/b = c/d$. Is it possible to define in such L a "distance function" with values in a distributive lattice?

IV.14. Characterize the congruence lattices of modular lattices. (Distributive algebraic lattices.)

IV.15. Find a common generalization of the Kuroš-Ore Theorem (Theorem 1.5) and the Ore Theorem (on direct representations). (For modular lattices of finite length this is done in E. T. Schmidt [1970].)

IV.16. For algebraic lattices, is M-symmetry equivalent to semimodularity?

IV.17. Investigate the construction of Exercise 2.12 from the point of view of M-symmetry.

IV.18. For infinite sets A and B, are Part(A) and Part(B) elementarily equivalent? If $|A| \leq |B|$, is there an elementary embedding of Part(A) into Part(B)?

IV.19. Characterize the congruence lattices of semimodular lattices. (Distributive algebraic lattices.)

IV.20. Characterize the congruence lattices of weakly modular lattices. (Distributive algebraic lattices.)

IV.21. Let the lattice L have a type 1 representation. Does L have a type 1 representation associated with a distance function?

IV.22. Let L be a lattice having a type 1 representation. Does every homomorphic image of L have a type 1 representation?

IV.23. Show that there is no finite set of identities defining the equational class generated by lattices having type 1 representation.

IV.24. Let L be an algebraic lattice having a type 1 representation. Does L have a type 1 complete representation?

IV.25. Let L be an algebraic lattice having a type 1 representation. Is there a (finitary) algebra $\mathfrak{A}$ with permutable congruences such that L is isomorphic to the congruence lattice of $\mathfrak{A}$?

IV.26. Let **K** be an equational class of algebras. If there is a nontrivial lattice identity holding in all congruence lattices of algebras in **K**, is then **K** congruence modular (R. N. McKenzie)?

IV.27. If the validity of a nontrivial lattice identity, ε, for congruence lattices of algebras in an equational class is a Mal'cev type condition, does then ε imply modularity (J. B. Nation)?

IV.28. Find a Mal'cev type condition equivalent to the validity of the arguesian identity for the congruence lattices of the members of an equational class of algebras.

IV.29. To what extent is an equational class **K** of lattices determined by $\{C(L) \mid L \in \mathbf{K}\}$? Check $\mathbf{K} = \mathbf{D}, \mathbf{N}_5, \mathbf{M}_3, \mathbf{M}_{3,3}$, etc.

IV.30. Let the lattice L have an embedding into a complemented modular lattice. Does every homomorphic image of L have such an embedding?

IV.31. Find an effective algorithm that turns configurational conditions for geometries into equivalent identities for the subspace lattices.

IV.32. Can every finite lattice be embedded in a finite partition lattice? In particular, does the dual of a finite partition lattice have such an embedding?[1]

---

[1] A letter received by the author in January 1977 from P. Pudlák and J. Tůma contains a remarkable proof that every finite lattice can be embedded in a finite partition lattice.

IV.33. For a finite lattice L, let $\mu(L)$ denote the smallest integer such that L can be embedded in Part(A) with $|A| \leq \mu(L)$. What is $\mu(L)$, where L is the dual of Part(A) with $|A| = 4$?

IV.34. Let $L(k_1, \ldots, k_n)$ denote the lattice with 0 and 1 consisting of n chains $c(i, 1) > c(i, 2) > \cdots > c(i, k_i)$, $i = 1, 2, \ldots, n$, such that if $i \neq j$, then $c(i, p)$ is a complement of $c(j, q)$. Compute $\mu(L(k_1, \ldots, k_n))$. (This is done for $n = 2$ in A. Ehrenfeucht, V. Faber, S. Fajtlowicz, and J. Mycielski [1973].)

IV.35. Is it true in Part(A), A finite, that any maximum sized antichain is of the form $E(i)$?

IV.36. Is every finite lattice isomorphic to the congruence lattice of a finite algebra?

IV.37. Is every finite type 1 lattice isomorphic to the congruence lattice of a finite algebra with permutable congruences?

IV.38. Investigate Problems IV.26–IV.28 for infinitary algebras.

IV.39. Is there an equational class **K** of infinitary algebras such that $\mathbf{M} = \{C(\mathfrak{A}) \mid \mathfrak{A} \in K\}$?

CHAPTER V

EQUATIONAL CLASSES OF LATTICES

1. Characterizations of Equational Classes

In this section we shall discuss the basic properties of equational classes of lattices. Of the four characterizations and descriptions given, three apply to arbitrary equational classes of universal algebras; the fourth is valid only for those equational classes of universal algebras that are *congruence distributive* (that is, the congruence lattice of any algebra in the class is distributive). For the sake of simplicity, all these results are stated and proved only for lattices.

For a class **K** of lattices, let Id(**K**) denote the set of all identities holding in all lattices of **K**; if $\mathbf{K}=\{L\}$ we write Id(L) for Id(**K**). (The same convention will be used for all "operators" as well; if **X** is an operator, that is, **X** is any one of **H**, **S**, **P**, **Equ**, $\mathbf{P_S}$, $\mathbf{P_U}$, **Si**, **I**, then we write **X**(L) for **X**(**K**) if $\mathbf{K}=\{L\}$.) For a class Σ of identities, let Mod(Σ) denote the class of all "lattice models of Σ"—lattices in which all the identities of Σ hold. By definition, **K** is an equational class of lattices iff $\mathbf{K}=\mathrm{Mod}(\Sigma)$ for some set Σ of identities, or equivalently, iff

$$\mathbf{K}=\mathrm{Mod}(\mathrm{Id}(\mathbf{K})).$$

Let $\varepsilon\colon p=q$ be an identity where p and q are n-ary lattice polynomials. Let **K** be an equational class and let $F_{\mathbf{K}}(n)$ be the free lattice over **K** with n free generators, $u_0,\ldots,u_{n-1}$; $F_{\mathbf{K}}(n)$ exists by Corollary I.5.6. Let $F_{\mathbf{K}}(\aleph_0)$ denote the free lattice over **K** with $\aleph_0$ free generators, $u_0, u_1, \ldots, u_n, \ldots$.

Let us suppose that ε holds for the free generators of $F_{\mathbf{K}}(n)$, that is,

$$p(u_0,\ldots,u_{n-1})=q(u_0,\ldots,u_{n-1}).$$

Let L be any lattice in **K** and let $a_0,\ldots,a_{n-1}\in L$. Then the map $u_i\to a_i$, $i=0,1,\ldots,n-1$, can be extended to a homomorphism φ of $F_{\mathbf{K}}(n)$ into L, and so

$$\begin{aligned}p(a_0,\ldots,a_{n-1})&=p(u_0\varphi,\ldots,u_{n-1}\varphi)=p(u_0,\ldots,u_{n-1})\varphi\\&=q(u_0,\ldots,u_{n-1})\varphi=q(u_0\varphi,\ldots,u_{n-1}\varphi)=q(a_0,\ldots,a_{n-1}),\end{aligned}$$

that is, ε holds in L. Thus

$$\varepsilon\in\mathrm{Id}(\mathbf{K})\quad\text{iff}\quad\varepsilon\in\mathrm{Id}(F_{\mathbf{K}}(n)).$$

If we make no restrictions on the arity of p and q we obtain:

$$\varepsilon\in\mathrm{Id}(\mathbf{K})\quad\text{iff}\quad\varepsilon\in\mathrm{Id}(F_{\mathbf{K}}(\aleph_0)).$$

This means that **K** is completely determined by $F_{\mathbf{K}}(\aleph_0)$.

Theorem 1.

(i) *There is a one-to-one correspondence between equational classes of lattices and (up to isomorphism) free lattices with $\aleph_0$ generators, wherein* **K** *corresponds to* $F_{\mathbf{K}}(\aleph_0)$.

(ii) *A lattice L is a free lattice with $\aleph_0$ generators over some equational class* **K** *of lattices iff L has a countable generating set U such that every map $U \to L$ can be extended to an endomorphism of L.*

Remark. (i) sets up the correspondence between equational classes and countably generated free lattices while (ii) describes which lattices occur in this correspondence. It follows from the results of Chapter VI that the U in (ii) is uniquely determined as the set of doubly irreducible elements of L.

Proof. (i) has already been proved. The "only if" part of (ii) is trivial. To prove the "if" part, let L and U be given as in (ii). We wish to construct an equational class **K** such that L is free over **K**. Let **K** be the class of all lattices A such that any map $U \to A$ can be extended to a homomorphism of L into A. Let Σ be the set of all identities $\varepsilon: p = q$ such that $p(u_0, u_1, \ldots) = q(u_0, u_1, \ldots)$ holds in L for all $u_0, u_1, \ldots \in U$.

We claim that $\mathbf{K} = \mathrm{Mod}(\Sigma)$. Indeed, arguing as at the beginning of this section, we obtain $\mathbf{K} \subseteq \mathrm{Mod}(\Sigma)$. Conversely, let $A \in \mathrm{Mod}(\Sigma)$ and let $\alpha: U \to A$ be a map. If $\varepsilon: p = q$ is an identity and $p(u_0, u_1, \ldots) = q(u_0, u_1, \ldots)$ for $u_0, u_1, \ldots \in U$ (with all u_i distinct), then $\varepsilon \in \Sigma$. Since $A \in \mathrm{Mod}(\Sigma)$ we obtain that ε holds in A and therefore $p(u_0\alpha, \ldots) = q(u_0\alpha, \ldots)$. Thus (see Exercise I.5.45) α can be extended to a homomorphism and so $L \in \mathbf{K}$.

This proves that $\mathbf{K} = \mathrm{Mod}(\Sigma)$, so **K** is an equational class. By the definition of **K**, L is free over **K**. ●

Let $u_0, u_1, \ldots$ be the free generators of $F(\aleph_0)$ and let $v_0, v_1, \ldots$ be the free generators of $F_{\mathbf{K}}(\aleph_0)$. Then $u_i \to v_i$, $i = 0, 1, \ldots$ extends to a homomorphism $\varphi = \varphi_{\mathbf{K}}$; let $\Theta = \Theta_{\mathbf{K}}$ be the kernel of φ. Since, by Theorem 1, the equational class **K** is determined by $F_{\mathbf{K}}(\aleph_0)$ and $F_{\mathbf{K}}(\aleph_0)$ is determined by Θ, we conclude that **K** is determined by Θ. If we can ascertain which congruences arise this way, we shall have another description of equational classes.

Call a congruence relation Θ of a lattice L *fully invariant* iff $a \equiv b\ (\Theta)$ implies that $a\alpha \equiv b\alpha\ (\Theta)$, for all $a, b \in L$ and for all endomorphisms α of L.

Theorem 2 (B. H. Neumann [1962]). *There is a one-to-one correspondence between equational classes of lattices and fully invariant congruence relations of $F(\aleph_0)$.*

Proof. Let **K** be an equational class and let φ, Θ $(= \Theta_{\mathbf{K}})$, u_i, v_i be as described above. We show that Θ is fully invariant. Let α be an endomorphism of $F(\aleph_0)$, let $a, b \in F(\aleph_0)$, and let $a \equiv b\ (\Theta)$.

Let β be the endomorphism of $F_{\mathbf{K}}(\aleph_0)$ extending $v_i \to u_i\alpha\varphi$. Since $a \in F(\aleph_0)$, $a =$

$p(u_0, \ldots, u_{n-1})$ for some integer n and for some n-ary polynomial p. We compute:

$$a\varphi\beta = p(u_0, \ldots, u_{n-1})\varphi\beta = p(u_0\varphi\beta, \ldots, u_{n-1}\varphi\beta) = p(v_0\beta, \ldots, v_{n-1}\beta)$$
$$= p(u_0\alpha\varphi, \ldots, u_{n-1}\alpha\varphi) = p(u_0, \ldots, u_{n-1})\alpha\varphi = a\alpha\varphi,$$

and similarly for b. Thus,

$$a\alpha\varphi = a\varphi\beta = b\varphi\beta = b\alpha\varphi,$$

and therefore $a\alpha \equiv b\alpha\ (\Theta)$.

Now let Θ be a fully invariant congruence relation of $F(\aleph_0)$. We shall show that $F(\aleph_0)/\Theta$ satisfies the condition of Theorem 1(ii). If $u_i \equiv u_j\ (\Theta)$ for some $i \neq j$, then it is easily shown that $\Theta = \iota$ and so $F(\aleph_0)/\Theta$ is the one-element lattice. So let us assume that $u_i \not\equiv u_j\ (\Theta)$, for all $i \neq j$. We take $U = \{[u_i]\Theta \mid i = 0, 1, \ldots\}$ and $\alpha\colon U \to F(\aleph_0)/\Theta$. Let $[u_i]\Theta\alpha = [a_i]\Theta$, $i = 0, 1, \ldots$ Then $u_i \to a_i$, $i = 0, 1, \ldots$, can be extended to an endomorphism β of $F(\aleph_0)$. Since $\Theta = \mathrm{Ker}(\varphi)$ is fully invariant, we have $\mathrm{Ker}(\varphi) \subseteq \mathrm{Ker}(\beta\varphi)$ and so $x\varphi \to x\beta\varphi$ extends α to an endomorphism of $F(\aleph_0)/\Theta$. ●

We have already proved in Section I.4 that an equational class is closed under the formation of homomorphic images, sublattices, and direct products. The converse, which is due to G. Birkhoff [1935], is the third description of equational classes.

Theorem 3. *A class* **K** *of lattices is an equational class iff* **K** *is closed under the formation of homomorphic images, sublattices, and direct products.*

Remark. The direct product of an empty family of lattices is the one-element lattice and therefore if **K** is closed under the formation of direct products, then **K** is not the empty class. Observe also, that if **K** is closed under the formation of homomorphic images, then **K** is closed under the formation of isomorphic copies.

Proof. Let **K** be closed under the formation of homomorphic images, sublattices, and direct products. If **K** consists of one-element lattices only, then **K** can be defined by the identity: $x_0 = x_1$ and so **K** is equational. Now we can assume that **K** contains a lattice of more than one element. Therefore, we conclude just as in Section I.5 that $F_{\mathbf{K}}(\mathfrak{m})$ exists for any cardinal $\mathfrak{m}$.

Let $L \in \mathrm{Mod}(\mathrm{Id}(\mathbf{K}))$ and take an $F_{\mathbf{K}}(\mathfrak{m}) \in \mathbf{K}$ with $|L| + \aleph_0 = \mathfrak{m}$. We denote by U the set of free generators of $F_{\mathbf{K}}(\mathfrak{m})$. Since $|L| + \aleph_0 = |U|$, there is a map α of U onto L. Let $p(u_0, u_1, \ldots, u_{n-1}) = q(u_0, u_1, \ldots, u_{n-1})$ hold in $F_{\mathbf{K}}(\mathfrak{m})$ with $u_0, \ldots, u_{n-1} \in U$. Without loss of generality we can assume that the u_i are all distinct. Therefore, $p = q \in \mathrm{Id}(F_{\mathbf{K}}(\mathfrak{m}))$ (as argued at the beginning of this section) and so, by the freeness of $F_{\mathbf{K}}(\mathfrak{m})$, $p = q \in \mathrm{Id}(\mathbf{K})$. Because $L \in \mathrm{Mod}(\mathrm{Id}(\mathbf{K}))$, $p = q \in \mathrm{Id}(L)$; in particular, $p(u_0\alpha, \ldots, u_{n-1}\alpha) = q(u_0\alpha, \ldots, u_{n-1}\alpha)$. This shows that α satisfies the hypothesis of Exercise I.5.45 and can therefore be extended to a homomorphism β of $F_{\mathbf{K}}(\mathfrak{m})$ onto L. Thus L is a homomorphic image of a member of **K**, and so $L \in \mathbf{K}$.

The converse was proved in Lemma I.4.8. ●

To obtain a slightly different version of this result, we introduce some notation. For a class **K** of lattices, let **H**(**K**), **S**(**K**), and **P**(**K**) denote the class of all homomorphic images, sublattices, and direct products of members of **K**, respectively.

Corollary 4 (A. Tarski [1946]). *Let* **K** *be a class of lattices. Then* **HSP(K)** *is the smallest equational class containing* **K**.

Proof. We start out by observing three formulas for any class **K** of lattices:

(i) $\mathbf{SH}(\mathbf{K}) \subseteq \mathbf{HS}(\mathbf{K})$.

(ii) $\mathbf{PH}(\mathbf{K}) \subseteq \mathbf{HP}(\mathbf{K})$.

(iii) $\mathbf{PS}(\mathbf{K}) \subseteq \mathbf{SP}(\mathbf{K})$.

To prove (i), let $L \in \mathbf{SH}(\mathbf{K})$. Then there is an $A \in \mathbf{K}$ and a homomorphism α of A onto a lattice B containing L as a sublattice. Set $C = \{x \mid x \in A \text{ and } x\alpha \in L\}$. Then $C \in \mathbf{S}(\mathbf{K})$ and $L \in \mathbf{H}(C)$. Hence $L \in \mathbf{HS}(\mathbf{K})$, proving (i).

To prove (ii), let $L \in \mathbf{PH}(\mathbf{K})$. Then there exist $A_i \in \mathbf{K}$ and a homomorphism α_i of A_i onto B_i, for $i \in I$, such that $L = \Pi(B_i \mid i \in I)$. It is clear that L is a homomorphic image of $\Pi(A_i \mid i \in I) \in \mathbf{P}(\mathbf{K})$, proving (ii).

The proof of (iii) follows that of (ii) with A_i a sublattice of B_i.

To show that $\mathbf{HSP}(\mathbf{K}) = \mathbf{K}_1$ is an equational class using Theorem 3, we have to verify that $\mathbf{H}(\mathbf{K}_1) \subseteq \mathbf{K}_1$, $\mathbf{S}(\mathbf{K}_1) \subseteq \mathbf{K}_1$, and $\mathbf{P}(\mathbf{K}_1) \subseteq \mathbf{K}_1$. Indeed (using (i)–(iii) and that $\mathbf{XX}(\mathbf{K}_2) = \mathbf{X}(\mathbf{K}_2)$ for $\mathbf{X} \in \{\mathbf{H}, \mathbf{S}, \mathbf{P}\}$ and any class $\mathbf{K}_2$):

$$\mathbf{H}(\mathbf{K}_1) = \mathbf{HHSP}(\mathbf{K}) = \mathbf{HSP}(\mathbf{K}) = \mathbf{K}_1,$$
$$\mathbf{S}(\mathbf{K}_1) = \mathbf{SHSP}(\mathbf{K}) \subseteq \mathbf{HSSP}(\mathbf{K}) = \mathbf{HSP}(\mathbf{K}) = \mathbf{K}_1,$$
$$\mathbf{P}(\mathbf{K}_1) = \mathbf{PHSP}(\mathbf{K}) \subseteq \mathbf{HPSP}(\mathbf{K}) \subseteq \mathbf{HSPP}(\mathbf{K}) = \mathbf{HSP}(\mathbf{K}) = \mathbf{K}_1.$$

If $\mathbf{K}_2$ is any equational class containing **K**, then

$$\mathbf{K}_2 = \mathbf{HSP}(\mathbf{K}_2) \supseteq \mathbf{HSP}(\mathbf{K}) = \mathbf{K}_1. \quad \bullet$$

Let $\mathbf{Equ}(\mathbf{K})$ denote the smallest equational class containing **K**. Then Corollary 4 can be summarized in the formula $\mathbf{Equ} = \mathbf{HSP}$.

For a class **K** of lattices, let $\mathbf{P}_S(\mathbf{K})$ denote the class of all lattices that are isomorphic to a subdirect product of members of **K**. The following variant of Corollary 4 is not especially useful, but the construction used in the proof has found some applications.

Corollary 5 (S. R. Kogalovskiĭ [1965]). *A class* **K** *of lattices is an equational class iff* **K** *is closed under the formation of homomorphic images and subdirect products. Moreover, for any class* **K** *of lattices,*

$$\mathbf{Equ}(\mathbf{K}) = \mathbf{HP}_S(\mathbf{K}).$$

Remark. In other words, $\mathbf{Equ} = \mathbf{HP}_S$.

Proof. We leave it to the reader to verify that both statements follow readily from the following inequality:

(iv) $\mathbf{S}(\mathbf{K}) \subseteq \mathbf{HP}_S(\mathbf{K})$.

To verify (iv), let $L \in \mathbf{S}(\mathbf{K})$, that is, let L be a sublattice of some $A \in \mathbf{K}$. Now take A^I with $|I| = \aleph_0$ and form a sublattice $B \subseteq A^I$ defined as follows: for $f \in A^I$, let $f \in B$

iff $f(i)=a$, for some $a\in L$ and for all but finitely many $i\in I$; a map $\varphi: B\to L$ is defined by mapping f to this a. It is trivial that $B\in \mathbf{P_S}(A)$ and $L\in \mathbf{H}(B)$, concluding that $L\in \mathbf{HP_S}(\mathbf{K})$. ●

We have obtained two formulas for **Equ**, namely, $\mathbf{Equ}=\mathbf{HSP}=\mathbf{HP_S}$. Their usefulness is, however, somewhat limited when applied to describing members of equational classes of lattices. A great improvement of these formulas is possible for lattices and we shall proceed to develop this.

Let L_i, $i\in I$, be lattices, let $I\neq\varnothing$, and let $\mathcal{D}$ be a dual ideal of the lattice $P(I)$ of all subsets of I. For $f, g\in \Pi(L_i \mid i\in I)$, we set

$$I(f,g)=\{i \mid i\in I \text{ and } f(i)=g(i)\}.$$

(I in $I(f,g)$ stands for "identical".)

We introduce a congruence relation $\Theta=\Theta(\mathcal{D})$ on $L=\Pi(L_i \mid i\in I)$. For $f, g\in L$, let

$$f\equiv g\ (\Theta) \quad \text{iff} \quad I(f,g)\in\mathcal{D}.$$

(View members of $\mathcal{D}$ as "sets of measure one". Then $f\equiv g\ (\Theta)$ iff f and g are identical on a set of measure one.) Θ is obviously reflexive and symmetric. If $f, g, h\in L$, $f\equiv g\ (\Theta)$, and $g\equiv h\ (\Theta)$, then $I(f,g)$ and $I(g,h)\in\mathcal{D}$ so that $I(f,h)\supseteq I(f,g)\cap I(g,h)\in\mathcal{D}$. Thus $I(f,h)\in\mathcal{D}$ and $f\equiv h\ (\Theta)$. If $f, g, h\in L$ and $f\equiv g\ (\Theta)$, then $I(f\wedge h, g\wedge h)\supseteq I(f,g)\in\mathcal{D}$ so that $I(f\wedge h, g\wedge h)\in\mathcal{D}$ and $f\wedge h\equiv g\wedge h\ (\Theta)$. Similarly, $f\vee h\equiv g\vee h\ (\Theta)$. Thus Θ is a congruence relation and we can form the lattice L/Θ, denoted by $\Pi_{\mathcal{D}}(L_i \mid i\in I)$, called a *reduced product of* $(L_i \mid i\in I)$. When $\mathcal{D}$ is a prime dual ideal (also called, ultra filter)–briefly, $\mathcal{D}$ *is prime over* I–, $\Pi_{\mathcal{D}}(L_i \mid i\in I)$ is called an *ultra product* (also called, *prime product*); see J. Łoś [1955]. For a class $\mathbf{K}$ of lattices, $\mathbf{P_U}(\mathbf{K})$ will denote the class of all lattices that are isomorphic to an ultra product of members of $\mathbf{K}$.

We start out by proving some elementary properties of these constructions:

Lemma 6. *Let I be a nonvoid set and let $\mathcal{D}$ be a dual ideal of $P(I)$. Let $J\in\mathcal{D}$, $J\neq\varnothing$, and define $\mathcal{E}=\{X\cap J \mid X\in\mathcal{D}\}$. Then $\mathcal{E}$ is a dual ideal of $P(J)$ and if $\mathcal{D}$ is prime so is $\mathcal{E}$. Furthermore, for any family $(L_i \mid i\in I)$ of lattices*

$$\Pi_{\mathcal{D}}(L_i \mid i\in I)\cong\Pi_{\mathcal{E}}(L_i \mid i\in J).$$

Proof. The statements concerning $\mathcal{E}$ are trivial. Let π denote the homomorphism $f\to f_J$, the restriction of f to J, and let $\varphi: \Pi(L_i \mid i\in I)\to\Pi_{\mathcal{D}}(L_i \mid i\in I)$ and $\psi: \Pi(L_i \mid i\in J)\to\Pi_{\mathcal{E}}(L_i \mid i\in J)$ be the natural homomorphisms. $\Theta=\Theta(\mathcal{D})$ is the kernel of φ. Let Φ be the kernel of $\pi\psi$. Then $f\equiv g\ (\Phi)$ iff $I(f_J, g_J)\in\mathcal{E}$, or equivalently, $J\cap I(f,g)\in\mathcal{D}$. In view of $J\in\mathcal{D}$, this is equivalent to $I(f,g)\in\mathcal{D}$. Thus $\Theta=\Phi$ and the isomorphism follows. ●

Corollary 7. *Let I be a nonvoid set and let $\mathcal{D}$ be a dual ideal of $P(I)$. If $\mathcal{D}$ is principal, $\mathcal{D}=[J)$, then for any family $(L_i \mid i\in I)$ of lattices,*

$$\Pi_{\mathcal{D}}(L_i \mid i\in I)\cong\Pi(L_i \mid i\in J);$$

in particular, if $\mathcal{D}$ is principal and prime, $\mathcal{D}=[\{j\})$, then

$$\Pi_{\mathcal{D}}(L_i \mid i\in I)\cong L_j.$$

Lemma 8. *Let $L_0, \ldots, L_{n-1}$ be finite lattices. Let $(L_i \mid i\in I)$, $I\neq\varnothing$, be a family of lattices each L_i being one of $L_0, \ldots, L_{n-1}$. Let $\mathcal{D}$ be prime over I. Then there is a j, $0\leq j<n$, such that*

$$\Pi_{\mathcal{D}}(L_i \mid i\in I)\cong L_j.$$

Proof. We can assume that the lattices $L_0, \ldots, L_{n-1}$ are pairwise nonisomorphic. Thus if we define $I_j=\{i \mid i\in I \text{ and } L_i=L_j\}$ for $0\leq j<n$, then $I_0, \ldots, I_{n-1}$ are pairwise disjoint and $I_0\cup\cdots\cup I_{n-1}=I$. Since $\mathcal{D}$ is prime, there is exactly one j with $0\leq j<n$ such that $I_j\in\mathcal{D}$. Applying Corollary 7 to I_j, we obtain

$$\Pi_{\mathcal{D}}(L_i \mid i\in I)\cong \Pi_{\mathcal{E}}(L_i \mid i\in I_j),$$

where $\mathcal{E}=\{X\cap I_j \mid X\in\mathcal{D}\}$ is prime over I_j. For $a\in L_j$, let $f_a\in\Pi(L_i \mid i\in I_j)$ be defined by $f_a(i)=a$, for all $i\in I$. Then for $a, b\in L_j$, $a\neq b$, we have $I(f_a, f_b)=\varnothing\notin\mathcal{E}$, and therefore $f_a\not\equiv f_b(\Theta(\mathcal{E}))$. Thus $\alpha: a\to[f_a]\Theta(\mathcal{E})$ embeds L_j in $\Pi_{\mathcal{E}}(L_i \mid i\in I_j)$. In fact, α is an isomorphism. To see that, let $f\in\Pi(L_i \mid i\in I_j)$. Then for each $a\in L$ we define $I_{j,a}=\{i \mid i\in I_j$ and $f(i)=a\}$. Since the $I_{j,a}$ are pairwise disjoint and $\bigcup(I_{j,a} \mid a\in L_j)=I_j$, we conclude that there exists exactly one $a\in L_j$ such that $I_{j,a}\in\mathcal{E}$. Now $I(f, f_a)=I_{j,a}\in\mathcal{E}$, thus $f\equiv f_a\,(\Theta(\mathcal{E}))$ and α is onto. Therefore, $\Pi_{\mathcal{E}}(L_i \mid i\in I_j)\cong L_j$. ●

Now we are ready to state the formula of B. Jónsson [1967] for **Equ**:

Theorem 9. *For a class* **K** *of lattices,*

$$\mathbf{Equ}(\mathbf{K})=\mathbf{P_S H S P_U}(\mathbf{K}).$$

Proof. By the Subdirect Product Representation Theorem it is sufficient to prove that if A is a subdirectly irreducible lattice and $A\in\mathbf{Equ}(\mathbf{K})$, then $A\in\mathbf{HSP_U}(\mathbf{K})$. By Corollary 4, $A\in\mathbf{HSP}(\mathbf{K})$ and therefore there exist $A_i\in\mathbf{K}$, $i\in I$, a sublattice B of $\Pi(A_i \mid i\in I)$, and a congruence relation Φ on B such that $B/\Phi\cong A$.

We claim that there is a $\mathcal{D}$ prime over I such that $\Theta(\mathcal{D})$ restricted to B is contained in Φ. Indeed, if we have such a $\mathcal{D}$, then by the Second Isomorphism Theorem, B/Φ is a homomorphic image of $B/\Theta(\mathcal{D})_B$, which is, in turn, a sublattice of $\Pi(A_i \mid i\in I)/\Theta(\mathcal{D})=\Pi_{\mathcal{D}}(A_i \mid i\in I)$. Thus, $A\cong B/\Phi\in\mathbf{HSP_U}(\mathbf{K})$, as required.

For $J\subseteq I$, set $\Theta_J=\Theta([J))$, that is, for $f, g\in\Pi(A_i \mid i\in I)$

$$f\equiv g\ (\Theta_J) \quad\text{iff}\quad I(f, g)\supseteq J.$$

Observe that $\Theta(\mathcal{D})=\bigcup(\Theta_J \mid J\in\mathcal{D})$; consequently, to find $\mathcal{D}$ we have to look for $\mathcal{D}$ in

$$\mathcal{E}=\{J \mid J\subseteq I \text{ and } (\Theta_J)_B\leq\Phi\}.$$

$\mathscr{E}$ has the following three properties:

(i) $I \in \mathscr{E}$ and $\varnothing \notin \mathscr{E}$;

(ii) $J_0 \in \mathscr{E}$ and $J_0 \subseteq J_1 \subseteq I$ imply that $J_1 \in \mathscr{E}$;

(iii) $M, N \subseteq I$ and $M \cup N \in \mathscr{E}$ imply that $M \in \mathscr{E}$ or $N \in \mathscr{E}$.

$\Theta_I = \omega$ and $\Theta_\varnothing = \iota$, so (i) is obvious. If $J_0 \subseteq J_1$, then $\Theta_{J_0} \geq \Theta_{J_1}$, proving (ii). To prove (iii), let M and $N \subseteq I$. It is trivial that $\Theta_{M \cup N} = \Theta_M \wedge \Theta_N$ and

$$(\Theta_{M \cup N})_B = (\Theta_M)_B \wedge (\Theta_N)_B.$$

Now let $M \cup N \in \mathscr{E}$. Then

$$(\Theta_{M \cup N})_B \leq \Phi,$$

that is,

$$(\Theta_M)_B \wedge (\Theta_N)_B \leq \Phi.$$

Since B/Φ is subdirectly irreducible, Φ is meet-irreducible in $C(B)$. So using that $C(B)$ is distributive we conclude that $(\Theta_M)_B \leq \Phi$ or $(\Theta_N)_B \leq \Phi$, that is, M or $N \in \mathscr{E}$, proving (iii).

Now let $\mathscr{D}$ be a dual ideal of $P(I)$ maximal with respect to the property $\mathscr{D} \subseteq \mathscr{E}$. We show that $\mathscr{D}$ is prime. By (i), $\varnothing \notin \mathscr{D}$ so $\mathscr{D}$ is proper. If $\mathscr{D}$ is proper but not prime, then there exists a $J \subseteq I$ such that $J \notin \mathscr{D}$ and $I - J \notin \mathscr{D}$. If $J \cap J' \in \mathscr{E}$ for every $J' \in \mathscr{D}$, then by (ii) $\mathscr{D}$ and J would generate a dual ideal contained in $\mathscr{E}$, contradicting $J \notin \mathscr{D}$ and the maximality of $\mathscr{D}$. Thus there exists a $J_0 \in \mathscr{D}$ such that $J \cap J_0 \notin \mathscr{E}$. Similarly, there exists a $J_1 \in \mathscr{D}$ such that $(I - J) \cap J_1 \notin \mathscr{E}$. Then

$$J_0 \cap J_1 = (J \cap (J_0 \cap J_1)) \cup ((I - J) \cap (J_0 \cap J_1)),$$

contradicting (iii). Thus $\mathscr{D}$ is prime. ●

Let $\mathbf{Si}(\mathbf{K})$ be the class of subdirectly irreducible lattices in $\mathbf{K}$. An equivalent form of Theorem 9 is the following:

Corollary 10. *For a class* $\mathbf{K}$ *of lattices,*

$$\mathbf{Si}\,\mathrm{Equ}(\mathbf{K}) \subseteq \mathbf{HSP_U}(\mathbf{K}).$$

We shall illustrate the power of Theorem 9 with two simple applications taken from B. Jónsson [1967].

If $\mathbf{K}$ is a finite set of finite lattices, then by Lemma 8 $\mathbf{P_U}(\mathbf{K})$ is, up to isomorphic copies, the same as $\mathbf{K}$, so we conclude:

Corollary 11. *Let* $\mathbf{K}$ *be a finite set of finite lattices. Then*

$$\mathbf{Si}\,\mathrm{Equ}(\mathbf{K}) \subseteq \mathbf{HS}(\mathbf{K}).$$

Observe that $\mathbf{HS}(\mathbf{K})$ is, up to isomorphic copies, a finite set of finite lattices.

COROLLARY 12. *Let A and B be finite nonisomorphic subdirectly irreducible lattices. If $|A| \leq |B|$, then there exists an identity ε holding in A but not holding in B.*

PROOF. $B \notin \mathbf{HS}(A)$ since $|B| \geq |A|$ and B is not isomorphic to A. Hence by Corollary 11, $B \notin \mathbf{Equ}(A)$ and so some identity holding in A must fail in B. ●

Exercises

1. Show that $\mathbf{K} \to \mathrm{Id}(\mathbf{K})$ and $\Sigma \to \mathrm{Mod}(\Sigma)$ set up a Galois connection.
2. Prove that $\mathbf{K}$ is an equational class iff $\mathbf{K} = \mathrm{Mod}(\mathrm{Id}(\mathbf{K}))$. For a set Σ of identities, $\Sigma = \mathrm{Id}(\mathbf{K})$ for some class of lattices iff $\Sigma = \mathrm{Id}(\mathrm{Mod}(\Sigma))$.
3. Characterize sets of identities Σ that are of the form $\mathrm{Id}(\mathbf{K})$ for some class $\mathbf{K}$ of lattices.
4. Find properties of equational classes $\mathbf{K}$ of lattices satisfying $\mathrm{Id}(\mathbf{K}) = \mathrm{Id}(F_{\mathbf{K}}(n))$ for some integer n.
5. Let L be a lattice and let U be a generating set of L. Show that if L is free over $\mathbf{K}$ with U as a free generating set, then the same holds over $\mathbf{Equ}(\mathbf{K})$.
6. Reprove Theorem 1(ii) using Exercise 5 and Theorem 3.
7. Prove that the elements of a free generating set are doubly irreducible (B. Jónsson [1971].)
8. Let $\mathbf{K}$ be an equational class of lattices. Let φ be the natural homomorphism of $F(\aleph_0)$ onto $F_{\mathbf{K}}(\aleph_0)$, and let Θ be the kernel of φ. We assign to Θ an equational class $\mathbf{K}_1$ over which $F(\aleph_0)/\Theta$ is free. Prove that $\mathbf{K} = \mathbf{K}_1$.
9. Let Θ be a fully invariant congruence relation of $F(\aleph_0)$. We assign to Θ an equational class $\mathbf{K}$, and to $\mathbf{K}$ a fully invariant congruence relation Θ_1 of $F(\aleph_0)$ as in the proof of Theorem 2. Prove that $\Theta = \Theta_1$.
10. Let L be a lattice. Show that the fully invariant congruence relations of L form a complete sublattice of $C(L)$.
11. Prove that the lattice of fully invariant congruence relations of a lattice L is a distributive algebraic lattice. Characterize the compact elements.
12. A congruence relation Θ of a lattice L is called *invariant* iff $a \equiv b\ (\Theta)$ implies that $a\alpha \equiv b\alpha\ (\Theta)$ for any automorphism α of L. Do the invariant congruences form a lattice, if so, is this lattice distributive or algebraic?
13. Let $\langle A; F\rangle$ and $\langle A; G\rangle$ be algebras and $F \subseteq G$. Show that the congruence lattice of $\langle A; G\rangle$ is a complete sublattice of the congruence lattice of $\langle A; F\rangle$.
14. Derive Exercises 10–12 from Exercise 13.
15. An equational class $\mathbf{K}$ of lattices is *generated by a lattice* A iff $\mathbf{K} = \mathbf{Equ}(A)$. Show that every equational class of lattices is generated by a lattice.
16. An equational class $\mathbf{K}$ of lattices is *locally finite* iff every finitely generated member of $\mathbf{K}$ is finite. Prove that an equational class generated by a finite lattice is locally finite. (Do not use Theorem 9 or any of its consequences. This result is true even in classes of algebras for which Theorem 9 fails.)
17. Is the converse of the statement of Exercise 16 true?
18. Prove that $\mathbf{P}_{\mathrm{S}}\mathbf{H}(\mathbf{K}) \subseteq \mathbf{HP}_{\mathrm{S}}(\mathbf{K})$ for any class $\mathbf{K}$ of lattices.
19. For a class $\mathbf{K}$ of lattices, let $\mathbf{P}_{\mathrm{R}}(\mathbf{K})$ denote the class of reduced products (up to isomorphism) of members of $\mathbf{K}$. Prove that $\mathbf{P}_{\mathrm{R}}(\mathbf{K}) \subseteq \mathbf{P}_{\mathrm{S}}\mathbf{P}_{\mathrm{U}}(\mathbf{K})$ (G. Grätzer and H. Lakser [1973]).
20. An *implication* is a sentence (all variables are universally quantified): $p_0(x_0, \ldots, x_{n-1}) =$

$q_0(x_0, \ldots, x_{n-1})$ and $\cdots$ and $p_{m-1}(x_0, \ldots, x_{n-1}) = q_{m-1}(x_0, \ldots, x_{n-1})$ imply that $p_m(x_0, \ldots, x_{n-1}) = q_m(x_0, \ldots, x_{n-1})$. (For instance, the semidistributive law of Chapter VI is an implication: $x \wedge y = x \wedge z$ implies that $x \wedge y = x \wedge (y \vee z)$. Any identity is an implication.) An *implicational class* is the class **K** of all lattices satisfying a set of implications. For a class $\mathbf{K}_1$, the class of all isomorphic copies of members of $\mathbf{K}_1$ is denoted $\mathbf{I}(\mathbf{K}_1)$. Prove that $\mathbf{ISP_R}(\mathbf{K})$ is the smallest implicational class containing **K** (A. I. Mal'cev [1966] and G. Grätzer and H. Lakser [1973]).

21. Let $\mathcal{D}$ be prime over I. An *ultra power* $L^I_{\mathcal{D}}$ of a lattice L is an ultra product $\Pi_{\mathcal{D}}(L_i \mid i \in I)$ such that $L_i = L$ for all $i \in I$. Prove that L can be embedded into $L^I_{\mathcal{D}}$.

*22. Prove that every lattice L can be embedded into an ultra product of all finitely generated sublattices of L. (Hint: let I be the set of all nonempty finite subsets of L. For $J \in I$, let L_J be the sublattice generated by J. Choose a $\mathcal{D}$ prime over I such that for each $J \in I$, $\{K \mid K \in I \text{ and } K \supseteq J\} \in \mathcal{D}$.)

*23. Let us define (first-order) formulas: (i) for n-ary polynomials p and q, $p = q$ is a formula; (ii) if Φ is a formula, so is $\neg\Phi$ (read: not Φ); (iii) if Φ_0 and Φ_1 are formulas, then so is $\Phi_0 \vee \Phi_1$ (read: Φ_0 or Φ_1); (iv) if Φ is a formula, so is $(\exists x_k)\,\Phi$. Then $\Phi_0 \wedge \Phi_1$ (Φ_0 and Φ_1) is introduced as $\neg((\neg\Phi_0) \vee (\neg\Phi_1))$, $(x_k)\,\Phi$ is introduced as $\neg((\exists x_k)\neg\Phi)$, and so on. Define inductively what it means for a formula Φ to be satisfied by certain elements of a lattice L.

24. Define precisely a *free variable* x which is not "bound" by a quantifier $\exists x$. A *sentence* is a formula without free variables. Find sentences expressing that the lattice L has at most or exactly n elements.

25. For a finite lattice K construct a sentence Φ such that Φ holds for a lattice L iff L has a sublattice isomorphic to K.

*26. Prove that a sentence Φ holds for an ultra product $\Pi_{\mathcal{D}}(L_i \mid i \in I)$ iff $\{i \mid \Phi \text{ holds in } L_i\} \in \mathcal{D}$ (J. Łoś [1955]).

27. Prove Lemma 8 using Exercise 26.

28. A *model* of a set Σ of sentences is a lattice L in which all sentences $\Phi \in \Sigma$ are satisfied. Prove the *Compactness Theorem*: Let Σ be a set of sentences. If every finite subset of Σ has a model, then Σ has a model. (Hint: for every nonempty finite $\Omega \subseteq \Sigma$ choose a model L_Ω. Let I be the set of all finite nonempty subsets of Σ and choose a $\mathcal{D}$ prime over I containing all sets of the form $\{X \mid X \in I \text{ and } X \supseteq J\}$, where $J \in I$. Then $L = \Pi_{\mathcal{D}}(L_\Omega \mid \Omega \in I)$ is a model of Σ.)

29. Let Σ and Σ_1 be sets of sentences. Σ *implies* Σ_1 iff every model of Σ is also a model of Σ_1. Σ *is equivalent to* Σ_1 iff Σ implies Σ_1 and Σ_1 implies Σ. Prove that if Σ is equivalent to $\{\Phi\}$, then there is a finite $\Sigma_1 \subseteq \Sigma$ that is equivalent to $\{\Phi\}$. (Use the Compactness Theorem.)

30. Call a lattice L *finitely subdirectly irreducible* iff ω is meet-irreducible in $C(L)$. Prove that Corollary 10 holds also for the finitely subdirectly irreducible members of $\mathbf{Equ}(\mathbf{K})$.

31. Let $\mathfrak{M}_4$ be the subspace lattice of a projective line with four points. Show that $\mathbf{M}_4 = \mathbf{Equ}(\mathfrak{M}_4) \supset \mathbf{M}_3 = \mathbf{Equ}(\mathfrak{M}_3)$ and for every equational class **K**, $\mathbf{M}_4 \supset \mathbf{K} \supseteq \mathbf{M}_3$ implies that $\mathbf{K} = \mathbf{M}_3$.

32. Find an equational class **N** of lattices such that $\mathbf{N} \supset \mathbf{N}_5 = \mathbf{Equ}(\mathfrak{N}_5)$, $\mathbf{N} \not\supseteq \mathbf{M}_3$ and $\mathbf{N} \supset \mathbf{K} \supseteq \mathbf{N}_5$ implies that $\mathbf{K} = \mathbf{N}_5$ for any equational class **K**.

33. Show that $\mathbf{Equ}(\mathbf{K}) = \mathbf{SHPS}(\mathbf{K})$ if **K** is one of the classes of lattices listed below:
   (i) $\mathbf{K} = \{L\}$, where L is a finite lattice;
   (ii) $\mathbf{K} = \{\mathfrak{M}_{\aleph_0}\}$, where $\mathfrak{M}_{\aleph_0}$ is the subspace lattice of a projective line with countably many points;

(iii) for each prime number p we choose a field F_p of characteristic p and $\mathbf{K} = \{L_p \mid p$ is a prime number$\}$, where L_p is the lattice of subspaces of the projective plane coordinatized by F_p.

34. Let $\mathbf{K}$ be an equational class of lattices. Prove that

$$F_{\mathbf{K}}(\mathfrak{m}) \in \mathbf{ISP}(F_{\mathbf{K}}(\aleph_0)),$$

for any cardinal $\mathfrak{m}$.

35. Prove the Compactness Theorem for any type of universal algebras. Find applications of this result to lattices that go beyond the Compactness Theorem of Exercise 28. (Hint: use a type with $\wedge$ and $\vee$ and infinitely many constants.)

2. The Lattice of Equational Classes of Lattices

Let $\mathbf{K}_0$ and $\mathbf{K}_1$ be equational classes of lattices. Then $\mathbf{K}_0 \cap \mathbf{K}_1$ is again an equational class and

$$\mathbf{K}_0 \cap \mathbf{K}_1 = \mathrm{Mod}(\mathrm{Id}(\mathbf{K}_0) \cup \mathrm{Id}(\mathbf{K}_1)).$$

There is also a smallest equational class $\mathbf{K}_0 \vee \mathbf{K}_1$ containing both $\mathbf{K}_0$ and $\mathbf{K}_1$, namely,

$$\mathbf{K}_0 \vee \mathbf{K}_1 = \mathrm{Mod}(\mathrm{Id}(\mathbf{K}_0) \cap \mathrm{Id}(\mathbf{K}_1)).$$

All equational classes of lattices form a lattice with the lattice operations $\mathbf{K}_0 \wedge \mathbf{K}_1 = \mathbf{K}_0 \cap \mathbf{K}_1$ and $\mathbf{K}_0 \vee \mathbf{K}_1$. (Axiomatic set theory does not permit the formation of a set whose elements are classes. Thus, formally, one cannot form the lattice of equational classes. It is easy to get around this difficulty. For instance, replace an equational class $\mathbf{K}$ by $\mathrm{Id}(\mathbf{K})$, which is a subset of the countable set of all lattice identities. Then we form the lattice of all subsets of the form $\mathrm{Id}(\mathbf{K})$ of the set of all lattice identities.)

The zero of this lattice is $\mathbf{T}$ (the trivial equational class) consisting of all one-element lattices. Let $\mathbf{K}$ be an equational class of lattices different from $\mathbf{T}$. Then there is a lattice L in $\mathbf{K}$ with $|L| > 1$. Therefore L has $\mathfrak{C}_2$ as a sublattice and so $\mathfrak{C}_2 \in \mathbf{K}$. By Corollary II.1.20, $\mathbf{Equ}(\mathfrak{C}_2) = \mathbf{D}$ the class of all distributive lattices. Thus $\mathbf{K} \supseteq \mathbf{D}$. We have just verified that $\mathbf{D}$ is the only atom of the lattice of equational classes of lattices and every nonzero member contains $\mathbf{D}$.

Now let $\mathbf{K}$ be an equational class of lattices properly containing $\mathbf{D}$. Then there is a nondistributive lattice L in $\mathbf{K}$. By Theorem II.1.1, $\mathfrak{N}_5$ or $\mathfrak{M}_3$ is a sublattice of L, hence $\mathfrak{N}_5$ or $\mathfrak{M}_3 \in \mathbf{K}$. Set

$$\mathbf{N}_5 = \mathbf{Equ}(\mathfrak{N}_5),$$

$$\mathbf{M}_3 = \mathbf{Equ}(\mathfrak{M}_3).$$

Then $\mathbf{K} \supseteq \mathbf{N}_5$ or $\mathbf{K} \supseteq \mathbf{M}_3$. Thus $\mathbf{D}$ is covered by exactly two equational classes, $\mathbf{N}_5$ and $\mathbf{M}_3$, and every equational class properly containing $\mathbf{D}$ contains $\mathbf{N}_5$ or $\mathbf{M}_3$, see Figure 1.

The correspondence set up in Theorem 1.2 between equational classes of lattices and fully invariant congruences of $F(\aleph_0)$ is an isomorphism between the dual of the lattice of equational classes of lattices and the lattice of fully invariant congruences of $F(\aleph_0)$.

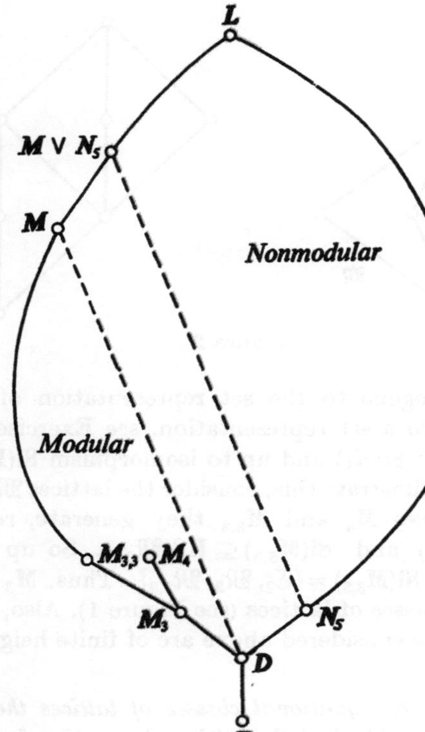

Figure 1

Theorem 1. ***The lattice of equational classes of lattices is distributive and dually algebraic. The dually compact elements are exactly those equational classes that can be defined by finitely many identities.***

Proof. Observe that an identity $p = q$ corresponds to forming the smallest fully invariant congruence under which $p(u_0, u_1, \ldots) \equiv q(u_0, u_1, \ldots)$. Thus the compact fully invariant congruence relations correspond to finite sets of identities. ●

The distributivity of the lattice also follows from the following formula (B. Jónsson [1967]):

Theorem 2. ***Let $\mathbf{K}_0$ and $\mathbf{K}_1$ be equational classes of lattices. Then,***

$$\mathbf{Si}(\mathbf{K}_0 \vee \mathbf{K}_1) = \mathbf{Si}(\mathbf{K}_0) \cup \mathbf{Si}(\mathbf{K}_1).$$

Proof. Apply Corollary 1.10 to $\mathbf{K} = \mathbf{K}_0 \cup \mathbf{K}_1$: if $L \in \mathbf{K}_0 \vee \mathbf{K}_1$ and L is subdirectly irreducible, then $L \in \mathbf{HSP}_{\mathrm{U}}(\mathbf{K}_0 \cup \mathbf{K}_1)$, that is, $L \in \mathbf{HS}(L_1)$, where L_1 is an ultra product of lattices from $\mathbf{K}_0 \cup \mathbf{K}_1$. Let $L_1 = \Pi_{\mathcal{D}}(L_i \mid i \in I)$, where $L_i \in \mathbf{K}_0 \cup \mathbf{K}_1$. Set $I_j = \{i \mid i \in I$ and $L_i \in K_j\}$, for $j = 0$ and 1. Then $I_0 \cup I_1 = I \in \mathcal{D}$, hence I_0 or $I_1 \in \mathcal{D}$. If $I_j \in \mathcal{D}$, then by Lemma 1.6, $L_1 \in \mathbf{P}_{\mathrm{U}}(\mathbf{K}_j) = \mathbf{K}_j$. Thus we conclude that $L_1 \in \mathbf{K}_0$ or $L_1 \in \mathbf{K}_1$. ●

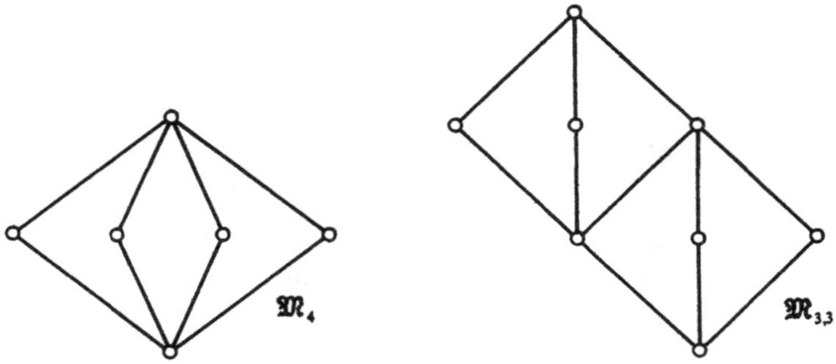

Figure 2

Thus $\mathbf{K} \to \mathbf{Si}(\mathbf{K})$ is analogous to the set representation of distributive lattices. (It is easy to turn this into a set representation, see Exercise 2.) In particular, we should note that if $\mathbf{Si}(\mathbf{K}_0) \subset \mathbf{Si}(\mathbf{K}_1)$ and up to isomorphism $\mathbf{Si}(\mathbf{K}_1)$ has only one more member, then $\mathbf{K}_0 \prec \mathbf{K}_1$. To illustrate this, consider the lattices $\mathfrak{M}_4$ and $\mathfrak{M}_{3,3}$ of Figure 2, and the equational classes $\mathbf{M}_4$ and $\mathbf{M}_{3,3}$ they generate, respectively. By Corollary 1.11, $\mathbf{Si}(\mathbf{M}_4) \subseteq \mathbf{HS}(\mathfrak{M}_4)$ and $\mathbf{Si}(\mathbf{M}_{3,3}) \subseteq \mathbf{HS}(\mathfrak{M}_{3,3})$. So up to isomorphic copies, $\mathbf{Si}(\mathbf{M}_4) = \{\mathfrak{C}_2, \mathfrak{M}_3, \mathfrak{M}_4\}$ and $\mathbf{Si}(\mathbf{M}_{3,3}) = \{\mathfrak{C}_2, \mathfrak{M}_3, \mathfrak{M}_{3,3}\}$. Thus, $\mathbf{M}_3 \prec \mathbf{M}_4$ and $\mathbf{M}_3 \prec \mathbf{M}_{3,3}$ in the lattice of equational classes of lattices (see Figure 1). Also, $\mathbf{M}_3 = \mathbf{M}_4 \wedge \mathbf{M}_{3,3}$.

All the equational classes considered above are of finite height.

Lemma 3. *The collection of equational classes of lattices that can be generated by a single finite lattice is an ideal of the lattice of equational classes of lattices. This ideal contains only elements of finite height.*

Proof. If the equational class $\mathbf{K}$ is generated by a finite lattice L, then by Corollary 1.11, up to isomorphism $\mathbf{Si}(\mathbf{K})$ is a finite set of finite lattices. Thus if $\mathbf{K}_0 \subset \mathbf{K}$, then up to isomorphism $\mathbf{Si}(\mathbf{K}_0)$ must be a subset of this finite set, hence there are only finitely many such $\mathbf{K}_0$. All the statements of this corollary now follow immediately. ●

From the observations made above it should be clear that the join-irreducible elements of the lattice of equational classes of lattices are connected with equational classes generated by a single subdirectly irreducible lattice. The following result states a number of connections of this sort (R. N. McKenzie [1972]):

Theorem 4. *Let $\mathbf{K}$ be an element of the lattice of equational classes of lattices.*

(i) *If* $[\mathbf{K})$ *is a* completely prime dual ideal *(that is,* $\bigvee(\mathbf{K}_i \mid i \in I) \in [\mathbf{K})$ *implies that* $\mathbf{K}_i \in [\mathbf{K})$, *for some* $i \in I$*), then* $\mathbf{K}$ *can be generated by a finite subdirectly irreducible lattice.*

(ii) *If* $\mathbf{K}$ *can be generated by a finite subdirectly irreducible lattice, then* $\mathbf{K}$ *is completely join-irreducible.*

(iii) *If* $\mathbf{K}$ *is completely join-irreducible, then* $\mathbf{K}$ *can be generated by a subdirectly irreducible lattice.*

(iv) *If* $\mathbf{K}$ *can be generated by a subdirectly irreducible lattice, then* $\mathbf{K}$ *is join-irreducible.*

Proof.

(i) Let $\mathbf{K}_n$ denote the equational class generated by the partition lattice on an n-element set. By Corollary IV.4.6, $\bigvee(\mathbf{K}_n \mid n = 1, 2, 3, \ldots) = \mathbf{L} \supseteq \mathbf{K}$; since $[\mathbf{K})$ is completely prime, $\mathbf{K}_n \supseteq \mathbf{K}$ for some integer n. Thus, by Lemma 3, $\mathbf{K}$ can be generated by finitely many finite subdirectly irreducible lattices. Since $\mathbf{K}$ is join-irreducible, $\mathbf{K}$ can be generated by a single finite subdirectly irreducible lattice. ◑

(ii) If $\mathbf{K}$ is generated by the finite subdirectly irreducible lattice L and $\mathbf{K} = \mathbf{K}_0 \vee \mathbf{K}_1$, then by Theorem 2, $L \in \mathbf{Si}(\mathbf{K}_0)$ or $L \in \mathbf{Si}(\mathbf{K}_1)$ implying $\mathbf{K} = \mathbf{K}_0$ or $\mathbf{K} = \mathbf{K}_1$. Thus $\mathbf{K}$ is join-irreducible. By Lemma 3, $\mathbf{K}$ is of finite height, hence $\mathbf{K}$ is completely join-irreducible. ◑

(iii) Any equational class $\mathbf{K}$ is of the form $\mathbf{K} = \bigvee(\mathbf{K}_0 \mid \mathbf{K}_0 \subseteq \mathbf{K}$ and $\mathbf{K}_0$ is generated by a subdirectly irreducible lattice). Thus if $\mathbf{K}$ is completely join-irreducible, $\mathbf{K}$ can be generated by a subdirectly irreducible lattice. ◑

(iv) We proceed as in (ii) by reference to Theorem 2. ●

The converse statements of (i)–(iv) all fail, see R. N. McKenzie [1972] and Exercises.

Figure 1 can also be used to illustrate the very important concept of *splitting* due to R. N. McKenzie [1972]. A pair of equational classes $\langle \mathbf{K}_0, \mathbf{K}_1 \rangle$ is said to be *splitting* iff for every equational class $\mathbf{K}_2$, either $\mathbf{K}_2 \subseteq \mathbf{K}_0$ or $\mathbf{K}_1 \subseteq \mathbf{K}_2$. For instance, $\langle \mathbf{M}, \mathbf{N}_5 \rangle$ is a splitting pair. ($\langle \mathbf{L}, \mathbf{T} \rangle$ is a trivial splitting pair.) Equivalently, $(\mathbf{K}_0]$ and $[\mathbf{K}_1)$ are prime, $\mathbf{K}_0 \not\supseteq \mathbf{K}_1$, and $(\mathbf{K}_0] \cup [\mathbf{K}_1) = (\mathbf{L}]$. Obviously, $\mathbf{K}_0$ determines $\mathbf{K}_1$, and conversely. Since $[\mathbf{K}_1)$ is a completely prime dual ideal, $\mathbf{K}_1$ can be generated by a finite subdirectly irreducible lattice. Finite subdirectly irreducible lattices that arise this way are called *splitting lattices* and they are characterized in R. N. McKenzie [1972].

How big is the lattice of equational classes of lattices? Since there are only $\aleph_0$ identities, there are at most $2^{\aleph_0}$ equational classes. Now we shall show by construction that there are exactly $2^{\aleph_0}$ equational classes of lattices.

Let Π be the set of prime numbers and, for $p \in \Pi$, let L_p be the subspace lattice of the projective plane coordinatized by the p-element field (that is, the Galois field, $GF(p)$). For a subset $S \subseteq \Pi$ set

$$\mathbf{K}(S) = \mathbf{Equ}(\{L_i \mid i \in S\}).$$

We claim that S can be recovered from $\mathbf{K}(S)$, in fact,

$$L_p \in \mathbf{K}(S) \quad \text{iff} \quad p \in S.$$

Obviously, if $p \in S$, then $L_p \in \mathbf{K}(S)$. Now let $L_p \in \mathbf{K}(S)$ and $p \notin S$. Since L_p is subdirectly irreducible we can apply Corollary 1.10: $L_p \in \mathbf{HSP}_{\mathrm{U}}(\{L_i \mid i \in S\})$, that is, $L_p \in \mathbf{HS}(L)$, where $L = \prod_{\mathscr{D}} (L_i \mid i \in I)$, $\mathscr{D}$ is prime over I, and each L_i, $i \in I$, is an L_j with $j \in S$. Each L_i is a complemented modular lattice of length three, in which any two atoms are perspective. Since these properties can be expressed by (first-order) sentences, by Exercise 1.26, L has the same property and so, by the results of Section IV.5, L is the subspace lattice of a nondegenerate projective plane. Each L_i satisfies Desargues' Theorem and, by Theorem IV.5.8, this property can be expressed by a sentence. Thus L satisfies Desargues' Theorem and, by the Coordinatization Theorem, L can be coordinatized by a division ring D. We assumed that $p \notin S$ so each L_i, $i \in I$,

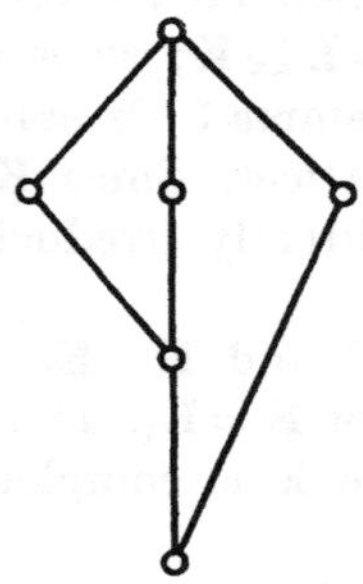

Figure 3

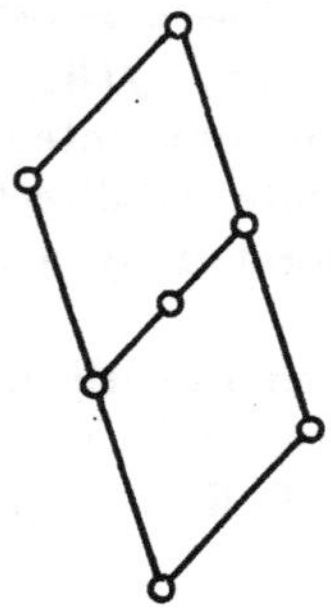

Figure 4

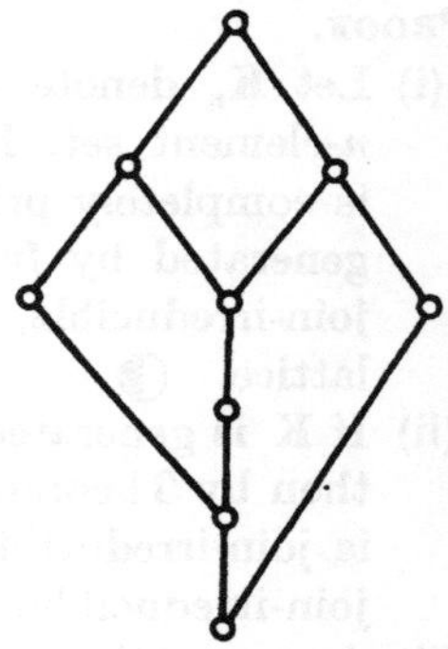

Figure 5

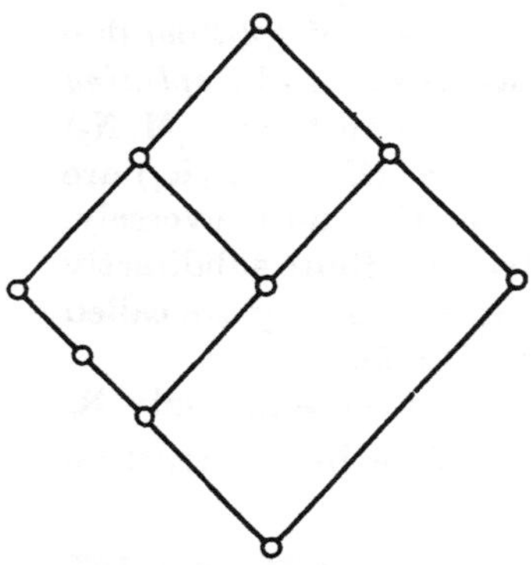

Figure 6

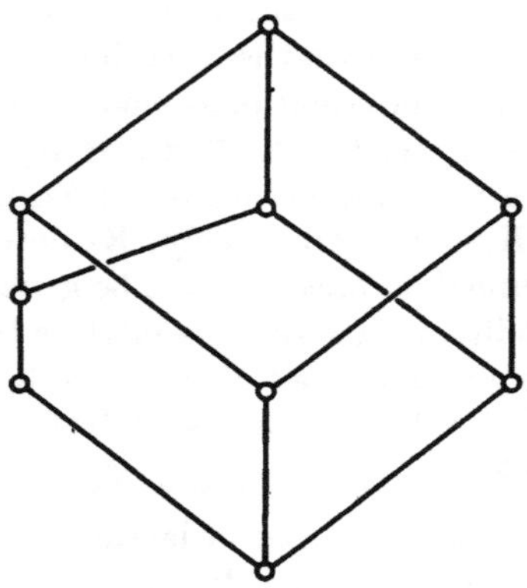

Figure 7

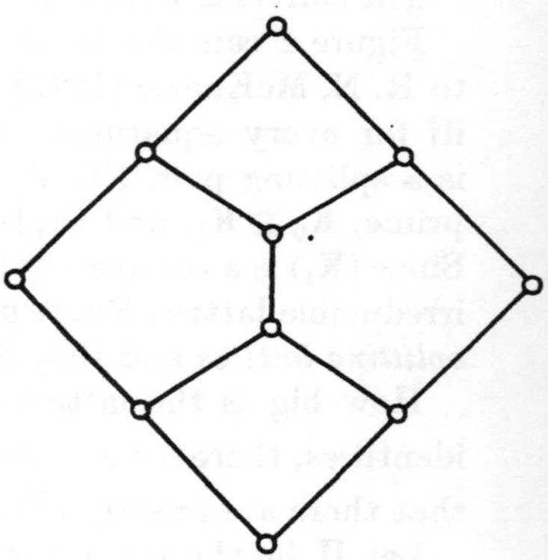

Figure 8

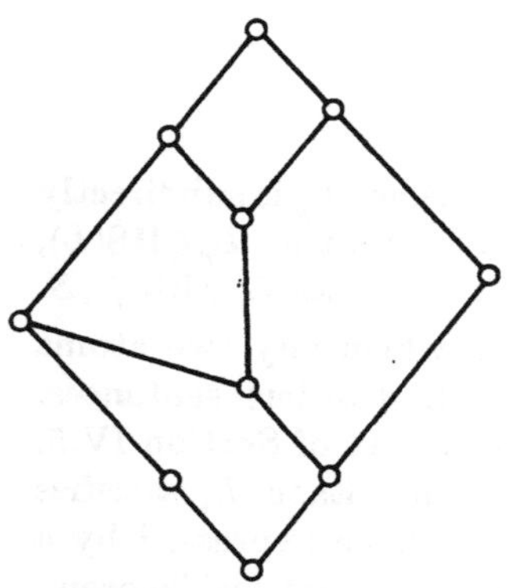

Figure 9

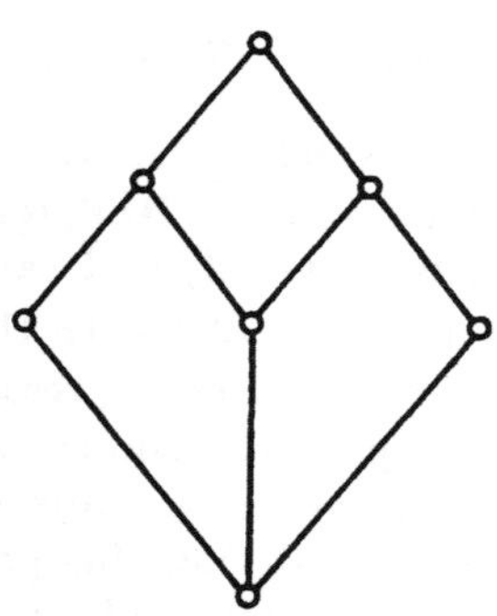

Figure 10

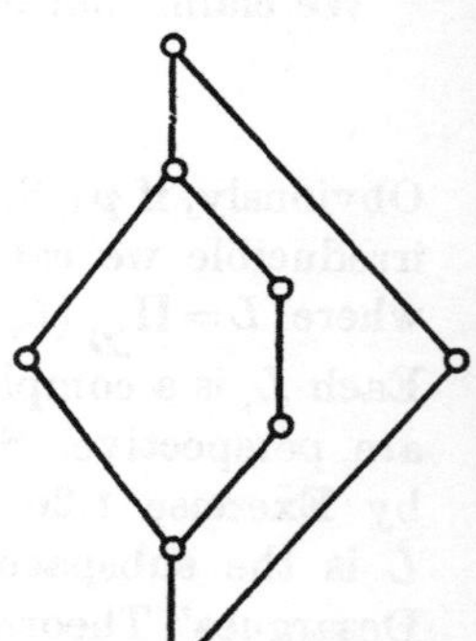

Figure 11

is coordinatized by a division ring not of characteristic p. This again can be expressed by a formula, hence D is not of characteristic p.

$L_p \in \mathbf{HS}(L)$, that is, L has a sublattice L' such that L_p is a homomorphic image of L'. Since both L and L_p are of length three and a proper homomorphic image of a modular lattice of length three is of length less than three, we conclude that we can assume that $L_p = L'$, that is, L_p is a sublattice of L. This is a clear contradiction: we can introduce $x+y$ for points x and y in L_p using only elements of L_p; thus for points x and y of L_p, $x+y$ in L_p is the same as $x+y$ in L. But in L_p we have $p \cdot x = 0$ while in L, $p \cdot x \neq 0$ if $x \neq 0$, which is impossible.

Thus there are at least as many equational classes of lattices as there are subsets of Π, which number $2^{\aleph_0}$.

Theorem 5. *There are $2^{\aleph_0}$ equational classes of (modular) lattices.*

This result was proved by K. A. Baker [1969a] (on whose example the above discussion was based), R. N. McKenzie [1970] (without modularity), and R. Wille [1972].

The following method of constructing equational classes of lattices is due to R. Wille [1972].

Let $\mathcal{P}$ be a set of finite posets. We denote by $\mathbf{Equ}(\mathcal{P})$ the class of all lattices that do *not* contain an isomorphic copy of a member of $\mathcal{P}$ as a primitive subset.

Theorem 6 (R. Wille [1972]). *For any set of finite posets $\mathcal{P}$, $\mathbf{Equ}(\mathcal{P})$ is an equational class of lattices.*

Proof. This is clear from Exercises III.1.22 and III.1.23. ●

The reader should verify that the construction in the proof of Theorem 5 could be verified using Theorem 6.

Exercises

1. Consider the lattices of Figures 3–11 and their duals (fifteen in all). Show that each one generates an equational class covering $\mathbf{N}_5$ (R. N. McKenzie [1972]).
2. Let A be a fixed countable set. For an equational class $\mathbf{K}$, let $\mathbf{Si}_A(\mathbf{K})$ denote the set of all subdirectly irreducible lattices L in $\mathbf{K}$ satisfying $L \subseteq A$. Prove that $\mathbf{K} \to \mathbf{Si}_A(\mathbf{K})$ is a set representation of the lattice of all equational classes of lattices. (Hint: $F_{\mathbf{K}}(\aleph_0)$ can be recovered from $\mathbf{Si}_A(\mathbf{K})$.)
3. Does the representation of Exercise 2 preserve infinite meets and joins of equational classes?
4. Prove that an equational class $\mathbf{K}$ of lattices can be generated by finitely many finite lattices iff $\mathbf{K}$ can be generated by a single finite lattice, which, in turn, is equivalent to the statement that all subdirectly irreducible lattices in $\mathbf{K}$ are finite and there are only finitely many nonisomorphic subdirectly irreducible lattices.
5. Show that $\mathbf{L}$ can be generated by a subdirectly irreducible lattice and $\mathbf{L}$ is completely join-reducible.

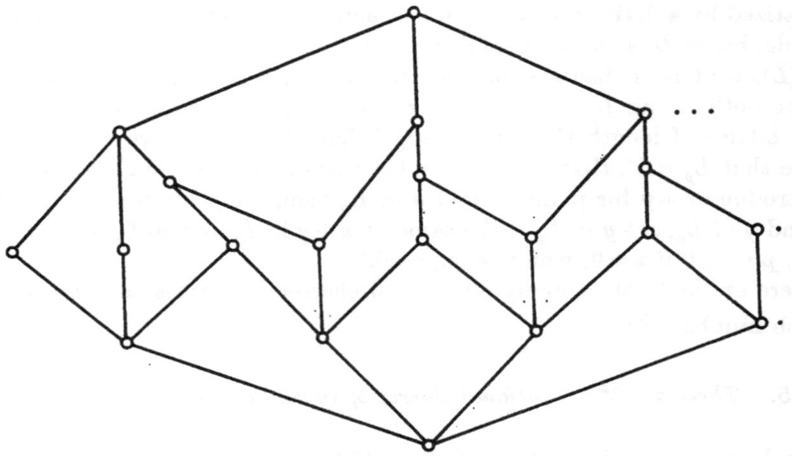

Figure 12

6. Prove that the equational class generated by the lattice of Figure 12 is join-irreducible but it cannot be generated by a single subdirectly irreducible lattice (R. N. McKenzie [1972]).
7. Let $\mathbf{K}_0$ and $\mathbf{K}_1$ be the equational classes of lattices generated by all projective planes satisfying Desargues' Theorem and Pappus' Theorem, respectively. Prove that $\mathbf{K}_0$ and $\mathbf{K}_1$ have the same finite members (K. A. Baker [1969a]).
8. Find an equational class of lattices that is not generated by its finite members (K. A. Baker [1969a] and R. Wille [1972]).
9. Find a sublattice of the lattice of equational classes of lattices that is isomorphic to the lattice of all subsets of a countable set (K. A. Baker [1969a]).
10. Let $\mathbf{K}$ be an equational class of lattices. If $\mathbf{K} \neq \mathbf{L}$, then $\mathbf{K} \prec \mathbf{K}_0$ for some equational class $\mathbf{K}_0$ (B. Jónsson [1967]). (Hint: Consider $\mathbf{K} \vee \mathbf{Equ}(L)$, where L is a finite lattice not in $\mathbf{K}$.)
11. Prove that $\mathbf{L}$ is join-irreducible (B. Jónsson [1967]). (Hint: If $\mathbf{K} \subset \mathbf{L}$, then the partition lattice on an infinite set does not belong to $\mathbf{K}$.)
12. There is no equational class of lattices covered by $\mathbf{L}$ (B. Jónsson [1967]).
13. Show that every proper interval of the lattice of equational classes of lattices contains a prime interval.
14. Let $\mathbf{K}_0$ and $\mathbf{K}_1$ be equational classes of lattices. If $\mathbf{K}_0 \subset \mathbf{K}_1$ and $\mathbf{K}_0$ can be defined by finitely many identities, then there exists an equational class $\mathbf{K}$ satisfying $\mathbf{K}_0 \prec \mathbf{K} \subseteq \mathbf{K}_1$. (Hint: use Theorems 1 and II.3.13.)
15. Prove that in Exercise 14 we cannot require that $\mathbf{K}_0 \subseteq \mathbf{K} \prec \mathbf{K}_1$.
16. Prove that $\mathbf{SiEqu}(\mathcal{P})$ consists of those subdirectly irreducible lattices in which no $P \in \mathcal{P}$ can be embedded.
17. Let $\mathcal{P} = \{\mathfrak{C}_5\}$. Describe $\mathbf{Equ}(\mathcal{P})$.
18. A lattice L is said to be of *breadth* n iff n is the smallest integer with the property that for every finite $X \subseteq L$ there exists a $Y \subseteq X$ such that $|Y| \leq n$ and $\bigvee X = \bigvee Y$. Choose a finite set of finite posets $\mathcal{P}$ such that $\mathbf{Equ}(\mathcal{P})$ is generated by all subdirectly irreducible lattices of breadth at most n.
19. Let $\mathcal{P}$ consist of the n-element unordered poset. Describe $\mathbf{Equ}(\mathcal{P})$ using the concept of width.
*20. Prove that $\mathbf{N}_5$ contains all lattices of width two (O. T. Nelson, Jr. [1968]).

3. Finding Equational Bases

An *equational basis* of a class $\mathbf{K}$ is a set of identities Σ such that $\mathbf{Equ}(\mathbf{K}) = \mathbf{Mod}(\Sigma)$. This Σ is of special interest if it is *irredundant* (that is, $\mathbf{Equ}(\mathbf{K}) \neq \mathbf{Mod}(\Sigma_1)$ for any $\Sigma_1 \subset \Sigma$) or finite. There is not much one can say about the problem of finding equational bases in general. However, if $\mathbf{K}$ has some special properties, then there is hope for some meaningful results.

A *universal disjunction of equations*, or briefly UDE, is a sentence of the form (recall, $\vee$ stands for " or ")

$$(x_0) \cdots (x_{n-1})\ (f_0(x_0, \ldots, x_{n-1}) = g_0(x_0, \ldots, x_{n-1}) \vee \cdots$$
$$\vee f_{m-1}(x_0, \ldots, x_{n-1}) = g_{m-1}(x_0, \ldots, x_{n-1})),$$

where $f_0, \ldots, f_{m-1}$ and $g_0, \ldots, g_{m-1}$ are polynomials. In the sequel, we shall omit the quantifiers and we shall assume that $f_i \leq g_i$ holds in any lattice. Examples abound: every identity is a UDE. The following lemma yields examples of a different kind. (In the formula $\bigvee$ stands for the disjunction of the terms.)

Lemma 1. *Let $P = \{a_0, \ldots, a_{n-1}\}$ be a poset. Set $g_i = \bigvee(x_j \mid a_j \leq a_i)$ and*

$$\Phi(P)\colon \bigvee(g_i = g_i \vee g_k \mid a_k \nleq a_i).$$

Then $\Phi(P)$ is a UDE which holds for a lattice L iff L has no subset isomorphic (as a poset) with P.

Proof. If $Q = \{b_0, \ldots, b_{n-1}\} \subseteq L$ is isomorphic with P and $b_i \to a_i$, $0 \leq i < n$, is an isomorphism, then setting $x_i = b_i$, $0 \leq i < n$, we find that $g_i(b_0, \ldots, b_{n-1}) = b_i$. Thus if $a_k \nleq a_i$ then $g_k \nleq g_i$; and so $g_i \neq g_i \vee g_k$. Since all the terms of $\Phi(P)$ fail, $\Phi(P)$ itself fails in L. Conversely, if $\Phi(P)$ fails in L, then there are elements $b_0, \ldots, b_{n-1} \in L$ such that $g_k(b_0, \ldots, b_{n-1}) \nleq g_i(b_0, \ldots, b_{n-1})$ whenever $a_k \nleq a_i$. Since $a_k \leq a_i$ obviously implies that $g_k(b_0, \ldots, b_{n-1}) \leq g_i(b_0, \ldots, b_{n-1})$ by the definition of g_i, we conclude that $a_i \to g_i(b_0, \ldots, b_{n-1})$ is an isomorphism of P with

$$\{g_0(b_0, \ldots, b_{n-1}), \ldots, g_{n-1}(b_0, \ldots, b_{n-1})\}. \quad \bullet$$

Another important UDE is

$$\Phi(n)\colon \bigvee(x_i = x_j \mid 0 \leq i < j \leq n)$$

which holds for L iff $|L| \leq n$. (To follow the inequality convention introduced above we should replace "$x_i = x_j$" by "$x_i \wedge x_j = x_i \vee x_j$".)

Now let us take an arbitrary UDE:

$$\Phi\colon \bigvee(f_i(x_0, \ldots, x_{n-1}) = g_i(x_0, \ldots, x_{n-1}) \mid 0 \leq i < m)$$

and consider the following statement:

$S(\Phi)$: for any integer k, the polynomial

$$r_k = p_{2m}(x, p_k(g_0(x_0, \ldots, x_{n-1}), y_0^0, \ldots, y_{k-1}^0), p_k(f_0(x_0, \ldots, x_{n-1}), y_0^0, \ldots, y_{k-1}^0), \ldots,$$
$$p_k(g_{m-1}(x_0, \ldots, x_{n-1}), y_0^{m-1}, \ldots, y_{k-1}^{m-1}), p_k(f_{m-1}(x_0, \ldots, x_{n-1}), y_0^{m-1}, \ldots, y_{k-1}^{m-1}))$$

does not depend on x, where p_k is the polynomial introduced in Lemma III.1.1.

In r_k the variables are $x, x_0, \ldots, x_{n-1}$ and y_j^i, $0 \leq i < m$, $0 \leq j < k$.

We claim that $S(\Phi)$ holds in a subdirectly irreducible lattice L iff Φ does.

Let Φ hold in L. Substituting $x_i = a_i$ for $0 \leq i < n$, we obtain the elements $f_i(a_0, \ldots, a_{n-1})$ and $g_i(a_0, \ldots, a_{n-1})$ for $0 \leq i < m$. By Φ, $f_i(a_0, \ldots, a_{n-1}) = g_i(a_0, \ldots, a_{n-1})$ for some i. Then two successive elements substituted into p_{2m} agree and so, by a trivial property of p_{2m} (see Exercise III.1.8), r_k does not depend on x, verifying $S(\Phi)$.

Now let Φ fail in L. Then there exist $a_0, \ldots, a_{n-1} \in L$ such that $f_i(a_0, \ldots, a_{n-1}) < g_i(a_0, \ldots, a_{n-1})$ for $0 \leq i < m$. Since L is subdirectly irreducible there are, by Corollary III.1.11, $a, b \in L$ satisfying $b < a$ and $a/b \approx_w g_i(a_0, \ldots, a_{n-1})/f_i(a_0, \ldots, a_{n-1})$. Thus for a suitable integer k and elements $c_0^i, \ldots, c_{k-1}^i \in L$ $(0 \leq i < m)$

$$p_k(g_i(a_0, \ldots, a_{n-1}), c_0^i, \ldots, c_{k-1}^i) = a,$$
$$p_k(f_i(a_0, \ldots, a_{n-1}), c_0^i, \ldots, c_{k-1}^i) = b.$$

Hence r_k with $x_i = a_i$ $(0 \leq i < n)$ and $y_j^i = c_j^i$ $(0 \leq i < m$ and $0 \leq j < k)$ takes the form

$$r_k = p_{2k}(x, a, b, \ldots, a, b).$$

But it is evident that $r_k = x$ on the interval $[b, a]$ (see Exercise III.1.9) and so r_k is dependent on x. Thus $S(\Phi)$ fails in L.

The statement that r_k does not depend on x is equivalent to the identity:

$$\varepsilon_k: r_k(x, \ldots) = r_k(z, \ldots)$$

where z is a variable distinct from x, x_i, and y_j^i. Set $\Sigma(\Phi) = \{\varepsilon_k \mid k = 1, 2, \ldots\}$. Now we are ready to state a solution to our problem:

Theorem 1 (K. A. Baker [1971]). *Let Φ be a* UDE. *Then $\Sigma(\Phi)$ is an equational basis for* Mod(Φ). *For a set Ω of* UDE-*s*, $\bigcup(\Sigma(\Phi) \mid \Phi \in \Omega)$ *is an equational basis for* Mod (Ω).

Proof. Let $\mathbf{K}_0$ be the equational class defined by $\bigcup(\Sigma(\Phi) \mid \Phi \in \Omega)$ and let $\mathbf{K}_1$ be the equational class generated by Mod(Ω). If $L \in \mathbf{Si}(\mathbf{K}_0)$, then by the discussion above $L \in \text{Mod}(\Omega) \subseteq \mathbf{K}_1$. Hence $\mathbf{K}_0 \subseteq \mathbf{K}_1$. Conversely, if $L \in \mathbf{Si}(\mathbf{K}_1)$, then by Corollary 1.10, $L \in \mathbf{HSP}_U(\text{Mod}(\Omega))$. But Ω is preserved under ultra products (Exercise 1.26) and, obviously, by the formation of sublattices and homomorphic images. Hence $L \in \text{Mod}(\Omega)$ and so, by the discussion above, $L \in \text{Mod}(\bigcup(\Sigma(\Phi) \mid \Phi \in \Omega)) = \mathbf{K}_0$. Thus $\mathbf{K}_1 \subseteq \mathbf{K}_0$, proving that $\mathbf{K}_0 = \mathbf{K}_1$. ●

We can recast this proof using the following concept: a set or sequence of quotients of a lattice L is *(k-)bounded* iff there is a proper quotient weakly projective (in k steps) into each quotient. Corollary III.1.11 and Exercise III.1.21 (slightly sharpened) then read:

Lemma 2.
(i) *In a subdirectly irreducible lattice any finite set of proper quotients is bounded.*
(ii) *Let φ be a homomorphism of the lattice L onto L', let $a_0/b_0, \ldots, a_{n-1}/b_{n-1}$ be quotients of L, and let $a_0\varphi/b_0\varphi, \ldots, a_{n-1}\varphi/b_{n-1}\varphi$ be k-bounded. Then $a_1/b_1, \ldots, a_n/b_n$ is $(k+2)$-bounded.*

Now for a UDE Φ: $\bigvee f_i = \overset{\bullet}{g_i}$, let $\alpha_k(\Phi) = \alpha_k$ denote the sentence stating that for no substitution is $g_0/f_0, \ldots, g_{m-1}/f_{m-1}$ k-bounded. We can write this down simply by requiring that if $u/v \overset{k}{\approx}_w g_i/f_i$ for all i, then $u = v$. Then observe:

Lemma 3.
(i) *Φ implies α_k and α_k implies α_t for any $t \leq k$.*
(ii) *If all α_k, $k = 1, 2, \ldots$, hold in a subdirectly irreducible lattice, then so does Φ.*
(iii) *α_k is preserved under the formation of direct products and sublattices.*
(iv) *For any set Ω of* UDE-*s*,

$$\mathbf{Equ}(\mathbf{Mod}(\Omega)) = \mathbf{Mod}(\{\alpha_k(\Phi) \mid \Phi \in \Omega \quad \textit{and} \quad k = 1, 2, \ldots\}).$$

Proof.
(i) and (iii) are trivial. ◑
(ii) restates Lemma 2(i). ◑
(iv) follows from Corollary 1.10 and (iii). ●
The reader should have no difficulty relating the sentence α_k to the identity ε_k.
We can use these ideas to give a simple proof of

Theorem 4 (R. N. McKenzie [1970]). *Any finite lattice has a finite equational basis.*

Remark. The present proof is due to C. Herrmann [1973] which is based on K. A. Baker [1974].

Proof. Let L have n elements. If Σ is a finite equational basis for $\mathbf{Mod}(\Phi(n))$ (the class of at most n element lattices), then we can easily find a finite equational basis for L: let A be a finite set of (up to isomorphism) all finite lattices N satisfying $N \notin \mathbf{Equ}(L)$ and $|N| \leq n$; for each $N \in A$ choose an identity ε_N holding in L but not in N; then $\Sigma \cup \{\varepsilon_N \mid N \in A\}$ is a finite equational basis for L.

Any lattice in the class $\mathbf{K} = \mathbf{Mod}(\Phi(n))$ has two properties: (i) it is defined by an at most n^2-termed UDE; (ii) every bounded set of at most n^2 quotients is n^2-bounded (because there are at most n^2 quotients in the lattice). Thus the following lemma completes the proof of Theorem 4:

Lemma 5. *Let* $\mathbf{K}$ *be a class of lattices and let m be an integer with the following two properties:*
(i) $\mathbf{K} = \mathbf{Mod}(\Sigma)$ *where Σ is a finite set of at most m-termed* UDE-*s*.

(ii) *There is an integer r such that in every subdirectly irreducible lattice $L \in \mathbf{K}$ every bounded set of m quotients is r-bounded.*

Then $\mathbf{K}$ *has a finite equational basis.*

PROOF. Let $\varrho_{m,r}$ denote the sentence expressing that any set of m quotients that is $(r+1)$-bounded is r-bounded. A trivial induction shows that $\varrho_{m,r}$ implies that any bounded set of m quotients is r-bounded.

Let $L \in \mathbf{Equ}(\mathbf{K})$. Then L is a subdirect product of subdirectly irreducible lattices L_i, $i \in I$, where each $L_i \in \mathbf{K}$ by Corollary 1.10, since UDE-s are preserved under the formation of ultra products, sublattices, and homomorphic images. Let $c \neq d$ and $c/d \approx_w a_j/b_j$, for $0 \leq j < m$ in L. Then $c\varphi_i \neq d\varphi_i$ and $c\varphi_i/d\varphi_i \approx_w a_j/b_j$, for $0 \leq j < m$, in L_i for some $i \in I$, where φ_i is the projection of L onto L_i. By the second hypothesis, $\varrho_{m,r}$ holds in L_i and so, by Lemma 2(ii), $\{a_j/b_j \mid 0 \leq j < m\}$ is $(r+2)$-bounded. Thus $\varrho_{m,r+2}$ holds in any $L \in \mathbf{Equ}(\mathbf{K})$.

By Lemma 3(iv), $\mathbf{Equ}(\mathbf{K})$ is defined by $\{\alpha_k(\Phi) \mid \Phi \in \Omega \text{ and } k = 1, 2, \ldots\}$. Hence this set of sentences implies $\varrho_{m,r+2}$ and so by the Compactness Theorem (Exercise 1.29) finitely many $\alpha_k(\Phi)$ imply $\varrho_{m,r+2}$. Let $A = \{\alpha_k(\Phi) \mid \Phi \in \Omega \text{ and } k = 1, 2, \ldots, t\}$ imply $\varrho_{m,r+2}$ and let us further assume that $t \geq r+2$. We claim that A defines $\mathbf{Equ}(\mathbf{K})$. Indeed, A holds in $\mathbf{Equ}(\mathbf{K})$ by Lemma 3, hence $\mathrm{Mod}(A) \supseteq \mathbf{Equ}(\mathbf{K})$. Conversely, A includes $\alpha_{r+2}(\Phi)$ for any $\Phi \in \Omega$, and A implies $\varrho_{m,r+2}$. But $\alpha_{r+2}(\Phi)$ and $\varrho_{m,r+2}$ imply $\alpha_i(\Phi)$ for any i; thus A implies $\alpha_i(\Phi)$ for any i and any $\Phi \in \Omega$, proving that $\mathrm{Mod}(A) \subseteq \mathbf{Equ}(\mathbf{K})$.

We have proved that A is equivalent to $\mathrm{Id}(\mathbf{K})$ and A is finite, hence, by the Compactness Theorem, A is equivalent to some finite $\Sigma \subseteq \mathrm{Id}(\mathbf{K})$. Σ is a finite equational basis for $\mathbf{K}$. ●

The proof of Theorem 4 as exhibited above does not give a finite equational basis; it only proves that there is one. In R. N. McKenzie [1970] and in K. A. Baker [1974] more complicated arguments are presented which actually construct a finite equational basis.

In some cases a finite equational basis for a finite lattice can be found using the following method. Let L be a finite lattice. Let us assume that we have constructed the lattices $L_0, L_1, \ldots$ such that (i) $\mathbf{Equ}(L)$ is covered by each $\mathbf{Equ}(L_i)$ and (ii) if $\mathbf{K}$ is an equational class of lattices and $\mathbf{Equ}(L) \subset \mathbf{K}$, then $L_i \in \mathbf{K}$ for some i. Find lattice identities $\varepsilon_0, \ldots, \varepsilon_{n-1}$ which hold in L but for which ε_i fails in L_i. Then $\{\varepsilon_0, \ldots, \varepsilon_{n-1}\}$ is a finite equational basis for L. We shall illustrate this method with an example.

THEOREM 6. *In the lattice of all equational classes of lattices, $\mathbf{M}_3$ is covered by $\mathbf{M}_4$ and $\mathbf{M}_{3,3}$. If $\mathbf{K}$ is any equational class of modular lattices and $\mathbf{M}_3 \subset \mathbf{K}$, then $\mathbf{M}_4 \subseteq \mathbf{K}$ or $\mathbf{M}_{3,3} \subseteq \mathbf{K}$.*

REMARK. This result was proved in G. Grätzer [1966] under the additional hypothesis that $\mathbf{K}$ is generated by a finite lattice. This hypothesis was removed in B. Jónsson [1968] where the following corollary was also stated.

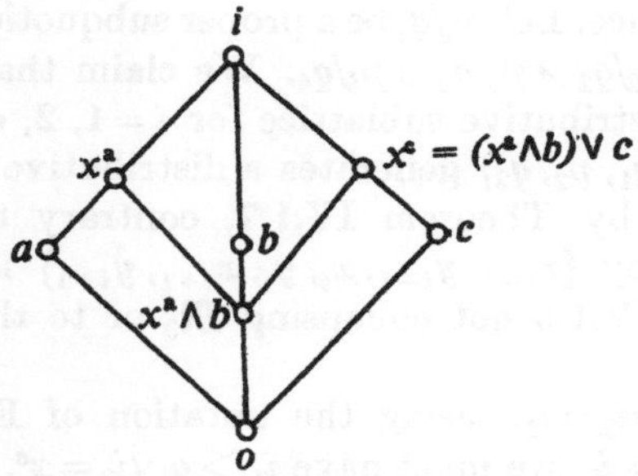

Figure 1

Corollary 7. *An equational basis for* $\mathbf{M}_3$ *is given by the modular identity and the identity*

$$x \wedge (y \vee z) \wedge (z \vee w) \wedge (w \vee y) \leq (x \wedge y) \vee (x \wedge z) \vee (x \wedge w).$$

By the discussion above, to prove this corollary it is sufficient to see that this identity holds in $\mathfrak{M}_3$ but fails in $\mathfrak{M}_4$ and $\mathfrak{M}_{3,3}$; this is left to the reader.

The proof of Theorem 6 is based on the following two lemmas.

Lemma 8. *Let L be a modular lattice and let $\{o, a, b, c, i\}$ be a diamond in L. Let $a \leq x^a < i$ and set* (see Figure 1)

$$x^b = b \vee (x^a \wedge c), \quad x^c = c \vee (x^a \wedge b), \quad \text{and} \quad o_1 = (x^a \wedge b) \vee (x^a \wedge c).$$

Then $\{o_1, x^a, x^b, x^c, i\}$ is a diamond.

Proof. Since $a \leq x^a < i$, $b \leq x^b < i$, and $c \leq x^c < i$, the relations $x^a \vee x^b = x^a \vee x^c = x^b \vee x^c = i$ obviously hold. Also,

$$x^a \wedge x^b = x^a \wedge (b \vee (x^a \wedge c)) = (x^a \wedge b) \vee (x^a \wedge c) = o_1,$$

and similarly, $x^a \wedge x^c = o_1$; finally,

$$\begin{aligned} x^b \wedge x^c &= (b \vee (x^a \wedge c)) \wedge (c \vee (x^a \wedge b)) \\ &= ((b \vee (x^a \wedge c)) \wedge c) \vee (x^a \wedge b) = (b \wedge c) \vee (x^a \wedge c) \vee (x^a \wedge b) = o \vee o_1 = o_1. \end{aligned}$$ ●

Lemma 9. *Let L be a subdirectly irreducible modular lattice which has no sublattice of which $\mathfrak{M}_{3,3}$ is a homomorphic image. Let $a/b \overset{n}{\approx} c/d$ in L such that a/b has no proper subquotient projective to a subquotient of c/d in less than n steps. Then $n \leq 3$.*

Proof. Let $x_0/y_0 \overset{4}{\approx} x_4/y_4$ such that no proper subquotient of x_0/y_0 is projective to a proper subquotient of x_4/y_4 in less than four steps. By duality we can assume that $x_0/y_0 \nearrow x_1/y_1 \searrow x_2/y_2 \nearrow x_3/y_3 \searrow x_4/y_4$. By (the proof of) Theorem IV.1.6, we can assume that this is a normal sequence. Let p_i/q_i be a proper subquotient of x_i/y_i, for $0 \leq i \leq 4$, such that $p_0/q_0 \nearrow p_1/q_1 \searrow p_2/q_2 \nearrow p_3/q_3 \searrow p_4/q_4$. We claim that $\{p_{i-1}, q_{i-1}, p_i, q_i, p_{i+1}, q_{i+1}\}$ cannot generate a distributive sublattice for $i = 1$, 2, or 3. Indeed, if one does, as for instance $\{p_0, q_0, p_1, q_1, p_2, q_2\}$ generates a distributive sublattice, then $p_0/q_0 \searrow p_0 \wedge p_2/q_0 \wedge q_2 \nearrow p_3/q_3 \searrow p_4/q_4$ by Theorem IV.1.7, contrary to the hypothesis. Thus the sublattice generated by $\{x_{i-1}, y_{i-1}, x_i, y_i, x_{i+1}, y_{i+1}\}$ is isomorphic to a homomorphic image of Figure IV.1.6 not collapsing $\mathfrak{M}_3$ or to the dual of such a lattice (see Figure 2).

Now we claim that $o_1 \vee i_2 = i_1$, using the notation of Figure 2. Let $o_1 \vee i_2 \neq i_1$. Since $i_1 > o_1$ and i_1 $(= x_1) > i_2$, we must have $i_1 > o_1 \vee i_2 = x^c$. Thus by Lemma 8, there are elements x^b, x^a such that $A_1 = \{x^b \wedge x^c, x^a, x^b, x^c, i_1\}$ is a diamond and $b_1 \leq x^b < i_1$, $a_1 \leq x^a < i_1$. Since $i_1/b_1 \searrow i_2/a_2$, there is an element y^a satisfying $i_1/x^b \searrow i_2/y^a$. Again by Lemma 8, we find the elements y^b and y^c satisfying $b_2 \leq y^b < i_2$, $c_2 \leq y^c < i_1$, and $A_2 = \{y^a \wedge y^b, y^a, y^b, y^c, i_2\}$ is a diamond. Thus $i_1/x^b \searrow i_2/y^a$ and since $i_2 \leq x^c < i_1$, we also have $x^c/x^b \wedge x^c \searrow i_2/y^a$. This last relation means that $A_1 \cup A_2 \cong \mathfrak{M}_{3,3}$ or if $x^c \neq i_2$ (and $x^b \wedge x^c \neq y^a$), then $(A_1 \cup A_2)/\Theta(x^c, i_2) \cong \mathfrak{M}_{3,3}$, a contradiction.

Therefore, $o_1 \vee i_2 = i_1$. By duality, $o_1 \wedge i_2 = o_2$, that is

$$i_1/o_1 \searrow i_2/o_2$$

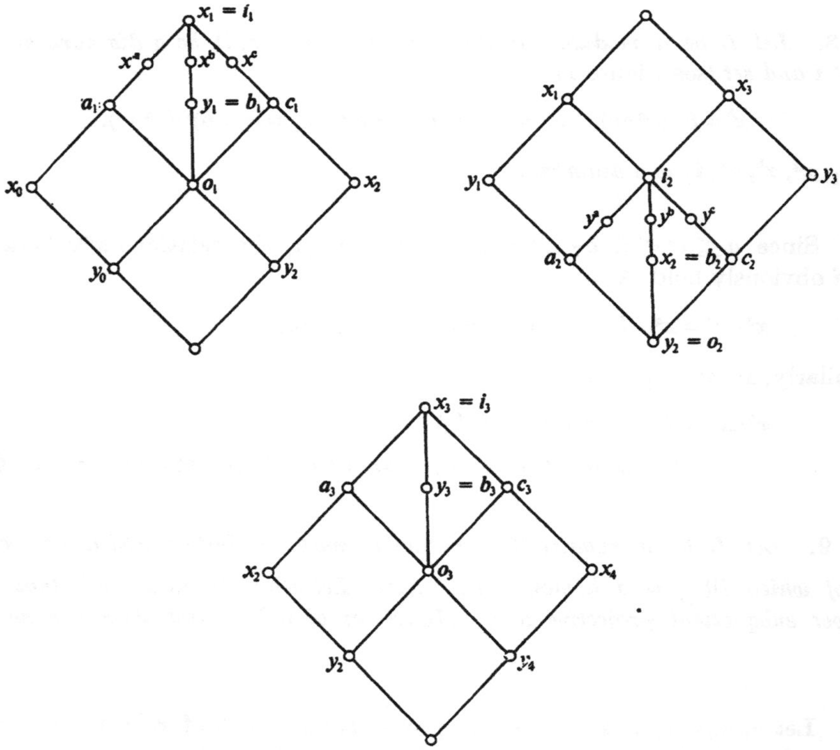

Figure 2

and similarly,

$$i_2/o_3 \nearrow i_3/o_3.$$

By normality $o_2 = o_1 \wedge o_3$, and by definition $i_2 = i_1 \wedge i_3$. Thus $\{o_1, o_2, o_3, i_1, i_2, i_3\}$ is contained in the sublattice generated by $\{o_1, i_1, o_3, i_3\}$ which is distributive by Theorem IV.1.13. We conclude, by Theorem IV.1.7, that

$$i_1/o_1 \nearrow i_1 \vee i_3/o_1 \vee o_3 \searrow i_3/o_3,$$

which trivially implies that

$$x_0/y_0 \nearrow x_1 \vee o_3/y_1 \vee o_3 \searrow x_4/y_4,$$

contrary to the hypothesis. ●

Proof of Theorem 6. Let $\mathbf{K}$ be an equational class of modular lattices such that $\mathbf{K} \supseteq \mathbf{M}_3$ and $\mathfrak{M}_4, \mathfrak{M}_{3,3} \notin \mathbf{K}$. In order to show that $\mathbf{K} = \mathbf{M}_3$, it is sufficient to verify that if $L \in \mathbf{Si}(\mathbf{K})$, then L is a sublattice of $\mathfrak{M}_3$. Assume to the contrary that $L \in \mathbf{Si}(\mathbf{K})$ but L is not a sublattice of $\mathfrak{M}_3$. Since $\mathfrak{M}_4, \mathfrak{M}_{3,3} \notin \mathbf{K}$, we must have $\mathfrak{M}_4, \mathfrak{M}_{3,3} \notin \mathbf{HS}(L)$.

Obviously, $|L| > 2$. If L is of length 2, then L must have $\mathfrak{M}_4$ as a sublattice, a contradiction. Thus L has a chain $c_0 < c_1 < c_2 < c_3$ of length three. Since L is subdirectly irreducible, $\Theta(c_0, c_1) \wedge \Theta(c_1, c_2) \neq \omega$ and so, by applying Theorem III.1.2 twice, we obtain a proper quotient x/y weakly projective into c_2/c_1 and into c_1/c_0. By Theorem IV.1.6 and by the symmetry of projectivity, there is a proper subquotient a/b of c_2/c_1 and a subquotient c/d of c_1/c_0 such that $a/b \approx c/d$. Choose these a, b, c, d so that n is minimal in $a/b \overset{n}{\approx} c/d$. Then a, b, c, d and L satisfy the conditions of Lemma 9 and therefore $n \leq 3$. We claim that $n = 3$. Indeed, in any lattice L we cannot have $a > b \geq c > d$ and $a/b \overset{n}{\approx} c/d$ with $n \leq 2$. (Proof. If $a/b \nearrow p/q \searrow c/d$, then $d = c \wedge q \geq c \wedge b = c$. If $a/b \searrow p/q \nearrow c/d$, then $a = b \vee p \leq b \vee c = b$. The case $n = 1$ is trivial.)

Now let $a/b = x_0/y_0 \sim x_1/y_1 \sim x_2/y_2 \sim x_3/y_3 = c/d$. If $a/b \nearrow x_1/y_1$, then $x_1 > y_1 \geq c > d$ and $x_1/y_1 \overset{2}{\approx} c/d$ which is impossible as noted in the previous paragraph. Thus $x_0/y_0 \searrow x_1/y_1 \nearrow x_2/y_2 \searrow x_3/y_3$.

Applying the same arguments to $\Theta(c_3, c_2) \wedge \Theta(a, b)$, we obtain $z_0/u_0 \searrow z_1/u_1 \nearrow z_2/u_2 \searrow z_3/u_3$, where z_0/u_0 and z_3/u_3 are proper subquotients of c_3/c_2 and a/b, respectively. By making the trivial replacements x_0/y_0 by z_3/u_3, x_1/y_1 by $x_1 \wedge z_3/x_1 \wedge u_3$, x_2/y_2 by $y_2 \vee (x_1 \wedge z_3)/y_2 \vee (x_1 \wedge u_3)$, and x_3/y_3 by $x_3 \wedge (y_2 \vee (x_1 \wedge z_3))/x_3 \wedge (y_2 \vee (x_1 \wedge u_3))$, we can assume that x_0/y_0 equals z_3/u_3. Thus by Theorem IV.1.7, there are diamonds $A_j = \{o_j, a_j, b_j, c_j, i_j\}$, $j = 0, 1$, such that $a/b \searrow i_0/a_0$ and $a/b \nearrow a_1/o_1$. We conclude that $a_1/o_1 \searrow i_0/a_0$ and so $A_0 \cup A_1$ is a sublattice of which $\mathfrak{M}_{3,3}$ is a homomorphic image, a contradiction. ●

Exercises

1. A *universal sentence* is a sentence Ψ of the form $(x_0) \cdots (x_{n-1})\, \Phi$, where there is in Φ no quantifier and no variable other than $x_0, \ldots, x_{n-1}$. Show that every universal sentence in which no negation or implication occurs is equivalent to a finite set of UDE-s.

2. Show that a UDE is preserved under the formation of sublattices and homomorphic images.
3. Are all UDE-s preserved under direct products?
4. Is every UDE equivalent to some $\Phi(P)$?
5. Let P be the n-element unordered set. Compare $\Phi(P)$ with $\Phi(n)$.
6. Let P be a finite poset. Show that **Equ**(**Mod**($\Phi(P)$)) = **Equ**($\{P\}$) as introduced in Section 2.
7. In the proof of Theorem 4 the statement is used that in an n-element lattice every bounded set of proper quotients is n^2-bounded. Can n^2 be improved in this statement?
8. Write out the formulas α_k and $\varrho_{m,r}$ to prove formally that these are indeed (first-order) formulas.
9. An algebra $\langle A; \wedge, \vee\rangle$ is called a *weakly associative lattice* (WA lattice) iff the following are satisfied: the two binary operations satisfy the idempotent, commutative, and absorption identities, and

$$x \leq z \quad \text{and} \quad y \leq z \quad \text{imply that} \quad x \vee y \leq z$$

and its dual hold, where $a \leq b$ means that $a = a \wedge b$ or equivalently, $a \vee b = b$ (E. Fried [1970] and H. L. Skala [1971]). Show that all the results of this section hold for WA lattices (K. A. Baker [1974]).
*10. Take the three-element WA-lattice $T = \{0, 1, 2\}$ defined by $0 \leq 1 \leq 2 \leq 0$. Find a finite equational basis for T (E. Fried and G. Grätzer [1973]).
*11. If L is a modular lattice of length n, then any bounded set of proper quotients is k-bounded, where $k \leq \left[\frac{3n}{2}\right] + 2$ (C. Herrmann [1973]). ($[x]$ stands for the largest integer $\leq x$.)
12. Let $\mathbf{M}^n$ denote the class of modular lattices of length at most n. Prove that $\mathbf{M}^n$ has a finite equational basis. (For $n = 2$ this is due to B. Jónsson [1968]; for $n = 3$ this is due to D. X. Hong [1972]. For general n this is due to K. A. Baker; a reference to this fact and a proof of this result based on Exercise 11 is due to C. Herrmann [1973].)
13. Show that a finite equational basis for $\mathbf{M}^2$, for notation see Exercise 12, is given by the modular identity and

$$x \wedge (y \vee (z \wedge u)) \vee (z \wedge u) \leq y \vee (x \wedge z) \vee (x \wedge u)$$

(B. Jónsson [1968]). (Hint: Use Lemma 9 and the reasoning in the proof of Theorem 6.)
14. Let L be a subdirectly irreducible lattice of length at most 3. Show that any bounded set of proper quotients is in fact 5-bounded (C. Herrmann [1973]).
15. Let $\mathbf{L}_3$ denote the class of all lattices of length at most 3. Prove that $\mathbf{L}_3$ has a finite equational basis (C. Herrmann [1973]).
16. Show that Lemma 8 can be derived from (and is implicit in) Theorem IV.1.7.
17. Using the notations of Lemma 8, project x^a into $[o, a]$, $[o, b]$, $[o, c]$ obtaining x_a, x_b, x_c, respectively. Show that $\{o, x_a, x_b, x_c, o_1\}$ is a diamond.
18. The results of this section depend heavily on Theorem 1.9. However, Theorems 4 and 6 use only Corollary 1.11. Show that the latter can be proved with no reference to ultra products, using only the primitive sets of Section III.1 (R. Wille [1972]).
*19. Let L be a subdirectly irreducible lattice of width not greater than four. Then either the lattice of Figure IV.3.4b or one of the eight lattices of Figures 3–10 is a homomorphic image of a sublattice of L (R. Freese [1972]).
20. Let $\mathbf{M}(4)$ denote the class of all modular lattices of width four. Prove that $\mathbf{M}(4)$ has a finite equational basis (R. Freese [1972]). (Hint: use Exercise 19.)

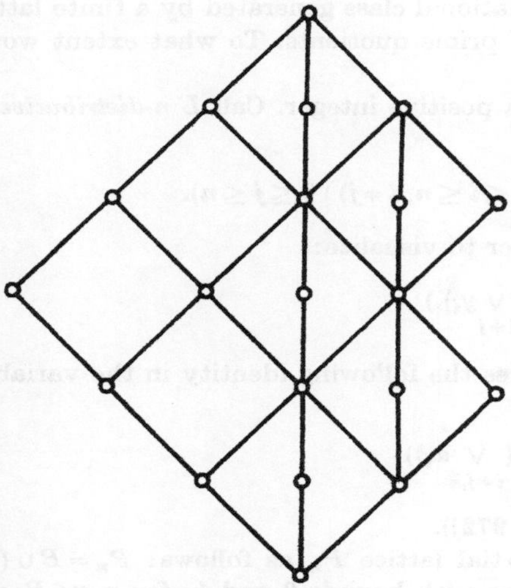

Figure 3

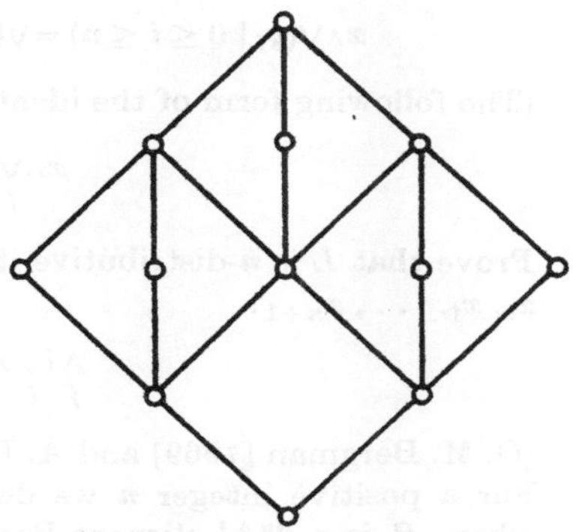

Figure 4

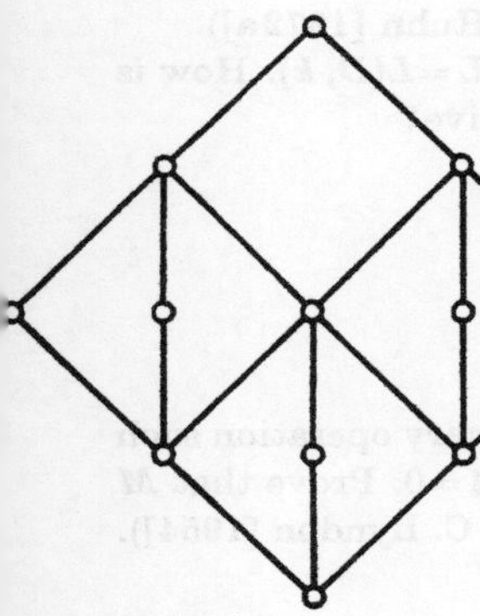

Figure 5

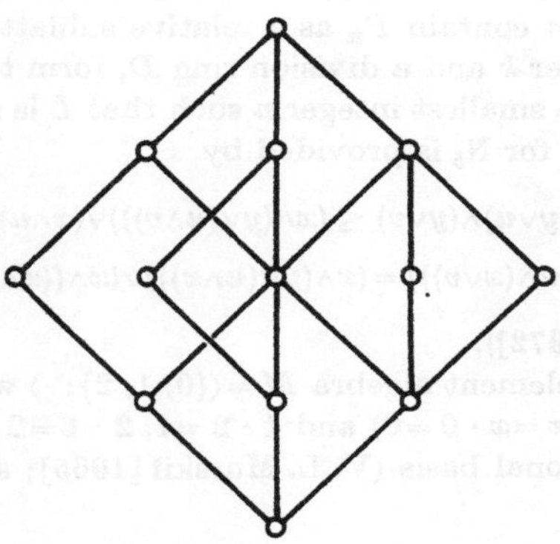

Figure 6

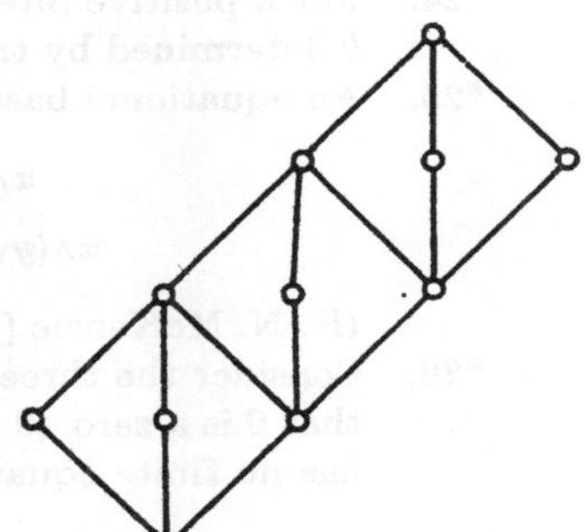

Figure 7

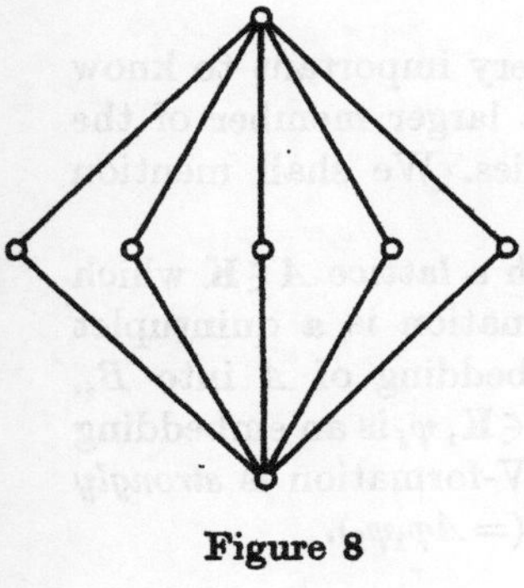

Figure 8

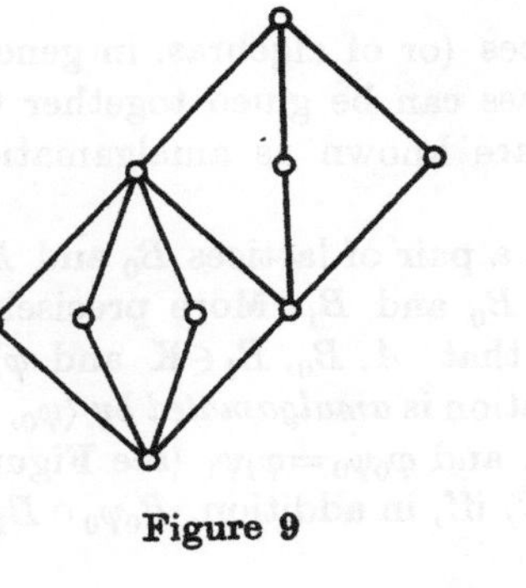

Figure 9

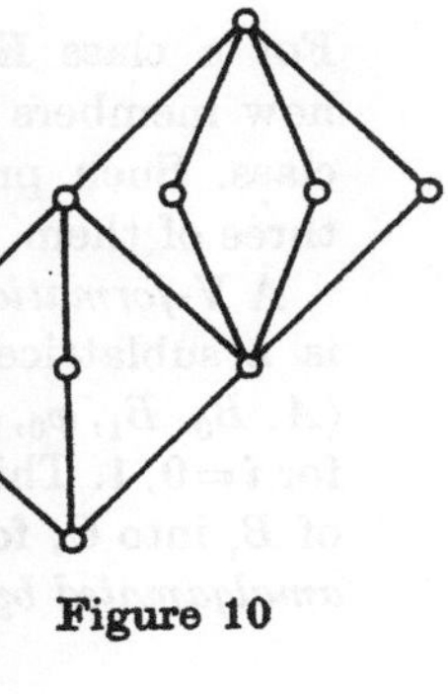

Figure 10

21. Show that to prove Theorem 6 for an equational class generated by a finite lattice it is sufficient to analyse projectivities of prime quotients. To what extent would this simplify the proof?

22. Let L be a modular lattice and let n be a positive integer. Call L *n-distributive* iff the following identity holds:

$$x \wedge \bigvee(y_i \mid 0 \leq i \leq n) = \bigvee(x \wedge \bigvee(y_i \mid 0 \leq i \leq n, i \neq j) \mid 0 \leq j \leq n).$$

(The following form of the identity is easier to visualize:

$$x \wedge \bigvee_i y_i = \bigvee_j (x \wedge \bigvee_{i \neq j} y_i).)$$

Prove that L is n-distributive iff it satisfies the following identity in the variables $x_0, x_1, \ldots, x_{n+1}$:

$$\bigwedge_j (\bigvee_i x_i) = \bigvee_k (\bigwedge_{j \neq k} (\bigvee_{i \neq j,k} x_i))$$

(G. M. Bergman [1969] and A. P. Huhn [1972]).

23. For a positive integer n we define a partial lattice P_n as follows: $P_n = B \cup \{w\}$, where B is a 2^{n+1}-element Boolean lattice with bounds 0 and 1; for $x, y \in B$, $x \wedge y$ and $x \vee y$ are defined as in B; $w \wedge 1 = 1 \wedge w = w$ and $w \wedge x = x \wedge w = 0$ for $x \in B - \{1\}$; $w \vee 0 = 0 \vee w = w$, $w \vee d = d \vee w = 1$ if d is a dual atom of B or if $d = 1$; $w \vee x$ and $x \vee w$ are not defined if $x \in B$ and $0 < h(x) < n$. Prove that a modular lattice L is n-distributive iff L does not contain P_n as a relative sublattice (A. P. Huhn [1972a]).

24. For a positive integer k and a division ring D, form the lattice $L = L(D, k)$. How is k determined by the smallest integer n such that L is n-distributive?

*25. An equational basis for $\mathbf{N}_5$ is provided by

$$x \wedge (y \vee u) \wedge (y \vee v) \leq (x \wedge (y \vee (u \wedge v))) \vee (x \wedge u) \vee (x \wedge v)$$

$$x \wedge (y \vee (u \wedge (x \vee v))) = (x \wedge (y \vee (u \wedge x))) \vee (x \wedge ((x \wedge y) \vee (u \wedge v)))$$

(R. N. McKenzie [1972]).

*26. Consider the three-element algebra $M = \langle\{0, 1, 2\}; \cdot\rangle$ with one binary operation such that 0 is a zero $(0 \cdot x = x \cdot 0 = 0)$ and $1 \cdot 2 = 1$, $2 \cdot 1 = 2 \cdot 2 = 2$, $1 \cdot 1 = 0$. Prove that M has no finite equational basis (V. L. Murskiĭ [1965]; see also R. C. Lyndon [1954]).

4. The Amalgamation Property

For a class $\mathbf{K}$ of lattices (or of algebras, in general) it is very important to know how members of the class can be glued together to obtain a larger member of the class. Such properties are known as amalgamation properties. We shall mention three of them.

A V-*formation in* $\mathbf{K}$ is a pair of lattices B_0 and B_1 in $\mathbf{K}$ with a lattice $A \in \mathbf{K}$ which is a sublattice of both B_0 and B_1. More precisely, a V-formation is a quintuplet $\langle A, B_0, B_1, \varphi_0, \varphi_1\rangle$ such that $A, B_0, B_1 \in \mathbf{K}$ and φ_i is an embedding of A into B_i, for $i = 0, 1$. This V-formation is *amalgamated by* $\langle\psi_0, \psi_1, C\rangle$ iff $C \in \mathbf{K}$, ψ_i is an embedding of B_i into C, for $i = 0, 1$, and $\varphi_0\psi_0 = \varphi_1\psi_1$ (see Figure 1). The V-formation is *strongly amalgamated by* $\langle\psi_0, \psi_1, C\rangle$ iff, in addition, $B_0\psi_0 \cap B_1\psi_1 = A\varphi_0\psi_0 (= A\varphi_1\psi_1)$.

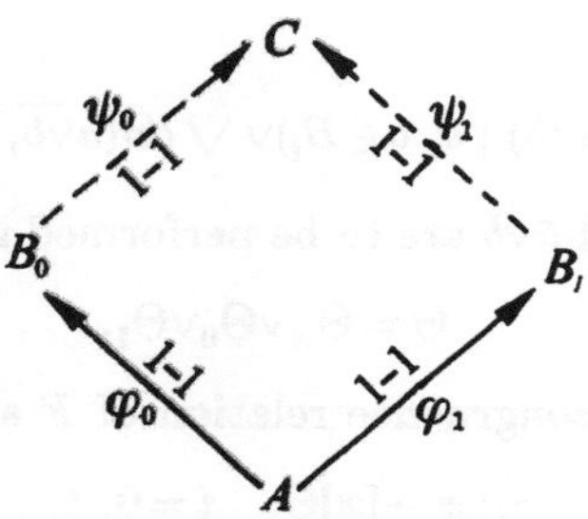

Figure 1

A class **K** is said to have the *(Strong) Amalgamation Property* iff every V-formation can be (strongly) amalgamated.

K is said to have the *Weak Amalgamation Property* (also called the *Embedding Property*) iff for any $B_0, B_1 \in \mathbf{K}$ there exists a $C \in \mathbf{K}$ into which both B_0 and B_1 can be embedded (this is the "special case" $A = \varnothing$ of the Amalgamation Property). This is of little interest for lattices: any equational class of lattices has the Weak Amalgamation Property. Indeed, $C = B_0 \times B_1$ will do with the embeddings $\psi_0: x \to \langle x, b_1 \rangle$, $\psi_1: x \to \langle b_0, x \rangle$, where b_i is a fixed element of B_i, for $i = 0, 1$.

L has the Strong Amalgamation Property. To see this take a V-formation $\langle A, B_0, B_1, \varphi_0, \varphi_1 \rangle$; we can assume that $B_0 \cap B_1 = A$ and A is a sublattice of B_0 and B_1. On the set $P = B_0 \cup B_1$ we define a partial ordering as follows:

For $a, b \in B_i$, $a \leq b$ in P iff $a \leq b$ in B_i $(i = 0, 1)$; for $a \in B_i$ and $b \in B_j$, $i \neq j$, $a \leq b$ in P iff $a \leq c$ in B_i and $c \leq b$ in B_j for some $c \in A$ $(i, j \in \{0, 1\})$.

It is easy to check that P is a poset. Then

$$a \wedge b = \inf\{a, b\}, \quad a \vee b = \sup\{a, b\}$$

turns P into a partial lattice (see Lemma I.5.21) of which A, B_0, and B_1 are sublattices. Thus by Theorem I.5.20, P can be embedded into a lattice, proving the Strong Amalgamation Property for **L**.

D does not have the Strong Amalgamation Property. Indeed, let $B_0 = B_1 = (\mathfrak{C}_2)^2$, $A = \mathfrak{C}_3$ $(A = \{0, a, 1\})$, and let $\varphi_0 = \varphi_1 = \varphi$ be given by $0\varphi = \langle 0, 0 \rangle$, $a\varphi = \langle 1, 0 \rangle$, $1\varphi = \langle 1, 1 \rangle$. Let $\langle \psi_0, \psi_1, C \rangle$ strongly amalgamate $\langle A, B_0, B_1, \varphi_0, \varphi_1 \rangle$ with $C \in \mathbf{D}$. Then $B_0\psi_0 \cap B_1\psi_1 = A$ and so $\langle 0, 1 \rangle\psi_0 \neq \langle 0, 1 \rangle\psi_1$. Both of these elements are relative complements of $\langle 1, 0 \rangle\psi_0$ $(= \langle 1, 0 \rangle\psi_1)$ in the interval $[\langle 0, 0 \rangle\psi_0, \langle 1, 1 \rangle\psi_0]$ of C, contradicting $C \in \mathbf{D}$ and Corollary II.1.3.

However, **D** has the Amalgamation Property. This we shall prove shortly after some general remarks.

Let **K** be an equational class of lattices (or of algebras, in general) and let $Q = \langle A, B_0, B_1, \varphi_0, \varphi_1 \rangle$ be a V-formation in **K**. We assume that $B_0 \cap B_1 = \varnothing$. We associate with Q a congruence relation $\Theta = \Theta(Q)$ as follows:

Let $F = F_{\mathbf{K}}(Q)$ be the free lattice over **K** freely generated by the set $\overline{B_0} \cup \overline{B_1}$ where $\overline{B_i} = \{\bar{x} \mid x \in B_i\}$, for $i = 0, 1$, and the sets B_0, B_1, $\overline{B_0}$, $\overline{B_1}$ are pairwise disjoint. We set

$$\Theta_A = \bigvee (\Theta(\overline{a\varphi_0}, \overline{a\varphi_1}) \mid a \in A)$$

and

$$\Theta_i = \bigvee(\Theta(\overline{a \wedge b}, \bar{a} \wedge \bar{b}) \mid a, b \in B_i) \vee \bigvee(\Theta(\overline{a \vee b}, \bar{a} \vee \bar{b}) \mid a, b \in B_i),$$

for $i = 0, 1$. Note that $\bar{a} \wedge \bar{b}$ and $\bar{a} \vee \bar{b}$ are to be performed in F. Now we define

$$\Theta = \Theta_A \vee \Theta_0 \vee \Theta_1.$$

Obviously, Θ is the smallest congruence relation of F such that

$$\alpha_i : x \to [x]\Theta, \quad i = 0, 1,$$

is a homomorphism of B_i into F/Θ (because $\Theta_i \leq \Theta$) and $\varphi_0 \alpha_0 = \varphi_1 \alpha_1$ (because $\Theta_A \leq \Theta$). Thus $\langle \alpha_0, \alpha_1, F/\Theta \rangle$ amalgamates Q iff α_0 and α_1 are one-to-one. Therefore, if α_0 and α_1 are one-to-one, then Q can be amalgamated. Conversely, if $\langle \psi_0, \psi_1, C \rangle$ ($C \in \mathbf{K}$) amalgamates Q, then map $\overline{B_0} \cup \overline{B_1}$ into C by

$$\bar{x} \to x\psi_i \quad \text{for} \quad x \in B_i, \quad i = 0, 1.$$

This map extends to a homomorphism β of F into C and obviously $\operatorname{Ker}(\beta) \geq \Theta$, and $\bar{x}\beta = x\psi_i$, for $x \in B_i$. Thus α_i followed by the natural homomorphism of F/Θ into C equals ψ_i which is one-to-one by assumption. So α_i is one-to-one, for $i = 0, 1$. We have proved (Theorem 1, Corollaries 2–5, and the application to the proof of Lemma 12 are from G. Grätzer [1975]):

Theorem 1. *Let $\mathbf{K}$ be an equational class, let $Q = \langle A, B_0, B_1, \varphi_0, \varphi_1 \rangle$ be a V-formation in $\mathbf{K}$, and let F and Θ be constructed as above. Then Q can be amalgamated in $\mathbf{K}$ iff, for $i = 0$ or 1 and $x, y \in B_i$,*

$$\bar{x} \equiv \bar{y} \ (\Theta) \quad \textit{implies that} \quad x = y.$$

Corollary 2. *An equational class $\mathbf{K}$ has the Amalgamation Property iff for any V-formation $\langle A, B_0, B_1, \varphi_0, \varphi_1 \rangle$ in $\mathbf{K}$ and $x, y \in B_0$, $x \neq y$, there exist a $C \in \mathbf{K}$ and homomorphisms $\psi_i : B_i \to C$ such that $\varphi_0 \psi_0 = \varphi_1 \psi_1$ and $x\psi_0 \neq y\psi_0$.*

Proof. In order to prove the α_i (introduced in the proof of Theorem 1) one-to-one it is sufficient to have homomorphisms ψ_i separating a fixed pair of distinct elements. The necessity is obvious. ●

Corollary 3. *Let $\mathbf{K}$ be an equational class and let $Q = \langle A, B_0, B_1, \varphi_0, \varphi_1 \rangle$ be a V-formation in $\mathbf{K}$. If Q cannot be amalgamated in $\mathbf{K}$, then there are finitely generated subalgebras A' of A and B_i' of B_i, $i = 0, 1$, such that $Q' = \langle A', B_0', B_1', \varphi_0', \varphi_1' \rangle$ cannot be amalgamated in $\mathbf{K}$, where φ_i' is the restriction of φ_i to A', $i = 0, 1$.*

Proof. Let F and Θ be given as in Theorem 1. If Q cannot be amalgamated in $\mathbf{K}$, then there are $i \in \{0, 1\}$ and $x, y \in B_i$ such that $x \neq y$ and $\bar{x} \equiv \bar{y} \ (\Theta)$. By Theorem III.1.2 and Lemma III.1.3, we can select finite subsets A^* of A and B_i^* of B_i, $i = 0, 1$, such that in computing $\bar{x} \equiv \bar{y}\ (\Theta)$ we use only elements of $A^* \cup B_0^* \cup B_1^*$. Thus we can set $A' = [A^*]$, $B_i' = [B_i^*]$, $i = 0, 1$. A trivial application of Theorem 1 shows that Q' cannot be amalgamated in $\mathbf{K}$. ●

Call a V-formation $\langle A, B_0, B_1, \varphi_0, \varphi_1 \rangle$ *finitely generated* (*finite*) iff A, B_0, and B_1 are finitely generated (finite).

Corollary 4. *An equational class* $\mathbf{K}$ *has the Amalgamation Property iff every finitely generated* V*-formation in* $\mathbf{K}$ *can be amalgamated in* $\mathbf{K}$.

An equational class $\mathbf{K}$ is called *locally finite* iff every finitely generated member of $\mathbf{K}$ is finite. It is easily seen that $\mathbf{K}$ is locally finite iff all $F_{\mathbf{K}}(n)$ are finite ($n < \omega$). This always holds if $\mathbf{K}$ is generated by a single finite lattice L. Let $\mathbf{K}_{\text{fin}}$ denote the class of finite members of $\mathbf{K}$.

Corollary 5. *Let* $\mathbf{K}$ *be a locally finite equational class. Then* $\mathbf{K}$ *has the Amalgamation Property iff all finite* V*-formations in* $\mathbf{K}$ *can be amalgamated in* $\mathbf{K}$, *or equivalently, iff* $\mathbf{K}_{\text{fin}}$ *has the Amalgamation Property.*

The last equivalence follows from the observation that if $Q = \langle A, B_0, B_1, \varphi_0, \varphi_1 \rangle$ is finite and can be amalgamated in $\mathbf{K}$, then some quotient of $F_{\mathbf{K}}(Q)$ will amalgamate Q; but $F_{\mathbf{K}}(Q) = F_{\mathbf{K}}(\bar{B}_0 \cup \bar{B}_1)$ is finite since K is locally finite.

Now we return to $\mathbf{D}$. By Theorem II.2.1, $|F_{\mathbf{D}}(n)| < 2^{2^n}$ and so $\mathbf{D}$ is locally finite.

Corollary 6. $\mathbf{D}$ *has the Amalgamation Property.*

Proof. Let $Q = \langle A, B_0, B_1, \varphi_0, \varphi_1 \rangle$ be a finite V-formation in $\mathbf{D}$ and let $x, y \in B_0$, $x \neq y$. Set $C = \mathfrak{C}_2$ and let ψ_0 be a homomorphism of B_0 into $\mathfrak{C}_2$ such that $x\psi_0 \neq y\psi_0$ (see Corollary II.1.11). Let P be the ideal kernel of $\varphi_0\psi_0$. If $P = A$, define $\psi_1 : B_1 \to \mathfrak{C}_2$ by $x\psi_1 = 0$ for all $x \in B_1$. If $P = \varnothing$, define $\psi_1 : B_1 \to \mathfrak{C}_2$ by $x\psi_1 = 1$ for all $x \in B_1$. If $P \neq A$ and $P \neq \varnothing$, then $P = (a]$ where $a \neq 1$ is a meet-irreducible element in A. Thus there is a unique $b \in A$ such that $b \succ a$. By Corollary II.1.13, there is a meet-irreducible element p in B_1 such that $a \leq p$ and $b \nleq p$. We then define $\psi_1 : B_1 \to \mathfrak{C}_2$ by $x\psi_1 = 0$ iff $x \leq p$ and $x\psi_1 = 1$ iff $x \nleq p$. It is obvious that $\langle \psi_0, \psi_1, \mathfrak{C}_2 \rangle$ satisfies the conditions of Corollary 2, thus Q can be amalgamated in $\mathbf{D}$. ●

To further illustrate the usefulness of Corollary 2, for a class $\mathbf{K}$ of lattices (or algebras, in general) define $A \in \mathbf{K}$ to be *injective* in $\mathbf{K}$ iff for any $B, C \in \mathbf{K}$, B a sublattice of C, any homomorphism of B into A can be extended to a homomorphism of C into A.

Corollary 7 (R. S. Pierce [1968]). *Let* $\mathbf{K}$ *be an equational class. If any member of* $\mathbf{K}$ *can be embedded in an injective member of* $\mathbf{K}$, *then* $\mathbf{K}$ *has the Amalgamation Property.*

PROOF. We apply Corollary 2. Let C be an injective member of $\mathbf{K}$ into which B_0 can be embedded; let ψ_0 be this embedding. Then $x\psi_0 \neq y\psi_0$. Set $A' = A\varphi_1$. Then $\varphi_1^{-1}\psi_0$ is a homomorphism (in fact, an embedding) of A' into C, thus this homomorphism can be extended to a homomorphism ψ_1 of B_1 into C. Obviously, $\varphi_0\psi_0 = \varphi_1\psi_1$. ●

A further application of Corollary 3 will be given at the end of this section.

A typical application of the Amalgamation Property is the sublattice theorem of free products of lattices discussed in Section VI.1.

The next two theorems are negative. They show that certain equational classes of lattices do not have the Amalgamation Property.

THEOREM 8.[1] *Let* $\mathbf{K}$ *be an equational class generated by a finite lattice. If* $\mathbf{K} \supset \mathbf{D}$, *then* $\mathbf{K}$ *does not have the Amalgamation Property.*

PROOF. If $\mathbf{K}$ is generated by a finite lattice L, then by Corollary 1.11, $\mathbf{Si}(\mathbf{K}) \subseteq \mathbf{HS}(L)$. Thus no subdirectly irreducible lattice in $\mathbf{K}$ has more than $n = |L|$ elements.

Since $\mathbf{K} \supset \mathbf{D}$, there is a nondistributive lattice in $\mathbf{K}$ and thus by Theorem II.1.1, $\mathfrak{N}_5$ or $\mathfrak{M}_3 \in \mathbf{K}$.

If $\mathfrak{N}_5 \in \mathbf{K}$, $\mathfrak{N}_5 = \{o, a, b, c, i\}$ as in Figure II.1.1, then consider the V-formation $\langle \mathfrak{C}_2, \mathfrak{N}_5, \mathfrak{N}_5, \varphi_0, \varphi_1 \rangle$, where $\mathfrak{C}_2 = \{0, 1\}$ and $0\varphi_0 = o$, $1\varphi_0 = i$, $0\varphi_1 = b$, $1\varphi_1 = a$. Let $\langle \psi_0, \psi_1, A \rangle$ amalgamate this V-formation. Then A will have the lattice A_1 of Figure 2

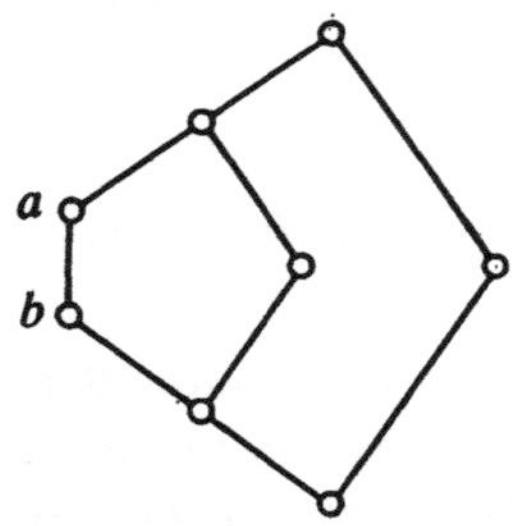

Figure 2

as a sublattice. In fact, A_1 is the union of the two images of $\mathfrak{N}_5$ in A. Observe that A_1 is again subdirectly irreducible and $|A_1| = 8$. Then we take the V-formation $\langle \mathfrak{C}_2, \mathfrak{N}_5, A_1, \varphi_0, \varphi_1 \rangle$ with $0\varphi_0 = o$, $1\varphi_0 = i$, $0\varphi_1 = b$, $1\varphi_1 = a$. The union of the images of $\mathfrak{N}_5$ and A_1 will form a sublattice A_2 which is again subdirectly irreducible and $|A_2| = 11$. Proceeding by induction we obtain the subdirectly irreducible lattice $A_k \in \mathbf{K}$ with $|A_k| = 5 + 3k$. If k is large enough so that $5 + 3k > n$, this is a contradiction.

If $\mathfrak{M}_3 \in \mathbf{K}$, $\mathfrak{M}_3 = \{o, a, b, c, i\}$, then we take $\langle \mathfrak{C}_2, \mathfrak{M}_3, \mathfrak{M}_3, \varphi_0, \varphi_1 \rangle$ where $0\varphi_0 = a$, $1\varphi_0 = i$, $0\varphi_1 = o$, $1\varphi_1 = c$. Then the union of the images is again a sublattice of the amalgam, thus obtaining $\mathfrak{M}_{3,3} \in \mathbf{K}$. Proceeding the same way we obtain a sequence of simple modular lattices of increasing size, leading to a contradiction as above. ●

[1] A. Day, S. D. Comer, and S. Fajtlowicz. See G. Grätzer, B. Jónsson, and H. Lakser [1973].

Theorem 9. **M** *does not have the Amalgamation Property.*

Remark. In fact, no equational class **K** of modular lattices has the Amalgamation Property unless **K** = **T** or **D**, as proved in G. Grätzer, B. Jónsson, and H. Lakser [1973].

Proof. Let us assume that **M** has the Amalgamation Property; under this assumption we prove a few embedding theorems.

1. Any modular lattice A can be embedded in a bounded modular lattice B which has a five-element chain.

This is trivial. If A has no five-element chain we add new zeros and ones until it has a five-element chain. If A is not bounded, add bounds. ◑

2. Any bounded modular lattice A can be embedded in a modular lattice B having the same bounds and having the property that for every $a \in A$, $a \neq 0, 1$, B has a diamond $\{0, a, b, c, 1\}$.

To prove this, let $A - \{0, 1\} = \{a_\gamma \mid \gamma < \alpha\}$ and we define an increasing sequence of modular lattices A_γ, $\gamma < \alpha$ with the same bounds 0 and 1. Set $\bar{A}_0 = A$. If A_γ, $\gamma < \delta$ $(\delta < \alpha)$, have already been defined, set $\bar{A}_\delta = \bigcup(A_\gamma \mid \gamma < \delta)$.

Let $\mathfrak{C}_3 = \{0, 1, 2\}$ and consider the V-formation $Q_\delta = \langle \mathfrak{C}_3, \mathfrak{M}_3, \bar{A}_\delta, \varphi_0, \varphi_1 \rangle$ defined by $0\varphi_0 = o$, $1\varphi_0 = a$, $2\varphi_0 = i$, $0\varphi_1 = 0$, $1\varphi_1 = a_\delta$, $2\varphi_1 = 1$. Let $\langle \psi_0, \psi_1, A \rangle$ amalgamate Q_δ. By forming the interval $[0\varphi_0\psi_0, 2\varphi_0\psi_0]$ we obtain A_δ. It is obvious that with $a = a_\delta$, A_δ has the property required of B. We define $B = \bigcup(A_\gamma \mid \gamma < \alpha)$, which obviously has the required properties. ◑

3. Any bounded modular lattice A can be embedded in a modular lattice B having the same bounds as A and having the following properties:

(i) For every $a \in B$, $a \neq 0, 1$, there is a diamond $\{0, a, b, c, 1\}$ in B.

(ii) B is a simple complemented modular lattice.

Indeed, set $A_0 = A$, and if A_n is defined, then define A_{n+1} as the lattice constructed from A_n in 2. Then $B = \bigcup(A_n \mid n < \omega)$ obviously satisfies (i).

Let Θ be a congruence relation of B, $\omega \neq \Theta$. Since B is a complemented modular lattice, B is relatively complemented; hence (see Theorem III.3.10) there is an $a \in B$, such that $a \neq 0$ and $a \equiv 0\ (\Theta)$. If $a \neq 1$, then by (i) there is a diamond $\{0, a, b, c, 1\}$. Since $\mathfrak{M}_3$ is simple, we obtain $0 \equiv 1\ (\Theta)$, that is, $\Theta = \iota$. If $a = 1$, again $\Theta = \iota$. Thus B is simple. ◑

Now we are ready to prove the theorem. Let $A \in$ **M**. By the embeddings 1–3 we can embed A into a simple complemented modular lattice B having a five-element chain. By Theorem IV.5.17, B can be $\{0, 1\}$-embedded into a modular geometric lattice C. C is directly indecomposable because it has a simple $\{0, 1\}$-sublattice (namely, B). C has a five-element chain because B has one. Thus Corollary IV.5.16 applies and $C \cong L(D, \mathfrak{m})$. Thus by Lemma IV.5.9 and Corollary IV.5.12, the arguesian identity holds in C. Since A is a sublattice of C, the arguesian identity holds in A.

Since A was an arbitrary member of **M** and the arguesian identity does not hold in **M**, this is a contradiction. ●

Since the Amalgamation Property fails for so many equational classes of lattices it seems reasonable to ask to what extent does it hold in general. The answer is rather surprising.

Following G. Grätzer and H. Lakser [1971], for a class $\mathbf{K}$ of lattices (or algebras, in general), let $\mathbf{Amal}(\mathbf{K})$ be the class of all those $A \in \mathbf{K}$ for which all V-formations in $\mathbf{K}$ of the form $\langle A, \ldots \rangle$ can be amalgamated in $\mathbf{K}$. Obviously, $\mathbf{K}$ has the Amalgamation Property iff $\mathbf{Amal}(\mathbf{K}) = \mathbf{K}$. For any equational class $\mathbf{K}$ of lattices $\mathbf{Amal}(\mathbf{K}) \neq \varnothing$. Indeed, $\mathfrak{C}_1 \in \mathbf{Amal}(\mathbf{K})$. (Argue similarly to the proof of the Weak Amalgamation Property for $\mathbf{K}$.)

Call a subclass $\mathbf{K}_1$ of $\mathbf{K}$ *cofinal in* $\mathbf{K}$ iff for all $A \in \mathbf{K}$ there is an extension $B \in \mathbf{K}_1$.

Theorem 10 (M. Yasuhara [1974]). *For an equational class* $\mathbf{K}$, $\mathbf{Amal}(\mathbf{K})$ *is cofinal in* $\mathbf{K}$.

The proof of this result will be given in the next two lemmas. Some of the ideas of Lemma 11 originate in A. Robinson [1971], [1971 a]. The present form of Lemma 11 is a slight variant of a lemma of M. Yasuhara and it is due to M. Makkai.

For a class $\mathbf{K}$ let $\mathbf{Ec}(\mathbf{K})$ stand for the class of all $A \in \mathbf{K}$ having the following property: For any extension $B \in \mathbf{K}$ of A and for any finite $X \subseteq A$ and $Y \subseteq B$, there is an embedding

$$\varphi : [X \cup Y] \to A$$

fixing X (that is, $x\varphi = x$ for all $x \in X$).

Lemma 11. *For an equational class* $\mathbf{K}$, $\mathbf{Ec}(\mathbf{K})$ *is cofinal in* $\mathbf{K}$.

Proof. To facilitate the proof we introduce some concepts. For $A \in \mathbf{K}$, $\langle \varphi, X, C \rangle$ is a *triple over* A *in* $\mathbf{K}$ iff X is a finite subset of A, C is a finitely generated member of $\mathbf{K}$, and $\varphi : X \to C$ is a homomorphism of the relative sublattice X of A into C. (This last condition means that if $x \wedge y = z$ or $x \vee y = z$ for $x, y, z \in X$, then $x\varphi \wedge y\varphi = z\varphi$ or $x\varphi \vee y\varphi = z\varphi$ in C, respectively.) The triple $\langle \varphi, X, C \rangle$ *is realized over* A *by* B iff $B \in \mathbf{K}$, B is an extension of A, and there is an embedding ψ of C into B such that $x\varphi\psi = x$ for all $x \in X$.

Now let $\langle \varphi_\gamma, X_\gamma, C_\gamma \rangle$, $\gamma < \alpha$, list all the triples over A. We define two increasing sequences of lattices as follows: $\bar{A}_0 = A$; if A_γ has been defined for $\gamma < \delta (< \alpha)$, then set $\bar{A}_\delta = \bigcup (A_\gamma \mid \gamma < \delta)$. Regard $\langle \varphi_\delta, X_\delta, C_\delta \rangle$ as a triple over $\bar{A}_\delta$. If it is realized in $\mathbf{K}$ define A_δ as any member of $\mathbf{K}$ realizing it. If it cannot be realized in $\mathbf{K}$, define $A_\delta = \bar{A}_\delta$. Then set $A^{(1)} = \bigcup (A_\gamma \mid \gamma < \alpha)$. We define

$$A^{(n+1)} = (A^{(n)})^{(1)}$$

and

$$A^* = \bigcup (A^{(n)} \mid n < \omega).$$

Since $\mathbf{K}$ is an equational class, $A^* \in \mathbf{K}$. We claim that $A^* \in \mathbf{Ec}(\mathbf{K})$. Indeed, let $B \in \mathbf{K}$ be an extension of A^*, and choose finite subsets $X \subseteq A^*$ and $Y \subseteq B$. Define $C = [X \cup Y]$

and let $\varphi: X \to A$ be the identity map. Then $\langle \varphi, X, C \rangle$ is a triple over A^* which can be realized by B. Since $X \subseteq A^*$ and X is finite, $X \subseteq A^{(n)}$ for some $n < \omega$. Hence $\langle \varphi, X, C \rangle$ can be regarded as a triple over $A^{(n)}$. Thus $\langle \varphi, X, C \rangle$ occurs as some $\langle \varphi_\delta, X_\delta, C_\delta \rangle$ in the list of all triples over $A^{(n)}$. In the δ-th step of the construction of $A^{(n+1)} = (A^{(n)})^{(1)}$, we view $\langle \varphi, X, C \rangle$ as a triple over $\overline{(A^{(n)})_\delta}$; we observe that $\langle \varphi, X, C \rangle$ can be realized by B, and so $(A^{(n)})_\delta$ realizes $\langle \varphi, X, C \rangle$, that is, there is an embedding $\psi: C \to (A^{(n)})_\delta$ such that $x\varphi\psi = x$ for $x \in X$. However, $x\varphi = x$ for all $x \in X$ and therefore, $x\varphi\psi = x\psi = x$ for all $x \in X$. Thus ψ is an embedding of $[X \cup Y]$ into A^* keeping X fixed, proving $A^* \in \mathbf{Ec}(\mathbf{K})$. ●

Lemma 12. *For an equational class* $\mathbf{K}$, $\mathbf{Ec}(\mathbf{K}) \subseteq \mathbf{Amal}(\mathbf{K})$.

Proof. Let $A \in \mathbf{Ec}(\mathbf{K})$ and consider a V-formation $Q = \langle A, B_0, B_1, \varphi_0, \varphi_1 \rangle$ in $\mathbf{K}$. We can assume that $A \subseteq B_0$, $A \subseteq B_1$, and $\varphi_0 = \varphi_1$ is the identity map on A. We wish to show that Q can be amalgamated in $\mathbf{K}$. Assume to the contrary, that Q cannot be amalgamated in $\mathbf{K}$. Then, by Corollary 3, there exist finite subsets $X \subseteq A$, $Y_0 \subseteq B_0$ and $Y_1 \subseteq B_1$, such that $Q' = \langle [X], [X \cup Y_0], [X \cup Y_1], \varphi_0', \varphi_1' \rangle$ cannot be amalgamated, where φ_i' is the restriction of φ_i to $[X \cup Y_i]$, for $i = 0, 1$. Since $A \in \mathbf{Ec}(\mathbf{K})$ there exist embeddings $\psi_i: [X \cup Y_i] \to A$ keeping X fixed $(i = 0, 1)$. Thus for $x \in X$, $x\varphi_i\psi_i = x$, for $i = 0, 1$, and so $a\varphi_i\psi_i = a$ for all $a \in [X]$. This shows that $\langle \psi_0, \psi_1, A \rangle$ amalgamates Q', which is a contradiction. ●

Exercises

1. Show that the class of all Boolean algebras does not have the Weak Amalgamation Property.
2. Show that the class of all groups has the Strong Amalgamation Property.
3. Investigate which equational classes of pseudocomplemented distributive lattices have which Amalgamation Property.
4. Show that $\mathbf{L}_{\text{fin}}$ has the Strong Amalgamation Property.
5. Define the **St-amal(K)** as the analogue of **Amal(K)** for the Strong Amalgamation Property. Determine **St-amal(D)**.
6. Show that "$C \in \mathbf{K}$" can be changed to "$C \in \mathbf{Si}(\mathbf{K})$" in Corollary 2.
7. Did we use the Axiom of Choice in verifying Corollary 5?
8. Let L be injective in $\mathbf{K}$. If A is an extension of L in $\mathbf{K}$, then L is a *retract* of A, that is, there is a homomorphism $\varphi: A \to L$ such that $x\varphi = x$ for all $x \in L$.
9. Show that any retract of a complete Boolean lattice is a complete Boolean lattice.
10. A distributive lattice L is injective in $\mathbf{D}$ iff L is a complete Boolean lattice (P. R. Halmos [1963]). (Hint: use Exercises 8, 9 and Corollary II.2.7.)
11. Prove Corollary 6 using Corollary 7.
12. Find further examples of the phenomenon observed twice in the proof of Theorem 8, namely that for some special V-formation, if $\langle \psi_0, \psi_1, C \rangle$ amalgamates $\langle A, B_0, B_1, \varphi_0, \varphi_1 \rangle$ then $B_0\psi_0 \cup B_1\psi_1$ is a uniquely determined sublattice of C.
13. Show that there are $2^{\aleph_0}$ equational classes of modular lattices not having the Amalgamation Property.

14. Let **K** be an equational class. Then for every $A \in \mathbf{K}$ we construct an extension $B \in \mathbf{Amal}(\mathbf{K})$. Give an upper bound for $|B|$.
15. Let **K** be an equational class having the Amalgamation Property. Let $A, B \in \mathbf{K}$, B an extension of A, and let α be an automorphism of A. Show that[1] there exist an extension $C \in \mathbf{K}$ of B and an automorphism β of C extending α.
16. Let **K** be an equational class of lattices. Then $\mathfrak{C}_2 \in \mathbf{Amal}(\mathbf{K})$ iff there is a $D \in \mathbf{Amal}(\mathbf{K})$ with $|D| > 1$.
17. Let **K** be an equational class of lattices. Prove that if $I(L) \in \mathbf{Amal}(\mathbf{K})$, then $L \in \mathbf{Amal}(\mathbf{K})$.
18. Let $\mathbf{K} = \mathbf{Equ}(L)$, where L is a finite subdirectly irreducible lattice. Then $L \in \mathbf{Amal}(\mathbf{K})$.

Further Topics and References

$\mathbf{H}, \mathbf{S}, \mathbf{P}, \mathbf{P}_{\mathbf{R}}, \mathbf{P}_{\mathbf{S}}$, and so on, are examples of operators: an *operator* **X** assigns to a class **K** of lattices (algebras) another class **X**(**K**) of lattices (algebras). If **X** and **Y** are operators, so is their product: **XY** defined by $\mathbf{XY}(\mathbf{K}) = \mathbf{X}(\mathbf{Y}(\mathbf{K}))$. Thus we can easily define the semigroup generated by a set of operators. The *standard semigroup* of a class **K** of algebras is the one generated by **H**, **I**, **S**, and **P**, equality defined with respect to **K**. For references see S. D. Comer and J. Johnson [1972] which also lists all the known results for lattices, distributive lattices, Boolean algebras, and so on; one problem they leave unresolved is our Problem V.1.

Theorem 2.1 is very trivial but it has a number of useful consequences: the lattice of equational classes of lattices satisfies the Meet Infinite Distributive Identity and every proper interval contains a prime interval; properties shared by all dually algebraic, distributive lattices. An additional property is that no element covers more than countably many elements, since the lattice then would have more than $2^{\aleph_0}$ elements.

For a more detailed discussion of reduced products and ultra products, see T. Frayne, A. C. Morel, and D. S. Scott [1962/63] and G. Grätzer [1968].

Splitting can be defined relative to an equational class. Apart from the general concept, only splittings in **M** and **E** have been studied to some extent, where **E** is the equational class generated by the subspace lattices of all projective planes. For instance, $\mathfrak{M}_4$ does not split in **M** but the subspace lattice of any projective plane splits in **E**. For a discussion of these and related results see R. Wille [b]; see also A. Day [1973] and D. X. Hong [1972].

Wille's method of constructing equational classes (Theorem 2.6) is based on his two earlier papers, R. Wille [1969] and [1969a]. For a refinement of Wille's method, see K. A. Baker [1974a].

By a result of R. N. McKenzie [1970], lattices can be defined by a single identity and obviously so can **T**; no other equational class of lattices can be defined by a single identity. By R. Padmanabhan [1969], any equational class of lattices defined by a finite set of identities can be defined by two identities.

---

[1] Exercises 15–18 are from G. Grätzer, B. Jónsson, and H. Lakser [1973]. Exercise 15 is implicit in B. H. Neumann and H. Neumann [1952].

Combining these results with a result of A. Tarski [1968] we obtain:

L and **T** have an irredundant equational basis of n elements for any $1 \leq n < \omega$. If the equational class **K** of lattices has a finite equational basis and $\mathbf{K} \neq \mathbf{L}, \mathbf{T}$, then **K** has an irredundant equational basis of n elements iff $2 \leq n < \omega$.

Lattices are quite exceptional in that Theorem 3.4 holds. A finite algebra with no finite equational basis was found by R. C. Lyndon [1954]; this was improved in V. L. Murskiĭ [1965] in which a three-element example is found (this is optimal, see E. L. Post [1921]) and P. Perkins [1969] in which the example is a six-element semigroup. For additional references and for the main result in the field (namely, that Theorem 3.4 holds in any congruence distributive equational class), see K. A. Baker [1974].

Let $\mathbf{K}_0$ and $\mathbf{K}_1$ be equational classes; the identities of $\mathbf{K}_0 \vee \mathbf{K}_1$ are those that are shared by $\mathbf{K}_0$ and $\mathbf{K}_1$. Thus $\mathbf{K}_0$ and $\mathbf{K}_1$ may have finite equational bases, while $\mathbf{K}_0 \vee \mathbf{K}_1$ may not have a finite equational basis. For lattices this negative result has been verified by K. Baker (unpublished) and in B. Jónsson [1974]. Some positive results are also given in B. Jónsson [1974]; for instance, $\mathbf{M} \vee \mathbf{N}_5$ has finite equational bases.

Equational bases are known for many equational classes of lattices (and other related algebras); see especially, K. A. Baker [1969 a], [1971], [1974], [1974 a], and [a], S. D. Comer and D. X. Hong [1972], A. Day [1973], R. Freese [1973] and [b], E. Fried and G. Grätzer [1973], C. Herrmann [1973] and [1973 a], C. Herrmann and W. Poguntke [1974] and [a], D. X. Hong [1972], G. Hutchinson [1973 b], B. Jónsson [1968], M. Makkai [1973], M. Makkai and G. McNulty [a], R. N. McKenzie [1970], [1972], and [1973], R. Padmanabhan [1968], [1969], [1972], and [1974 a], M. Schützenberger [1945], and R. Wille [b]; see also R. Padmanabhan [1966] and D. H. Potts [1965].

K. A. Baker's method of finding equational bases discussed in Section 3 is exploited in great detail in K. A. Baker [1974]; for further results and applications, see K. A. Baker [1974 a] and [a]. C. Herrmann [1973] and M. Makkai [1973] are further variants on the same theme; see also G. Grätzer and H. Lakser [1971 a].

Theorem 3.6 has been extended in B. Jónsson [1968] and D. X. Hong [1972]. Related questions are considered in C. Herrmann [1973] and in the papers of R. Freese.

The development of the Amalgamation Property is described in B. Jónsson [1965]; probably, it was B. Jónsson's work more than any other influence that convinced the algebraists of the importance of this property. The Amalgamation Property for lattices and posets is noted in B. Jónsson [1956]; it is observed for Boolean algebras in A. Daigneault [1959].

The results of Section 4 on the Amalgamation Property are mostly negative. A further negative result of G. Grätzer, B. Jónsson, and H. Lakser [1973] states that if **K** is an equational class of lattices, $\mathbf{K} \subseteq \mathbf{M}$, and **K** is not arguesian, then $\mathfrak{C}_2 \notin \mathbf{Amal}(\mathbf{K})$. This was strengthened in G. Grätzer and H. Lakser [1973 a].

The existentially complete algebras of M. Yasuhara [1974] are different from members of $\mathbf{Ec}(\mathbf{K})$. Thus Lemma 4.11 is slightly stronger and Lemma 4.12 is slightly weaker than the corresponding results in M. Yasuhara [1974].

The connections of the amalgamation properties and the interpolation theorems of logic are examined in P. D. Bacsich [1975]; see also P. C. Eklof [1974].

Finite members of $\mathbf{Amal}(\mathbf{M}_n)$ are described in E. Fried, G. Grätzer, and H. Lakser [1971].

Problems

V. 1. Is $\mathbf{Equ}(\mathbf{K}) = \mathbf{SHPS}(\mathbf{K})$ for any class $\mathbf{K}$ of lattices?

V. 2. Describe those sets of operators introduced in this chapter that generate finite semigroups for lattices.

V. 3. Characterize splitting lattices in $\mathbf{M}$.

V. 4. The method of R. N. McKenzie [1972] to give an equational basis for $\mathfrak{N}_5$ has been modified in S. D. Comer and D. X. Hong [1972] to obtain an equational basis for $\mathfrak{M}_3$. Explain the similarities and determine whether the method has other applications (for instance, to yield equational bases for covers of $\mathbf{N}_5$).

V. 5. For an equational class $\mathbf{K}$ of lattices, define

$i(\mathbf{K})$ $(i_g(\mathbf{K}))$ as the smallest integer n such that $F_{\mathbf{K}}(n)$ is infinite (and generates $\mathbf{K}$); and if there is no such integer, then $i(\mathbf{K}) = \infty$ $(i_g(\mathbf{K}) = \infty)$;

$e(\mathbf{K})$ $(e_f(\mathbf{K}))$ as the smallest integer n such that every finite member of $\mathbf{K}$ can be embedded into an n-generated (finite) member of $\mathbf{K}$; and if there is no such integer, then $e(\mathbf{K}) = \infty$ $(e_f(\mathbf{K}) = \infty)$;

$c(\mathbf{K})$ as the smallest integer n such that every countable member of $\mathbf{K}$ can be embedded in an n-generated member of $\mathbf{K}$; and if there is no such integer, then $c(\mathbf{K}) = \infty$;

$f(\mathbf{K})$ $(f_i(\mathbf{K}))$ as the smallest integer n such that $F_{\mathbf{K}}(n)$ contains $F_{\mathbf{K}}(n+1)$ $(F_{\mathbf{K}}(\aleph_0))$ as a sublattice; and $f(\mathbf{K}) = \infty$ $(f_i(\mathbf{K}) = \infty)$ if there is no such integer.

Are there any relationships amongst these apart from the obvious ones $(e(\mathbf{K}) \leq e_f(\mathbf{K}), c(\mathbf{K}) \leq f_i(\mathbf{K})$, and so on)?

V. 6. Compute the functions introduced in Problem V.5 for particular equational classes of lattices: $\mathbf{N}_5$, $\mathbf{M}_3$, $\mathbf{M}_n$, $\mathbf{M}$, and so on. ($\mathfrak{M}_n$ is the subspace lattice of a projective line with n points and $\mathbf{M}_n = \mathbf{Equ}(\mathfrak{M}_n)$.)

V. 7. For an equational class $\mathbf{K}$ of lattices and positive integer n, define $\mathbf{K}_n = \mathbf{Equ}(F_{\mathbf{K}}(n))$. Characterize $I = \{n \mid \mathbf{K}_n \neq \mathbf{K}_{n+1}\}$. (Conjecture: $I = \{1, 2, 3, \dots, k\}$ or $I = \{1, 2, \dots, k, \dots\}$. For universal algebras in general, see B. Jónsson, G. McNulty, and R. W. Quackenbush [1975].)

V. 8. Find an $L \in \mathbf{Amal}(\mathbf{M})$, $|L| > 1$.

V. 9. Does there exist an equational class $\mathbf{K}$ satisfying $\mathbf{D} \subset \mathbf{K} \subseteq \mathbf{M}$ and $\mathfrak{C}_2 \in \mathbf{Amal}(\mathbf{K})$?

V.10. Does there exist an equational class $\mathbf{K} \subset \mathbf{L}$ having the Strong Amalgamation Property? How many such equational classes exist? (Conjecture: finite or countably infinite.)

V.11. Show that there are only countably many equational classes of lattices having the Amalgamation Property.

V.12. Describe $\mathbf{Amal}(\mathbf{M}_n)$.

V.13. Investigate $\mathbf{Amal}(\mathbf{Equ}$(complemented modular lattices)).

V.14. Is it true that in the lattice of equational classes of lattices every proper interval contains an atom? [1]

---

[1] Yes, see A. Day [b].

V.15. In the lattice of equational classes of lattices, are the elements of finite height exactly the equational classes of the form **Equ**(L), where L is a finite lattice?

V.16. Find all covers of **Equ**(L) where L is finite planar.

V.17. Find all the covers of $\mathbf{N}_5$.[1]

V.18. Is there an equational class of lattices with uncountably many covers?[2]

V.19. Is there a nontrivial lattice identity holding in the lattice of equational classes of commutative semigroups?

V.20. Find a finite equational basis for $\mathbf{M} \vee \mathbf{N}_5$.[3]

V.21. Let $\mathbf{K}_0$, $\mathbf{K}_1$ be equational classes of modular lattices. Is it true that if $\mathbf{K}_0$ and $\mathbf{K}_1$ have finite equational bases, then so does $\mathbf{K}_0 \vee \mathbf{K}_1$?

---

[1] The covers of $\mathbf{N}_5$ are those listed in Exercise 2.1. This was proved for equational classes generated by a finite lattice by I. Rival and in the general case by B. Jónsson.

[2] Yes, there are such equational classes. An example was given by R. Wille.

[3] Solved in B. Jónsson, The variety covering the variety of modular lattices. Manuscript.

CHAPTER VI

FREE PRODUCTS

1. Free Products of Lattices

The formation of a free product of a family of lattices is one of the most fundamental constructions of lattice theory. This specializes to the construction of free lattices, which form a class of lattices that is probably the closest rival of the class of distributive lattices in the richness of its structure. Also, free products provide a very useful tool for the construction of pathological lattices.

This is made possible, in part, by the Structure Theorem of Free Products (Theorem 10 below). Since the theorem is based on a large number of long inductive definitions, we shall present first a short intuitive description of its contents. We ask the reader to follow it up with a careful reading of the theorem bearing in mind that the final result is very simple and most efficient in use.

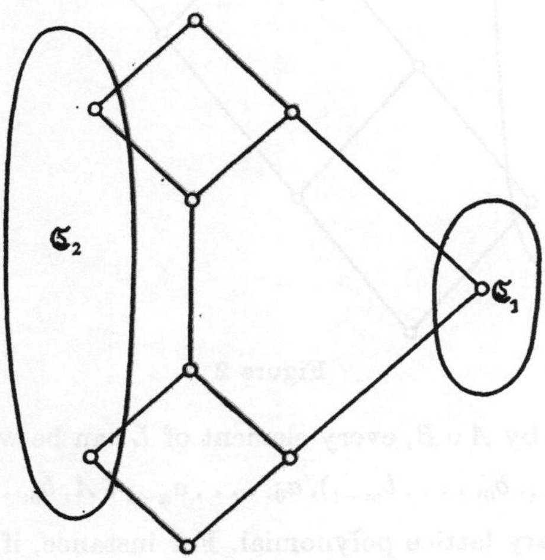

Figure 1

Let A and B be lattices and let L be a free product of A and B, in symbols, $L = A * B$; for the present discussion this should mean that A and B are sublattices of L, $A \cup B$ generates L, and L is the "most general" lattice with these properties, as in Section I.5. Figures 1 and 2 provide two simple examples: Figure 1 is the diagram of $\mathfrak{C}_2 * \mathfrak{C}_1$ and Figure 2 that of $\mathfrak{C}_3 * \mathfrak{C}_1$.

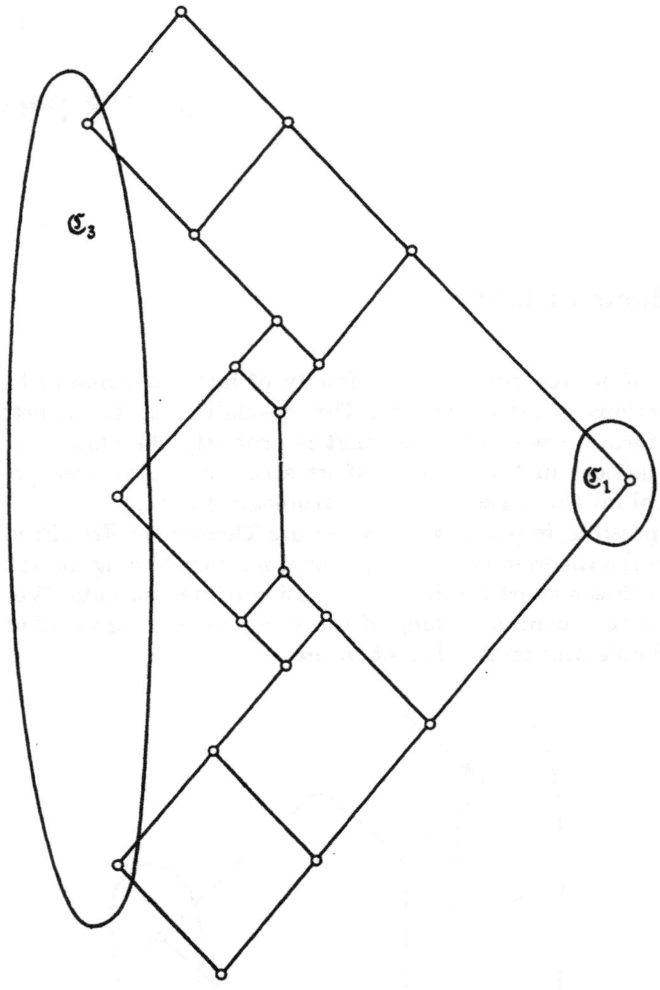

Figure 2

Since L is generated by $A \cup B$, every element of L can be written in the form

$$p(a_0, \ldots, a_{n-1}, b_0, \ldots, b_{m-1}),\ a_0, \ldots, a_{n-1} \in A,\ b_0, \ldots, b_{m-1} \in B,$$

where p is an $(n+m)$-ary lattice polynomial. For instance, if A and B are as given in Figure 3, then

$$a_1,\ b_2,\ a_3 \vee b_1,\ (a_3 \wedge b_2) \vee (a_0 \wedge b_0),\ ((a_5 \wedge b_0) \vee a_3) \wedge ((a_2 \wedge b_0) \vee a_2 \vee b_3)$$

are examples of elements of L. Observe, however, that expressions of this sort that appear to be very different may, in fact, denote the same element of L. For instance,

$$a_1 \vee ((a_3 \vee b_3) \wedge (a_4 \vee b_3)) = a_5 \vee b_3.$$

Theorem 10 will help the reader to construct much more complicated examples.

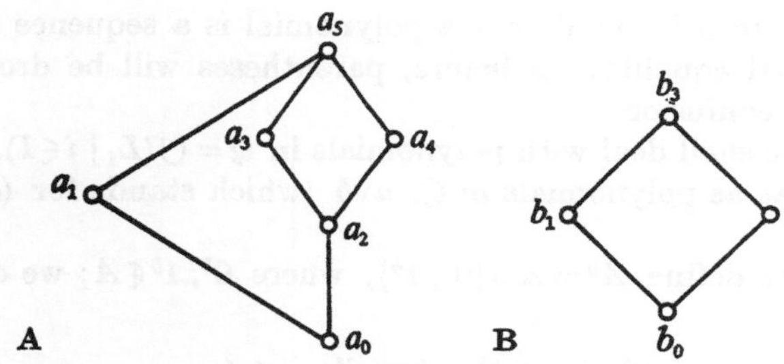

Figure 3

The problem then is how do we decide when two expressions represent the same element of L. The key to this problem is the observation that if

$$p = p(a_0, \ldots, a_{n-1}, b_0, \ldots, b_{m-1}) \leq a$$

for some $a \in A$, then there is a smallest element of A with this property. In fact, this smallest element (called the upper cover of p in A) is easy to compute knowing only p, A, and B.

So now our plan is the following: define formally the free product, expressions of the form $p(a_0, \ldots)$, upper and (dually) lower covers, present the algorithm deciding $p(a_0, \ldots) \leq q(a_0, \ldots)$, and finally prove that this algorithm works.

For this whole section, let L_i, $i \in I$, $I \neq \varnothing$, be a fixed family of lattices; we assume that L_i and L_j are disjoint for $i, j \in I$, $i \neq j$. We set $Q = \bigcup (L_i \mid i \in I)$ and we consider Q a poset under the following partial ordering:

For $a, b \in Q$, let $a \leq b$ iff $a, b \in L_i$ for some $i \in I$ and $a \leq b$ in L_i.

A free product L of the L_i, $i \in I$, is a free lattice $F(Q)$ ($= F_{\mathbf{L}}(Q)$) generated by Q, in the sense of Definition I.5.2. Or, equivalently,

Definition 1. *Let* $\mathbf{K}$ *be an equational class of lattices and let* $L, L_i \in \mathbf{K}$ *for* $i \in I$. *The lattice* L *is a* **free K-product** *of the lattices* L_i, $i \in I$, *iff the following conditions are satisfied*:

(i) *Each* L_i *is a sublattice of* L *and for* $i, j \in I$, $i \neq j$, L_i *and* L_j *are disjoint.*
(ii) L *is generated by* $\bigcup (L_i \mid i \in I)$.
(iii) *For any lattice* $A \in \mathbf{K}$ *and for any family of homomorphisms* $\varphi_i\colon L_i \to A$, $i \in I$, *there exists a homomorphism* $\varphi\colon L \to A$ *such that* φ *on* L_i *agrees with* φ_i, *for all* $i \in I$.

For $\mathbf{K} = \mathbf{L}$, L is called a *free product*.

The next definition is a slight adaptation of Definition I.4.1.

Definition 2. *Let* X *be an arbitrary set. The set* $\mathbf{P}(X)$ *of* **polynomials** *in* X *is the smallest set satisfying* (i) *and* (ii):

(i) $X \subseteq \mathbf{P}(X)$.
(ii) *If* $p, q \in \mathbf{P}(X)$, *then* $(p \wedge q)$, $(p \vee q) \in \mathbf{P}(X)$.

The reader should keep in mind that a polynomial is a sequence of symbols and equality means formal equality. As before, parentheses will be dropped whenever there is no danger of confusion.

In what follows, we shall deal with polynomials in $Q = \bigcup(L_i \mid i \in I)$. Let $a, b, c \in L_i$, $a \vee b = c$. Observe that as polynomials in Q, $a \vee b$ (which stands for $(a \vee b)$) and c are distinct.

For a lattice A, we define $A^b = A \cup \{0^b, 1^b\}$, where $0^b, 1^b \notin A$; we order A^b by the rules:

$$0^b < x < 1^b, \quad \text{for all} \quad x \in A.$$

$$x \leq y \quad \text{in} \quad A^b \quad \text{iff} \quad x \leq y \quad \text{in } A, \quad \text{for} \quad x, y \in A.$$

Thus A^b is a bounded lattice. Note, however, that $A^b \neq A$ even if A is itself bounded. It is important to observe that 0^b is meet-irreducible and 1^b is join-irreducible. Thus if $a \wedge b = 0^b$ then either a or b is 0^b, and dually. This will be quite important in subsequent computations.

Definition 3. *Let $p \in \mathbf{P}(Q)$ and $i \in I$. The* upper i-cover of p, *in notation, $p^{(i)}$, is an element of $(L_i)^b$ defined as follows:*

(i) *For $a \in Q$, we have $a \in L_j$ for exactly one j; if $j = i$, then $a^{(i)} = a$; if $j \neq i$, then $a^{(i)} = 1^b$.*

(ii) *$(p \wedge q)^{(i)} = p^{(i)} \wedge q^{(i)}$ and $(p \vee q)^{(i)} = p^{(i)} \vee q^{(i)}$, where $\wedge$ and $\vee$ on the right side of these equations are to be taken in $(L_i)^b$.*

The definition of the *lower i-cover* of p, in notation, $p_{(i)}$, is analogous, with 0^b replacing 1^b in (i).

If there is some danger of confusion, we shall write p^{L_i} and p_{L_i} for $p^{(i)}$ and $p_{(i)}$, respectively.

An upper cover or a lower cover is *proper* if it is not 0^b or 1^b. Observe that, however, no upper cover is 0^b and no lower cover is 1^b.

Corollary 4. *For any $p \in \mathbf{P}(Q)$ and $i \in I$ we have that*

$$p_{(i)} \leq p^{(i)},$$

and if $p_{(i)}$ and $p^{(j)}$ are proper, then $i = j$.

Proof. If $p \in Q$, then $p = p_{(i)} = p^{(i)}$ for $p \in L_i$ and $p_{(i)} = 0_b < 1_b = p^{(i)}$ for $p \notin L_i$, so the first statement is true. If the first statement holds for p and q, then

$$(p \wedge q)_{(i)} = p_{(i)} \wedge q_{(i)} \leq p^{(i)} \wedge q^{(i)} = (p \wedge q)^{(i)},$$

and so the first statement holds for $p \wedge q$ and similarly for $p \vee q$. To prove the second statement it is sufficient to verify that if $p_{(i)}$ is proper, then $p^{(j)}$ is not proper for any $j \neq i$. This is obvious for $p \in Q$ by 3(i). If $p = q \wedge r$ and $p_{(i)}$ is proper, then both $q_{(i)}$ and $r_{(i)}$ are proper, hence $q^{(j)} = r^{(j)} = 1^b$, and so $p^{(j)} = 1^b$. Finally, if $p = q \vee r$

and $p_{(i)}$ is proper, then $q_{(i)}$ or $r_{(i)}$ is proper, hence $q^{(j)} = 1^b$ or $r^{(j)} = 1^b$, ensuring $p^{(j)} = q^{(j)} \vee r^{(j)} = 1^b$. ●

Finally, we introduce a quasi-ordering of $\mathbf{P}(Q)$.

Definition 5. *For $p, q \in \mathbf{P}(Q)$, set $p \sqsubseteq q$ iff it follows from the following rules:*

(C) *For some $i \in I$, $p^{(i)} \leq q_{(i)}$.*

($_\wedge$W) $p = p_0 \wedge p_1$, *where* $p_0 \sqsubseteq q$ *or* $p_1 \sqsubseteq q$.

($_\vee$W) $p = p_0 \vee p_1$, *where* $p_0 \sqsubseteq q$ *and* $p_1 \sqsubseteq q$.

(W$_\wedge$) $q = q_0 \wedge q_1$, *where* $p \sqsubseteq q_0$ *and* $p \sqsubseteq q_1$.

(W$_\vee$) $q = q_0 \vee q_1$, *where* $p \sqsubseteq q_0$ *or* $p \sqsubseteq q_1$.

Remarks. In (C), C stands for Cover; W stands for Ph. M. Whitman. Each Whitman condition assumes a $\wedge$ or $\vee$ on the left or right of $\sqsubseteq$; at most two of these conditions may be applicable to a particular $p \sqsubseteq q$. Note also, that if p and $q \in Q$, then only (C) can apply.

Definition 5 gives essentially the algorithm we have been looking for. For $p, q \in \mathbf{P}(Q)$, it will be shown that p and q represent the same element of the free product iff $p \sqsubseteq q$ and $q \sqsubseteq p$. We shall show this by actually exhibiting the free product as the set of equivalence classes of $\mathbf{P}(Q)$ under this relation. To be able to do this we have to establish a number of properties of the relation $\sqsubseteq$. All the proofs are by induction and will use the *rank* $r(p)$ of a polynomial $p \in \mathbf{P}(Q)$: for $p \in Q$, $r(p) = 1$; $r(p \wedge q) = r(p \vee q) = r(p) + r(q)$.

Lemma 6. *Let $p, q, r \in \mathbf{P}(Q)$ and $i \in I$.*

(i) $p \sqsubseteq p$.

(ii) $p \sqsubseteq q$ *implies that* $p_{(i)} \leq q_{(i)}$ *and* $p^{(i)} \leq q^{(i)}$.

(iii) $p \sqsubseteq q$ *and* $q \sqsubseteq r$ *imply that* $p \sqsubseteq r$.

Proof.

(i) Proof by induction on $r(p)$. If $p \in Q$, then $p \in L_i$ for some $i \in I$. Hence $p = p_{(i)} = p^{(i)}$ by 3(i), and so $p \sqsubseteq p$ by (C). Let $p = q \wedge r$. Then $q \sqsubseteq q$ and $r \sqsubseteq r$ by induction hypothesis. By ($_\wedge$W), $q \wedge r \sqsubseteq q$ and $q \wedge r \sqsubseteq r$, hence by (W$_\wedge$), $q \wedge r \sqsubseteq q \wedge r$, that is, $p \sqsubseteq p$. We proceed similarly if $p = q \vee r$. ◑

(ii) If $p \sqsubseteq q$ by (C), then $p^{(j)} \leq q_{(j)}$ for some $j \in I$. We conclude on the one hand, that $p_{(j)} \leq q_{(j)}$ and $p^{(j)} \leq q^{(j)}$ by Corollary 4, and on the other that $p^{(j)}$ and $q_{(j)}$ are proper; thus again by Corollary 4, $p_{(i)}$ and $q^{(i)}$ are not proper for $i \neq j$, hence $p_{(i)} = 0^b$, $q^{(i)} = 1^b$, and so $p_{(i)} \leq q_{(i)}$ and $p^{(i)} \leq q^{(i)}$.

Now we induct on $r(p) + r(q)$. If $p, q \in Q$, then only (C) applies to $p \sqsubseteq q$; in this case and also in the induction step if (C) is applied, we obtain the result by the last paragraph. Now if $p = p_0 \wedge p_1$ and ($_\wedge$W) applies, then $p_0 \sqsubseteq q$ or $p_1 \sqsubseteq q$, say $p_0 \sqsubseteq q$. Then $(p_0)_{(i)} \leq q_{(i)}$ and $(p_0)^{(i)} \leq q^{(i)}$, hence

$$p_{(i)} = (p_0)_{(i)} \wedge (p_1)_{(i)} \leq (p_0)_{(i)} \leq q_{(i)}$$

and similarly for upper covers. ◑

(iii) If $p \subseteq q$ by (C), then $p^{(i)} \leq q_{(i)}$ for some $i \in I$. By (ii), $q_{(i)} \leq r_{(i)}$, hence $p^{(i)} \leq r_{(i)}$ and so by (C), $p \subseteq r$. This takes care of the base of the induction, $p, q, r \in Q$ since then only (C) applies. We induct on the sum of the ranks of p, q, and r. If $p \subseteq q$ by (C), $p \subseteq r$ has already been proved.

If $p \subseteq q$ follows from $({}_\wedge W)$, then $p = p_0 \wedge p_1$ and $p_0 \subseteq q$ or $p_1 \subseteq q$. Thus, by the induction hypotheses, $p_0 \subseteq r$ or $p_1 \subseteq r$, and so by $({}_\wedge W)$, $p_0 \wedge p_1 = p \subseteq r$.

If $p \subseteq q$ follows from $({}_\vee W)$, then $p = p_0 \vee p_1$, $p_0 \subseteq q$, and $p_1 \subseteq q$, and so again $p_0 \subseteq r$ and $p_1 \subseteq r$, implying $p_0 \vee p_1 = p \subseteq r$ by $({}_\vee W)$.

If $q \subseteq r$ follows from (C), $(W_\wedge)$, or $(W_\vee)$, we can proceed dually (that is, by interchanging $\wedge$ and $\vee$). Only two cases remain; since the second is the dual of the first, we shall state only one: $q = q_0 \wedge q_1$, $(W_\wedge)$ applies to $p \subseteq q$, and $({}_\wedge W)$ is applied to $q \subseteq r$ (observe that $({}_\vee W)$ is not applicable). Then $p \subseteq q_0$ and $p \subseteq q_1$, and $q_0 \subseteq r$ or $q_1 \subseteq r$. Hence $p \subseteq q_i \subseteq r$ for $i = 0$ or 1, hence by the induction hypotheses, $p \subseteq r$. ●

By Lemma 6, the relation $\subseteq$ is a quasi-ordering and so (Exercise I.1.28) we can define:

$$p \equiv q \quad \text{iff} \quad p \subseteq q \quad \text{and} \quad q \subseteq p \qquad (p, q \in \mathbf{P}(Q)).$$
$$R(p) = \{q \mid q \in \mathbf{P}(Q) \text{ and } p \equiv q\} \qquad (p \in \mathbf{P}(Q)).$$
$$R(Q) = \{R(p) \mid p \in \mathbf{P}(Q)\}.$$
$$R(p) \leq R(q) \quad \text{iff} \quad p \subseteq q.$$

In other words, we split $\mathbf{P}(Q)$ into blocks under the equivalence relation $p \equiv q$; $R(Q)$ is the set of blocks which we partially order by $\leq$.

Lemma 7. *$R(Q)$ is a lattice, in fact,*

$$R(p) \wedge R(q) = R(p \wedge q) \quad \text{and} \quad R(p \vee q) = R(p) \vee R(q).$$

Furthermore, if $a, b, c, d \in L_i$, $i \in I$, and $a \wedge b = c$, $a \vee b = d$ in L_i, then

$$R(a) \wedge R(b) = R(c) \quad \text{and} \quad R(a) \vee R(b) = R(d);$$

and if $x, y \in Q$, $x \neq y$, then $R(x) \neq R(y)$.

Proof. $p \wedge q \subseteq p$ and $p \wedge q \subseteq q$ by $p \subseteq p$, $q \subseteq q$, and $({}_\wedge W)$. If $r \subseteq p$ and $r \subseteq q$, then $r \subseteq p \wedge q$ by $(W_\wedge)$; this argument and its dual give the first statement.

$c \subseteq a$ and $c \subseteq b$ is obvious by (C), hence $R(c) \leq R(a)$ and $R(c) \leq R(b)$. Now let $R(p) \leq R(a)$ and $R(p) \leq R(b)$ for some $p \in \mathbf{P}(Q)$. Then $p \subseteq a$ and $p \subseteq b$, and so, by Lemma 6, $p^{(i)} \leq a^{(i)} = a$ and $p^{(i)} \leq b^{(i)} = b$. Therefore, $p^{(i)} \leq c = c_{(i)}$ and thus $p \subseteq c$ by (C). The second part follows by duality. Finally, if $R(x) = R(y)$ $(x, y \in Q)$, then $x \subseteq y$. Since only (C) applies, $x^{(i)} \leq y_{(i)}$ for some $i \in I$. Thus $x^{(i)}$ and $y_{(i)}$ are proper and so $x = x^{(i)}$, $y = y_{(i)}$. We conclude that $x \leq y$; similarly, $y \leq x$. Thus $x = y$. ●

By Lemma 7

$$a \to R(a), \quad a \in L_i,$$

is an embedding of L_i into $R(Q)$, for $i \in I$. Therefore, by identifying $a \in L_i$ with $R(a)$ we get each L_i as a sublattice of $R(Q)$ and hence $Q \subseteq R(Q)$. It is also obvious that the partial ordering induced by $R(Q)$ on Q agrees with the original partial ordering.

Theorem 8. *$R(Q)$ is a free product of the L_i, $i \in I$.*

Proof. 1(i) and 1(ii) have already been observed. Let $\varphi_i \colon L_i \to A$ be given for all $i \in I$. We define inductively a map

$$\psi \colon \mathbf{P}(Q) \to A$$

as follows: for $p \in Q$ there is exactly one $i \in I$ with $p \in L_i$; set $p\psi = p\varphi_i$; if $p = p_0 \wedge p_1$ or $p = p_0 \vee p_1$, and if $p_0\psi$ and $p_1\psi$ have already been defined, then set $p\psi = p_0\psi \wedge p_1\psi$ or $p\psi = p_0\psi \vee p_1\psi$, respectively. Now we prove:

Lemma 9. *Let $p \in \mathbf{P}(Q)$ and $i \in I$.*

(i) *If $p_{(i)}$ is proper, then $p_{(i)}\psi \leq p\psi$.*
(ii) *If $p^{(i)}$ is proper, then $p\psi \leq p^{(i)}\psi$.*
(iii) *$p \subseteq q$ implies that $p\psi \leq q\psi$ for $p, q \in \mathbf{P}(Q)$.*

Proof.

(i) If $p \in Q$ and $p_{(i)}$ is proper, then $p \in L_i$. Hence $p = p_{(i)}$ and so $p_{(i)}\psi \leq p\psi$ is obvious. If $p = q \wedge r$, then $p_{(i)} = q_{(i)} \wedge r_{(i)}$ so $q_{(i)}$ and $r_{(i)}$ are proper. Thus $q_{(i)}\psi \leq q\psi$ and $r_{(i)}\psi \leq r\psi$ by induction; hence $p_{(i)}\psi = q_{(i)}\psi \wedge r_{(i)}\psi \leq q\psi \wedge r\psi = (q \wedge r)\psi = p\psi$. If $p = q \vee r$, then $q_{(i)}$ or $r_{(i)}$ is proper. If both $q_{(i)}$ and $r_{(i)}$ are proper, we proceed as in the previous case. If, say, $q_{(i)}$ is proper and $r_{(i)} = 0^b$, then $p_{(i)}\psi = (q_{(i)} \vee r_{(i)})\psi = q_{(i)}\psi \leq q\psi \leq p\psi$. ◑

(ii) This follows by duality from (i). ◑

(iii) If $p \subseteq q$ follows from (C), then, for some $i \in I$, $p^{(i)} \leq q_{(i)}$. Thus $p^{(i)}$ and $q_{(i)}$ are proper. Therefore, $p\psi \leq p^{(i)}\psi$ by (ii), $p^{(i)}\psi \leq q_{(i)}\psi$ because $p^{(i)}$ and $q_{(i)} \in Q$, and $q_{(i)}\psi \leq q\psi$ by (i), implying $p\psi \leq q\psi$.
This takes care of $p, q \in Q$ and the first case in the induction step.
If $p \subseteq q$ follows from $({}_\wedge\mathrm{W})$, then $p = p_0 \wedge p_1$, where $p_0 \subseteq q$ or $p_1 \subseteq q$. Hence $p_0\psi \leq q\psi$ or $p_1\psi \leq q\psi$, therefore $p\psi = p_0\psi \wedge p_1\psi \leq q\psi$.
If $p \subseteq q$ follows from $({}_\vee\mathrm{W})$, $(\mathrm{W}_\wedge)$, or $(\mathrm{W}_\vee)$, the proof is analogous to the proof in the last case. ●

Now take a $p \in \mathbf{P}(Q)$ and define

$$R(p)\varphi = p\psi.$$

φ is well-defined since if $R(p) = R(q)$ $(p, q \in \mathbf{P}(Q))$, then $p \subseteq q$ and $q \subseteq p$. Hence, by Lemma 9, $p\psi \leq q\psi$ and $q\psi \leq p\psi$, and so $p\psi = q\psi$. Since

$$(R(p) \wedge R(q))\varphi = R(p \wedge q)\varphi = (p \wedge q)\psi = p\psi \wedge q\psi = R(p)\varphi \wedge R(q)\varphi$$

and similarly for $\vee$, we conclude that φ is a homomorphism. Finally, for $p \in L_i$, $i \in I$,

$$R(p)\varphi = p\psi = p\varphi_i$$

by the definition of ψ, hence φ restricted to L_i agrees with φ_i. ●

Lemma 6(ii) implies that if $p \equiv q\,(p, q \in \mathbf{P}(Q))$, then, for all $i \in I$, $p_{(i)} = q_{(i)}$ and $p^{(i)} = q^{(i)}$. Hence we can define

$$(R(p))_{(i)} = p_{(i)} \quad \text{and} \quad (R(p))^{(i)} = p^{(i)}.$$

All our results will now be summarized (G. Grätzer, H. Lakser, and C. R. Platt [1970]):

Theorem 10 (The Structure Theorem of Free Products). *Let L_i, $i \in I$, be lattices and let L be a free product of the L_i, $i \in I$. Then for every $a \in L$ and $i \in I$, if some element of L_i is contained in a, then there is a largest one with this property, $a_{(i)}$. If $a = p(a_0, \ldots, a_{n-1})$, where p is an n-ary polynomial and $a_0, \ldots, a_{n-1} \in \bigcup(L_i \mid i \in I)$, then $a_{(i)}$ can be computed by the algorithm given in Definition 3. Dually, $a^{(i)}$ can be computed. For $a, b \in L$, $a = p(a_0, \ldots, a_{n-1})$, $b = q(b_0, \ldots, b_{m-1})$, $a_0, \ldots, a_{n-1}$, $b_0, \ldots, b_{m-1} \in \bigcup(L_i \mid i \in I)$, we can decide whether $a \leq b$ using the algorithm of Definition 5.*

The idea of the proof of Theorem 10 goes back to Ph. M. Whitman [1941] and R. P. Dilworth [1945].

We should comment on the use of the term "algorithm" in Theorem 10 and elsewhere in this section. (C) of Definition 5 brings in covers and so the procedure described in Definition 3. This procedure is an algorithm insofar as the structure of L_i is described in an effective way. Thus if we consider the free product of finitely many finite lattices then we really deal with an algorithm. In the general case, algorithm should be interpreted intuitively and not be given the precise meaning assigned to it in mathematical logic.

The existence of covers is not special to **L**, as it was observed in B. Jónsson [1971]. To prove their existence let L be a free **K**-product of L_i, $i \in I$. Fix an $i \in I$ and for $j \in I$ define $\psi_j \colon L_j \to (L_i)^b$ as follows: ψ_i is the identity map on L_i; for $j \neq i$, $a\psi_j = 0^b$ for all $a \in L_j$. By Exercise I.4.14, $(L_i)^b \in \mathbf{K}$ and so by the definition of free **K**-product, there is a homomorphism $\psi \colon L \to (L_i)^b$ extending all ψ_j, $j \in I$. For $a \in L$, $a_{(i)} = a\psi$. Now if $b \in L_i$, $b \leq a$, then $b\psi \leq a\psi$, that is, $b \leq a_{(i)}$. Thus if there is $b \in L_i$, $b \leq a$, then $a_{(i)}$ is the largest such element. Otherwise $a\psi = 0_b$; indeed, if $a\psi \neq 0_b$, then $b = a\psi \in L_i$, a contradiction in view of $b = a\psi \leq a$. (We use the fact that $a\psi \leq a$ for all $a \in L$ since this holds for all $a \in \bigcup(L_i \mid i \in I)$.)

Free products (in fact, free **K**-products) of two lattices satisfy a special condition which is due to G. Grätzer, H. Lakser, and C. R. Platt [1970]:

Theorem 11 (The Splitting Theorem). *Let the lattice L be a free product of the lattices L_0 and L_1. For every $a \in L$, if $a_{(0)}$ is not proper then $a^{(1)}$ is proper and conversely.*

Thus

$$L = (L_0] \cup [L_1),$$

where the union is a disjoint union. In other words, for every element a of L either a is contained in some element of L_0 or it contains an element of L_1.

Proof. $(L_0] \cup [L_1)$ is a sublattice of L containing a generating set, namely, $L_0 \cup L_1$. Hence $(L_0] \cup [L_1) = L$. Now Theorem 10 yields the statement. ●

Corollary 12. *$L_0 * L_1$ can be represented as a disjoint union of four convex sublattices: the convex sublattices generated by L_0 and L_1, respectively, $(L_0] \cap (L_1]$, and $[L_0) \cap [L_1)$.*

The next result is a trivial application of Theorem 10.

Corollary 13. *Let K_i be a sublattice of L_i, for $i \in I$, and let L be a free product of the L_i, $i \in I$. Let K be the sublattice of L generated by $\bigcup(K_i \mid i \in I)$. Then K is a free product of the K_i, $i \in I$.*

Proof. This is obvious since for $p, q \in \mathbf{P}(\bigcup(K_i \mid i \in I))$ we have $p \subseteq q$ iff it follows from Definition 5 and in applying Definition 5 we use only elements of the K_i, $i \in I$. Thus, by the proof of Theorem 8, K is a free product of the K_i, $i \in I$. ●

This result is a special case of a result of B. Jónsson [1961] and [1965]:

Theorem 14. *Let $\mathbf{K}$ be an equational class satisfying the Amalgamation Property. Let $A, B, C \in \mathbf{K}$ and let C be a free $\mathbf{K}$-product of A and B. Let A_1 be a sublattice of A, let B_1 be a sublattice of B, and let C' be the sublattice of C generated by $A_1 \cup B_1$. Then C' is a free $\mathbf{K}$-product of A_1 and B_1.* (See Figure 4.)

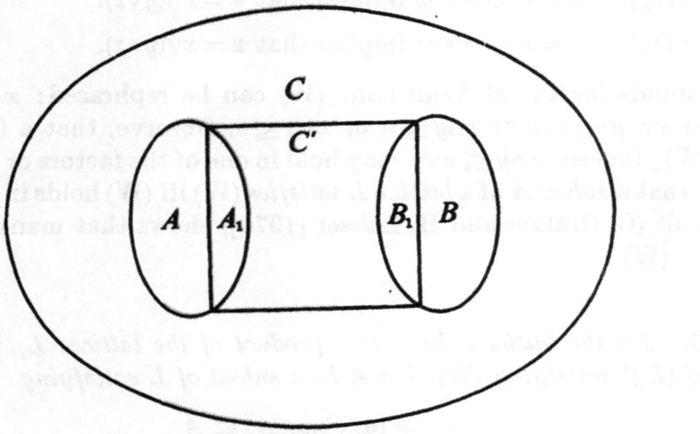

Figure 4

Proof. Let C'' be a free $\mathbf{K}$-product of A_1 and B_1. Thus there exists a homomorphism χ of C'' into C' such that χ is the identity on A_1 and B_1. A_1 is a sublattice of A and of C''; thus by the Amalgamation Property there is a lattice D in $\mathbf{K}$ containing A and C'' as sublattices, $A_1 \subseteq A \cap C''$. Similarly, B_1 is a sublattice of D and B, and thus there exists a lattice E in $\mathbf{K}$ containing B and D as sublattices, $B_1 \subseteq B \cap D$

(see Figure 5). Since C is a free product of A and B, there exists a homomorphism φ_1 of C into E such that φ_1 is the identity on A and B. Let φ be the restriction of φ_1 to C'. Then φ maps C' into C'', $\varphi\chi$ is the identity on A_1 and B_1, and $\varphi\chi$ is thus the identity on C'. Similarly, $\chi\varphi$ is the identity on C'', so φ is an isomorphism between C' and C''. ●

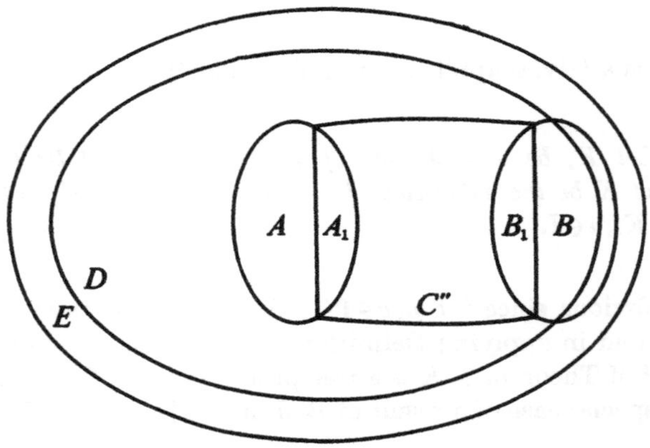

Figure 5

There are three properties that play an important role for free lattices. They are the following:

(W) $x\wedge y \leq u\vee v$ implies that $[x\wedge y, u\vee v] \cap \{x, y, u, v\} \neq \varnothing$.

($SD_\wedge$) $u = x\wedge y = x\wedge z$ implies that $u = x\wedge(y\vee z)$.

($SD_\vee$) $u = x\vee y = x\vee z$ implies that $u = x\vee(y\wedge z)$.

W again stands for Ph. M. Whitman. (W) can be rephrased: $x\wedge y \leq u\vee v$ implies that $x \leq u\vee v$ or $y \leq u\vee v$ or $x\wedge y \leq u$ or $x\wedge y \leq v$. Observe, that a free product need not satisfy (W); indeed, $x\wedge y \leq u\vee v$ may hold in one of the factors or on account of (C).

Let us say that a *subset* A of a lattice L *satisfies* (W) iff (W) holds in L for $x, y, z, u \in A$. The next result (G. Grätzer and H. Lakser [1974]) shows that many subsets of a free product have (W).

THEOREM 15. *Let the lattice L be a free product of the lattices L_i, $i \in I$. Let A_i, $i \in I$, be a subset of $(L_i)^b$ satisfying* (W). *Let A be a subset of L satisfying*

$$A_{(i)} = \{a_{(i)} \mid a \in A\} \subseteq A_i$$

and

$$A^{(i)} = \{a^{(i)} \mid a \in A\} \subseteq A_i,$$

for all $i \in I$. Then A satisfies (W) *in L.*

PROOF. Let $x, y, u, v \in A$, $x = R(p)$, $y = R(q)$, $u = R(r)$, $v = R(s)$, where $p, q, r, s \in \mathbf{P}(Q)$, Let $x\wedge y \leq u\vee v$. Then $p\wedge q \subseteq r\vee s$. If (C) applies, then, for some $i \in I$, $(p\wedge q)^{(i)} \leq$

$(r \vee s)_{(i)}$, hence $p^{(i)} \wedge q^{(i)} \leq r_{(i)} \vee s_{(i)}$ in A_i. Therefore, $p^{(i)} \leq r_{(i)} \vee s_{(i)}$, or $q^{(i)} \leq r_{(i)} \vee s_{(i)}$, or $p^{(i)} \wedge q^{(i)} \leq r_{(i)}$, or $p^{(i)} \wedge q^{(i)} \leq s_{(i)}$. Again by (C) we conclude that $p \subseteq r \vee s$, or $q \subseteq r \vee s$, or $p \wedge q \subseteq r$, or $p \wedge q \subseteq s$. $({}_\wedge\mathrm{W})$ and $(\mathrm{W}_\vee)$ give exactly the same conclusions. We conclude therefore that (W) holds in A. ●

Observe how strong (W) really is. For instance, it follows immediately, that there is no element in A which is doubly reducible.

Naturally, every $a \in L$ has infinitely many representations of the form $p = p(a_0, \ldots, a_{n-1}) \in \mathbf{P}(Q)$ $(a_0, \ldots, a_{n-1} \in Q)$. If $r(p)$ is minimal we call p a *minimal representation of a* and we call p a *minimal polynomial*. Using the notation of Theorem 10, the next result (H. Lakser [1970a]) tells us how to recognize a minimal representation:

Theorem 16. *Let $p \in \mathbf{P}(Q)$. Then p is a minimal representation iff $p \in Q$, or if $p = p_0 \vee \cdots \vee p_{n-1}$, $n > 1$, where no p_j is a join of more than one polynomial and conditions* (i)–(v) *below hold, or the dual of the preceding case holds.*

(i) *Each p_j is minimal, $0 \leq j < n$.*
(ii) *For each $0 \leq j < n$, $p_j \not\subseteq p_0 \vee \cdots \vee p_{j-1} \vee p_{j+1} \vee \cdots \vee p_{n-1}$.*
(iii) *If $0 \leq j < n$, $r(p_j) > 1$, $i \in I$, then $(p_j)^{(i)} \nleq p_{(i)}$ in $(L_i)^b$.*
(iv) *If $p_j = p_j' \wedge p_j''$ $(0 \leq j < n$ and $p_j', p_j'' \in \mathbf{P}(Q))$, then $p_j' \not\subseteq p$ and $p_j'' \not\subseteq p$.*
(v) *If $p_j, p_k \in L_i$ $(0 \leq j \leq k < n$ and $i \in I)$, then $j = k$.*

Proof. Consider the polynomials

$$\begin{aligned}
q_i &= p_0 \vee \cdots \vee p_{j-1} \vee p_j' \vee p_{j+1} \vee \cdots \vee p_{n-1}, \text{ where } p_j' \equiv p_j \text{ and } r(p_j') < r(p_j),\\
q_{ii} &= p_0 \vee \cdots \vee p_{j-1} \vee p_{j+1} \vee \cdots \vee p_{n-1},\\
q_{iii} &= p_0 \vee \cdots \vee p_{j-1} \vee (p_j)^{(i)} \vee p_{j+1} \vee \cdots \vee p_{n-1},\\
q_{iv} &= p_0 \vee \cdots \vee p_{j-1} \vee p_j' \vee p_{j+1} \vee \cdots \vee p_{n-1},\\
q_v &= p_0 \vee \cdots \vee p_{j-1} \vee p_{j+1} \vee \cdots \vee p_{k-1} \vee p_{k+1} \vee \cdots \vee p_{n-1} \vee q,
\end{aligned}$$

where $q \in L_i$ and $q = p_j \vee p_k$ in L_i.

It is obvious, that if condition (x) fails $(i \leq x \leq v)$, then $p \equiv q_x$ and $r(q_x) < r(p)$, contradicting the minimality of p.

To prove the converse we show that if $p, q \in \mathbf{P}(Q)$, $p \equiv q$, p satisfies (i)–(v), and q is minimal, then $r(p) = r(q)$. This gives us an algorithm to reduce any polynomial to a minimal one and at the same time verifies the converse. Indeed, if p was not minimal, there would be a q with $p \equiv q$, $r(p) > r(q)$, and q minimal, contradicting the above statement.

So let p and q be given as specified, $p = p_0 \vee \cdots \vee p_{n-1}$, $n > 1$.

Firstly, we claim that $r(q) > 1$. Indeed, if $r(q) = 1$, then $q \in L_i$, for some $i \in I$. Thus, by Lemma 6(ii), $p_{(i)} = p^{(i)} = q$. At most one component of p is in Q. Indeed, if $p_j, p_k \in Q$, $j \neq k$, $p_j \in L_{i_j}$, $p_k \in L_{i_k}$, then by (v), $i_j \neq i_k$, and so one of them is not i, say $i \neq i_j$. But then $p_{i_j} \leq p$, hence $p_{(i_j)}$ is proper, which contradicts by Corollary 4 the fact that $p^{(i)}$ is proper. Thus there is a j with $0 \leq j < n$ and $r(p_j) > 1$. Then $(p_j)^{(i)} \leq p^{(i)} = q$ and so $(p_j)^{(i)}$ is proper, and $(p_j)^{(i)} \leq q = p_{(i)}$, contradicting the fact that p satisfies (iii).

Secondly, we claim that q is not of the form $q_0 \wedge q_1$. Indeed, let $q = q_0 \wedge q_1$. Consider $q \subseteq p$. If (C) applies, then $q^{(i)} \leq p_{(i)} = q_{(i)}$, hence $q_{(i)} = q^{(i)} \equiv q$, contradicting $r(q) > 1$

and the minimality of q. If $({}_\wedge W)$ applies, then $q_0 \subseteq p$ or $q_1 \subseteq p$. Obviously, $p \subseteq q_i$, thus if, say, $q_0 \subseteq p$, then $p \equiv q_0$, contradicting the minimality of q. Finally, if $(W_\vee)$ applies, then $q \subseteq p_0 \vee \cdots \vee p_{n-2}$ or $q \subseteq p_{n-1}$. The first possibility yields $p_{n-1} \subseteq p_0 \vee \cdots \vee p_{n-2}$ while the second gives $p_{n-1} \equiv p$ (and therefore $p_0 \subseteq p_1 \vee \cdots \vee p_{n-1}$), both contradicting the fact that p satisfies (ii).

Thus q is of the form

$$q = q_0 \vee \cdots \vee q_{m-1}, \quad m > 1,$$

where no q is the join of two polynomials.

Next we show that there are functions

$$f: \{0, 1, \ldots, n-1\} \to \{0, 1, \ldots, m-1\}, \quad g: \{0, 1, \ldots, m-1\} \to \{0, 1, \ldots, n-1\}$$

satisfying the following conditions:

(a) $g(f(j)) = j$ for $0 \leq j < n$ and $f(g(j)) = j$ for $0 \leq j < m$.

(b) If $0 \leq j < n$ and $r(p_j) > 1$, then $q_{f(j)} \equiv p_j$ and $r(q_{f(j)}) = r(p_j)$, and similarly for any $0 \leq j < m$ with $r(q_j) > 1$.

(c) If $p_j \in L_i$ $(0 \leq j < n$ and $i \in I)$, then $q_{f(j)} \in L_i$, and similarly for $q_j \in L_i$.

Let $0 \leq j < n$ and $r(p_j) > 1$. Then $p_j \subseteq q$. If (C) is applicable, then $(p_j)^{(i)} \leq q_{(i)}$, for some $i \in I$; since $q_{(i)} = p_{(i)}$, we obtain $(p_j)^{(i)} \leq p_{(i)}$, contradicting condition (iii) for p. $({}_\wedge W)$ is not applicable either because it would contradict that p satisfies (iv). Hence only $(W_\vee)$ is applicable. Thus $p_j \subseteq q_0 \vee \cdots \vee q_{m-2}$ or $p_j \subseteq q_{m-1}$. Continuing this argument we conclude that $p_j \subseteq q_{f(j)}$ for some $0 \leq f(j) < m$. If $q_{f(j)} \in L_i$, for some $i \in I$, then $p_j \subseteq q_{f(j)}$ implies that $(p_j)^{(i)} \leq q_{f(j)}$ in L_i, thus $(p_j)^{(i)} \leq q_{f(j)} \leq q_{(i)} = p_{(i)}$, contradicting the fact that condition (iii) holds for p. Therefore, $r(q_{f(j)}) > 1$.

Since q is minimal it satisfies (i)–(v), hence we can define the function g. Thus

$$p_j \subseteq q_{f(j)} \subseteq p_{g(f(j))},$$

and so by (ii), $g(f(j)) = j$ and $p_j \equiv q_{f(j)}$. Similarly, $f(g(j)) = j$ and $q_j \equiv p_{g(j)}$.

Now let $0 \leq j < n$ and $r(p_j) = 1$. Then $p_j \in L_i$ for some $i \in I$. Since $p_j \subseteq q$ we get that $q_{(i)}$ is proper. Since $q_{(i)} = (q_0)_{(i)} \vee \cdots \vee (q_{m-1})_{(i)}$, some $(q_j)_{(i)}$ must be proper. By renumbering $q_0, \ldots, q_{m-1}$ we get that $(q_k)_{(i)}$ is proper iff $0 \leq k < t$, where $t \leq m$. Thus

$$q_{(i)} = (q_0)_{(i)} \vee \cdots \vee (q_{t-1})_{(i)}.$$

If $r(q_s) > 1$ for all $0 \leq s < t$, then $r(p_{g(s)}) > 1$ and $(p_{g(s)})_{(i)} = (q_s)_{(i)}$. Therefore, $g(s) \neq j$ for all $0 \leq s < t$, and so

$$p_j \leq (q_0)_{(i)} \vee \cdots \vee (q_{t-1})_{(i)} \leq (p_0 \vee \cdots \vee p_{j-1} \vee p_{j+1} \vee \cdots \vee p_{n-1})_{(i)}.$$

Thus $p_j \subseteq p_0 \vee \cdots \vee p_{j-1} \vee p_{j+1} \vee \cdots \vee p_{n-1}$, contradicting the fact that (ii) holds for p.

Consequently, we can choose $0 \leq f(j) < n$ such that $r(q_{f(j)}) = 1$ and $(q_{f(j)})_{(i)}$ is proper, that is, such that $q_{f(j)} \in L_i$. Similarly, we define $g(j)$ for $0 \leq j < m$ and $r(q_j) = 1$. It is obvious that (a), (b), and (c) are satisfied.

Now (a) implies that f and g are one-to-one, hence $n = m$. Since $r(p_j) = r(q_{f(j)})$, we conclude that $r(p) = r(q)$. ●

A *subset* A of a lattice L is said to *satisfy* $(SD_\vee)$ iff $(SD_\vee)$ holds for all $x, y, z \in A$. The following result (G. Grätzer and H. Lakser [1974]) establishes for $(SD_\vee)$ what was done for (W) in Theorem 15.

THEOREM 17. *Let the lattice L be a free product of the lattices L_i, $i \in I$. For each $i \in I$, let A_i be a subset of $(L_i)^b$ satisfying* (SD$_\vee$). *Let A be a subset of L such that $A_{(i)} \subseteq A_i$, for all $i \in I$. Then A satisfies* (SD$_\vee$).

PROOF. Let $x = R(p)$, $y = R(q)$, $z = R(s) \in A$, $x \vee y = x \vee z$, and let $u = u_0 \vee \cdots \vee u_{n-1}$, $n \geq 1$, be a representation of $p \vee q \equiv p \vee s$ satisfying (i)–(v) of Theorem 16. We show that for each j with $0 \leq j < n$ and $r(u_j) > 1$ we have $u_j \subseteq p$ or $u_j \subseteq q$. Consider $u_j \subseteq p \vee q$. If (C) applies, then $(u_j)^{(i)} \leq (p \vee q)_{(i)} = u_{(i)}$, contradicting 16(iii). If ($_\wedge$W) applies, that is, if $u_j = u_j' \wedge u_j''$ and $u_j' \subseteq p \vee q$, we get a contradiction with 16(iv). Thus only (W$_\vee$) can apply, yielding $u_j \subseteq p$ or $u_j \subseteq q$. Similarly, $u_j \subseteq p$ or $u_j \subseteq s$, hence $u_j \subseteq p \vee (q \wedge s)$. Now take a j with $0 \leq j < n$ and $r(u_j) = 1$. Then $u_j = u_{(i)}$ for some $i \in I$. Therefore, $u_{(i)} = p_{(i)} \vee q_{(i)}$ and $u_{(i)} = p_{(i)} \vee s_{(i)}$ in $(L_i)^b$ and $p_{(i)}, q_{(i)}, s_{(i)} \in A_i$, and so $u_j \leq p_{(i)} \vee (q_{(i)} \wedge s_{(i)})$ by (SD$_\vee$). Thus, $u_j \subseteq p \vee (q \wedge s)$ by (C). Since $u_j \subseteq p \vee (q \wedge s)$ is proved for all j with $0 \leq j < n$, we conclude that $u_0 \vee \cdots \vee u_{n-1} \subseteq p \vee (q \wedge s)$, and so $u = x \vee (y \wedge z)$, as claimed. ●

By dualizing Theorem 17 we can obtain the analogous result for (SD$_\wedge$).

We conclude this section by proving that the *Common Refinement Property* holds for free products (G. Grätzer and J. Sichler [1975]).

THEOREM 18. *Let L be a free product of A_0 and A_1 and also of B_0 and B_1. Then L is a free product of $(A_i \cap B_j \mid i, j = 0, 1, A_i \cap B_j \neq \varnothing)$.*

PROOF. Let $a \in A_0$. We claim that

$$a_{B_0} \in (A_0 \cap B_0) \cup \{0^b\}.$$

Indeed, since L is generated by $B_0 \cup B_1$,

$$a = p(b_{0,0}, b_{0,1}, \ldots, b_{1,0}, b_{1,1}, \ldots),$$

where $b_{0,0}, b_{0,1}, \ldots \in B_0$ and $b_{1,0}, b_{1,1}, \ldots \in B_1$. Computing the lower A_0-covers and observing that $a_{A_0} = a$, we obtain

$$a = p((b_{0,0})_{A_0}, (b_{0,1})_{A_0}, \ldots, (b_{1,0})_{A_0}, (b_{1,1})_{A_0}, \ldots).$$

Forming lower B_0-covers in the original expression for a we get

$$a_{B_0} = p(b_{0,0}, b_{0,1}, \ldots, 0^b, 0^b, \ldots),$$

and from the previous formula:

$$a_{B_0} = (a_{A_0})_{B_0} = p(((b_{0,0})_{A_0})_{B_0}, \ldots, ((b_{1,0})_{A_0})_{B_0}, \ldots).$$

But $b_{1,m} \geq (b_{1,m})_{A_0}$ and so $0^b = (b_{1,m})_{B_0} \geq ((b_{1,m})_{A_0})_{B_0}$, hence $((b_{1,m})_{A_0})_{B_0} = 0^b$. Thus

$$\begin{aligned} a_{B_0} &= p(b_{0,0}, b_{0,1}, \ldots, 0^b, 0^b, \ldots) \\ &= p(((b_{0,0})_{A_0})_{B_0}, ((b_{0,1})_{A_0})_{B_0}, \ldots, 0^b, 0^b, \ldots). \end{aligned}$$

Since p is isotone and $b_{0,m} \geq (b_{0,m})_{A_0} \geq ((b_{0,m})_{A_0})_{B_0}$, we obtain

$$a_{B_0} = p(b_{0,0}, b_{0,1}, \ldots, 0^b, 0^b, \ldots) \geq p((b_{0,0})_{A_0}, (b_{0,1})_{A_0}, \ldots, 0^b, 0^b, \ldots)$$
$$\geq p(((b_{0,0})_{A_0})_{B_0}, ((b_{0,1})_{A_0})_{B_0}, \ldots, 0^b, 0^b, \ldots) = a_{B_0},$$

and so

$$a_{B_0} = p((b_{0,0})_{A_0}, (b_{0,1})_{A_0}, \ldots, 0^b, 0^b, \ldots) \in A_0 \cup \{0^b\}.$$

By definition, $a_{B_0} \in B_0 \cup \{0^b\}$, hence $a_{B_0} \in (A_0 \cap B_0) \cup \{0^b\}$. Similarly, for $i, j \in \{0, 1\}$, $a_{B_j} \in (A_i \cap B_j) \cup \{0^b\}$. It follows immediately, that

$$a = p((b_{0,0})_{A_0}, (b_{0,1})_{A_0}, \ldots, (b_{1,0})_{A_0}, (b_{1,1})_{A_1}, \ldots) \in (A_0 \cap B_0) \cup (A_0 \cap B_1) \cup \{0^b\}.$$

Now a simple induction on the rank of a polynomial proves that for $a \in A_0$ and the polynomial p of smallest rank representing a in the form

$$a = p(a_0, \ldots, a_{n-1}),\ a_0, \ldots, a_{n-1} \in (A_0 \cap B_0) \cup (A_0 \cap B_1) \cup \{0^b\},$$

no a_i is 0^b. We conclude that

$$A_0 \subseteq [(A_0 \cap B_0) \cup (A_0 \cap B_1)].$$

Thus

$$L = [A_0 \cup A_1] = [\bigcup (A_i \cap B_j \mid i, j = 0, 1)].$$

Applying Corollary 13 twice we get

$$A_0 = (A_0 \cap B_0) * (A_0 \cap B_1) \quad \text{and} \quad A_1 = (A_1 \cap B_0) * (A_1 \cap B_1),$$
$$B_0 = (A_0 \cap B_0) * (A_1 \cap B_0) \quad \text{and} \quad B_1 = (A_0 \cap B_1) * (A_1 \cap B_1),$$

hence

$$L = (A_0 \cap B_0) * (A_0 \cap B_1) * (A_1 \cap B_0) * (A_1 \cap B_1)$$

(to be more precise, drop all $A_i \cap B_j = \varnothing$), and this is the common refinement of $A_0 * A_1 = B_0 * B_1$. ●

Exercises

1. Prove that $\mathfrak{C}_2 * \mathfrak{C}_1$ is indeed the lattice of Figure 1.
2. Show that $\mathfrak{C}_3 * \mathfrak{C}_1$ is the lattice of Figure 2.
3. Find an infinite descending chain in $\mathfrak{C}_4 * \mathfrak{C}_1$. (Hint: Let $a_0 < a_1 < a_2 < a_3$ and b be the two chains. Define

$$c_1 = ((a_2 \wedge (a_1 \vee b)) \vee (a_3 \wedge b)) \wedge ((a_3 \wedge (a_0 \vee b)) \vee a_1)$$

and, for $n > 1$,

$$c_n = ((c_{n-1} \wedge a_2) \vee (a_3 \wedge b)) \wedge ((c_{n-1} \wedge (a_0 \vee b)) \vee a_1).$$

Then $c_1 > c_2 > c_3 > \cdots$.)

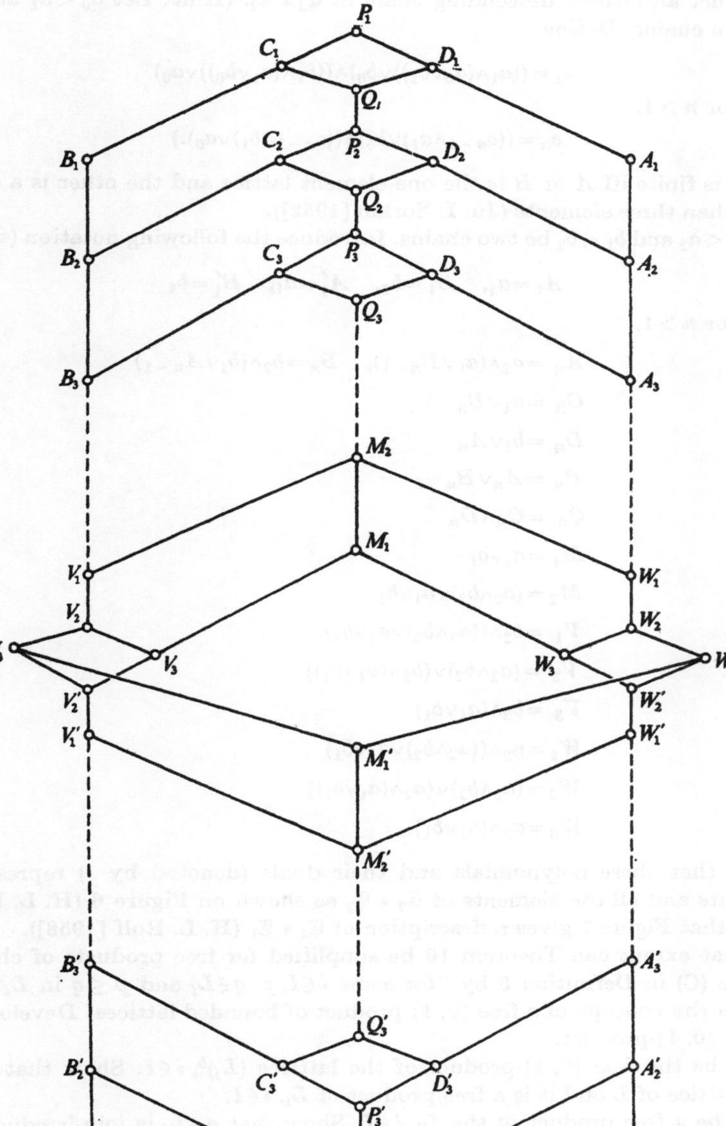
P_1
C_1
D_1
Q_1
P_2
C_2
D_2
B_1
A_1
Q_2
P_3
C_3
D_3
B_2
A_2
Q_3
B_3
A_3
M_2
M_1
V_1
W_1
V_2
W_2
V_3
V_3
W_3
W_3'
V_2'
W_2'
V_1'
W_1'
M_1'
M_2'
B_3'
A_3'
Q_3'
C_3'
D_3'
B_2'
A_2'
P_3'
Q_2'
B_1'
A_1'
C_2'
D_2'
P_2'
Q_1'
C_1'
D_1'
P_1'

Figure 6

4. Construct an infinite descending chain in $\mathfrak{C}_2 * \mathfrak{C}_2$. (Hint: Let $a_0 < a_1$ and $b_0 < b_1$ be the two chains. Define

$$c_1 = ((a_1 \wedge (a_0 \vee b_1)) \vee b_0) \wedge ((b_1 \wedge (a_1 \vee b_0)) \vee a_0)$$

and, for $n > 1$,

$$c_n = ((c_{n-1} \wedge a_1) \vee b_0) \wedge ((c_{n-1} \wedge b_1) \vee a_0).)$$

5. $A * B$ is finite iff A or B is the one-element lattice and the other is a chain of not more than three elements (Ju. I. Sorkin [1952]).
6. Let $a_1 < a_2$ and $b_1 < b_2$ be two chains. Introduce the following notation (see Figure 6):

$$A_1 = a_2, \quad B_1 = b_2, \quad A_1' = a_1, \quad B_1' = b_1$$

and, for $n > 1$,

$$A_n = a_2 \wedge (a_1 \vee B_{n-1}), \quad B_n = b_2 \wedge (b_1 \vee A_{n-1})$$
$$C_n = a_1 \vee B_n$$
$$D_n = b_1 \vee A_n$$
$$P_n = A_n \vee B_n$$
$$Q_n = C_n \wedge D_n$$
$$M_1 = a_1 \vee b_1$$
$$M_2 = (a_2 \wedge b_2) \vee a_1 \vee b_1$$
$$V_1 = b_2 \wedge ((a_2 \wedge b_2) \vee a_1 \vee b_1)$$
$$V_2 = (a_2 \wedge b_2) \vee (b_2 \wedge (a_1 \vee b_1))$$
$$V_3 = b_2 \wedge (a_1 \vee b_1)$$
$$W_1 = a_2 \wedge ((a_2 \wedge b_2) \vee a_1 \vee b_1)$$
$$W_2 = (a_2 \wedge b_2) \vee (a_2 \wedge (a_1 \vee b_1))$$
$$W_3 = a_2 \wedge (a_1 \vee b_1).$$

Prove that these polynomials and their duals (denoted by ′) represent distinct elements and all the elements of $\mathfrak{C}_2 * \mathfrak{C}_2$ as shown on Figure 6 (H. L. Rolf [1958]).
7. Show that Figure 7 gives a description of $\mathfrak{C}_4 * \mathfrak{C}_1$ (H. L. Rolf [1958]).
8. To what extent can Theorem 10 be simplified for free products of chains. (Hint: replace (C) in Definition 5 by "for some $i \in I$, $p, q \in L_i$ and $p \leq q$ in L_i".)
9. Define the concept of a free {0, 1}-product of bounded lattices. Develop the theory of free {0, 1}-product.
10. Let L be the free {0, 1}-product of the lattices $(L_i)^b$, $i \in I$. Show that $L - \{0, 1\}$ is a sublattice of L and it is a free product of L_i, $i \in I$.
11. Let L be a free product of the L_i, $i \in I$. Show that $a \in L_i$ is join-irreducible in L iff a is join-irreducible in L_i. (Hint: use only covers in the proof.)
12. Show that Exercise 11 holds for free **K**-products in general (B. Jónsson [1971]).
13. Let L be a free product of the L_i, $i \in I$, and let $\varphi_i\colon L_i \to A$ be an isotone map of L_i into a lattice A, for $i \in I$. Show that there is an isotone map $\varphi\colon L \to A$ extending the φ_i (Ju. I. Sorkin [1952]).
14. In a free product of chains, the minimal representation of an element is uniquely determined up to commutativity and associativity.
15. Define the *canonical representation* p of an element of a free product in the following way: $p \in Q$; or p is as in Theorem 16, all p_i are canonical and if $p_i \in L_j$, then $p_i = p_{(j)}$;

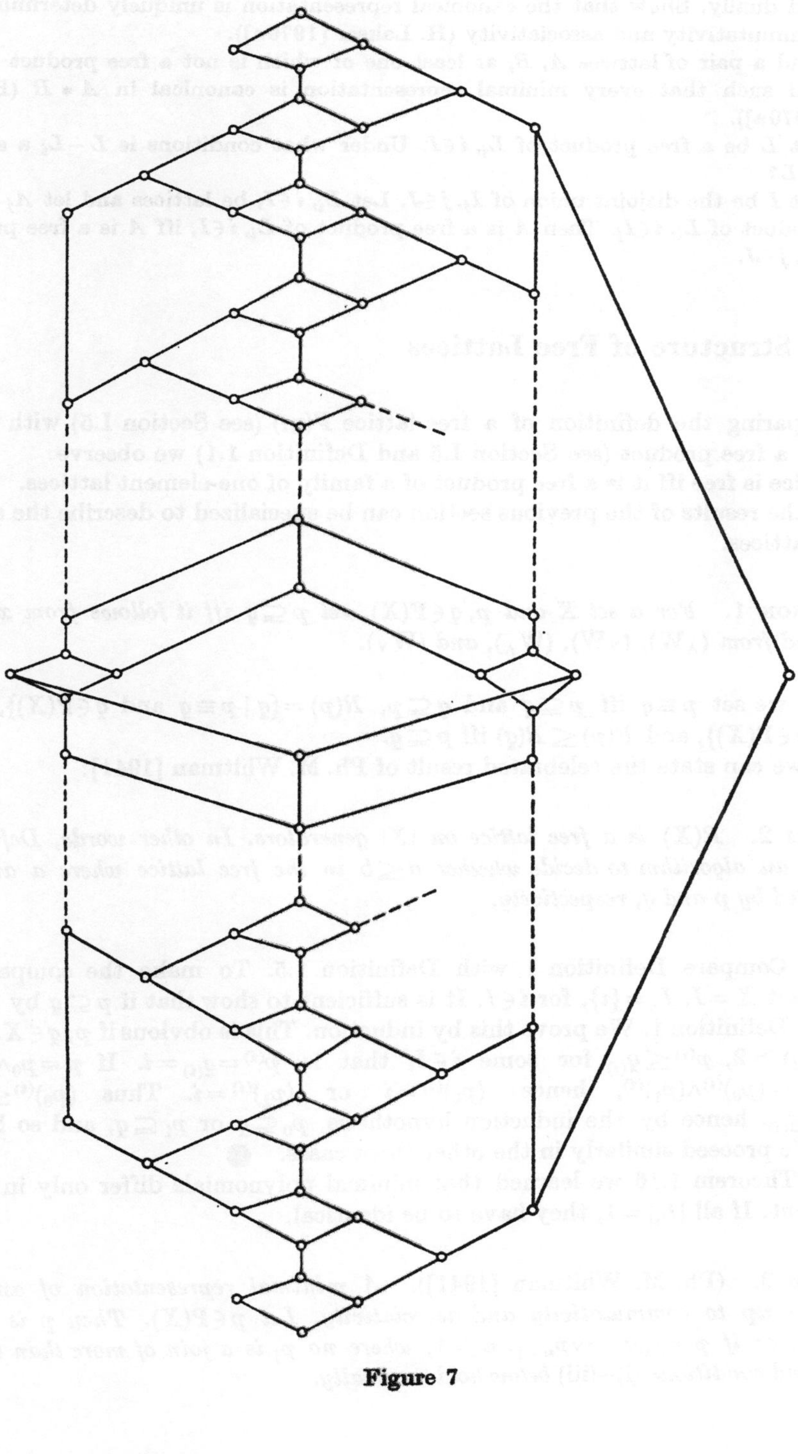

Figure 7

and dually. Show that the canonical representation is uniquely determined up to commutativity and associativity (H. Lakser [1970a]).

16. Find a pair of lattices A, B, at least one of which is not a free product of chains and such that every minimal representation is canonical in $A * B$ (H. Lakser [1970a]).
17. Let L be a free product of L_i, $i \in I$. Under what conditions is $L - L_i$ a sublattice of L?
18. Let I be the disjoint union of I_j, $j \in J$. Let L_i, $i \in I$, be lattices and let A_j be a free product of L_i, $i \in I_j$. Then A is a free product of L_i, $i \in I$, iff A is a free product of A_j, $j \in J$.

2. The Structure of Free Lattices

By comparing the definition of a free lattice $F(\mathfrak{m})$ (see Section I.5) with the definition of a free product (see Section I.5 and Definition 1.1) we observe:

A lattice is free iff it is a free product of a family of one-element lattices.

Thus the results of the previous section can be specialized to describe the structure of free lattices.

DEFINITION 1. *For a set X and $p, q \in \mathbf{P}(X)$, set $p \subseteq q$ iff it follows from $x \subseteq x$, for $x \in X$, and from* $(_\wedge\mathrm{W})$, $(_\vee\mathrm{W})$, $(\mathrm{W}_\wedge)$, *and* $(\mathrm{W}_\vee)$.

Again we set $p \equiv q$ iff $p \subseteq q$ and $q \subseteq p$, $R(p) = \{q \mid p \equiv q$ and $q \in \mathbf{P}(X)\}$, $R(X) = \{R(p) \mid p \in \mathbf{P}(X)\}$, and $R(p) \leq R(q)$ iff $p \subseteq q$.

Now we can state the celebrated result of Ph. M. Whitman [1941]:

THEOREM 2. *$R(X)$ is a free lattice on $|X|$ generators. In other words, Definition 1 provides an algorithm to decide whether $a \leq b$ in the free lattice where a and b are represented by p and q, respectively.*

PROOF. Compare Definition 1 with Definition 1.5. To make the comparison we have to set $X = I$, $L_i = \{i\}$, for $i \in I$. It is sufficient to show that if $p \subseteq q$ by (C), then $p \subseteq q$ by Definition 1. We prove this by induction. This is obvious if $p, q \in X$. Now let $r(p) + r(q) > 2$, $p^{(i)} \leq q_{(i)}$ for some $i \in I$, that is, $p^{(i)} = q_{(i)} = i$. If $p = p_0 \wedge p_1$, then $(p_0 \wedge p_1)^{(i)} = (p_0)^{(i)} \wedge (p_1)^{(i)}$, hence $(p_0)^{(i)} = i$ or $(p_1)^{(i)} = i$. Thus $(p_0)^{(i)} \leq q_{(i)}$ or $(p_1)^{(i)} \leq q_{(i)}$, hence by the induction hypothesis, $p_0 \subseteq q$ or $p_1 \subseteq q$, and so by $(_\wedge\mathrm{W})$, $p \subseteq q$. We proceed similarly in the other three cases. ●

From Theorem 1.16 we learned that minimal polynomials differ only in some L_i component. If all $|L_i| = 1$, they have to be identical.

THEOREM 3. (Ph. M. Whitman [1941]). *A minimal representation of an element is unique up to commutativity and associativity. Let $p \in \mathbf{P}(X)$. Then p is minimal iff $p \in X$, or if $p = p_0 \vee \cdots \vee p_{n-1}$, $n > 1$, where no p_j is a join of more than one polynomial and conditions* (i)–(iii) *below hold, or dually.*

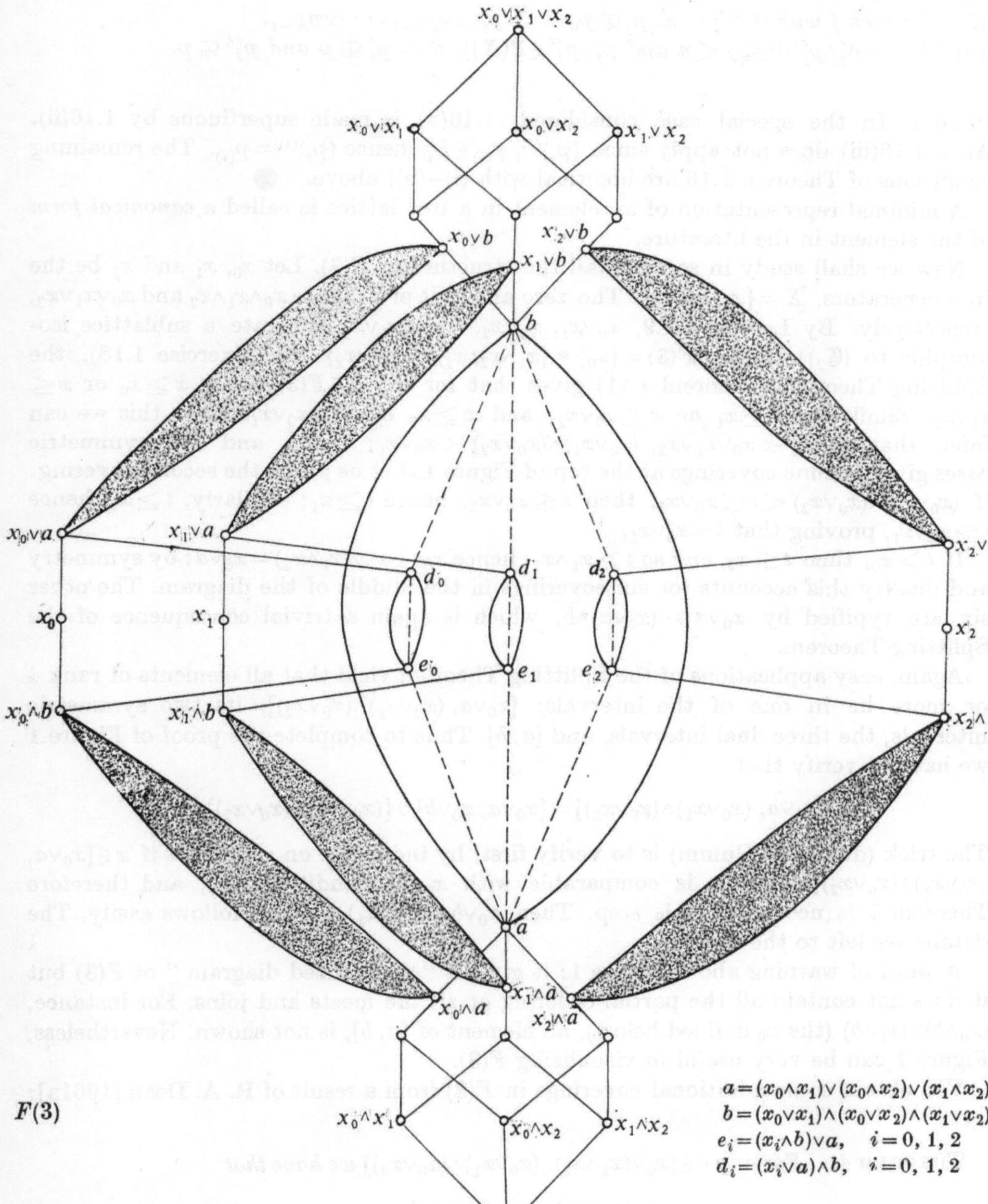

$x_0 \vee x_1 \vee x_2$
$x_0 \vee x_1$
$x_0 \vee x_2$
$x_1 \vee x_2$
$x_0 \vee b$
$x_2 \vee b$
$x_1 \vee b$
b
$x_0 \vee a$
$x_1 \vee a$
$x_2 \vee a$
d_0
d_1
d_2
x_0
x_1
x_2
e_0
e_1
e_2
$x_0 \wedge b$
$x_1 \wedge b$
$x_2 \wedge b$
a
$x_1 \wedge a$
$x_0 \wedge a$
$x_2 \wedge a$
$F(3)$
$x_0 \wedge x_1$
$x_0 \wedge x_2$
$x_1 \wedge x_2$
$x_0 \wedge x_1 \wedge x_2$
$a = (x_0 \wedge x_1) \vee (x_0 \wedge x_2) \vee (x_1 \wedge x_2)$
$b = (x_0 \vee x_1) \wedge (x_0 \vee x_2) \wedge (x_1 \vee x_2)$
$e_i = (x_i \wedge b) \vee a, \quad i = 0, 1, 2$
$d_i = (x_i \vee a) \wedge b, \quad i = 0, 1, 2$

Figure 1

(i) *Each* p_j *is minimal*, $0 \leq j < n$.
(ii) *For each* j *with* $0 \leq j < n$, $p_j \nsubseteq p_0 \vee \cdots \vee p_{j-1} \vee p_{j+1} \vee \cdots \vee p_{n-1}$.
(iii) *If* $p_j = p_j' \wedge p_j''$ $(0 \leq j < n$ *and* $p_j', p_j'' \in \mathbf{P}(X))$, *then* $p_j' \nsubseteq p$ *and* $p_j'' \nsubseteq p$.

PROOF. In the special case considered, 1.16(v) is made superfluous by 1.16(ii). Also, 1.16(iii) does not apply since $(p_j)^{(i)}$, $p_{(i)} \in L_i$, hence $(p_j)^{(i)} = p_{(i)}$. The remaining conditions of Theorem 1.16 are identical with (i)–(iii) above. ●

A minimal representation of an element in a free lattice is called a *canonical form* of the element in the literature.

Now we shall study in some detail the structure of $F(3)$. Let x_0, x_1 and x_2 be the free generators, $X = \{x_0, x_1, x_2\}$. The zero and unit of $F(3)$ are $x_0 \wedge x_1 \wedge x_2$ and $x_0 \vee x_1 \vee x_2$, respectively. By Lemma I.5.9, $x_0 \vee x_1$, $x_0 \vee x_2$, and $x_1 \vee x_2$ generate a sublattice isomorphic to $(\mathfrak{C}_2)^3$. Since $F(3) = \{x_0\} * \{x_1 \wedge x_2, x_1, x_2, x_1 \vee x_2\}$ (see Exercise 1.18), the Splitting Theorem (Theorem 1.11) gives that for any $x \in F(3)$ either $x \geq x_0$ or $x \leq x_1 \vee x_2$. Similarly, $x \geq x_1$ or $x \leq x_0 \vee x_2$; and $x \geq x_2$ or $x \leq x_0 \vee x_1$. From this we can infer that $x_0 \vee x_1 \prec x_0 \vee x_1 \vee x_2$, $(x_0 \vee x_1) \wedge (x_0 \vee x_2) \prec x_0 \vee x_1$; these, and the symmetric cases give the nine coverings at the top of Figure 1. Let us prove the second covering: if $(x_0 \vee x_1) \wedge (x_0 \vee x_2) < t \leq x_0 \vee x_1$, then $t \nleq x_0 \vee x_2$, hence $t \geq x_1$; similarly, $t \geq x_0$, hence $t \geq x_0 \vee x_1$, proving that $t = x_0 \vee x_1$.

If $t > x_0$, then $t \nleq x_0$, and so $t \geq x_1 \wedge x_2$; hence $x_0 \prec x_0 \vee (x_1 \wedge x_2) = x_0 \vee a$; by symmetry and duality this accounts for six coverings in the middle of the diagram. The other six are typified by $x_0 \vee a \succ (x_0 \vee a) \wedge b$, which is again a trivial consequence of the Splitting Theorem.

Again, easy applications of the Splitting Theorem yield that all elements of rank 4 or more lie in one of the intervals: $[x_0 \vee a, (x_0 \vee x_1) \wedge (x_0 \vee x_2)]$, its two symmetric intervals, the three dual intervals, and $[a, b]$. Thus to complete the proof of Figure 1 we have to verify that

$$[x_0 \vee a, (x_0 \vee x_1) \wedge (x_0 \vee x_2)] = [x_0 \vee a, x_0 \vee b] \cup \{(x_0 \vee x_1) \wedge (x_0 \vee x_2)\}.$$

The trick (due to P. Gumm) is to verify first, by induction on $r(x)$, that if $x \in [x_0 \vee a, (x_0 \vee x_1) \wedge (x_0 \vee x_2)]$, then x is comparable with $x_0 \vee b$. Condition (W), and therefore Theorem 5 is needed in this step. Then $x_0 \vee b \prec (x_0 \vee x_1) \wedge (x_0 \vee x_2)$ follows easily. The details are left to the reader.

A word of warning about Figure 1: it gives a " generalized diagram " of $F(3)$ but it does not contain all the partial ordering or all the meets and joins. For instance, $(x_0 \wedge b) \vee (x_1 \wedge b)$ (the u_0 defined below), an element of $[a, b]$, is not shown. Nevertheless, Figure 1 can be very useful in visualizing $F(3)$.

We obtain some additional coverings in $F(3)$ from a result of R. A. Dean [1961a]:

THEOREM 4. *For any* $c \in [x_0 \vee (x_1 \wedge x_2), (x_0 \vee x_1) \wedge (x_0 \vee x_2)]$ *we have that*

$$c \wedge (x_1 \vee x_2) \prec (c \wedge (x_1 \vee x_2)) \vee x_0.$$

PROOF. These two elements are obviously distinct since they belong to disjoint intervals. If $c \wedge (x_1 \vee x_2) \leq t \leq (c \wedge (x_1 \vee x_2)) \vee x_0$ and $t \geq x_0$, then obviously $t = (c \wedge$

$(x_1 \vee x_2)) \vee x_0$. Otherwise, $t \leq x_1 \vee x_2$, hence $t \leq ((c \wedge (x_1 \vee x_2)) \vee x_0) \wedge (x_1 \vee x_2) \leq c \wedge (x_1 \vee x_2)$, and so $t = c \wedge (x_1 \vee x_2)$. ●

Apart from the Splitting Theorem the most important properties of a free lattice are given in the next result, which is due to Ph. M. Whitman [1941] and B. Jónsson [1961]:

Theorem 5. *Conditions* (W), $(\mathrm{SD}_\wedge)$, *and* $(\mathrm{SD}_\vee)$ *hold in any free lattice.*

Proof. If $|A_i| = 1$, then $(A_i)^b = \mathfrak{C}_3$, hence it satisfies all three conditions. So choose $A_i = (L_i)^b$ and the result follows from Theorems 1.15 and 1.17. ●

We illustrate the power of these conditions by two results.

Theorem 6 (B. Jónsson [1961]). *A sublattice A of finite length of a free lattice* (*of any lattice satisfying* $(\mathrm{SD}_\wedge)$) *is finite.*

Proof. Let A be of length n. We prove the statement by induction on n. If $n = 1$, the statement is trivial. Now let A be of length n and let a be an atom of A. Then

$$A = [a) \cup \{x \mid a \wedge x = 0\}.$$

By $(\mathrm{SD}_\wedge)$, $\{x \mid a \wedge x = 0\}$ is a sublattice. Hence A is a union of two sublattices of length less than n, and so A is a union of two finite sets. Thus A is finite. ●

A sublattice of a free lattice need not be free. The next result shows how free sublattices can be found.

Theorem 7. *Let the lattice L be generated by X. If X is* irredundant, *that is,*

$$x \leq \bigvee Y \textit{ implies } x \in Y \textit{ for any } x \in X \textit{ and finite } Y \subseteq X,$$

and dually, then L is a free lattice iff L satisfies (W).

Proof. The necessity of these conditions follows from the Splitting Theorem and from Theorem 5. Now let these conditions hold. With every $a \in L$ associate the set $S(p)$ of all $p \in \mathrm{P}(X)$ which represent a. If $a = S(p)$, $b = S(q)$, then $a \leq b$ iff $p \subseteq q$ by Definition 1. This is easily seen except if $p = p_0 \wedge p_1$ and $q = q_0 \vee q_1$ in which case it follows from (W). Thus L is a free lattice. ●

Now consider in $F(3)$ the elements

$$u_0 = (x_0 \wedge (x_1 \vee x_2)) \vee (x_1 \wedge (x_0 \vee x_2)),$$

$$u_1 = (x_0 \wedge (x_1 \vee x_2)) \vee (x_2 \wedge (x_0 \vee x_1)),$$

$$u_2 = (x_0 \vee (x_1 \wedge x_2)) \wedge (x_1 \vee (x_0 \wedge x_2)),$$

$$u_3 = (x_0 \vee (x_1 \wedge x_2)) \wedge (x_2 \vee (x_0 \wedge x_1)).$$

It is an easy computation to show that $U=\{u_0, u_1, u_2, u_3\}$ is irredundant. Hence $F(3)\supset F(4)$ as a sublattice. Now define $y_0=u_0$, $a_0=u_1$, $b_0=u_2$, $c_0=u_3$. If y_n, a_n, b_n, and c_n are defined, construct y_{n+1}, a_{n+1}, b_{n+1}, and c_{n+1} from a_n, b_n, and c_n, just as the u_i were constructed from x_0, x_1, and x_2.

We claim that $Y=\{y_0, y_1, y_2, \ldots\}$ is an irredundant set. To see this we prove by induction that $y_0, \ldots, y_n, a_n, b_n, c_n$ generate a free lattice on $n+4$ generators. This is true for $n=0$. Assume it for n. Then

$$[y_0, \ldots, y_n, y_{n+1}, a_{n+1}, b_{n+1}, c_{n+1}]$$

is the sublattice generated by $[y_0, \ldots, y_n]$ and $[y_{n+1}, a_{n+1}, b_{n+1}, c_{n+1}]$ in $[y_0, \ldots, y_n, a_n, b_n, c_n]=[y_0, \ldots, y_n] * [a_n, b_n, c_n]$. Thus by Theorem 1.14,

$$\begin{aligned}[y_0, \ldots, y_n, y_{n+1}, a_{n+1}, b_{n+1}, c_{n+1}] &\cong [y_0, \ldots, y_n] * [y_{n+1}, a_{n+1}, b_{n+1}, c_{n+1}]\\ &\cong F(n+1) * F(4) \cong F(n+5).\end{aligned}$$

This proves the statement. Thus Y is irredundant and so, by Theorem 7, $[Y]\cong F(\aleph_0)$. So we have proved:

Theorem 8 (Ph. M. Whitman [1941]). *$F(3)$ contains $F(\aleph_0)$ as a sublattice.*

A free generator of a free lattice is doubly irreducible, and of course no other element is such. Hence we obtain:

Theorem 9. *There is a one-to-one correspondence between automorphisms of a free lattice and permutations of its free generating set.*

This is used in proving the following result of F. Galvin and B. Jónsson [1961].

Theorem 10. *Every chain of a free lattice is countable.*

Proof. Let X be the free generating set of L. We can assume that X is not countable. Let X_0 be a countable subset of L, let $L_0=[X_0]$, and let C be a chain in L.

For $a, b \in L$ let $a\equiv b$ mean that $a\varphi=b$ for some automorphism φ of L. Obviously, $\equiv$ is an equivalence relation. Now if $a\in L$, then $a\in[Y]$ for some finite $Y\subseteq X$. Take a permutation p of X with $Yp\subseteq X_0$, and let φ be the automorphism of L extending p. Then $a\varphi\in L_0$. Hence there are no more equivalence classes than $|L_0|=\aleph_0$. Therefore the proof will be complete if we show that for any $a, b\in C$, $a\neq b$, we have $a\not\equiv b$. Indeed, let $a<b$ and $a\varphi=b$ for some automorphism φ. For some finite $Y\subseteq X$, $a\in[Y]$. There is a permutation p of X such that $xp=x\varphi$ for $x\in Y$ and $xp=x$ for $x\in X-(Y\cup Y\varphi)$. Let ψ be the automorphism of L extending p. Then $a\psi=b$ and ψ is of some finite order n (as a group element). Thus,

$$a<a\psi<a\psi^2<\cdots<a\psi^n=a,$$

a contradiction. ●

A somewhat unexpected dividend of the study of the structure of free lattices is the following result of R. P. Dilworth [1945] and R. A. Dean [1956]:

Theorem 11. *There exists a three-generated partial lattice with infinitely many maximal elements.*

Proof. Let x_0, x_1, x_2 be the free generators of $F(3)$ and define $y_0 = x_0$,

$$y_n = x_0 \vee (x_1 \wedge (x_2 \vee (x_0 \wedge (x_1 \vee (x_2 \wedge y_{n-1}))))),$$

$$y_{-n} = x_0 \wedge (x_1 \vee (x_2 \wedge (x_0 \vee (x_1 \wedge (x_2 \vee y_{-n+1}))))),$$

and

$$a_n = x_1 \vee (y_n \wedge (y_{-n} \vee x_2)).$$

It is easy to check that $[a_1, a_2, \ldots] \cong F(\aleph_0)$.

For a polynomial p, define the *component subset of* p, $\mathrm{Komp}(p)$, as follows: $\mathrm{Komp}(x_i) = \{x_i\}$; $\mathrm{Komp}(p_0 \wedge p_1) = \mathrm{Komp}(p_0) \cup \mathrm{Komp}(p_1) \cup \{p_0 \wedge p_1\}$ and similarly for $p_0 \vee p_1$.

If $a \in F(3)$, let $a = R(p)$ where p is the canonical polynomial representing a. Define $\mathrm{Komp}(a) = \{R(q) \mid q \in \mathrm{Komp}(p)\}$. Then $\mathrm{Komp}(a) \subseteq F(3)$. Regard $\mathrm{Komp}(a)$ as a relative sublattice of $F(3)$. Then $\mathrm{Komp}(a)$ becomes a partial lattice. $\mathrm{Komp}(a)$ is generated by $\{x_0, x_1, x_2\} \cap \mathrm{Komp}(a)$.

Now define $A = \bigcup(\mathrm{Komp}(a_n) \mid n = 1, 2, \ldots)$. All the elements of A are easily enumerated and we observe that $a_1, a_2, \ldots$ are maximal elements of A. ●

Corollary 12 (Ju. I. Sorkin [1954] and R. A. Dean [1956]). *Every countable lattice can be embedded in a three-generated lattice.*

Proof. Let $L = \{b_0, b_1, b_2, b_3, \ldots\}$ and, with the partial lattice A of Theorem 11, form the set $B = L \cup A$ (we assume that L and A are disjoint). We define a partial ordering $\leq$:

for $u, v \in L$ or $u, v \in A$, $u \leq v$ iff $u \leq v$ in L or A, respectively;
$a_{2n+1} \leq b_n$ and $a_{2n+2} \leq b_n$; for $u \in A$ and $v \in L$, $u \leq v$ iff there are $v_0, \ldots, v_{n-1} \in L$ such that (i) $v_0 \wedge \cdots \wedge v_{n-1} \leq v$ in L; (ii) for each i, $0 \leq i < n$, there is a b_j such that $b_j \leq v_i$ in L and $u \leq a_{2j+1}$, $u \leq a_{2j+2}$ in A;
for $u \in A$ and $v \in L$, $v \leq u$ never holds.

Now it is easy to check that B is a partial lattice, $b_n = a_{2n+1} \vee a_{2n+2}$, for $n = 0, 1, \ldots$, and so B is three-generated. Every partial lattice B can be embedded in a lattice C and obviously, $[B]$ in C is three-generated and contains L as a sublattice. ●

Exercises

1. Let L be a lattice completely freely generated by the poset P. Show that an algorithm is provided for deciding whether $a \leq b$ in L by the rules: if $a, b \in P$, $a \sqsubseteq b$ iff $a \leq b$ in P; and by the rules $({}_\wedge\mathrm{W})$, $({}_\vee\mathrm{W})$, $(\mathrm{W}_\wedge)$, and $(\mathrm{W}_\vee)$ (R. P. Dilworth [1945]).
2. Let L be a free product of L_i, $i \in I$, and let $Q = \bigcup(L_i \mid i \in I)$ (disjoint union). Show that L is completely freely generated by Q iff all L_i are chains.

3. Prove that $(\mathfrak{C}_2)^n$ is a sublattice of a free lattice iff $n \leq 3$.
4. Let L be a modular lattice. Verify that if L is a sublattice of a free lattice, then L is distributive.
5. Let the lattice L be of length n. Show that if L is a sublattice of a free lattice, then $|L| \leq 2^n$.
6. Let the lattice L be generated by a finite set X and let us assume that $x \nleq \bigvee(X-\{x\})$ for some $x \in X$. Then for any $a \geq x$ we have that

$$a \wedge \bigvee(X-\{x\}) \prec (a \wedge \bigvee(X-\{x\})) \vee x.$$

7. Let $\mathfrak{m} \geq \aleph_0$. Show that there is no covering in $F(\mathfrak{m})$.
8. Let $\mathfrak{m} \geq \aleph_0$ and let $a, b \in F(\mathfrak{m})$, $a < b$. Show that $[a, b]$ has a sublattice isomorphic to $F(\mathfrak{m})$.
9. Let $f(n) = \max\{|L| \mid L$ is a sublattice of a free lattice and L is of length $n\}$. Then $(\sqrt{2})^n \leq f(n)$ (B. Jónsson and J. E. Kiefer [1962]).
10. A *linear decomposition* A_i, $i \in I$, of a lattice A consists of a chain I, sublattices A_i of A, for $i \in I$, such that if $i, j \in I$, $i < j$, then $a < b$ in A for all $a \in A_i$ and $b \in B_j$ and $A = \bigcup(A_i \mid i \in I)$. Show that if all A_i are sublattices of a free lattice and I is countable, then A is also a sublattice of a free lattice.
11. A distributive lattice D is a sublattice of a free lattice iff D has a linear decomposition A_i, $i \in I$, such that $|I| \leq \aleph_0$ and each A_i is either $\mathfrak{C}_1$, or $(\mathfrak{C}_2)^3$, or of the form $\mathfrak{C}_2 \times C$ where C is a countable chain (F. Galvin and B. Jónsson [1961]).
12. Let A be an algebra with the following properties: a partial ordering $\leq$ is defined on A; A has a generating set X such that every permutation on X can be extended to an isotone automorphism of A. Prove that every chain in A is countable.
13. Enumerate in canonical form all the elements of A in the proof of Theorem 11.
14. Prove that B in the proof of Corollary 12 is a partial lattice.
15. Is there a general lemma about "gluing together" two partial lattices which contains the construction of Corollary 12?
16. Use component subsets to prove that if $a, b \in F(\aleph_0)$, $a \neq b$, then there is a homomorphism φ of $F(\aleph_0)$ onto a finite lattice such that $a\varphi \neq b\varphi$ (R. A. Dean [1956]).
17. Derive Exercise 16 from Corollary IV.4.6.
18. Prove that every finite lattice can be embedded in a finite three-generated lattice.
19. Using the notation of Figure 1, let $y \in F(3)$, $e_i \leq y_i \leq d_i$, $i = 0, 1, 2$. Prove that $[y_0, y_1, y_2]$ is a proper sublattice of $F(3)$ isomorphic to $F(3)$.

3. Reduced Free Products

Let L_i, $i \in I$, be bounded lattices and let L be a free $\{0, 1\}$-product of the L_i, $i \in I$ (see Definition II.5.14 and the discussion following it). As we shall see, a pair of elements x, y is *complementary in* L (that is, $x \wedge y = 0$ and $x \vee y = 1$) iff either they belong to some L_i and they are complementary in L_i or if there exist elements x_0, x_1, y_0, y_1 in some L_i such that $x_0 \leq x \leq y_0$, $x_1 \leq y \leq y_1$, and $\{x_0, x_1\}$, $\{y_0, y_1\}$ are complementary in L_i. We shall describe a construction in which there are many more complements than in the free $\{0, 1\}$-product, but in which we can still keep track of the complements. We call this construction the reduced free product. Several applications will be given in this section and the next.

In the discussion below let L_i, $i \in I$, $I \neq \varnothing$, be pairwise disjoint bounded lattices and $Q = \bigcup(L_i \mid i \in I)$.

Definition 1. *A* C-relation $\mathscr{C}$ *on* L_i, $i \in I$, *is a set of two element subsets of* Q *such that if* $\{a, b\} \in \mathscr{C}$, *then there exist* $i, j \in I$, $i \neq j$, *satisfying* $a \in L_i - \{0_i, 1_i\}$ *and* $b \in L_j - \{0_j, 1_j\}$.

Definition 2. *Let* $\mathscr{C}$ *be a C-relation on* L_i, $i \in I$. *A lattice* L *is a* $\mathscr{C}$-reduced free product of the L_i, $i \in I$, *iff the following conditions hold*:

(i) *Each* L_i, $i \in I$, *is a* $\{0, 1\}$*-sublattice of* L *and* $L = [\bigcup(L_i \mid i \in I)]$.

(ii) *If* $\{a, b\} \in \mathscr{C}$, *then* a, b *is a complementary pair in* L.

(iii) *If, for* $i \in I$, φ_i *is a* $\{0, 1\}$*-homomorphism of* L_i *into the bounded lattice* A, *and* $\{a, b\} \in \mathscr{C}$ $(a \in L_i, b \in L_j)$ *implies that* $a\varphi_i$, $b\varphi_j$ *are complementary in* A, *then there is a homomorphism* φ *of* L *into* A *extending all the* φ_i, $i \in I$.

Using the technique of Section I.5, the existence of a $\mathscr{C}$-reduced free product is proved if we find a lattice satisfying (i) and (ii). Such a lattice is

$$\{0, 1\} \cup \bigcup(L_i - \{0_i, 1_i\} \mid i \in I)$$

with the obvious partial ordering. The uniqueness of $\mathscr{C}$-reduced free products can be established as in Section I.5. The next result shall give a description of a $\mathscr{C}$-reduced free product based on the Structure Theorem of Free Products. This description is then used to describe the complementary pairs in a $\mathscr{C}$-reduced free product.

Theorems 4 and 5 appeared in their present form in G. Grätzer [1973], but earlier versions can be found in G. Grätzer [1971 a] and C. C. Chen and G. Grätzer [1969]. The germ of the idea can be traced back to R. P. Dilworth [1945].

Definition 3. *We define a subset* S *of* $\mathbf{P}(Q)$; *for* $p \in \mathbf{P}(Q)$, $p \in S$ *is defined by induction on the rank of* p:

(i) *For* $p \in Q$, $p \in S$ *iff* $p \in L_i - \{0_i, 1_i\}$ *for some* $i \in I$.

(ii) *For* $p = q \wedge r$, $p \in S$ *iff* $q, r \in S$ *and the following two conditions hold*:

(ii_1) $p \subseteq 0_i$ *for no* $i \in I$;

(ii_2) $p \subseteq x \wedge y$ *for no* $\{x, y\} \in \mathscr{C}$.

(iii) *For* $p = q \vee r$, $p \in S$ *iff* $q, r \in S$ *and the following two conditions hold*:

(iii_1) $1_i \subseteq p$ *for no* $i \in I$.

(iii_2) $x \vee y \subseteq p$ *for no* $\{x, y\} \in \mathscr{C}$.

Now we set

$$L = \{0, 1\} \cup \{R(p) \mid p \in S\},$$

and partially order L by

$$0 < R(p) < 1 \quad \text{for} \quad p \in S,$$

$$R(p) \leq R(q) \quad \text{iff} \quad p \subseteq q.$$

If we identify $a \in L_i$ with $R(a)$, then we get the setup we need:

THEOREM 4. *L is a $\mathscr{C}$-reduced free product of the L_i, $i \in I$.*

PROOF. L is obviously a poset. To show that L is a lattice we have to find the meet and the join of $R(p)$ and $R(q)$ in L $(p, q \in S)$. We claim that $R(p) \wedge R(q) = R(p \wedge q)$ if $p \wedge q \in S$ and otherwise $R(p) \wedge R(q) = 0$.

To verify this claim it is sufficient to prove that for any $u \in \mathbf{P}(Q)$, if $u \subseteq 0_i$ for some $i \in I$ or $u \subseteq x \wedge y$ for some $\{x, y\} \in \mathscr{C}$, then $u \notin S$. We prove this by induction on the rank of u. If $u \in Q$ and $u \subseteq 0_i$, then $u = 0_i \notin S$. If $u \in Q$ and $u \subseteq x \wedge y$ for some $\{x, y\} \in \mathscr{C}$, then $u \in L_i$, $x \in L_k$, $y \in L_n$, $k \neq n$, $u \leq x$ and $u \leq y$; these imply that $i = k$ and $i = n$, a contradiction. If $u = u_0 \wedge u_1$, and u_0 or $u_1 \notin S$, then $u \notin S$ by 3(ii); if $u_0, u_1 \in S$, then $u \notin S$ by 3(ii$_1$) or 3(ii$_2$). Finally, if $u = u_0 \vee u_1$, then $u_0 \subseteq 0_i$ or $u_1 \subseteq 0_i$ in the first case, and $u_0 \subseteq x \wedge y$ or $u_1 \subseteq x \wedge y$ in the second case, and so u_0 or $u_1 \notin S$ implying $u \notin S$ by 3(iii).

Dually, $R(p) \vee R(q) = R(p \vee q)$ iff $p \vee q \in S$; $R(p) \vee R(q) = 1$ otherwise.

Now it is obvious that $a \to R(a)$ is a $\{0, 1\}$-embedding of L_i into L. So after the identification 2(i) becomes obvious. 2(ii) is clear in view of 3(ii$_1$), 3(ii$_2$), and our description of meet and join in L.

Let K be the free product of the L_i, $i \in I$, as constructed in Section 1. Then $L - \{0, 1\} \subseteq K$. We define a congruence Θ on K:

$$\Theta = \bigvee(\Theta(x, 0_i) \mid i \in I, x \leq 0_i) \vee \bigvee(\Theta(x, 1_i) \mid i \in I, x \geq 1_i)$$
$$\vee \bigvee(\Theta(x, u \wedge v) \mid x \leq u \wedge v, \{u, v\} \in \mathscr{C}) \vee \bigvee(\Theta(x, u \vee v) \mid x \geq u \vee v, \{u, v\} \in \mathscr{C}).$$

In other words, Θ is the smallest congruence relation under which all 0_i and $u \wedge v$ $(u, v \in \mathscr{C})$ are in the congruence class which is the zero of K/Θ and dually. We claim that

$$K/\Theta \cong L.$$

To see this, it is sufficient to prove that every congruence class modulo Θ except the two extremal ones contains one and only one element of S.

Let ε_i be the identity map as a map of L_i into L. Then there is a map φ extending all ε_i, $i \in I$, into a homomorphism of K into L. Observe, that $R(p)\varphi = R(p)$ for all $p \in S$. Indeed, $p = p(a_0, \ldots, a_{n-1})$, where $a_0, \ldots, a_{n-1} \in S \cap Q$; hence

$$R(p)\varphi = R(p\varphi) = R(p(a_0, \ldots, a_{n-1})\varphi)$$
$$= R(p(a_0\varphi, \ldots, a_{n-1}\varphi)) = R(p(a_0, \ldots, a_{n-1})) = R(p),$$

since $a_0\varphi = a_0, \ldots, a_{n-1}\varphi = a_{n-1}$.

Let Φ be the congruence kernel of φ. Since L satisfies 2(i) and 2(ii), $\Theta \leq \Phi$. Now if $p, q \in S$, and $R(p)\varphi = R(q)\varphi$, then $R(p) = R(q)$. In other words, $R(p) \equiv R(q)$ (Φ) implies that $R(p) = R(q)$. Therefore, the same holds for Θ. This proves that there is at most one $R(p)$ in the nonextremal congruence classes of Θ. To show "at least one" take a $p \in \mathbf{P}(Q)$ such that $R(p) \not\equiv 0_i$ (Θ) and $R(p) \not\equiv 1_i$ (Θ) (for any/all $i \in I$); we prove that there exists a $q \in S$ such that $R(p) \equiv R(q)$ (Θ).

Let $p \in L_i$, for some $i \in I$. Then, by assumption, $p \neq 0_i$ and 1_i; hence we can take $q = p$. Let $p = p_0 \wedge p_1$, $R(p_0) \equiv R(q_0)$ (Θ), $R(p_1) \equiv R(q_1)$ (Θ), where $q_0, q_1 \in S$. If $q_0 \wedge q_1 \in S$, take $q = q_0 \wedge q_1$. Otherwise, by 3(ii), $q_0 \wedge q_1 \equiv 0_i$ (Θ), hence $p \equiv 0_i$ (Θ), contrary to our assumption. The dual argument completes the proof.

Thus we have proved that $K/\Theta \cong L$.

Now we are ready to verify 2(iii). For each $i \in I$, let φ_i be a $\{0, 1\}$-homomorphism of L_i into the bounded lattice A. Since K is the free product of the L_i, $i \in I$, there is a homomorphism ψ of K into A extending all the φ_i, $i \in I$. Let Ψ be the congruence kernel of ψ. It obviously follows from the definition of Θ that $\Theta \leq \Psi$. Therefore, by the Second Isomorphism Theorem,

$$[x]\Theta \to x\psi$$

is a homomorphism of K/Θ into A. Combining this with the isomorphism $L \cong K/\Theta$ as described above, we get a $\{0, 1\}$-homomorphism φ of L into A extending all the φ_i, $i \in I$. ●

Theorem 5. *Let a, b be a complementary pair in a $\mathscr{C}$-reduced free product L of the L_i, $i \in I$. Then there exist a_0, b_0 and a_1, b_1 such that*

$$a_0 \leq a \leq a_1 \quad \textit{and} \quad b_0 \leq b \leq b_1$$

such that either $\{a_0, b_0\}, \{a_1, b_1\} \in \mathscr{C}$ or, for some $i \in I$, a_0, b_0 and a_1, b_1 are complementary pairs in L_i, and conversely.

Proof. The converse is, of course, obvious. In either case, by Definition 2, a_0, b_0 and a_1, b_1 are complementary in L, hence

$$a \wedge b \leq a_1 \wedge b_1 = 0, \quad a \vee b \geq a_0 \vee b_0 = 1,$$

and so a and b are complementary in L.

Now to prove the main part of the theorem, take $p, q \in S$ such that $a = R(p)$ and $b = R(q)$ are complementary in L. Then $p \wedge q$ violates 3(ii$_1$) or 3(ii$_2$) and $p \vee q$ violates 3(iii$_1$) or 3(iii$_2$). The four cases will be handled separately.

Case 1. *$p \wedge q$ violates* 3(ii$_1$) *and $p \vee q$ violates* 3(iii$_1$). Hence, for some $i, j \in I$, $p \wedge q \subseteq 0_i$ and $1_j \subseteq p \vee q$. Thus in the free product K of the L_i, $i \in I$, $(p \wedge q)^{(i)} = 0_i$ and $(p \vee q)_{(j)} = 1_j$. Note that $q^{(i)}$ is proper, because otherwise $p^{(i)} = 0_i$, that is, $p \subseteq 0_i$, contradicting $p \in S$. Similary, $q_{(j)}$ is proper. This is a contradiction unless $i = j$, in which case we can put $a_0 = p_{(i)}$, $b_0 = q_{(i)}$, $a_1 = p^{(i)}$, $b_1 = q^{(i)}$ and these obviously satisfy the requirements of the theorem.

Case 2. *$p \wedge q$ violates* 3(ii$_1$) *and $p \vee q$ violates* 3(iii$_2$). Hence there exist $i \in I$ and $\{x, y\} \in \mathscr{C}$ such that

$$p \wedge q \subseteq 0_i \quad \text{and} \quad x \vee y \subseteq p \vee q.$$

Let $x \in L_j$ and $y \in L_k$, $j, k \in I$, $j \neq k$. Then $i \neq j$ or $i \neq k$; let us assume that $i \neq j$. Since $(p \wedge q)^{(i)} = 0_i$ and $p^{(i)} \wedge q^{(i)} = 0_i$, we conclude, just as in Case 1, that $p^{(i)}$ and $q^{(i)}$ are proper. From $i \neq j$ we conclude that $p_{(j)}$ and $q_{(j)}$ are not proper, that is, $p_{(j)} = q_{(j)} = 0^b$. Thus $(p \vee q)_{(j)} = p_{(j)} \vee q_{(j)} = 0^b$, contradicting that $p \vee q \supseteq x \in L_j$. Case 2 cannot occur.

CASE 3. *$p\wedge q$ violates* 3(ii$_2$) *and $p\vee q$ violates* 3(iii$_1$). This leads to a contradiction just as Case 2 does.

CASE 4. *$p\wedge q$ violates* 3(ii$_2$) *and $p\vee q$ violates* 3(iii$_2$). Then there exist $\{a_0, b_0\}\in\mathscr{C}$ and $\{a_1, b_1\}\in\mathscr{C}$ such that $a_0\in L_i$, $b_0\in L_j$, $a_1\in L_k$, $b_1\in L_n$, $i, j, k, n\in I$, $i\neq j$, $k\neq n$,

$$a_0\vee b_0\subseteq p\vee q \quad\text{and}\quad p\wedge q\subseteq a_1\wedge b_1.$$

We conclude, as above, that

$$a_0\leq p_{(i)}\vee q_{(i)},\quad b_0\leq p_{(j)}\vee q_{(j)},$$
$$p^{(k)}\wedge q^{(k)}\leq a_1,\quad p^{(n)}\wedge q^{(n)}\leq b_1.$$

Therefore, $p_{(i)}$ or $q_{(i)}$ is proper, $p_{(j)}$ or $q_{(j)}$ is proper, $p^{(k)}$ or $q^{(k)}$ is proper, and $p^{(n)}$ or $q^{(n)}$ is proper.

We cannot have both $p_{(i)}$ and $q_{(i)}$ proper, because then neither $p^{(k)}$ nor $q^{(k)}$ could be proper unless $i=k$; and similarly, $i=n$, contradicting $k\neq n$.

So we can assume that $p_{(i)}$ is proper and $q_{(i)}=0^b$, and therefore $p_{(i)}=(p\vee q)_{(i)}\geq a_0$. We cannot have $p_{(j)}$ proper because then $q_{(j)}=0^b$ and $p_{(j)}\geq b_0$; thus $p\supseteq a_0\vee b_0$. Now $k\neq i$ and $k\neq j$ yields a contradiction (neither $p^{(k)}$ nor $q^{(k)}$ could be proper), hence $k=i$ or $k=j$. Let $k=i$ (the case $k=j$ is similar). Since $i\neq j$, $q^{(i)}$ is not proper, hence $p^{(i)}$ is proper and $p^{(i)}=(p\wedge q)^{(i)}\leq a_1$. But $i\neq n$, hence $p^{(n)}$ is not proper and so $q^{(n)}$ must be proper. Since $q_{(j)}$ is proper we conclude that $n=j$ and $q^{(j)}=(p\wedge q)^{(j)}\leq b_1$. To sum up, we obtained

$$a_0\leq p_{(i)},\quad p^{(i)}\leq a_1,\quad b_0\leq q_{(i)},\quad q^{(i)}\leq b_1,$$

hence

$$a_0\subseteq p\subseteq a_1,\quad b_0\subseteq q\subseteq b_1,$$

as required. ●

Let us say that a bounded lattice A *has no comparable complements* iff A contains no pentagon $\{0, a, b, c, 1\}$. Let Comp(A) stand for the set of complementary pairs in A.

A C-relation $\mathscr{C}$ is said to have *no comparable complements* iff $\{a_0, b_0\}, \{a_1, b_1\}\in\mathscr{C}$, $a_0\leq a_1$ and $b_0\leq b_1$ imply that $a_0=a_1$ and $b_0=b_1$. The following result is immediate from Theorem 5.

COROLLARY 6. *Let L_i, $i\in I$, be a family of lattices with no comparable complements and let $\mathscr{C}$ be a* C-*relation on the L_i, $i\in I$, with no comparable complements. Let L be a $\mathscr{C}$-reduced free product of the L_i, $i\in I$. Then*

$$\mathrm{Comp}(L)=\mathscr{C}\cup\bigcup(\mathrm{Comp}(L_i)\mid i\in I),$$

and L has no comparable complements.

Now we are ready for our first application.

THEOREM 7 (C. C. Chen and G. Grätzer [1969]). *Let L be a bounded lattice in which every element has at most one complement. Then L has a $\{0, 1\}$-embedding into a* uniquely complemented lattice *K (that is, into a lattice K in which every element has exactly one complement).*

PROOF. If L is complemented, then set $K=L$. Otherwise let $L=L_0$. We define by induction the lattice L_n. If L_{n-1} is defined let I_{n-1} be the set of noncomplemented elements of L_{n-1}. For $i \in I_{n-1}$ let $L_i = \{a_i\}^b$. Define the C-relation $\mathscr{C}_{n-1}$ on the family $\{L_{n-1}\} \cup (L_i \mid i \in I_{n-1})$ by the rule

$$\{a, b\} \in \mathscr{C}_{n-1} \quad \text{iff} \quad \{a, b\} = \{i, a_i\} \quad \text{for some} \quad i \in I_{n-1}.$$

Let L_n be the $\mathscr{C}_{n-1}$-reduced free product. Since

$$L = L_0 \subseteq L_1 \subseteq L_2 \subseteq \cdots$$

and all these containments are $\{0, 1\}$-embeddings, we can form $K = \bigcup(L_i \mid i \in I)$.

Now observe that L_0 is a lattice with no comparable complements. By induction, if this is known of L_{n-1}, then it is true for L_n since $\mathscr{C}_{n-1}$ has no comparable complements and so Corollary 6 applies. Therefore, again by Corollary 6,

$$\operatorname{Comp}(L_n) = \operatorname{Comp}(L_{n-1}) \cup \mathscr{C}_{n-1}.$$

Thus L_n is at most uniquely complemented and every element of L_{n-1} has a complement in L_n. It is now obvious that K is uniquely complemented. ●

Every lattice L can be embedded in a lattice in which every element has at most one complement, namely into L^b. Thus as a special case of Theorem 7 we get the celebrated result of R. P. Dilworth [1945]:

COROLLARY 8. *Every lattice can be embedded into a uniquely complemented lattice.*

For a graph G (or, more precisely, $\langle G; E\rangle$) let $(L_a \mid a \in G)$ be a family of lattices, where $L_a = \{0_a, a, 1_a\}$ is a three-element lattice. Define the C-relation $\mathscr{C}$ on $(L_a \mid a \in G)$ by

$$\{x, y\} \in \mathscr{C} \quad \text{iff} \quad \{x, y\} \in E.$$

Let $L(G)$ denote the $\mathscr{C}$-reduced free product of $(L_a \mid a \in G)$. Some examples are given in Figures 1–3.

For a bounded lattice L, let $\operatorname{End}_{0,1}(L)$ denote the *monoid* (that is, semigroup with identity) of all $\{0, 1\}$-endomorphisms of L. For a graph G, $\operatorname{End}(G)$ denotes the monoid of endomorphisms of G; a map $\varphi\colon G \to G$ is an *endomorphism* iff $\langle a, b\rangle \in E$ implies that $\langle a\varphi, b\varphi\rangle \in E$. Observe that $G \subseteq L(G)$ and, in fact, G generates $L(G)$ as a $\{0, 1\}$-sublattice (that is, $G \cup \{0, 1\}$ generates $L(G)$). Therefore, every $\varphi \in \operatorname{End}(G)$ has at most one extension $\bar{\varphi}$ to a $\{0, 1\}$-endomorphism of $L(G)$.

COROLLARY 9. *Every endomorphism φ of a graph G has exactly one extension $\bar{\varphi}$ to a $\{0, 1\}$-endomorphism of $L(G)$. If φ is onto so is $\bar{\varphi}$.*

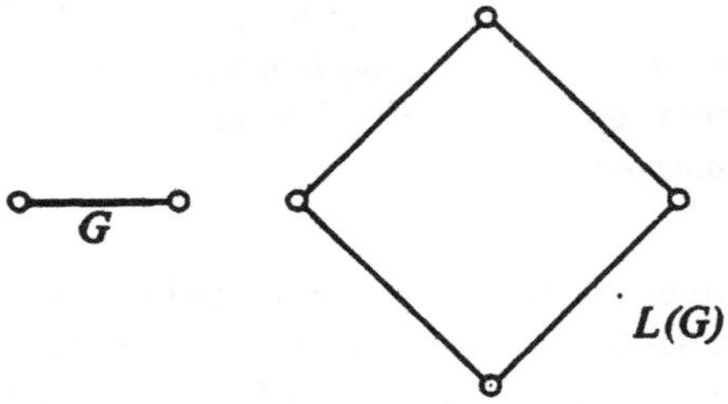

Figure 1

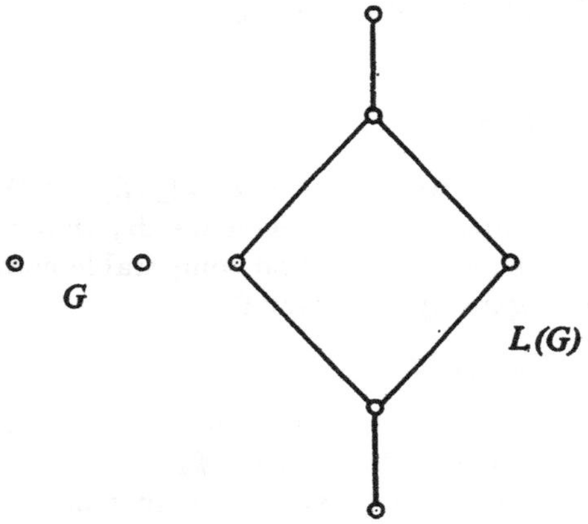

Figure 2

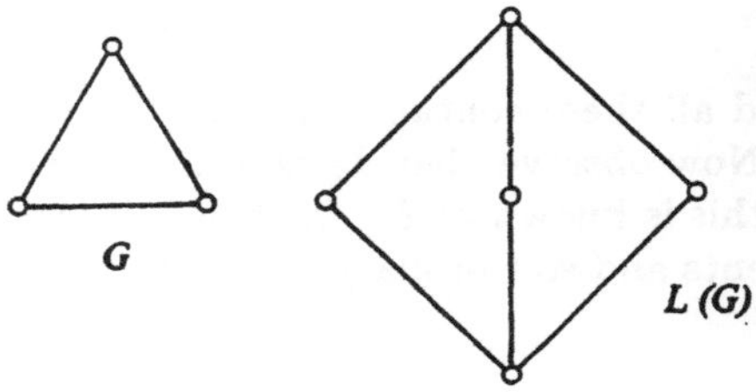

Figure 3

Proof. This is clear from Definition 2(iii) with $A = L(G)$. ●

For an integer $n \geq 2$, an *n-cycle* of a graph G is an n-tuple of elements $\langle a_0, \ldots, a_{n-1}\rangle$ with $\{a_0, a_1\}, \ldots, \{a_{n-2}, a_{n-1}\}, \{a_{n-1}, a_0\} \in E$.

Theorem 10. *Let G be a graph satisfying the property that every element of G is contained in some cycle of odd length. Then* $\mathrm{End}(G)$ *is isomorphic with* $\mathrm{End}_{0,1}(L(G))$.

Proof. It is obvious that the lattices and the C-relation used in forming $L(G)$ satisfy the hypotheses of Corollary 6. Therefore,

$$\mathrm{Comp}(L(G)) = E \cup \{0, 1\}.$$

By our assumption, every element of G is the endpoint of an edge, and G is recognized in $L(G)$ as the set of complemented elements, other than 0 and 1. Since a $\{0, 1\}$-endomorphism ψ takes a complementary pair into a complementary pair, we conclude that $G\psi \subseteq G \cup \{0, 1\}$. Let $g\psi \in \{0, 1\}$ for $g \in G$, for instance $g\psi = 0$. By assumption, there is a cycle of odd length $\langle g_0, \ldots, g_{2n}\rangle$ with $g = g_0$. This means that $\{g_0, g_1\}, \ldots, \{g_{2n-1}, g_{2n}\}$, and $\{g_{2n}, g_0\}$ are complementary pairs. Thus

$$g\psi = g_0\psi = 0, \quad g_1\psi = 1, \quad g_2\psi = 0, \quad g_3\psi = 1, \ldots, g_{2n-1}\psi = 1, \quad g_{2n}\psi = 0, \quad \text{and} \quad g_0\psi = 1,$$

a contradiction. This shows that every $\{0, 1\}$-endomorphism ψ is the unique extension of a map φ of G into itself; this φ is obviously a graph endomorphism. Thus the map $\varphi \to \bar{\varphi}$ is the required isomorphism of $\mathrm{End}(G)$ with $\mathrm{End}_{0,1}(L(G))$. ●

As it is shown in the Exercises, every monoid is the endomorphism semigroup of a graph in which every element lies on a cycle of odd length. Thus we conclude a result of G. Grätzer and J. Sichler [1970]:

Theorem 11. *Every monoid can be represented as the* $\{0, 1\}$*-endomorphism semigroup of a bounded lattice.*

Exercises

The following exercises should provide the reader with the necessary background in category theory (theory of concrete categories, graph theory) necessary for Theorem 11 and for the results of the next section. References will be given in the Further Topics and References. The present sequence of exercises is based on unpublished lecture notes of J. Sichler.

In what follows, we shall deal with systems $\langle A; R\rangle$, $\langle A; R_0, \ldots, R_{n-1}\rangle$, $\langle A; R_0, \ldots, R_n, \ldots\rangle$, where A is a nonempty set and R and the R_i are binary relations. Given two systems of the same type, say $\langle A; R_0, \ldots, R_n, \ldots\rangle$ and $\langle B; S_0, \ldots, S_n, \ldots\rangle$ a map $\varphi: A \to B$ is called a *homomorphism* iff

$$\langle a, b\rangle \in R_i \quad \text{implies that} \quad \langle a\varphi, b\varphi\rangle \in S_i, \quad \text{for all} \quad i=0, 1, \ldots$$

An *endomorphism* of $\langle A; R_0, \ldots\rangle$ is a homomorphism of $\langle A; R_0, \ldots\rangle$ into itself. The endomorphism semigroup $\mathrm{End}(\langle A; R_0, \ldots\rangle)$ is defined as before. $\langle A; R_0, \ldots\rangle$ is *rigid* iff the only endomorphism is the identity map.

1. Let $\langle A; \leq\rangle$ be a well-ordered set with unit. Let A_c consist of all $a \in A$ that are cofinal with ω, that is, for which there is a sequence $a_1 < \cdots < a_n < \cdots$ such that $a = \bigvee(a_i \mid i=1, 2, \ldots)$. For each $a \in A_c$ fix such a sequence $\langle a_1, a_2, \ldots\rangle$. Let R_0 be the relation $<$, and let $\langle x, y\rangle \in R_i$, for $i=1, 2, \ldots$, iff $y \in A_c$ and $x = y_i$ (that is, x is the i-th member of the sequence associated with y). Let φ be an endomorphism of $\langle A; R_0, R_1, \ldots\rangle$. Prove that $x \leq x\varphi$ for all $x \in A$.
2. Let $a < a\varphi$ for some $a \in A$. Set $a_1 = a\varphi$, $a_2 = a_1\varphi$, $\ldots$, and $b = \bigvee(a_n \mid n=1, 2, \ldots)$. Prove that $b\varphi = b$.
3. Prove that $b_n\varphi = b_n$, for all $n=1, 2, \ldots$ Conclude that $\langle A; R_0, R_1, \ldots\rangle$ is rigid.
4. Given a system $\langle A; R_0, R_1, \ldots\rangle$ we construct a new one $\langle B; S_0, S_1\rangle$ as follows: $B = A \times N$ ($N = \{0, 1, 2, \ldots\}$),

$$\langle\langle x, n\rangle, \langle y, m\rangle\rangle \in S_0 \quad \text{iff} \quad x=y \quad \text{and} \quad m=n+1, \quad \text{or}$$
$$x=y \quad \text{and} \quad n=0,\ m=2;$$
$$\langle\langle x, n\rangle, \langle y, m\rangle\rangle \in S_1 \quad \text{iff} \quad n=m \quad \text{and} \quad \langle x, y\rangle \in R_n.$$

Prove that $\mathrm{End}(\langle A; R_0, R_1, \ldots\rangle) \cong \mathrm{End}(\langle B; S_0, S_1\rangle)$ and if A is infinite, then $|A| = |B|$.

5. Given a system $\langle A; R_0, R_1\rangle$ we construct a new one $\langle B; S\rangle$ as follows: $B = A \times \{0, 1, 2, 3, 4\}$,

$$\langle\langle a, i\rangle, \langle b, j\rangle\rangle \in S \quad \text{iff} \quad a = b \quad \text{and} \quad j = i + 1, \quad \text{or}$$
$$a = b \quad \text{and} \quad i = 0, j = 4, \quad \text{or}$$
$$i = 0, j = 2, \quad \text{and} \quad \langle a, b\rangle \in R_0, \quad \text{or}$$
$$i = 2, j = 4, \quad \text{and} \quad \langle a, b\rangle \in R_1.$$

Prove that $\mathrm{End}(\langle A; R_0, R_1\rangle) \cong \mathrm{End}(\langle B; S\rangle)$ and if A is infinite, then $|A| = |B|$.

6. Prove that for each infinite cardinal $\mathfrak{m}$ there exists a rigid $\langle A; R\rangle$ with $|A| = \mathfrak{m}$.
7. Prove the statement of Exercise 6 for finite cardinals.
8. Let $\langle A; (R_i \mid i \in I)\rangle$ be a system without any restriction on $|I|$. Let $\langle I; R\rangle$ be a connected rigid graph. We define a new system $\langle B; R_0, R_1\rangle$ as follows: $B = A \times I$,

$$\langle\langle a, i\rangle, \langle b, j\rangle\rangle \in R_0 \quad \text{iff} \quad a = b \quad \text{and} \quad \langle i, j\rangle \in R,$$
$$\langle\langle a, i\rangle, \langle b, j\rangle\rangle \in R_1 \quad \text{iff} \quad i = j \quad \text{and} \quad \langle a, b\rangle \in R_i.$$

Prove that $\mathrm{End}(\langle A; (R_i \mid i \in I)\rangle) \cong \mathrm{End}(\langle B; R_0, R_1\rangle)$.

9. Let M be a monoid. For every $a \in M$ define $R_a = \{\langle x, ax\rangle \mid x \in M\}$. Show that $\mathrm{End}(\langle M; (R_a \mid a \in M)\rangle) \cong M$.
10. For any monoid M, find a system $\langle A; R\rangle$ such that $\mathrm{End}(\langle A; R\rangle) \cong M$.
11. For a system $\langle A; R\rangle$ and $B \subseteq A$ we construct a new system $\langle A; R, S_B\rangle$ as follows; $\langle x, y\rangle \in S_B$ iff $x \neq y$ and $x, y \in B$ or $x, y \notin B$. Show that if $\langle A; R\rangle$ is rigid, then the new systems are *mutually rigid*, that is, if φ is a one-to-one homomorphism of one system to another, then the two systems are the same and φ is the identity map.
12. Prove that there are $2^{\mathfrak{m}}$ mutually rigid systems of cardinality $\mathfrak{m}$, where $\mathfrak{m}$ is any infinite cardinal.
13. Consider the graph $\langle G; E\rangle$ of Figure 4. Let $\langle A; R\rangle$ be a system satisfying $\langle a, a\rangle \in R$ for no $a \in A$. We construct a new graph $\langle H; F\rangle$ by "replacing each $\langle a, b\rangle \in R$ by a copy of $\langle G; E\rangle$ with the pair $\langle 4, 8\rangle$ signifying $\langle a, b\rangle$".
Formally,

$$H = (R \times \{1, 2, 3, 5, 6, 7, 9, 10\}) \cup A,$$

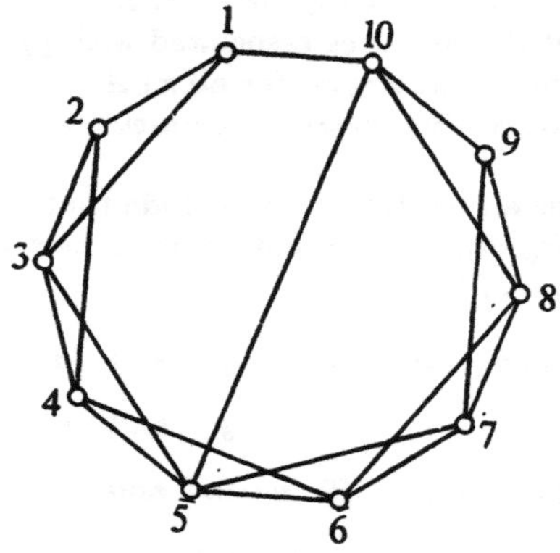

Figure 4

and for each $e = \langle a, b\rangle \in R$, $\langle e, 1\rangle$, $\langle e, 2\rangle$, $\langle e, 3\rangle$, a, $\langle e, 5\rangle$, $\langle e, 6\rangle$, $\langle e, 7\rangle$, b, $\langle e, 9\rangle$, $\langle e, 10\rangle$ should form a subgraph isomorphic with $\langle G; E\rangle$. Discuss the connections between $\mathrm{End}(\langle A; R\rangle)$ and $\mathrm{End}(\langle H; F\rangle)$.

14. A *triangle* $\{a, b, c\}$ in a graph $\langle G; E\rangle$ is a set of three elements of G such that $\{a, b\}$, $\{b, c\}$, and $\{c, a\} \in E$. A graph $\langle G; E\rangle$ is *triangle connected* iff for any $a, b \in G$, $a = b$ or there is a sequence $T_0, \ldots, T_{n-1}$ of triangles such that $a \in T_0$, $b \in T_{n-1}$, and $T_i \cap T_{i+1} \neq \varnothing$, for $i = 0, \ldots, n-2$. Prove that for each infinite cardinal $\mathfrak{m}$ there are $2^{\mathfrak{m}}$ mutually rigid triangle connected graphs.
15. Prove that for each infinite cardinal $\mathfrak{m}$, there are $2^{\mathfrak{m}}$ mutually rigid connected graphs in which every element lies on a cycle of length 7.
16. Let M be a monoid and let $\mathfrak{m}$ be an infinite cardinal satisfying $|M| \leq \mathfrak{m}$. Prove that there are $2^{\mathfrak{m}}$ pairwise nonisomorphic graphs $\langle G; E\rangle$ of cardinality $\mathfrak{m}$ satisfying $\mathrm{End}(\langle G; E\rangle) \cong M$.
17. Prove that the L of Theorem 4 is a $\mathscr{C}$-reduced free product of the L_i, $i \in I$, by verifying directly 2(iii). (Hint: pattern the proof after the proof of Theorem 1.8.)
18. A *bi-uniquely complemented lattice* L is a bounded lattice in which every element $x \neq 0, 1$ has exactly two complements. Define the concept of a free bi-uniquely complemented lattice and prove the existence and uniqueness (up to isomorphism) of a free bi-uniquely complemented lattice on $\mathfrak{m}$ generators.[1]
19. Let F_0 and F_1 be free bi-uniquely complemented lattices and let X_0 and X_1 be the free generating sets of F_0 and F_1, respectively. Let α be a one-to-one map of X_0 onto X_1. Let $|X_0| = |X_1| = \aleph_0$. Show that there are $2^{\aleph_0}$ isomorphisms of F_0 and F_1 extending α.
20. In a bounded lattice L the *complementation is transitive* iff whenever x, y and y, z are complementary pairs, then either $x = z$ or x, z is also a complementary pair. Prove that every lattice can be embedded into a bi-uniquely complemented lattice with transitive complementation.
21. Let L be a bounded lattice. Under what condition is there a $\{0, 1\}$-embedding of L into a bi-uniquely complemented lattice with transitive complementation?
22. Show that a relatively complemented uniquely complemented lattice is Boolean. (Hint: use Exercise I.6.5.)
23. Prove that a modular uniquely complemented lattice is Boolean.
24. Show that an atomic uniquely complemented lattice is Boolean (T. Ogasawara and U. Sasaki [1949]). (Hint: For an element a, let $R(a)$ denote the set of atoms $\leq a$; then $a \to R(a)$ is a set representation.)
25. Let L be a uniquely complemented lattice and let $a \succ b$ in L. Prove that $a \wedge b'$ is an atom, where b' is the complement of b (R. Padmanabhan). (Hint: if $a \wedge b' = 0$, then b' has two complements; if $0 < x < a \wedge b'$, then $b \vee a'$ has two complements.)
26. A lattice L is called *relatively atomic* iff, for all $a, b \in L$, if $a < b$, then there exist x and y satisfying $a \leq x \prec y \leq b$. Show that a relatively atomic uniquely complemented lattice is Boolean. (This includes the result of V. N. Salii [1972] that every uniquely complemented algebraic lattice is Boolean.)
27. A *lattice with complementation* $\langle L; \wedge, \vee, ', 0, 1\rangle$ is a bounded lattice in which for each $a \in L$, a' is a complement of L. An *endomorphism* φ is a lattice endomorphism satisfying $(a\varphi)' = a'\varphi$ for all $a \in L$. Prove that for any monoid M there is a lattice with complementation whose endomorphisms semigroup is isomorphic to M (G. Grätzer and J. Sichler [1970]).
28. In Theorem 11 and in Exercise 27, how many pairwise nonisomorphic lattices of a given cardinality can be constructed satisfying the requirements?

---

[1] Exercises 18–21 are based on C. C. Chen and G. Grätzer [1969].

4. Hopfian Lattices

A lattice L is *hopfian* iff every onto endomorphism is an automorphism. Equivalently, a lattice L is hopfian iff $L \cong L/\Theta$ implies that $\Theta = \omega$, that is, L is not isomorphic to a proper quotient of itself. The hopfian property is similarly defined for groups, rings, and algebras, in general. The first definition of the hopfian property also applies for graphs; one should note however, that for a graph a one-to-one and onto endomorphism need not be an automorphism. For graphs, an automorphism has to be defined as an invertible endomorphism.

Obviously, every finite lattice (algebra, graph) is hopfian. In fact, the hopfian property is a generalization of one of the most important characteristics of finiteness.

Closest to finite lattices are *finitely presented lattices*, that is, lattices of the form $F(Q)$ (see Section I.5) where Q is a finite partial lattice. (It is easily seen that these are the lattices that can be described by finitely many generators and finitely many relations, so this concept agrees with "finitely presented" as it is used for groups and semigroups.)

THEOREM 1 (T. Evans [1969]). *Every finitely presented lattice is hopfian.*

The proof of Theorem 1 is contained in the following three lemmas.

LEMMA 2. *A finitely presented lattice can be represented as a subdirect product of finite lattices.*

PROOF. It is sufficient to prove that if Q is a finite partial lattice, $a, b \in F(Q)$, and $a \neq b$, then there exists a finite lattice K and a homomorphism $\varphi: F(Q) \to K$ such that $a\varphi \neq b\varphi$. Just as in Section 2, we can define the component subsets of a and b and set

$$Q' = Q \cup \mathrm{Komp}(a) \cup \mathrm{Komp}(b).$$

Regarding Q' as a partial lattice, obviously $F(Q) = F(Q')$. By (the proof of) Theorem I.5.20, Q' can be embedded into some finite lattice K. This embedding extends to a homomorphism φ of $F(Q') = F(Q)$ into K. Finally, $a\varphi \neq b\varphi$, since φ is one-to-one on Q'. ●

LEMMA 3. *Let L be a finitely generated lattice and let Θ be a congruence relation of L. If L/Θ is finite, then there exists a fully invariant congruence relation Φ of L such that $\Phi \leq \Theta$ and L/Φ is finite.*

PROOF. Set $\mathbf{K} = \mathbf{Equ}(L/\Theta)$. We define

$$\Phi = \bigvee(\Theta(p(a_0, \ldots, a_{n-1}), q(a_0, \ldots, a_{n-1})) \mid p = q \in \mathrm{Id}(\mathbf{K}), a_0, \ldots, a_{n-1} \in L).$$

It is easily seen that Φ is fully invariant. Since the identities used to construct Φ all hold in L/Θ, we conclude that $\Phi \leq \Theta$. Finally, L/Φ is finitely generated and $L/\Phi \in \mathbf{K}$ which is a locally finite class, hence L/Φ is finite. ●

A subdirect representation of a lattice L is associated with a family $(\Theta_i \mid i \in I)$ of congruence relations of L such that $\bigwedge (\Theta_i \mid i \in I) = \omega$. If all Θ_i, $i \in I$, are fully invariant, we call this subdirect representation *fully invariant.*

Lemma 4. *Let the lattice L have a representation as a fully invariant subdirect product of hopfian lattices. Then L is hopfian.*

Proof. Let $(\Theta_i \mid i \in I)$ be a family of fully invariant congruence relations of L satisfying $\bigwedge (\Theta_i \mid i \in I) = \omega$. Let φ be an onto endomorphism of L. Since Θ_i is fully invariant for $i \in I$, we can define

$$\varphi_i \colon [a]\Theta_i \to [a\varphi]\Theta_i,$$

which is an endomorphism of L/Θ_i. Obviously, all φ_i are onto, hence they all are automorphisms. Now if $a, b \in L$ and $a\varphi = b\varphi$, then, for all $i \in I$, $([a]\Theta_i)\varphi_i = ([b]\Theta_i)\varphi_i$, hence $[a]\Theta_i = [b]\Theta_i$. Since $\bigwedge (\Theta_i \mid i \in I) = \omega$, we conclude that φ is one-to-one. ●

R. Wille [1975] exhibits a finitely generated lattice that is not hopfian. An earlier example of a finitely generated lattice that is not finitely presented (or presentable, to be more precise) is $F_{\mathbf{M}}(4)$ (see T. Evans and D. X. Hong [1972]); of course, $F_{\mathbf{M}}(4)$ is hopfian.

The free product of two finitely presented lattices is again finitely presented. What about hopfian lattices?

Theorem 5 (G. Grätzer and J. Sichler [1974]). *There exist hopfian lattices L_0 and L_1 whose free product is not hopfian.*

The proof of Theorem 5 is based on the construction of the lattice $L(G)$ from the graph G introduced in Section 3 and a lattice theoretic result (Theorem 6) which is due to J. Sichler [1972] and H. Lakser [1972]. In fact, the result we shall prove will be stronger than Theorem 5 and it will be based on J. Sichler [1975].

For graphs $G_i(=\langle G_i; E_i\rangle)$, $i \in I$, we form the lattices $L(G_i)$ (with bounds 0_i, 1_i) and the free product L of the $L(G_i)$, $i \in I$.

A *triangle* of G_i is a three-element set $\{a, b, c\}$ such that $\{a, b\}, \{a, c\}, \{b, c\} \in E_i$. If $\{a, b, c\}$ is a triangle of G_i, then $\{0_i, a, b, c, 1_i\}$ is a diamond. We call this sublattice the *diamond associated with a triangle.*

Theorem 6. *Any diamond in L is associated with a triangle of some G_i, $i \in I$.*

Proof. Let $M = \{o, a, b, c, i\}$ be a diamond in L. Let us first assume that $M \subseteq L(G_j)$, for some $j \in I$.

CASE 1. *$o=0_j$ and $i=1_j$.* Then $\{a, b\}, \{a, c\}, \{b, c\} \in \mathrm{Comp}(L(G_j))$, hence by Corollary 3.6, $\{a, b\}, \{a, c\}, \{b, c\} \in E_i$, and indeed, M is associated with the triangle $\{a, b, c\}$ of G_j.

CASE 2. *$o=0_j$ and $i<1_j$.* Recall that $L(G_j)-\{0_j, 1_j\} \subseteq F(G_j)$ (the free lattice generated by the set G_j), hence $a, b, c, i \in F(G_j)$, and there is a congruence relation Θ on $F(G_j)$ such that $F(G_j)/\Theta \cong L(G_j)$ and $[x]\Theta \to x$ under this isomorphism for all $x \in L(G_j)-\{0_j, 1_j\}$. Since $i=a \vee b=a \vee c$ in $F(G_j)$, by $(\mathrm{SD}_\vee)$, $i=a \vee (b \wedge c)$. But $b \wedge c=0_j$, that is, $[b \wedge c]\Theta$ is the zero of $F(G_j)/\Theta$. Thus

$$[i]\Theta=[a]\Theta \vee [b \wedge c]\Theta=[a]\Theta,$$

and we conclude $a=i$, a contradiction. Hence this case is not possible.

CASE 3. *$o>0_j$ and $i=1_j$.* This is impossible just as Case 2.

CASE 4. *$o>0_j$ and $i<1_j$.* Then $M \subseteq F(G_j)$ which is impossible since there is no diamond in a free lattice.

Thus Theorem 6 is proved for sublattices of $L(G_j)$. Now let $M \subseteq L$ and consider, for each $j \in J$,

$$M_{(j)}=\{x_{(j)} \mid x \in M\}.$$

Since $x \to x_{(j)}$ is a homomorphism, $M_{(j)} \cong M$ or $M_{(j)}$ is a singleton. If, for all $j \in J$, $M_{(j)}$ is a singleton, then Theorem 1.17 yields that $(\mathrm{SD}_\vee)$ holds in M, a contradiction. Hence there exists a $j \in I$ such that $M_{(j)}$ is a diamond and it is therefore associated with a triangle of G_j, in particular,

$$o_{(j)}=0_j \quad \text{and} \quad i_{(j)}=1_j.$$

By duality, there exists a $k \in I$ such that $M^{(k)}$ is a diamond; in particular,

$$o^{(k)}=0_k \quad \text{and} \quad i^{(k)}=1_k.$$

This yields immediately that $j=k$ and $o=0_j$, $i=1_j$. Since $a_{(j)} \vee b_{(j)}=1_j$, $a^{(j)} \wedge b^{(j)}=0_j$, $a_{(j)} \leq a^{(j)}$, and $b_{(j)} \leq b^{(j)}$, we conclude that $a_{(j)}$ and $a^{(j)}$ are complements of $b_{(j)}$ in $L(G_j)$. But $L(G_j)$ is a lattice with no comparable complements, hence $a_{(j)}=a^{(j)}=a \in L(G_j)$. Similarly, $b, c \in L(G_j)$. Hence $M \subseteq L(G_j)$ and M is associated with a triangle of G_j. ●

Let I be a set of at least two elements. For each $i \in I$, a lattice L_i will be constructed such that the free product of all L_i, $i \in I$, is not hopfian but the free product of L_i, $i \in I'$, is hopfian for all $\varnothing \neq I' \subset I$. The special case $|I|=2$ is Theorem 5.

Let $N=\{1, 2, 3, \ldots\}$ and we consider maps $\varphi: N \to I$. The map φ is *eventually constant* iff there exists an integer n such that $n\varphi=(n+1)\varphi=\cdots$. Let $M(I)$ denote the set of all eventually constant maps of N into I. By $M_n(I)$ we denote the set of all maps $\varphi: \{1, \ldots, n\} \to I$ and $M_\omega(I)=\bigcup(M_n(I) \mid n=1, 2, \ldots)$. For $\varphi \in M(I)$, $n \in N$, and for $\varphi \in M_m(I)$, $n \leq m$, let φ_n be the restriction of φ to $\{1, \ldots, n\}$. Finally, for $\varphi \in M_n(I)$ we write $n=D(\varphi)$ and $FV(\varphi)=n\varphi$ (D for domain and FV for final value).

For a graph G $(=\langle G; E\rangle)$, $a, b \in G$, we say that a and b are *triangle connected* iff $a=b$ or there is a sequence $T_0, \ldots, T_n$ of triangles of G such that $a \in T_0$, $b \in T_n$, and $T_i \cap T_{i+1} \neq \varnothing$, for $i=0, \ldots, n-1$. Any graph can be decomposed into a disjoint union of triangle connected components. G is *triangle connected* iff it has a single component.

Now choose a cardinal $\mathfrak{m} > |M(I)|$, a triangle connected graph G_0, and for each $\alpha \in M(I)$ choose a triangle connected graph G_α such that $|G_0| = |G_\alpha| = \mathfrak{m}$ and the graphs G_0 and G_α, $\alpha \in M(I)$, are pairwise disjoint and mutually rigid (Exercise 3.14). We fix the elements $a_0 \in G_0$ and $a_\alpha \in G_\alpha$ for all $\alpha \in M(I)$.

For every $n \in N$ and $\varphi \in M_n(I)$ we define a graph G^φ:

$$G^\varphi = \bigcup(G_\alpha \mid \alpha \in M(I) \text{ and } \alpha_n = \varphi),$$
$$E^\varphi = \bigcup(E_\alpha \mid \alpha \in M(I) \text{ and } \alpha_n = \varphi) \cup \{\{a_0, a_\alpha\} \mid \alpha \in M(I)\}.$$

In words, G^φ is a disjoint union of all G_α such that $\alpha_n = \varphi$ and we add the edges connecting the distinguished element of G_0 with the distinguished element of the G_α. Observe that G^φ is connected and that G_0 and the G_α are the triangle connected components of G^φ.

Let χ be a homomorphism of G^φ into G^ψ ($\varphi \in M_n(I)$ and $\psi \in M_m(I)$). Then the triangle connected components of G^φ have to be mapped into the triangle connected components of G^ψ. Thus if $\alpha \in M(I)$ with $\alpha_n = \varphi$ or $\alpha = 0$ then $G_\alpha \chi \subseteq G_\beta$, where $\beta \in M(I)$ and $\beta_m = \psi$ or $\beta = 0$. In view of the mutual rigidity of these graphs we must have $\alpha = \beta$ and χ is the identity on G_α. So we conclude, that if $\alpha \in M(I)$ and $\alpha_n = \varphi$, then $\alpha_m = \psi$; this is possible iff $m \leq n$ and $\alpha_m = \beta$. Thus there is a homomorphism of G^φ into G^ψ iff $D(\varphi) \geq D(\psi)$ and $\varphi_{D(\psi)} = \psi$. Moreover, in this case there is exactly one homomorphism which is the identity map on G_0 and on all G_α $(\alpha_n = \varphi)$.

Now we are ready to define L_i, for $i \in I$: let L_i be a free product of all $L(G^\varphi)$ satisfying $FV(\varphi) = i$.

Let L be a free product of all L_i, $i \in I$; in other words, L is a free product of all $L(G^\varphi)$, $\varphi \in M_\omega(I)$. We verify that L is not hopfian. We define a map β:

β restricted to G^φ is the homomorphism of G^φ into $G^{\varphi_{n-1}}$, where $n = D(\varphi) > 1$; β on G^φ is the identity if $D(\varphi) = 1$. Thus β extends to a homomorphism γ of $L(G^\varphi)$ into $L(G^{\varphi_{n-1}})$ if $n = D(\varphi) > 1$ and to the identity map γ on $L(G^\varphi)$ if $D(\varphi) = 1$. Since L is a free product, γ extends to an endomorphism δ of L.

Observe that the image of β covers all G^φ. Indeed, if $a \in G^\varphi$, then $a \in G_\alpha$ for some $\alpha \in M(I)$ with $\alpha_n = \varphi$, where $n = D(\varphi)$. Then β maps $a \in G_\alpha \subseteq G^{\alpha_{n+1}}$ onto $a \in G_\alpha \subseteq G^\varphi$.

Thus δ is an onto endomorphism of L. β is not one-to-one since every element of any G^φ with $D(\varphi) = 1$ is the image of two elements: of itself and of a suitable G^ψ with $D(\psi) = 2$. Thus δ is not one-to-one. We have proved that L is not hopfian.

Now let $\varnothing \neq I' \subset I$ and let L' be the free product of the L_i, $i \in I'$. We have to show that L' is hopfian.

Observe that L' is the free product of all $L(G^\varphi)$ satisfying $FV(\varphi) \in I'$.

Let δ be an onto endomorphism of L'. We shall verify that δ is the identity map, implying that L' is hopfian.

Assume to the contrary that δ is not the identity map. Then there is an $a \in G^\varphi$ $(FV(\varphi) \in I', D(\varphi) = n)$ such that $a \neq a\delta$. Since a is an element of a triangle, it is an atom of the associated diamond M. If $|M\varphi| = 1$, then the bounds of $L(G^\varphi)$ are collapsed by δ, hence all of $L(G^\varphi)$ is mapped by δ onto a single element. Otherwise, $M\varphi \cong M$, hence, by Theorem 6, $a\delta \in G^\psi$ for some $\psi \in M_\omega(I)$. Since G^φ is connected, all of G^φ is mapped by δ into G^ψ. Hence $D(\varphi) \geq D(\psi)$ and $\psi = \varphi_m$ where $m = D(\psi) < D(\varphi)$, since $D(\psi) = D(\varphi)$ implies that $\varphi = \psi$, contradicting the rigidity of G^φ.

In either case we see that $G^\varphi \cap G^\varphi\delta$ has at most one element. Let $i \in I - I'$ and consider the map $\beta\colon N \to I$ defined by $\beta_n = \varphi$ and $k\beta = i$ for $k > n$. Then $\beta \in M(I)$ and $G_\beta \subseteq G^\varphi$. Let $G_\beta^* = G_\beta - (G^\varphi \cap G^\varphi\delta)$. Since δ is onto, every $a \in G_\beta^*$ must be in the image of some G^ψ; but it cannot come from the homomorphism of G^ψ into G^φ since it would imply $D(\psi) = m > n = D(\varphi)$ and $\beta_m = \psi$, contradicting the definition of L'. (G^ψ is not in L' since $FV(\psi) = i \notin I'$.) Thus $a \in G_\beta^*$ must come from a G^ψ collapsed by δ onto a. However, $|G_\beta^*| = \mathfrak{m}$ and $|M(I)| < \mathfrak{m}$, a contradiction. Thus δ must be the identity map. ●

Exercises

1. Prove that for any nontrivial equational class $\mathbf{K}$ and infinite cardinal $\mathfrak{m}$, $F_{\mathbf{K}}(\mathfrak{m})$ is not hopfian.
2. Prove that for any equational class $\mathbf{K}$ of lattices and natural number n, $F_{\mathbf{K}}(n)$ is hopfian.
3. Find a one-to-one and onto endomorphism of a graph that is not an automorphism.
4. Can the graph in Exercise 3 be chosen to be finite?
5. Define a finitely presented lattice as the "most free" lattice generated by $x_0, \ldots, x_{n-1}$ satisfying the "relations" $p_i(x_0, \ldots, x_{n-1}) = q_i(x_0, \ldots, x_{n-1})$, $i = 1, \ldots, m$, where p_i and q_i are n-ary polynomials. Show that this is equivalent to forming an $F(Q)$ with a suitable partial lattice Q.
6. Let $F(Q) \supseteq Q' \supseteq Q$ where Q and Q' are partial lattices. Show that $F(Q) \cong F(Q')$ provided that Q' is generated by Q.
7. Formulate and prove the converse of Exercise 6.
8. For a lattice L and equational class $\mathbf{K}$ of lattices, we constructed in the proof of Lemma 3 a congruence relation $\Phi = \Phi(\mathbf{K})$. Prove that $\Phi(\mathbf{K})$ is fully invariant.
9. Let Φ be a fully invariant congruence relation of a lattice L. Prove that $\Phi = \Phi(\mathbf{K})$ for some equational class $\mathbf{K}$.
10. Let L be a fully invariant subdirect product of the L_i, $i \in I$. For an onto endomorphism φ write down a formula for $\operatorname{Ker}\varphi$ from which Lemma 4 can be derived.
11. What is the analogue of Lemma 4 for graphs?
12. Show that the free product of two finitely presented lattices is finitely presented.
13. Prove that $L(G_0 \cup G_1)$ is a free $\{0, 1\}$-product of $L(G_0)$ and $L(G_1)$, where $G_0 \cup G_1$ is the disjoint union of the graphs G_0 and G_1.
14. Find two bounded hopfian lattices whose free $\{0, 1\}$-product is not hopfian (G. Grätzer and J. Sichler [1974]).
15. Let G_0 and G_1 be hopfian graphs. Prove that a free product of $L(G_0)$ and $L(G_1)$ is hopfian.

Further Topics and References

The idea of the covers of elements in a free product goes back to R. P. Dilworth [1945]. It becomes more explicit in C. C. Chen and G. Grätzer [1969] in which $L * F(\aleph_0)$ is considered. The four W conditions, are, of course, from Ph. M. Whitman [1941]; this should be apparent from Theorem 2.2.

The Structure Theorem has its limitations. Despite many attempts, it could not be used to verify the Common Refinement Property for free products.

The Splitting Theorem is first formulated for $F(n)$ in Ph. M. Whitman [1941]: for $a \in F(n)$, if $a \geq x_1$ does not hold, then $a \leq x_2 \vee \cdots \vee x_n$; proof by induction on $r(a)$ using Theorem 2.2. Theorem 1.11 is much more general; still, it is a very special case of a complete triviality:

The Splitting Theorem. Let L be a lattice and $L = [X]$. Then the prime ideals of L are in one-to-one correspondence with subsets Y of X satisfying $Y \neq \varnothing$, $X - Y \neq \varnothing$, and

$$\bigvee Y_1 \ngeq \bigwedge X_1,$$

for any finite nonvoid $Y_1 \subseteq Y$ and $X_1 \subseteq X - Y$. If X is finite, this condition reduces to $\bigvee Y \ngeq \bigwedge (X - Y)$.

A very special case of the Common Refinement Property for free products was proved in A. Kostinsky [1971]. The Common Refinement Property holds for free **K**-products for any equational class **K** of lattices by G. Grätzer and J. Sichler [1975]. B. Jónsson and E. Nelson [1974] verify the same result for *regular equational classes* of algebras, that is, equational classes defined by identities in which the same variables occur on both sides. In G. Grätzer and J. Sichler [1975] examples are exhibited of lattices that cannot be represented as free products of freely indecomposable lattices; the same result holds also relative to any equational class **K** of lattices $\mathbf{K} \neq \mathbf{T}$.

Let

$$g(L) = \min \{|X| \mid X \text{ generates } L\},$$

by G. Grätzer and J. Sichler [1974a], the formula

$$g(A * B) = g(A) + g(B)$$

holds; the same formula holds in any equational class $\mathbf{K} \neq \mathbf{T}$ of lattices.

Many of the observations made about $F(3)$ hold for all $F_{\mathbf{K}}(3)$, where $\mathbf{K} \neq \mathbf{T}$ is any equational class of lattices. An interesting example is $x_0 \vee a \succ x_0 \succ x_0 \wedge b$. This even holds for $\mathbf{K} = \mathbf{D}$, where $a = b$. Many intervals of $F(3)$ contain isomorphic copies of $F(3)$. The same problem for $F_{\mathbf{K}}(n)$ is considered in J. Berman [1972a]; this problem is closely connected with the Amalgamation Property.

Computing $F_{\mathbf{K}}(n)$ may be very difficult even when it is finite. For instance, $F_{\mathbf{N}_5}(3)$ has 99 elements and a rather complicated structure, see A. G. Waterman [1967]. A rather effective method is worked out in R. Wille [a] which lends itself well to computer programming.

A curious consequence of Theorem 2.7 is that if **K** is an equational class of lattices, $\mathbf{K} \neq \mathbf{T}$, and $F_{\mathbf{K}}(\aleph_0)$ satisfies (W), then $\mathbf{K} = \mathbf{L}$.

Corollary 2.12 suggests that there are very many three-generated lattices. Indeed, in P. Crawley and R. A. Dean [1959], it is proved that there are $2^{\aleph_0}$ three-generated lattices. Call a lattice L $\mathfrak{m}$-*universal* iff $|L| = \mathfrak{m}$ and every lattice of cardinality at most $\mathfrak{m}$ is isomorphic to a sublattice of L. Thus there are no $\aleph_0$-universal lattices. By B. Jónsson [1956], $\aleph_\alpha$-universal lattices exist for all $\alpha > 0$.

Corollary 2.12 has been generalized to WA lattices in E. Fried and G. Grätzer [a]; in this case "two-generated" suffices.

For an empty C-relation a $\mathscr{C}$-reduced free product becomes a free $\{0, 1\}$-product; the structure theorem for this special case is quite easy. This suggests an alternative development for this chapter. Start with establishing the structure theorem for free $\{0, 1\}$-products. Derive from this the development of Section 1 by observing that the free product of the L_i, $i \in I$, is the sublattice $[\bigcup(L_i \mid i \in I)]$ of the free $\{0, 1\}$-product of $(L_i)^b$, $i \in I$. Still one more approach to Section 1 is the one in B. Jónsson [1971]: Once we know the Structure Theorem, it is possible to give a direct proof without the use of polynomials, equivalence classes, and so on. This approach is the most economical but not necessarily the most illuminating.

The basic idea of Section 3 is in Definition 3. The credit for this should go to R. P. Dilworth [1945] which contains a significant special case of this construction. This was generalized to $L * F(\aleph_0)$ in C. C. Chen and G. Grätzer [1969] under the constraints of Corollary 3.6.

An alternative proof of Corollary 3.8 is given in P. Crawley and R. P. Dilworth [1973].

The background of Corollary 3.8 is quite interesting. Results, such as the one in E. V. Huntington [1904], suggested that Boolean lattices can be characterized as uniquely complemented lattices. There was a widely accepted conjecture in the thirties supported by a growing body of results of the type, that if a lattice L is uniquely complemented and has property X, then L is Boolean. For X = modular, and X = relatively complemented, this appears to have been known already in the thirties to G. Birkhoff and J. von Neumann; for X = finite dimensional, see R. P. Dilworth [1940a]; for X = complete, atomic, and dually atomic, see G. Birkhoff and M. Ward [1939]. Then came R. P. Dilworth's result showing that without some additional property the conjecture is false. Thus the search for weaker and weaker properties X continued. For X = atomic, see T. Ogasawara and U. Sasaki [1949], see also J. E. McLaughlin [1956]; for X = the map $a \to a'$ is order-inverting, see G. Birkhoff [1948]; for X = algebraic, see V. N. Salii [1972]; see also Exercises 3.25 and 3.26. R. Padmanabhan has some related unpublished results: for X we can take $L \in \mathbf{M} \vee \mathbf{N}_5$; for X we can also take $L \in \mathbf{Equ}(K)$, where K is a finite lattice satisfying $(\mathrm{SD}_\wedge)$ or an implication which is due to E. Fried and G. Grätzer:

$$x \wedge y = x \wedge z \quad \text{implies that} \quad x \vee (y \wedge z) = (x \vee y) \wedge (x \vee z).$$

Many of the results of this chapter can be extended to $\mathfrak{m}$-complete lattices (lattices in which $\bigwedge X$ and $\bigvee X$ exist for all nonvoid subsets of X with $|X| < \mathfrak{m}$). Some of these results can be found in P. Crawley and R. A. Dean [1959] and B. Jónsson [1962]. The Structure Theorem of Free Products and its applications have not been published for the $\mathfrak{m}$-complete case.[1]

---

[1] Now, however, see the abstracts by G. Grätzer and D. Kelly, Free $\mathfrak{m}$-products of lattices. Notices Amer. Soc. January, February, April 1977.

Some covering relations in $F(3)$ are shown in Figure 1; some more can be found in Ph. M. Whitman [1941] and [1942] and in R. A. Dean [1961 a]. In R. N. McKenzie [1972] an algorithm was found deciding whether $p \succ q$ holds in $F(3)$. Some of McKenzie's results combined with a new theorem of A. Day [b] yield the astonishing fact that every proper interval of $F(3)$ contains a prime interval!

The solution to the word problem for $CF(P)$, the completely free lattice generated by a poset P, is very similar to Theorem 2.2; except, of course, that the relations we have to start with are $a \leq b$ where $a, b \in P$ and $a \leq b$ in P. M. E. Adams and D. Kelly [a] prove that the free product of L_i, $i \in I$, can be embedded into $CF(\bigcup (L_i \mid i \in I))$. This is the crucial step in proving that free products preserve chain conditions.

For various generalizations of the free product construction see R. Balbes and A. Horn [1967], R. A. Dean [1964], Z. Ladzianska [1974], and H. Lakser [1968].

Condition (W) is implicit in Ph. M. Whitman [1941]; it was first explicitly stated in B. Jónsson [1961]. A remarkable result of R. Freese [1975] states that a lattice L with no infinite chains satisfies (W) iff L is a retract of some $\mathcal{D}(I(F(n)))$.

Condition (W) plays a role in the characterization of finite (R. N. McKenzie [1972]) and finitely generated (A. Kostinsky [1972]) projective lattices.[1] In B. A. Davey and B. Sands [a] it is proved that for a lattice L with no infinite chains (W) is equivalent to the projectivity of L in the class of all lattices with no infinite chains.

B. Sands has examples showing that without any chain condition this result does not hold.

T. Kucera and B. Sands have recently considered finite lattices L such that for all finite lattices K, the set of all homomorphisms of L into K form a lattice under pointwise ordering. Again (W) plays a role in this investigation.

On the connections of (W) and transferability, see the Concluding Remarks. An easy, but illuminating property of (W) is pointed out in K. A. Baker and A. W. Hales [1974]: (W) holds for a lattice L iff it holds for $I(L)$.

The characterization problem of finite sublattices of a free lattice goes back, at least, to Ph. M. Whitman [1941]. In B. Jónsson [1961] and B. Jónsson and J. E. Kiefer [1962] it became apparent that anything that can be proved for finite sublattices of a free lattice follows already from (W), $(SD_\vee)$, and $(SD_\wedge)$. The obvious converse, conjectured by B. Jónsson, is still open. For an early review article on this field, see R. A. Dean [1961]. The distributive case is settled in F. Galvin and B. Jónsson [1961]. For some additional comments on recent results see the Concluding Remarks and H. S. Gaskill [a].

$F(3)$ has many nonfinite sublattices. It was observed in P. Crawley and R. A. Dean [1959] that for every countable poset P, $CF(P)$ is a sublattice of $F(3)$. R. N. McKenzie [1973] discusses some open problems related to free lattices.

Elementary equivalence and free products are considered in B. Jónsson and Ph. Olin [a]. They prove that if $\mathbf{K} \neq \mathbf{T}$ is any equational class of lattices, then free $\mathbf{K}$-products do not preserve elementary equivalence. This contrasts with Boolean algebras, see Ph. Olin [a].

Automorphism groups of lattices are, up to isomorphism, arbitrary groups. This

---

[1] Projective lattices are characterized in R. Freese and J. B. Nation, Projective lattices. Abstract. Notices Amer. Math Soc. 23 (1976): A–478.

special case of Theorem 3.11 goes back to G. Birkhoff [1946]; in fact, the same result is proved there for distributive lattices. Moreover, for finite groups the lattice constructed is finite. Small lattices with given automorphism group are considered in R. Frucht [1948] and [1950]. R. N. McKenzie and J. Sichler have some related results for lattices of finite length. Two sample results: every group is the automorphism group of a lattice of finite length; for every lattice L, there exists a lattice K, such that $\mathrm{End}(L) \cong \mathrm{End}_{0,1}(K)$ and if L is finite or finite length, then so is K, where $\mathrm{End}(L)$ is the endomorphism semigroup of L. See also J. Sichler [1972].

Theorem 3.11 is one of a large body of results representing monoids as endomorphism monoids of various types of algebras. All these results are based on P. Vopěnka, A. Pultr, and Z. Hedrlín [1965] proving the existence of rigid relations and on Z. Hedrlín and A. Pultr [1964] proving the representation for graphs. See also Z. Hedrlín and A. Pultr [1966] for the case of algebras with two unary operations; Z. Hedrlín and J. Lambek [1969] for the case of semigroups and for an alternate proof of the existence of rigid graphs. The result on triangle connected graphs is a special case of a result of P. Hell [1972].

Theorem 4.6 combines a result of J. Sichler [1972] with a very special case of a result of H. Lakser [1972]. The result of J. Sichler states that the diamonds in an $L(G)$ are all associated with the triangles of G. H. Lakser investigates the simple sublattices of a free product L of lattices L_i, $i \in I$; in particular, he proves that if S is a simple lattice, and S cannot be embedded in any L_i, $i \in I$, then S cannot be embedded in L. In addition, he obtains that L contains a diamond iff some L_i contains a diamond and L contains a diamond which is not in one of the L_i, $i \in I$, iff some L_i contains $\mathfrak{M}_3 \times \mathfrak{C}_2$ as a sublattice.

Problems

VI. 1. Let X be a finite smallest generating set of $A * B$. Is it true that $X \subseteq A \cup B$? [1]

VI. 2. Does some form of the Common Refinement Property hold for free $\{0, 1\}$-products? Investigate this problem also in an arbitrary equational class of lattices.

VI. 3. For which equational classes of lattices does *Sorkin's Theorem* (Exercise 1.13) hold?

VI. 4. Find a condition implying the Common Refinement Property for free K-products for an equational class K of algebras including the case of equational classes of lattices and of regular equational classes.

VI. 5. Which properties of homomorphisms $B_1 \to B_2$ are preserved by the induced homomorphism $A * B_1 \to A * B_2$? (See B. Jónsson and Ph. Olin [a].)

VI. 6. For lattices A and B, does $A * A \cong B * B$ imply that $A \cong B$? More generally does $A_1 * \cdots * A_n \cong B_1 * \cdots * B_n$, $A_1 \cong \cdots \cong A_n$, and $B_1 \cong \cdots \cong B_n$ imply that $A_1 \cong B_1$?

---

[1] No, it is not true. See M. E. Adams, Generators of free products of lattices. Algebra Universalis 7 (1977): 409–410.

VI. 7. Does $A_1 * \cdots * A_n * C \cong B_1 * \cdots * B_n * C$, $A_1 \cong \cdots \cong A_n$, and $B_1 \cong \cdots \cong B_n$ imply that $A_1 * C \cong B_1 * C$?

VI. 8. Investigate Problems VI.6 and VI.7 for free $\{0, 1\}$-products and free **K**-products.

VI. 9. Give a structure theorem of free $\mathbf{M}_3$-products. Same problem for $\mathbf{N}_5$.

VI.10. Let **K** be an equational class of lattices and let us assume that there exists an algorithm deciding identities in **K**. Is there an algorithm (relative to A and B) deciding the structure of the free **K**-product of A and B?

VI.11. Let **K** be an equational class of lattices for which the word problem for $F_{\mathbf{K}}(\aleph_0)$ is solved. When can we conclude that Theorem 1.14 holds for **K**?

VI.12. Is there any equational class $\mathbf{K} \supset \mathbf{D}$ of modular lattices for which the conclusion of Theorem 1.14 holds?

VI.13. For which equational classes **K** of lattices is it true that free **K**-products preserve the $\mathfrak{m}$-chain condition? ($\mathfrak{m}$ is an infinite uncountable regular cardinal; the $\mathfrak{m}$-*chain condition* means that all chains are of cardinality $<\mathfrak{m}$.)

VI.14. Investigate equational classes **K** of lattices in which the following holds: let $L_i \in \mathbf{K}$, $i \in I$, be a disjoint family of lattices and let L be a free **K**-product of L_i, $i \in I$; then L has an order embedding (that is, one-to-one and isotone map) into $F_{\mathbf{K}}(\bigcup(L_i \mid i \in I)^m)$, where $\bigcup(L_i \mid i \in I)$ is regarded as a poset.

VI.15. For which equational classes **K** is $\mathbf{S}(F_{\mathbf{K}}(\aleph_0))$ closed under the formation of free **K**-products?

VI.16. Give necessary and sufficient conditions for an interval $[p, q]$ of $F(3)$ to contain a copy (infinitely many copies) of $F(3)$.

VI.17. Does $F(3)$ have intervals of length n, for $n = 3, 4, \ldots$?

VI.18. Is a finite lattice isomorphic to a sublattice of $F(3)$ iff it satisfies (W), $(\mathrm{SD}_\wedge)$, and $(\mathrm{SD}_\vee)$?

VI.19. For $p, q \in F(\aleph_0)$, is it decidable whether there is an endomorphism φ of $F(\aleph_0)$ satisfying $p\varphi = q$?

VI.20. For which equational class **K** of lattices does $F_{\mathbf{K}}(\aleph_0)$ satisfy $(\mathrm{SD}_\vee)$?

VI.21. Does $F_{\mathbf{M}}(4)$ have a proper sublattice isomorphic to $F_{\mathbf{M}}(4)$? In general, if **K** is an equational class of lattices, $n < \aleph_0$, and $F_{\mathbf{K}}(n)$ is infinite, does $F_{\mathbf{K}}(n)$ then have a proper sublattice isomorphic with $F_{\mathbf{K}}(n)$?

VI.22. Find a common generalization of the McKenzie-Kostinsky Theorem and the Davey-Sands Theorem.

VI.23 (FC 1). Characterize the endomorphism semigroups of lattices.

VI.24. Describe the endomorphism semigroups and $\{0, 1\}$-endomorphism semigroups of complemented lattices.

VI.25. Characterize the endomorphism semigroups and the $\{0, 1\}$-endomorphism semigroups of complete lattices. What about the semigroups of complete endomorphisms?

VI.26. Characterize the $\{0,1\}$- endomorphism semigroups of finite lattices, of lattices of finite length, and of lattices without infinite chains. (Consider also the categorical versions of problems VI.23–VI.26.) [1]

[1] Every finite monoid is the $\{0, 1\}$-endomorphism semigroup of a finite lattice. For this result and others concerning the second and third part of this problem, see M. E. Adams and J. Sichler, Bounded endomorphisms of lattices of finite height. Manuscript.

VI.27. Let M be a monoid, let $\mathfrak{m}$ be an infinite cardinal, and let $|M| < \mathfrak{m} \leq 2^{|M|}$. How many pairwise nonisomorphic bounded lattices exist whose $\{0, 1\}$-endomorphism semigroup is isomorphic to M? (If $\mathfrak{m} \leq |M|$, the answer is $2^{\mathfrak{m}}$; see G. Grätzer and J. Sichler [1970].)

VI.28 (FC 25). Give a concrete example of a nondistributive uniquely complemented lattice.

VI.29 (FC 26). Does there exist a complete nondistributive uniquely complemented lattice?

VI.30. Does every uniquely complemented nondistributive lattice contain $F(3)$ as a sublattice? (An affirmative answer would also settle the next problem.)

VI.31. Does there exist a uniquely complemented nondistributive lattice satisfying a nontrivial lattice identity?

VI.32. For any infinite cardinal $\mathfrak{m}$, prove the existence of $\mathfrak{m}$-complete uniquely complemented nondistributive lattices.[1] Can they be made join- or meet-continuous, or both?

VI.33. Investigate finitely presented lattices. (See G. Grätzer and H. Lakser [1975].)

VI.34. Is it true that the cardinality of a basis of a finitely presented lattice is uniquely determined?

VI.35 (FC 12). Is the automorphism group of a finitely presented lattice finite?

VI.36. Let L be an infinite lattice. Show that with finitely many exceptions if L is finitely presented, then it contains $F(3)$ as a sublattice.

VI.37. Describe sublattices of finitely presented lattices.

VI.38. Describe all finite partial lattices $\mathfrak{A}$ for which $F(\mathfrak{A})$ is finite.

VI.39. For an equational class $\mathbf{K}$ of lattices, define finitely $\mathbf{K}$-presented. Investigate Problems VI.33–VI.38 for finitely $\mathbf{K}$-presented lattices.

VI.40. Let $\mathbf{K}$ be an equational class of lattices and let $n < \omega$. Show that if $\mathbf{K} \neq \mathbf{L}$ and $F_{\mathbf{K}}(n)$ is infinite, then $F_{\mathbf{K}}(n)$ is not finitely presented. (For $\mathbf{K} = \mathbf{M}$ and $n = 4$ this has already been verified; see T. Evans and D. X. Hong [1972].)

VI.41. Under what conditions does Theorem 4.1 hold for finitely $\mathbf{K}$-presented lattices?

VI.42. Describe hopfian distributive lattices.

VI.43. Improve Theorem 4.5 by requiring that one or both of L_0 and L_1 be finite, finitely presented, finitely generated, or countable. (It follows from Theorem 4.1 that both of L_0 and L_1 cannot be finite or finitely presented.)

VI.44. Prove that the free product of two bounded hopfian lattices is hopfian again. (If this is true, then in Problem VI.43, both L_0 and L_1 cannot be finitely generated.)

VI.45. In which equational classes $\mathbf{K}$ does Theorem 4.5 hold (L_0, $L_1 \in \mathbf{K}$ and we form the free $\mathbf{K}$-product).[2]

---

[1] This much has been proved by G. Grätzer and D. Kelly, Free $\mathfrak{m}$-products of lattices. Abstract. Notices Amer. Math. Soc. January, February, April 1977.

[2] Theorem 4.5 has been proven to hold in all nontrivial equational classes, see M. E. Adams and J. Sichler, $\mathbf{V}$-free products of hopfian lattices. Math. Z. 151 (1976): 259–262.

VI.46. Under what operators is the class of hopfian lattices closed? For instance, is the direct product of hopfian lattices hopfian?

VI.47. For what equational classes **K**, do there exist two cohopfian lattices in **K** whose direct product is not cohopfian? (A lattice L is *cohopfian* iff L is not isomorphic to any of its proper sublattices.)

VI.48. Investigate first-order properties Φ which are preserved when passing from a lattice to its ideal lattice, or when passing from the ideal lattice to the lattice, or both ways. (See Concluding Remarks.)

VI.49. Let L be a lattice satisfying ($SD_\wedge$), ($SD_\vee$), and (W). Prove that if $|L| > 1$, then L contains a prime ideal.

VI.50. Find an equational class of lattices $\mathbf{K} \neq \mathbf{L}$ such that a finite lattice is projective in **K** iff it is embeddable in $F_{\mathbf{K}}(\aleph_0)$.

[Note added in January 1978. During the last ten months solutions were found for a number of problems proposed in this book.

Recent results of C. R. Platt, B. Sands, and the author show that, for finite lattices, transferable = sharply transferable, solving the first half of I.26. This also yields an affirmative solution to I.23. A recent paper of A. Day contains the proof demanded by I.25. All equational classes have the property described in I.40, as proved by M. Adams and J. Sichler. W. A. Lampe solved II.20 in the negative. A negative solution to IV.26 was given by V. Polin. R. Freese and B. Jónsson proved that the arguesian identity is equivalent to modularity for congruence lattices of an equational class; IV.28 follows from this. IV.35 was disproved by E. Rodney Canfield.]

CONCLUDING REMARKS

What is lattice theory? Where to draw the line between lattice theory proper and its allied fields? Does a theorem characterizing a lattice associated with some object outside of lattice theory belong to lattice theory?

These questions are rather academic but they require some response and, in fact, decisions in a book of this kind.

The first question is really answered by this book just like the question, "What is an animal?" is best answered by a tour of a zoo.

The second question is more difficult. For instance, the theory of Boolean algebras develops by an interaction of lattice theory, topology, axiomatic set theory, and logic. A complete bibliography of lattice theory would be rather difficult to compile on account of jurisdictional disputes.

The third question, however, is a simple one for me: a theorem characterizing the subgroup lattices of abelian groups belongs to the theory of abelian groups and not to lattice theory.

A contentious case in point is the representation theory of algebraic lattices (complete lattices) as subalgebra lattices and congruence lattices of universal algebras (infinitary universal algebras). In my opinion, these results belong to universal algebra and not to lattice theory. However, in view of the close affinity between the workers in lattice theory and in universal algebra, I digress to outline this field.

Let L be an algebraic lattice. Then by Theorem II.3.13, $L \cong I(C)$, where C is a join-semilattice with 0. For $a \in C$, introduce on C the binary operation

$$f_a(x, y) = a \wedge (x \vee y).$$

Then the subalgebras of $\langle C; \{f_a \mid a \in C\}\rangle$ are exactly the ideals of C, hence the subalgebra lattice is isomorphic to L. Thus we obtain that every algebraic lattice is isomorphic to the subalgebra lattice of some algebra. The converse is even easier. Thus we have obtained a result of G. Birkhoff and O. Frink [1948]. In the same paper they conjecture that, up to isomorphism, congruence lattices of algebras can also be characterized as algebraic lattices. This is proved in G. Grätzer and E. T. Schmidt [1962]. For the briefest known proof, see P. Pudlák [b]. A rather detailed history of this field, including the case of infinitary algebras and the connections between congruence lattices of algebras and type 2 and type 3 representations of lattices, can be found in the paper G. Grätzer and W. A. Lampe [a]. A detailed discussion of this material with proofs would require a monograph of about 400 pages.

The congruence lattice of the algebra $\langle A; F\rangle$ is a sublattice of Part(A); if a lattice has the property that whenever it is represented as a $\{0, 1\}$-sublattice of some Part(A), then that sublattice is the congruence lattice of some algebra $\langle A; F\rangle$, then the lattice is called *strongly representable.* Such lattices are described in R. W. Quackenbush and B. Wolk [1971] and P. Pudlák [a].

Another case in point is the investigation of the lattice $\Sigma(X)$ of all topological spaces defined on a set, where for the topologies S_1 and S_2 on X we set $S_1 \leq S_2$ iff every S_1 open set is also S_2 open. Sample results: $\Sigma(X)$ is a complemented lattice; see A. K. Steiner [1966] and [1966a].

This point of view was introduced by G. Birkhoff [1936] and it permeates the topology book of R. Vaidyanathaswamy [1947] (reprinted in 1960).

Along with $\Sigma(X)$, about a dozen lattices are being considered in topology. For an excellent survey of this field (with 124 references) the reader should consult R. E. Larson and S. J. Andima [1975]; see also J. Rosický [1975].

The closed (and also the open) subsets of a topological space X form a lattice $L(X)$, and this lattice determines the topological space (provided it is T_1). These lattices were characterized by H. Wallman [1938]. Thus certain parts of topology can be studied within lattice theory–the theory of compactifications being one example (H. Wallman [1938], O. Frink [1964]). The set of all continuous real functions on a topological space X also forms a lattice $C(X)$ under pointwise ordering. The lattice $C(X)$ determines X if X is compact Hausdorff (I. Kaplansky [1947]; see also L. Gillman and M. Jerison [1960]).

A *topological lattice* L is a lattice equipped with a topology such that $\wedge$ and $\vee$ are continuous functions. The theory of topological lattices, the study of which was started by A. D. Wallace and his students, is quite extensive. The methods used are mostly topological rather than lattice theoretical in nature. The reader should consult L. W. Anderson [1961] for an early review of this field.

In contrast with topological lattices, intrinsic topologies are considered on arbitrary lattices. *Intrinsic topologies* are topologies defined on a lattice in terms of the partial ordering. For instance, the interval topology takes the intervals as a subbase for closed sets. For more details, see O. Frink [1942], G. Birkhoff [1962], and B. C. Rennie [1951].

The study of *partially ordered topological vector spaces* was pioneered by the works of F. Riesz, H. Freudenthal, S. Kakutani, L. V. Kantorovič, M. G. Krein, H. Nakano, and G. Birkhoff. The motivation came from a variety of sources–the theory of Hermitian operators in Hilbert spaces, C^* algebra theory, choquet boundary theory, and the inescapable fact that partially ordered topological vector spaces abound in analysis. The goal is to study how the presence of a partial order relation compatible with the algebraic structure of a vector space over the field of real numbers can substantially influence the topology. The earliest expositions of the subject are L. V. Kantorovič, B. Z. Vulih, and A. G. Pinkser [1950], B. Z. Vulih [1967], and H. Nakano [1950]. The modern theory of partially ordered topological vector spaces is characterized by the major role played by the concept of duality (just as in the modern theory of topological vector spaces) and there are a number of recent books stressing this aspect of the subject: A. L. Peressini [1967], G. Jameson [1970], Yau-chuen Wong and Kung-fu Ng [1973] and M. Dohoux [1974]. These books have extensive bibliographies for the interested reader.

The important work C. Carathéodory [1956] (English translation: C. Carathéodory [1963]) initiated a study of finitely additive functions and measures, and the integrals they generate on rings of elements of Boolean algebras. The concept of integral was made completely algebraic. Thus he created what may, in analogy to J. von Neumann's study of Continuous Geometries, be called "pointless" integration. Rediscovering Caratheodory's work and using the potent tools of *topological Riesz spaces* (partially ordered linear topological spaces which are also lattices with respect to the partial order), D. H. Fremlin [1974] gives a thoroughly modern account of this field.

There are many publications in quantum mechanics that border on lattice theory. A typical question is whether the lattice of propositions of quantum mechanics is modular, as postulated by G. Birkhoff and J. von Neumann [1936]. For some of the recent problems discussed in the literature, see the books: J. M. Jauch [1968] and G. W. Mackey [1963] and the papers: P. D. Finch [1969], J. M. Jauch and C. Piron [1963] and [1969], P. Mittelstaedt [1972], and C. Piron [1964]. It may be of some interest to note that of the two volume work V. S. Varadarajan [1968] and [1970] on quantum mechanics, a substantial part of the first volume deals with lattices.

Lattices and posets appear in other areas alongside with an algebraic structure and some or all the operations relate in some way to the partial ordering (for instance, they are isotone). For a survey of this field, see L. Fuchs [1963].

Many generalizations of the lattice concept appear in the literature. However, only one has been developed to some depth: weakly associative lattices, introduced independently by E. Fried [1970] and H. L. Skala [1971]. Here is a partial bibliography of this new field: P. Erdös, E. Fried, A. Hajnal, and E. C. Milner [1972]; E. Fried [1973], [1974], [a], [b], and [c]; E. Fried and G. Grätzer [1973], [1973a], and [a]; E. Fried, G. Grätzer, and R. W. Quackenbush [a] and [b]; E. Fried and V. T. Sós [1975]; K. Gladstien [1973]; G. Grätzer and H. Lakser [1971a]; H. L. Skala [1972]; R. W. Quackenbush [1972].

There is one topic that I would have liked to see included in this book, unfortunately, its late development made it impossible to discuss it in detail. This is the theory of transferability.

Let us call a lattice K *transferable* iff, for any lattice L, if K has an embedding φ into $I(L)$, then K has an embedding ψ into L. If, in addition, this ψ can always be chosen to satisfy

$$a\psi \in a\varphi \quad \text{but} \quad a\psi \notin b\varphi \quad \text{for any} \quad b < a$$

then K is called *sharply transferable.*

These concepts originate in G. Grätzer [1970] (which is the text of a lecture given in 1966). There are only a few results on transferability:

A transferable lattice has no doubly reducible element (G. Grätzer [1970] and [1974]).

Every finite projective lattice is transferable (K. A. Baker and A. W. Hales [1974]).

A finite[1] transferable lattice satisfies $(SD_\wedge)$ and $(SD_\vee)$ (H. S. Gaskill, G. Grätzer, and C. R. Platt [1975]).

---

[1] This result is true even when „finite“ is dropped, see G. Grätzer and C. R. Platt, Two embedding theorems for lattices. Abstract. Notices Amer. Math. Soc. April 1977. See also footnote to Problem I.24.

It is obvious from Corollary IV.4.3, that if K is transferable and $a \in K$, then (a] is finite.

T. Tan [1975] found the first example of a transferable lattice that is not sharply transferable: $\mathfrak{C}_2 \times \mathfrak{C}_\omega$, where $\mathfrak{C}_\omega$ is the chain of natural numbers. No such finite lattice is known.

It is primarily due to H. S. Gaskill's work that sharp transferability has been much further developed. To describe the new results, conditions $(T_\wedge)$ and $(T_\vee)$ have to be introduced.

Let K be a finite lattice, $p \in K$, and $J \subseteq K$. We call $\langle p, J \rangle$ a *pair* iff $p \notin J$, $|J| \geq 2$, and $p \leq \bigvee J$. A pair $\langle p, J \rangle$ is *minimal* iff whenever $\langle p, J' \rangle$ is also a pair and for every $a \in J'$ there exists a $b \in J$ with $a \leq b$, then $J \subseteq J'$. (Minimal pairs are the most economical way of describing the join-structure of a lattice.) A finite lattice K satisfies the condition $(T_\vee)$ iff a rank function r can be defined on K (with integer values) such that if $\langle p, J \rangle$ is a minimal pair of K and $q \in J$, thus $r(p) < r(q)$. The condition $(T_\wedge)$ is the dual of $(T_\vee)$.

THEOREM. *For a finite lattice K the following conditions are equivalent*:

(i) *K is sharply transferable.*

(ii) *K satisfies* $(T_\wedge)$, $(T_\vee)$, *and* (W).

(iii) *K can be embedded in* $F(3)$.

This result was discovered in many parts. H. S. Gaskill introduced the conditions $(T_\wedge)$ and $(T_\vee)$ in his Ph. D. Thesis, see also H. S. Gaskill [1972], and he proved that (ii) implies (i) and that (i) implies $(T_\wedge)$ and $(T_\vee)$. I suggested a construction for lattices failing $(T_\vee)$ to show that they are not sharply transferable; in a related vein, C. R. Platt came up with a construction that applies to all finite lattices. This is the construction given in H. S. Gaskill, G. Grätzer, and C. R. Platt [1975]; as a corollary of Platt's construction one obtains one of the results on transferability mentioned above. Meanwhile, H. S. Gaskill and C. R. Platt [1975] proved that sharp transferability and (W) are together equivalent to embeddability in $F(3)$. Finally, I proved that sharp transferability implies (W). Thus the equivalence of (i)–(iii) is established.

H. Lakser observed that the result of K. A. Baker and A. W. Hales can be sharpened: every finite projective lattice is sharply transferable.

All distributive sharply transferable and all distributive transferable lattices have been described by T. Tan [1975].

If all lattices considered are restricted to an equational class **K** we get the concepts of **K**-*transferable* and *sharply* **K**-*transferable* lattices.

According to H. S. Gaskill [1972a] (see also E. Nelson [1974]) every finite distributive lattice is sharply **D**-transferable.

G. Grätzer and T. Tan proved that if K is **D**-transferable, then for every $a \in K$, (a] and $\{b \mid b \succ a\}$ are finite; moreover, if K is sharply **D**-transferable, then K does not contain $\mathfrak{C}_2 \times \mathfrak{C}_\omega$ as a sublattice. It is natural to conjecture that the linear sums of type ω of finite distributive lattices are the infinite sharply **D**-transferable lattices.

For a finite lattice K, the embeddability of K into a lattice L can be expressed as a first-order property. Thus K is transferable iff this first-order property is preserved when passing from $I(L)$ to L. The general problem of investigating and classifying such properties is discussed in G. Grätzer [1970]. For some interesting contributions to this problem, see K. A. Baker and A. W. Hales [1974]. An important first-order property of this type is investigated in W. Poguntke and I. Rival [1974].

Many problems proposed in FC are still open. As Piet Hein [1969] writes: Problems worthy—of attack—prove their worth—by hitting back.

The following is an account of the work done in relation to problems in FC; it is not claimed to be comprehensive.

3. Solved in J. Sichler [1972].

9. Solved in C. R. Platt [1976] and D. Kelly and I. Rival [1975a].

12. and 13. A negative answer is provided for the case $\mathbf{K} = \mathbf{M}$ in A. P. Huhn [1975]. The problem is still open for $\mathbf{K} = \mathbf{L}$.

14. and 15. These are discussed above.

16. $n = 2$ and 3 are done in R. Padmanabhan [1972].

18. An affirmative solution is given in R. W. Quackenbush [1974].

21. and 23. Some relevant references are given in the Further Topics and References of Chapter I.

29. Solved in R. Padmanabhan [1974a].

30. See the references given to Problem II.3.

31. According to J. D. Monk this can be solved affirmatively under the Generalized Continuum Hypothesis.

32. An affirmative solution was found independently by B. Jónsson (unpublished) and K. D. Magill [1970].

33. References to various solutions are given in the Further Topics and References of Chapter II.

34. The question raised in the problem was answered in the negative by T. Tan. The problem remains unsolved.

35. See V. Vilhelm [1955], J. Jakubik [1975], and F. Šik [a].

39. Solved by M. E. Adams [1974] and R. Balbes [1972], independently.

47. A contribution to this problem was made in H. Lakser [1973a]. The problem is solved in M. E. Adams and D. Kelly [b].

48. Solved in M. E. Adams and D. Kelly [a].

51. Answered in the negative in G. Grätzer, B. Jónsson, and H. Lakser [1973].

54. Solved in T. Katriňák [1972].

55. Solutions are offered in T. Katriňák [1973a] and H. A. Priestley [1974].

56. A description is given in A. Urquhart [1973].

57. See M. E. Adams [1973a], T. Katriňák [c], and H. A. Priestley [1975].

63. This is verified in M. E. Adams [1976]. (The result of M. E. Adams was announced in Notices Amer. Math. Soc. 20 (1973): A-560. A new proof is announced by A. Wrónski [1976].)

65. Solved by T. Katriňák [1973c].

66. The case $n = 1$ is considered in T. Katriňák [1966] and [1968].

I hope I shall be able to keep the reader informed about developments related to the problems proposed in this book through the journal Algebra Universalis.

BIBLIOGRAPHY

S. Abian and A. B. Brown

1961. A theorem on partially ordered sets, with applications to fixed point theorems. Canad. J. Math. 13: 78–82.

M. E. Adams

1973. The Frattini sublattice of a distributive lattice. Algebra Universalis 3: 216–228.

1973a. The structure of distributive lattices. Ph. D. Thesis, Bristol University.

1974. A problem of A. Monteiro concerning relative complementation of lattices. Colloq. Math. 30: 61–67.

1975. The poset of prime ideals of a distributive lattice. Algebra Universalis 5: 141–142.

1976. Implicational classes of pseudocomplemented distributive lattices. J. London Math. Soc. (2) 13: 381–384.

M. E. Adams and D. Kelly

1975. Homomorphic images of free distributive lattices that are not sublattices. Algebra Universalis 5: 143–144.

a. Chain conditions in free products of lattices. Algebra Universalis 7 (1977): 235–244.

b. Disjointness conditions in free products of lattices. Algebra Universalis 7 (1977): 245–258.

M. Aigner

1969. Graphs and partial orderings. Monatsh. Math. 73: 385–396.

J. W. Alexander

1939. Ordered sets, complexes and the problem of compactification. Proc. Nat. Acad. Sci. U.S.A. 25: 296–298.

M. Altwegg

1950. Zur Axiomatik der teilweise geordneten Mengen. Comment. Math. Helv. 24: 149–155.

I. Amemiya

1957. On the representation of complemented modular lattices. J. Math. Soc. Japan 9: 263–279.

I. Amemiya and I. Halperin

1959. Coordinatization of complemented modular lattices. Nederl. Akad. Wetensch. Proc. Ser. A. 62: 70–78.

I. Amemiya and I. Halperin

1959a. Complemented modular lattices derived from nonassociative rings. Acta Sci. Math. (Szeged) 20: 181–201.

F. W. Anderson and R. L. Blair

1961. Representations of distributive lattices as lattices of functions. Math. Ann. 143: 187–211.

L. W. Anderson

1961. Locally compact topological lattices. In *Lattice Theory*, Proc. Symp. Pure Math., II, pp. 195–197. Amer. Math. Soc., Providence, R. I.

S. J. Andima

(See R. E. Larson and S. J. Andima.)

R. Antonius and I. Rival

1974. A note on Whitman's property for free lattices. Algebra Universalis 4: 271–272.

G. Ja. Areškin

1953. On congruence relations in distributive lattices with zero. (Russian.) Dokl. Akad. Nauk SSSR N. S. 90: 485–486.

1953a. Free distributive lattices and free bicompact T_0-spaces. (Russian.) Mat. Sb. N.S. 33(75): 133–156.

E. Artin

1940. Coordinates in affine geometry. Rep. Math. Colloq. (2) 2: 15–20.

B. Artmann

1969. Hjelmslev planes derived from modular lattices. Canad. J. Math. 21: 76–83.

S. P. Avann

1958. A numerical condition for modularity of a lattice. Pacific J. Math. 8: 17–22.

1958a. Dual symmetry of projective sets in a finite modular lattice. Trans. Amer. Math. Soc. 89: 541–558.

1960. Upper and lower complementation in a modular lattice. Proc. Amer. Math. Soc. 11: 17–22.

1960/61. Application of the join-irreducible excess function to semi-modular lattices. Math. Ann. 142: 345–354.

1961. Metric ternary distributive semi-lattices. Proc. Amer. Math. Soc. 12: 407–414.

1961a. Distributive properties in semi-modular lattices. Math. Z. 76: 283–287.

1964. Increases in the join-excess function in a lattice. Math. Ann. 154: 420–426.

1964a. Dependence of finiteness conditions in distributive lattices. Math. Z. 85: 245–256.

L. Babai

1972. Automorphism groups of planar graphs. I. Discrete Math. 2: 295–307.

1973. Automorphism groups of planar graphs. II. In *Coll. Math. Soc. J. Bolyai. 10. Infinite and Finite Sets*, pp. 29–84. North-Holland, Amsterdam.

P. D. Bacsich

1975. Amalgamation properties and interpolation theorems for equational theories. Algebra Universalis 5: 45–55.

R. Baer

1952. *Linear Algebra and Projective Geometry.* Academic Press, New York, N.Y.

K. A. Baker

1969. A generalization of Sperner's lemma. J. Combinatorial Theory 6: 224–225.

1969a. Equational classes of modular lattices. Pacific J. Math. 28: 9–15.

1971. Equational axioms for classes of lattices. Bull. Amer. Math. Soc. 77: 97–102.

1973. Inside free semilattices. In *Proc. Univ. Houston Lattice Theory Conf.*, pp. 306–331.

1974. Finite equational bases for finite algebras in a congruence distributive equational class. Advances in Math.

1974a. Primitive satisfaction and equational problems for lattices and other algebras. Trans. Amer. Math. Soc. 190: 125–150.

a. Equational axioms for classes of Heyting algebras. Algebra Universalis 6 (1966): 105–120.

K. A. Baker, P. C. Fishburn, and F. S. Roberts

1970. Partial orders of dimension 2, interval orders, and interval graphs. Rand Corp. P-4376.

K. A. Baker and A. W. Hales

1970. Distributive projective lattices. Canad. J. Math. 22: 472–475.

1974. From a lattice to its ideal lattice. Algebra Universalis 4: 250–258.

K. A. Baker and A. F. Pixley

a. Polynomial interpolation and the Chinese Remainder Theorem for algebraic systems. Math. Z. 43 (1975): 165–174.

J. W. de Bakker

1967. On convex sublattices of distributive lattices. Math. Centrum Amsterdam Afd. Zuivere Wisk. ZW-003.

R. Balbes

1967. Projective and injective distributive lattices. Pacific J. Math. 21: 405–420.

1969. A note on distributive sublattices of a modular lattice. Fund. Math. 65: 219–222.

1971. On the partially ordered set of prime ideals of a distributive lattice. Canad. J. Math. 23: 866–874.

1972. Solution to a problem concerning the intersection of maximal filters and maximal ideals in a distributive lattice. Algebra Universalis 2: 389–392.

R. Balbes and Ph. Dwinger

1971. Coproducts of Boolean algebras and chains with applications to Post algebras. Colloq. Math. 24: 15–25.

R. Balbes and Ph. Dwinger
1971 a Uniqueness of representation of a distributive lattice as a free product of a Boolean algebra and a chain.
Colloq. Math. 24: 27–35.
1974. *Distributive Lattices.*
Univ. Missouri Press, Columbia, Miss.

R. Balbes and G. Grätzer
1971. Injective and projective Stone algebras.
Duke Math. J. 38: 339–347.

R. Balbes and A. Horn
1967. Order sums of distributive lattices.
Pacific J. Math. 21: 421–435.
1970. Stone lattices.
Duke Math. J. 37: 537–545.
1970 a. Injective and projective Heyting algebras.
Trans. Amer. Math. Soc. 148: 549–559.
1970 b. Projective distributive lattices.
Pacific J. Math. 33: 273–279.

J. G. Basterfield and L. M. Kelly
1968. A characterization of sets of *n* points which determine *n* hyperplanes.
Proc. Cambridge Philos. Soc. 64: 585–588.

R. Beazer
1975. Post-like algebras and injective Stone algebras.
Algebra Universalis 5: 16–23.
1975 a. Hierarchies of distributive lattices satisfying annihilator conditions.
J. London Math. Soc. (2) 11: 216–222.

E. A. Behrens
1960. Distributiv darstellbare Ringe I.
Math. Z. 73: 409–432.
1961. Distributiv darstellbare Ringe II.
Math. Z. 76: 367–384.

N. D. Belnap and J. H. Spencer
1966. Intensionally complemented distributive lattices.
Portugal. Math. 25: 99–104.
(See also J. M. Dunn and N. D. Belnap.)

G. M. Bergman
1969. On 2-firs (weak Bezout rings) with distributive divisor lattices.
Manuscript.

J. Berman
1972. On the length of the congruence lattice of a lattice.
Algebra Universalis 2: 18–19.
1972 a. Sublattices of intervals in relatively free lattices.
Algebra Universalis 2: 174–176.
1973. Congruence relations of pseudocomplemented distributive lattices.
Algebra Universalis 3: 288–293.

G. Birkhoff
1933. On the combination of subalgebras.
Proc. Cambridge Philos. Soc. 29: 441–464.
1935. On the structure of abstract algebras.
Proc. Cambridge Philos. Soc. 31: 433–454.

G. Birkhoff

1935a. Abstract linear dependence and lattices.
Amer. J. Math. 57: 800–804.

1935b. Combinatorial relations in projective geometries.
Ann. of Math. 36: 743–748.

1936. On the combination of topologies.
Fund. Math. 26: 156–166.

1940. *Lattice Theory*. First edition.
Amer. Math. Soc., Providence, R. I.

1940a. Neutral elements in general lattices.
Bull. Amer. Math. Soc. 46: 702–705.

1944. Subdirect unions in universal algebra.
Bull. Amer. Math. Soc. 50: 764–768.

1946. On groups of automorphisms. (Spanish.)
Rev. Un. Mat. Argentina 11: 155–157.

1948. *Lattice Theory*. Second edition.
Amer. Math. Soc., Providence, R. I.

1962. A new interval topology for dually directed sets.
Univ. Nac. Tucumán Rev. Ser. A. 14: 325–331.

1967. *Lattice Theory*. Third edition.
Amer. Math. Soc., Providence, R. I.

1970. What can lattices do for you?
In *Trends in Lattice Theory*, pp. 1–40.
Van Nostrand-Reinhold, New York, N.Y.

G. Birkhoff and O. Frink

1948. Representations of lattices by sets.
Trans. Amer. Math. Soc. 64: 299–316.

G. Birkhoff and J. von Neumann

1936. The logic of quantum mechanics.
Ann. of Math. 37: 823–843.

G. Birkhoff and M. Ward

1939. A characterization of Boolean algebras.
Ann. of Math. 40: 609–610.

R. L. Blair

(See F. W. Anderson and R. L. Blair.)

L. Blumental and K. Menger

1970. *Studies in Geometry*.
Freeman, San Francisco, Cal.

A. B. Brown

(See S. Abian and A. B. Brown.)

T. J. Brown

1969. A recursion formula for finite partition lattices.
Proc. Amer. Math. Soc. 22: 124–126.

G. Bruns

1959. Verbandstheoretische Kennzeichnung vollständiger Mengenringe.
Arch. Math. (Basel) 10: 109–112.

1961. Distributivität und subdirekte Zerlegbarkeit vollständiger Verbände.
Arch. Math. (Basel) 12: 61–66.

1962. Darstellungen und Erweiterungen geordneter Mengen. I, II.
J. Reine Angew. Math. 209: 167–200, 210: 1–23.

G. Bruns
1962a. On the representation of Boolean algebras.
Canad. Math. Bull. 5: 37–41.
1965. Ideal representations of Stone lattices.
Duke Math. J. 32: 555–556.

G. Bruns and G. Kalmbach
1971. Varieties of orthomodular lattices.
Canad. J. Math. 23: 802–810.
1972. Varieties of orthomodular lattices. II.
Canad. J. Math. 24: 328–337.

G. Bruns and H. Lakser
1970. Injective hulls of semilattices.
Canad. Math. Bull. 13: 115–118.

J. R. Büchi
1952. Representation of complete lattices by sets.
Portugal. Math. 11: 151–167.

E. Buttafuoco
(See R. Musti and E. Buttafuoco.)

R. D. Byrd and R. A. Mena
1975. Chains in generalized Boolean lattices.
J. Austral. Math. Soc. (A) 21 (1976): 231–240.

R. D. Byrd, R. A. Mena, and L. A. Troy
1975. Generalized Boolean lattices.
J. Austral. Math. Soc. 19: 225–237.

L. Byrne
1946. Two brief formulations of Boolean algebra.
Bull. Amer. Math. Soc. 52: 269–272.

A. D. Campbell
1943. Set-coordinates for lattices.
Bull. Amer. Math. Soc. 49: 395–398.

C. Carathéodory
1956. *Mass und Integral und ihre Algebraisierung.*
Birkhäuser Verlag, Basel und Stuttgart.
1963. *Algebraic Theory of Measure and Integration.*
Chelsea, New York, N.Y.

L. Carroll
1865. *Alice's Adventures in Wonderland.*
McMillan and Co., London.

C. C. Chang and A. Horn
1961. Prime ideal characterization of generalized Post algebra.
In *Lattice Theory. Proc. Symp. Pure Math.*, II, pp. 43–48.
Amer. Math. Soc., Providence, R.I.
1962. On the representation of α-complete lattices.
Fund. Math. 51: 253–258.

C. C. Chen and G. Grätzer
1969. On the construction of complemented lattices.
J. Algebra 11: 56–63.
1969a. Stone lattices I. Construction theorems.
Canad. J. Math. 21: 884–894.

C. C. Chen and G. Grätzer

1969b. Stone lattices II. Structure theorems.
Canad. J. Math. 21: 895–903.

C. C. Chen and K. M. Koh

1972. On the lattice of convex sublattices of a finite lattice.
Nanta Math. 5 No. 3: 93–95.

C. C. Chen, K. M. Koh, and S. K. Tan

1973. Frattini sublattices of distributive lattices.
Algebra Universalis 3: 294–303.

1975. On the Frattini sublattice of a finite distributive lattice.
Algebra Universalis 5: 88–97.

R. Church

1940. Numerical analysis of certain free distributive structures.
Duke Math. J. 6: 732–734.

1965. Enumeration by rank of the elements of the free distributive lattice with seven generators.
Abstract. Notices Amer. Math. Soc. 12: 724.

J. M. Cibulskis

1969. A characterization of the lattice orderings on a set which induce a given betweenness.
J. London Math. Soc. (2)1: 480–482.

I. S. Cohen

1950. Commutative rings with restricted minimum conditions.
Duke Math. J. 17: 27–42.

P. M. Cohn

1971. *Free Rings and Their Relations.*
Academic Press, New York, N.Y.

S. D. Comer and D. X. Hong

1972. Some remarks concerning the varieties generated by the diamond and the pentagon.
Trans. Amer. Math. Soc. 174: 45–54.

S. D. Comer and J. Johnson

1972. The standard semigroup of operators of a variety.
Algebra Universalis 2: 77–79.

W. H. Cornish

1972. Normal lattices.
J. Austral. Math. Soc. 14: 200–215.

1974. Crawley's completion of a conditionally upper continuous lattice.
Pacific J. Math. 51: 397–405.

1974a. n-normal lattices.
Proc. Amer. Math. Soc. 45: 48–54.

1974b. Quasicomplemented lattices.
Comment. Math. Univ. Carolinae. 15: 501–511.

M. Cotlar

1944. A method of construction of structures and its application to topological spaces and abstract arithmetic.
Univ. Nac. Tucumán Rev. Ser. A. 4: 105–157.

H. H. Crapo

1965. Single-element extensions of matroids.
J. Res. Nat. Bur. Standards Sect. B. 69B: 55–65.

H. H. Crapo

1966. The Möbius function of a lattice.
J. Combinatorial Theory 1: 126–131.

1968/69. Möbius inversion in lattices.
Arch. Math. (Basel) 19: 595–607.

H. H. Crapo and G.-C. Rota

1970. Geometric lattices.
In *Trends in Lattice Theory*, pp. 127–172.
Van Nostrand-Reinhold, New York, N.Y.

1971. *Combinatorial Geometries.*
M.I.T. Press, Boston, Mass.

P. Crawley

1959. The isomorphism theorem in compactly generated lattices.
Bull. Amer. Math. Soc. 65: 377–379.

1960. Lattices whose congruences form a Boolean algebra.
Pacific J. Math. 10: 787–795.

1961. Decomposition theory for nonsemimodular lattices.
Trans. Amer. Math. Soc. 99: 246–254.

1962. Regular embeddings which preserve lattice structure.
Proc. Amer. Math. Soc. 13: 748–752.

1962a. Direct decompositions with finite dimensional factors.
Pacific J. Math. 12: 457–468.
(See also R. P. Dilworth and P. Crawley.)

P. Crawley and R. A. Dean

1959. Free lattices with infinite operations.
Trans. Amer. Math. Soc. 92: 35–47.

P. Crawley and R. P. Dilworth

1973. *Algebraic Theory of Lattices.*
Prentice-Hall, Englewood Cliffs, N.J.

R. Croisot

1951. Contribution à l'étude des treillis semi-modulaires de longueur infinie.
Ann. Sci. École Norm. Sup. (3) 68: 203–265.

1952/53. Quelques applications et propriétés des treillis semimodulaires de longueur infinie.
Ann. Fac. Sci. Univ. Toulouse (4) 16: 11–74.

1957/58. Sur l'irréductibilité dans les treillis géométriques et les géométries projectives.
Convegno internazionale: Reticoli e geometrie proiettive, Palermo, Messina, pp. 3–10.
Gremonese, Rome.
(See also M. L. Dubreil-Jacotin, L. Lesieur, and R. Croisot.)

H. B. Curry

1963. *Foundations of Mathematical Logic.*
McGraw-Hill, New York, N.Y.

M. Curzio

1953. Alcune limitazioni sul minimo ordine dei reticoli modulari di lunghezza 3 contenenti sottoreticoli d'ordine dato.
Ricerche Mat. 2: 140–147.

A. Daigneault

1959. Products of polyadic algebras and of their representations.
Ph. D. Thesis, Princeton University.

B. A. Davey
1973. A note on representable posets.
Algebra Universalis 3: 345–347.
1974. Free products of bounded distributive lattices.
Algebra Universalis 4: 106–107.
1974a. Some annihilator conditions on distributive lattices.
Algebra Universalis 4: 316–322.
B. A. Davey, W. Poguntke, and I. Rival
1975. A characterization of semi-distributivity.
Algebra Universalis 5: 72–75.
B. A. Davey and I. Rival
a. Finite sublattices of three-generated lattices.
J. Austral. Math. Soc. (A) 1976: 171–178.
B. A. Davey and B. Sands
a. An application of Whitman's condition to lattices with no infinite chains.
Algebra Universalis 7 (1977): 171–178.
A. C. Davis
1955. A characterization of complete lattices.
Pacific J. Math. 5: 311–319.
A. Day
1969. A characterization of modularity for congruence lattices of algebras.
Canad. Math. Bull. 12: 167–173.
1970. A simple solution of the word problem for lattices.
Canad. Math. Bull. 13: 253–254.
1970a. Injectivity in congruence distributive equational classes.
Ph. D. Thesis, McMaster University.
1970b. Injectives in non-distributive equational classes of lattices are trivial.
Arch. Math. (Basel) 21: 113–115.
1973. Splitting algebras and a weak notion of projectivity.
In *Proc. Univ. Houston Lattice Theory Conf.*, pp. 466–485 and also Algebra Universalis 5 (1975): 153–162.
a. Lattice conditions implying congruence modularity.
Algebra Universalis 6 (1976): 291–303.
b. Splitting lattices generate all lattices.
Algebra Universalis 7 (1977): 163–170.
A. Day, C. Herrmann, and R. Wille
1972. On modular lattices with four generators.
Algebra Universalis 2: 317–323.
G. W. Day
1970. Maximal chains in atomic Boolean algebras.
Fund. Math. 67: 293–296.
R. A. Dean
1956. Component subsets of the free lattice on *n* generators.
Proc. Amer. Math. Soc. 7: 220–226.
1956a. Completely free lattices generated by partially ordered sets.
Trans. Amer. Math. Soc. 83: 238–249.
1961. Sublattices of free lattices.
In *Lattice Theory, Proc. Symp. Pure Math.*, II, pp. 31–42.
Amer. Math. Soc. Providence, R. I.
1961a. Coverings in free lattices.
Bull. Amer. Math. Soc. 67: 548–549.

R. A. Dean

1964. **Free lattices generated by partially ordered sets and preserving bounds.** Canad. J. Math. 16: 136–148.
(See also P. Crawley and R. A. Dean.)

R. A. Dean and T. Evans

1969. **A remark on varieties of lattices and semigroups.** Proc. Amer. Math. Soc. 21: 394–396.

R. A. Dean and R. H. Oehmke

1964. **Idempotent semigroups with distributive right congruence lattices.** Pacific. J. Math. 14: 1187–1209.

R. Dedekind

1900. **Über die von drei Moduln erzeugte Dualgruppe.** Math. Ann. 53: 371–403.

R. Demarr

1964. **Common fixed points for isotone mappings.** Colloq. Math. 13: 45–48.

V. Devidé

1963. **On monotonous mappings of complete lattices.** Fund. Math. 53: 147–154.

A. H. Diamond and J. C. C. McKinsey

1947. **Algebras and their subalgebras.** Bull. Amer. Math. Soc. 53: 959–962.

R. P. Dilworth

1940. **Lattices with unique irreducible decompositions.** Ann. of Math. (2) 41: 771–777.

1940a. **On complemented lattices.** Tôhoku Math. J. 47: 18–23.

1941. **Ideals in Birkhoff lattices.** Trans. Amer. Math. Soc. 49: 325–353.

1941a. **The arithmetical theory of Birkhoff lattices.** Duke Math. J. 8: 286–299.

1944. **Dependence relations in a semi-modular lattice.** Duke Math. J. 11: 575–587.

1945. **Lattices with unique complements.** Trans. Amer. Math. Soc. 57: 123–154.

1946. **Note on the Kurosch-Ore theorem.** Bull. Amer. Math. Soc. 52: 659–663.

1950. **A decomposition theorem for partially ordered sets.** Ann. of Math. (2) 51: 161–166.

1950a. **The structure of relatively complemented lattices.** Ann. of Math. (2) 51: 348–359.

1954. **Proof of a conjecture on finite modular lattices.** Ann. of Math. (2) 60: 359–364.

1960. **Some combinatorial problems on partially ordered sets.** In *Proc. Symp. Appl. Math.*, 10, pp. 85–90. Amer. Math. Soc., Providence, R.I.

1961. **Structure and decomposition theory of lattices.** In *Lattice Theory, Proc. Symp. Pure Math.*, II, pp. 3–16. Amer. Math. Soc., Providence, R.I.
(See also P. Crawley and R. P. Dilworth.)

R. P. Dilworth and P. Crawley
1960. Decomposition theory for lattices without chain conditions.
Trans. Amer. Math. Soc. 96: 1–22.

R. P. Dilworth and R. Freese
a. Generators of lattice varieties.
Algebra Universalis 6 (1966): 263–267.

R. P. Dilworth and C. Greene
1971. A counterexample to the generalization of Sperner's theorem.
J. Combinatorial Theory 10: 18–21.

R. P. Dilworth and M. Hall
1944. The embedding problem for modular lattices.
Ann. of Math. (2) 45: 450–456.

R. P. Dilworth and J. E. McLaughlin
1952. Distributivity in lattices.
Duke Math. J. 19: 683–693.

M. Dohoux
1974. *Espaces vectoriels topologiques préordonnés.*
Inst. de Math., Université Catholique de Louvain.

T. A. Dowling and R. M. Wilson
1974. The slimmest geometric lattices.
Trans. Amer. Math. Soc. 196: 203–215.
1975. Whitney number inequalities for geometric lattices.
Proc. Amer. Math. Soc. 47: 504–512.

D. A. Drake
1971. Another note on Sperner's lemma.
Canad. Math. Bull. 14: 255–256.

H. Draškovičová
1974. On a representation of lattices by congruence relations.
Mat. Časopis Sloven. Akad. Vied. 24: 69–75.

M. L. Dubreil-Jacotin, L. Lesieur, and R. Croisot
1953. *Leçons sur la théorie des treillis des structures algébriques ordonnées et des treillis géométriques.*
Gauthier-Villars, Paris.

D. G. Duncan
(See R. M. Thrall and D. G. Duncan.)

J. M. Dunn and N. D. Belnap
1968. Homomorphisms of intensionally complemented distributive lattices.
Math. Ann. 176: 28–38.

W. D. Duthie
1942. Segments of ordered sets.
Trans. Amer. Math. Soc. 51: 1–14.

Ph. Dwinger
(See R. Balbes and Ph. Dwinger.)

M. Edelberg
(See D. Kleitman, M. Edelberg, and D. Lubell.)

A. Ehrenfeucht, V. Faber, S. Fajtlowicz, and J. Mycielski
1973. Representations of finite lattices as partition lattices on finite sets.
In *Proc. Univ. Houston Lattice Theory Conf.*, pp. 17–35.

P. C. Eklof
1974. Algebraic closure operators and strong amalgamation bases.
Algebra Universalis 4: 89–98.

R. Engelking
1965. Cartesian products and dyadic spaces.
Fund. Math. 57: 287–304.

G. Epstein
1960. The lattice theory of Post algebras.
Trans. Amer. Math. Soc. 95: 300–317.

P. Erdös, E. Fried, A. Hajnal, and E. C. Milner
1972. Some remarks on simple tournaments.
Algebra Universalis 2: 238–245.

T. Evans
1969. Finitely presented loops, lattices, etc. are hopfian.
J. London Math. Soc. 44: 551–552.
1969a. Some connections between residual finiteness, finite embeddability, and the word problem.
J. London Math. Soc. (2) 1: 399–403.
(See also R. A. Dean and T. Evans.)

T. Evans and D. X. Hong
1972. The free modular lattice on four generators is not finitely presentable.
Algebra Universalis 2: 284–285.

V. Faber
(See A. Ehrenfeucht, V. Faber, S. Fajtlowicz, and J. Mycielski.)

S. Fajtlowicz
(See A. Ehrenfeucht, V. Faber, S. Fajtlowicz, and J. Mycielski.)

N. D. Filippov
1966. Projectivity of lattices. (Russian.)
Mat. Sb. 70(112): 36–54.

P. D. Finch
1969. On the structure of quantum logic.
J. Symbolic Logic 34: 275–282.

D. T. Finkbeiner
1951. A general dependence relation for lattices.
Proc. Amer. Math. Soc. 2: 756–759.
1960. Irreducible congruence relations on lattices.
Pacific J. Math. 10: 813–821.
1960a. A semimodular imbedding of lattices.
Canad. J. Math. 12: 582–591.

P. C. Fishburn
(See K. A. Baker, P. C. Fishburn, and F. S. Roberts.)

I. Fleischer
1976. Embedding a semilattice in a distributive lattice.
Algebra Universalis 6: 85–86.

T. Frayne, A. C. Morel, and D. S. Scott
1962/63. Reduced direct products.
Fund. Math. 51: 195–228.

R. Freese
1972. Varieties generated by modular lattices of width four.
Bull. Amer. Math. Soc. 78: 447–450.

R. Freese
1973. Breadth two modular lattices.
In *Proc. Univ. Houston Lattice Theory Conf.*, pp. 409–451.
1974. An application of Dilworth's lattice of maximal antichains.
Discrete Math. 7: 107–109.
1975. Ideal lattices of lattices.
Pacific J. Math. 57: 125–133.
1975a. Congruence modularity. Abstract.
Notices Amer. Math. Soc. 22: A-301.
a. Planar sublattices of $FM(4)$.
Algebra Universalis 6 (1976): 69–72.
b. *The structure of modular lattices of width four with applications to varieties of lattices.*
Memoirs Amer. Math. Soc. (1977) No. 181.
(See also R. P. Dilworth and R. Freese.)

R. Freese and J. B. Nation
1973. Congruence lattices of semilattices.
Pacific J. Math. 49: 51–58.

P. A. Freĭdman
1967. Rings with a distributive lattice of subrings. (Russian.)
Mat. Sb. 73(4): 513–534.

D. H. Fremlin
1974. *Topological Riesz Spaces and Measure Theory.*
Cambridge University Press, Cambridge.

E. Fried
1970. Tournaments and nonassociative lattices.
Ann. Univ. Sci. Budapest. Eötvös Sect. Math. 13: 151–164.
1973. Weakly associative lattices with join and meet of several elements.
Ann. Univ. Sci. Budapest. Eötvös Sect. Math. 16: 93–98.
1974. Weakly associative lattices with congruence extension property.
Algebra Universalis 4: 151–162.
a. Subdirect irreducible weakly associative lattices with congruence extension property.
Ann. Univ. Sci. Budapest. Eötvös Sect. Math.
b. Prime factorization, algebraic extension, and construction of weakly associative lattices.
Ann. Univ. Sci. Budapest. Eötvös Sect. Math.
c. Equational classes which cover the class of distributive lattices.
Acta Sci. Math. (Szeged) 37 (1975): 37–40.
(See also P. Erdös, E. Fried, A. Hajnal, and E. C. Milner.)

E. Fried and G. Grätzer
1973. A nonassociative extension of the class of distributive lattices.
Pacific J. Math. 49: 59–78.
1973a. Some examples of weakly associative lattices.
Colloq. Math. 27: 215–221.
a. Free and partial weakly associative lattices.
Houston J. Math. 2 (1976): 501–512.

E. Fried, G. Grätzer, and H. Lakser
1971. Amalgamation and weak injectives in the equational class of modular lattices $\mathbf{M}_n$. Abstract.
Notices Amer. Math. Soc. 18: 624.

E. Fried, G. Grätzer, and R. W. Quackenbush

a. Uniform congruence schemes.

b. The equational class generated by weakly associative lattices with the unique bound property.

E. Fried and E. T. Schmidt

1975. Standard sublattices.
Algebra Universalis 5: 203–211.

E. Fried and V. T. Sós

1975. Weakly associative lattices and projective planes.
Algebra Universalis 5: 114–119.

H. Friedman and D. Tamari

1967. Problèmes d'associativité: une structure de treillis finis induite par une loi demi-associative.
J. Combinatorial Theory 2: 215–242.

O. Frink

1942. Topology in lattices.
Trans. Amer. Math. Soc. 51: 569–582.

1946. Complemented modular lattices and projective spaces of infinite dimension.
Trans. Amer. Math. Soc. 60: 452–467.

1962. Pseudo-complements in semi-lattices.
Duke Math. J. 29: 505–514.

1964. Compactifications and semi-normal spaces.
Amer. J. Math. 86: 602–607.
(See also G. Birkhoff and O. Frink.)

B. Frontera Marqués

1964. Generalization of the concept of equivalence relation in a set. (Spanish.)
In *Proc. Fifth Annual Reunion of Spanish Mathematicians*, pp. 203–220.
Publ. Inst. "Jorge Juan" Mat., Madrid.

R. Frucht

1948. On the construction of partially ordered systems with a given group of automorphisms.
Rev. Un. Mat. Argentina 13: 12–18.

1950. Lattices with a given abstract group of automorphisms.
Canad. J. Math. 2: 417–419.

K. D. Fryer and I. Halperin

1956. The von Neumann coordinatization theorem for complemented modular lattices.
Acta Sci. Math. (Szeged) 17: 450–456.

1958. On the construction of coordinates for non-Desarguesian complemented modular lattices. I., II.
Nederl. Akad. Wetensch. Proc. Ser. A. 61: 142–161.

L. Fuchs

1949. Über die Ideale arithmetischer Ringe.
Comment. Math. Helv. 23: 334–341.

1963. *Partially Ordered Algebraic Systems.*
Pergamon Press, New York, N.Y.

S. Fujiwara

(See U. Sasaki and S. Fujiwara.)

T. Fujiwara and K. Murata

1953. On the Jordan-Hölder-Schreier theorem.
Proc. Japan Acad. 29: 151–153.

N. Funayama
1942. On the congruence relations on lattices.
Proc. Imp. Acad. Tokyo 18: 530–531.
1944. On the completion by cuts of distributive lattices.
Proc. Imp. Acad. Tokyo 20: 1–2.
1953. Notes on lattice theory. IV. On partial (semi)lattices.
Bull. Yamagata Univ. Natur. Sci. 2: 171–184.
1959. Imbedding infinitely distributive lattices completely isomorphically into Boolean algebras.
Nagoya Math. J. 15: 71–81.

N. Funayama and T. Nakayama
1942. On the distributivity of a lattice of lattice-congruences.
Proc. Imp. Acad. Tokyo 18: 553–554.

H. Gaifman
1964. Infinite Boolean polynomials I.
Fund. Math. 54: 229–250.

F. Galvin and B. Jónsson
1961. Distributive sublattices of a free lattice.
Canad. J. Math. 13: 265–272.

B. Ganter and I. Rival
1973. Dilworth's covering theorem for modular lattices: a simple proof.
Algebra Universalis 3: 348–350.
a. An arithmetical theorem for modular lattices.
Algebra Universalis 5 (1975): 395–396.

H. S. Gaskill
1972. On transferable semilattices.
Algebra Universalis 2: 303–316.
1972a. On the relation of a distributive lattice to its lattice of ideals.
Bull. Austral. Math. Soc. 7: 377–385.
a. Constructing finite sublattices of free lattices.

H. S. Gaskill, G. Grätzer, and C. R. Platt
1975. Sharply transferable lattices.
Canad. J. Math. 27: 1247–1262.

H. S. Gaskill and C. R. Platt
1975. Sharp transferability and finite sublattices of a free lattice.
Canad. J. Math. 27: 1036–1041.

I. M. Gelfand and V. A. Ponomarev
1970. Problems of linear algebra and classification of quadruples of subspaces in a finite dimensional vector space.
In *Coll. Math. Soc. J. Bolyai* 5. *Hilbert Space Operators*, pp. 163–237.
North Holland, Amsterdam.

A. Ghouilà-Houri
1962. Caractérisation des graphes non orientés dont on peut orienter les arêtes de manière à obtenir le graphe d'une relation d'ordre.
C. R. Acad. Sci. Paris. Sér. A–B 254: 1370–1371.

E. N. Gilbert
1954. Lattice theoretic properties of frontal switching functions.
J. Mathematical Phys. 33: 57–67.

L. Gillman and M. Jerison
1960. *Rings of Continuous Functions.*
University Series in Higher Mathematics.
Van Nostrand, Princeton, N.J.

P. C. Gilmore and A. J. Hoffman
1964. A characterization of comparability graphs and of interval graphs.
Canad. J. Math. 16: 539–548.

K. Gladstien
1973. A characterization of complete trellises of finite length.
Algebra Universalis 3: 341–344.

V. Glivenko
1929. Sur quelques points de la logique de M. Brouwer.
Bull. Acad. des Sci. de Belgique 15: 183–188.

M. M. Gluhov
1960. On the problem of isomorphism of lattices. (Russian.)
Dokl. Akad. Nauk SSSR 132: 254–256.

G. Grätzer
1959. Standard ideals. (Hungarian.)
Magyar Tud. Akad. Mat. Fiz. Oszt. Közl. 9: 81–97.

1962. A characterization of neutral elements in lattices. (Notes on Lattice Theory. I.)
Magyar Tud. Akad. Mat. Kutató Int. Közl. 7: 191–192.

1962a. On Boolean functions. (Notes on Lattice Theory. II.)
Rev. Roumaine Math. Pures Appl. 7: 693–697.

1963. A generalization of Stone's representation theorem for Boolean algebras.
Duke J. Math. 30: 469–474.

1963a. On semi-discrete lattices whose congruence relations form a Boolean algebra.
Acta Math. Acad. Sci. Hungar. 14: 441–445.

1964. Boolean functions on distributive lattices.
Acta Math. Acad. Sci. Hungar. 15: 195–201.

1966. Equational classes of lattices.
Duke J. Math. 33: 613–622.

1968. *Universal Algebra.*
University Series in Higher Mathematics.
Van Nostrand, Princeton, N.J.

1969. Stone algebras form an equational class. (Notes on Lattice Theory. III.)
J. Austral. Math. Soc. 9: 308–309.

1970. Universal algebra.
In *Trends in Lattice Theory*, pp. 173–215.
Van Nostrand-Reinhold, New York, N.Y.

1970a. Two Mal'cev type theorems in universal algebras.
J. Combinatorial Theory 8: 334–342.

1971. *Lattice Theory: First Concepts and Distributive Lattices.*
Freeman, San Francisco, Cal.

1971a. A reduced free product of lattices.
Fund. Math. 73: 21–27.

1973. Free products and reduced free products of lattices.
In *Proc. Univ. Houston Lattice Theory Conf.*, pp. 539–563.

1974. A property of transferable lattices.
Proc. Amer. Math. Soc. 43: 269–271.

1975. A note on the Amalgamation Property. Abstract.
Notices Amer. Math. Soc. 22: A–453.

G. Grätzer

(See also R. Balbes and G. Grätzer; C. C. Chen and G. Grätzer; E. Fried and G. Grätzer; E. Fried, G. Grätzer, and H. Lakser; E. Fried, G. Grätzer, and R. W. Quackenbush; H. S. Gaskill, G. Grätzer, and C. R. Platt.)

G. Grätzer, B. Jónsson, and H. Lakser

1973. The Amalgamation Property in equational classes of modular lattices. Pacific J. Math. 45: 507–524.

G. Grätzer, K. M. Koh, and M. Makkai

1972. On the lattice of subalgebras of a Boolean algebra. Proc. Amer. Math. Soc. 36: 87–92.

G. Grätzer and H. Lakser

1968. Extension theorems on congruences of partial lattices. Abstract. Notices Amer. Math. Soc. 15: 732, 785.

1969. Chain conditions in the distributive free product of lattices. Trans. Amer. Math. Soc. 144: 301–312.

1969a. Some applications of free distributive products. Abstract. Notices Amer. Math. Soc. 16: 405.

1969b. Equationally compact semilattices. Colloq. Math. 20: 27–30.

1971. The structure of pseudocomplemented distributive lattices. II. Congruence extension and amalgamation. Trans. Amer. Math. Soc. 156: 343–358.

1971a. Identities for equational classes generated by tournaments. Abstract. Notices Amer. Math. Soc. 18: 794.

1972. The structure of pseudocomplemented distributive lattices. III. Injective and absolute subretracts. Trans. Amer. Math. Soc. 169: 475–487.

1973. A note on the implicational class generated by a class of structures. Canad. Math. Bull. 16: 603–605.

1973a. Three remarks on the arguesian identity. Abstract. Notices Amer. Math. Soc. 20: A-253, A-313.

1974. Free lattice like sublattices of free products of lattices. Proc. Amer. Math. Soc. 44: 43–45.

1975. Finitely presented lattices. I. Abstract. Notices Amer. Math. Soc. 22: A-380.

G. Grätzer, H. Lakser, and C. R. Platt

1970. Free products of lattices. Fund. Math. 69: 233–240.

G. Grätzer and W. A. Lampe

a. Representation theorems related to congruence lattices of algebras. Advances in Math.

G. Grätzer and R. N. McKenzie

1967. Equational spectra and reduction of identities. Abstract. Notices Amer. Math. Soc. 14: 697.

G. Grätzer and E. T. Schmidt

1957. On the Jordan-Dedekind chain condition. Acta Sci. Math. (Szeged) 18: 52–56.

1957a. On a problem of M. H. Stone. Acta Math. Acad. Sci. Hungar. 8: 455–460.

G. Grätzer and E. T. Schmidt
1958. On the lattice of all join-endomorphisms of a lattice.
Proc. Amer. Math. Soc. 9: 722–726.
1958a. Characterizations of relatively complemented distributive lattices.
Publ. Math. Debrecen 5: 275–287.
1958b. Two notes on lattice-congruences.
Ann. Univ. Sci. Budapest. Eötvös Sect. Math. 1: 83–87.
1958c. On ideal theory for lattices.
Acta Sci. Math. (Szeged) 19: 82–92.
1958d. Ideals and congruence relations in lattices.
Acta. Math. Acad. Sci. Hungar. 9: 137–175.
1958e. On the generalized Boolean algebra generated by a distributive lattice.
Indag. Math. 20: 547–553.
1961. Standard ideals in lattices.
Acta Math. Acad. Sci. Hungar. 12: 17–86.
1962. On congruence lattices of lattices.
Acta Math. Acad. Sci. Hungar. 13: 179–185.

G. Grätzer and J. Sichler
1970. On the endomorphism semigroup (and category) of bounded lattices.
Pacific J. Math. 35: 639–647.
1974. Free products of hopfian lattices.
J. Austral. Math. Soc. 17: 234–245.
1974a. On generating free products of lattices.
Proc. Amer. Math. Soc. 46: 9–14.
1975. Free decompositions of a lattice.
Canad. J. Math. 27: 276–285.

G. Grätzer and B. Wolk
1970. Finite projective distributive lattices.
Canad. Math. Bull. 13: 139–140.

G. Grätzer and M. J. Wonenburger
1962. Some examples of complemented modular lattices.
Canad. Math. Bull. 5: 111–121.

M. Grayev
1947. Isomorphisms of direct decompositions in Dedekind structures. (Russian.)
Izv. Akad. Nauk SSSR Ser. Mat. 11: 33–46.

C. Greene
1970. A rank inequality for finite geometric lattices.
J. Combinatorial Theory 9: 357–364.
(See also R. P. Dilworth and C. Greene.)

O. Hajek
1965. Representation of finite-length modular lattices.
Czechoslovak Math. J. (90) 15: 503–520.
1965a. Characteristics of modular finite-length lattices.
Czechoslovak Math. J. (90) 15: 521–525.

A. Hajnal
(See P. Erdös, E. Fried, A. Hajnal, and E. C. Milner.)

A. W. Hales
1964. On the non-existence of free complete Boolean algebras.
Fund. Math. 54: 45–66.

A. W. Hales
1970. Partition representations of free lattices.
Proc. Amer. Math. Soc. 24: 517–520.
(See also K. A. Baker and A. W. Hales.)

M. Hall
1954. *Projective Planes and Related Topics.*
California Institute of Technology.
(See also R. P. Dilworth and M. Hall.)

Ph. Hall
1934. A contribution to the theory of groups of prime power order.
Proc. London Math. Soc. (2) 36: 29–95.

P. R. Halmos
1961. Injective and projective Boolean algebras.
In *Lattice Theory, Proc. Symp. Pure Math.*, II, pp. 114–122.
Amer. Math. Soc., Providence, R.I.
1963. *Lectures on Boolean Algebras.*
Mathematical Studies No. 1.
Van Nostrand, Princeton, N.J.

I. Halperin
1961. Complemented modular lattices.
In *Lattice Theory, Proc. Symp. Pure Math.*, II, pp. 51–64.
Amer. Math. Soc., Providence, R.I.
(See also I. Amemiya and I. Halperin; K. D. Fryer and I. Halperin.)

W. Hanf
1976. Representing real numbers in denumerable Boolean Algebras.
Fund. Math. 91: 167–170.

G. Hansel
1967. Problèmes de dénombrement et d'évaluation de bornes concernant les éléments du treillis distributif libre.
Publ. Inst. Statist. Univ. Paris 16: 159–300.

L. H. Harper
1974. The morphology of partially ordered sets.
J. Combinatorial Theory Ser. A. 17: 44–58.

M. A. Harrison
1965. *Introduction to Switching and Automata Theory.*
McGraw-Hill, New York, N.Y.

J. Hartmanis
1956. Two embedding theorems for finite lattices.
Proc. Amer. Math. Soc. 7: 571–577.
1959. Lattice theory of generalized partitions.
Canad. J. Math. 11: 97–106.
1961. Generalized partitions and lattice embedding theorems.
In *Lattice Theory, Proc. Symp. Pure Math.*, II, pp. 22–30.
Amer. Math. Soc., Providence, R.I.

J. Hashimoto
1952. Ideal theory for lattices.
Math. Japon. 2: 149–186.
1957. Direct, subdirect decompositions and congruence relations.
Osaka Math. J. 9: 87–112.
1963. Congruence relations and congruence classes in lattices.
Osaka Math. J. 15: 71–86.

J. Hashimoto and S. Kinugawa
1963. On neutral elements in lattices.
Proc. Japan Acad. 39: 162–163.
G. Havas and M. Ward
1969. Lattices with sublattices of a given order.
J. Combinatorial Theory 7: 281–282.
V. Havel
1955. Remark on the uniqueness of direct decompositions of elements in modular lattices of finite length. (Czech.)
Mat. Časopis Sloven. Akad. Vied 5: 90–93.
T. Hecht and T. Katriňák
1972. Equational classes of relative Stone algebras.
Notre Dame J. Formal Logic 13: 248–254.
Z. Hedrlín
(See P. Vopěnka, A. Pultr, and Z. Hedrlín.)
Z. Hedrlín and J. Lambek
1969. How comprehensive is the category of semigroups?
J. Algebra 11: 195–212.
Z. Hedrlín and A. Pultr
1964. Relations (graphs) with given infinite semigroups.
Monatsch. Math. 68: 421–425.
1966. On full embeddings of categories of algebras.
Illinois J. Math. 10: 392–406.
P. Hein
1969. *Grooks* 1.
Borgens Forlag, Copenhagen.
P. Hell
1972. Full embeddings into some categories of graphs.
Algebra Universalis 2: 129–141.
C. Herrmann
1973. Weak (projective) radius and finite equational bases for classes of lattices.
Algebra Universalis 3: 51–58.
1973a. On the equational theory of submodule lattices.
In *Proc. Univ. Houston Lattice Theory Conf.*, pp. 105–118.
1973b. Modulare Verbände von Länge ≤ 6.
In *Proc. Univ. Houston Lattice Theory Conf.*, pp. 119–146.
1973c. *S*-verklebte Summen von Verbänden.
Math. Z. 130: 225–274.
1975. Concerning M. M. Gluhov's paper on the word problem for free modular lattices.
Algebra Universalis 5: 445.
(See also A. Day, C. Herrmann, and R. Wille.)
C. Herrmann, M. Kindermann, and R. Wille
a. On modular lattices generated by $1+2+2$.
Algebra Universalis 5 (1975): 243–251.
C. Herrmann and W. Poguntke
1974. The class of sublattices of normal subgroup lattices is not elementary.
Algebra Universalis 4: 280–286.
a. Axiomatic classes of lattices of normal subgroups.
C. Herrmann, C. M. Ringel, and R. Wille
a. On modular lattices with four generators.

D. Higgs
1971. Lattices isomorphic to their ideal lattices.
Algebra Universalis 1: 71–72.

A. J. Hoffman
(See P. C. Gilmore and A. J. Hoffman.)

D. X. Hong
1972. Covering relations among lattice varieties.
Pacific J. Math. 40: 575–603.
(See also S. D. Comer and D. X. Hong; T. Evans and D. X. Hong.)

A. Horn
1962. On α-homomorphic images of α-rings of sets.
Fund. Math. 51: 259–266.
1968. A property of free Boolean algebras.
Proc. Amer. Math. Soc. 19: 142–143.
1969. Logic with truth values in a linearly ordered Heyting algebra.
J. Symbolic Logic 34: 395–408.
1969a. Free L-algebras.
J. Symbolic Logic 34: 475–480.
(See also R. Balbes and A. Horn; C. C. Chang and A. Horn.)

A. Horn and N. Kimura
1971. The category of semilattices.
Algebra Universalis 1: 26–38.

S. Huang and D. Tamari
1972. Problems of associativity: A simple proof for the lattice property of systems ordered by a semi-associative law.
J. Combinatorial Theory Ser. A 13: 7–13.

D. Huguet
1975. La Structure du treillis des polyèdres de paranthésages.
Algebra Universalis 5: 82–87.

A. P. Huhn
1972. Ph. D. Thesis, University of Szeged.
1972a. Schwach distributive Verbände. I.
Acta Sci. Math. (Szeged) 33: 297–305.
1975. On G. Grätzer's problem concerning automorphisms of a finitely presented lattice.
Algebra Universalis 5: 65–71.

E. V. Huntington
1904. Sets of independent postulates for the algebra of logic.
Trans. Amer. Math. Soc. 5: 288–309.

G. Hutchinson
1971. Modular lattices and abelian categories.
J. Algebra 19: 156–184.
1973. The representation of lattices by modules.
Bull. Amer. Math. Soc. 79: 172–176.
1973a. Recursively unsolvable word problems of modular lattices and diagram-chasing.
J. Algebra 26: 385–399.
1973b. On classes of lattices representable by modules.
In *Proc. Univ. Houston Lattice Theory Conf.*, pp. 69–94.
a. Embedding and unsolvability theorems for modular lattices.
Algebra Universalis 7 (1977): 47–84.

Iqbalunnisa
1963/64. Maximal congruences on a lattice.
J. Madras Univ. B. 33: 113–128.
1964. On neutral elements in a lattice.
J. Indian Math. Soc. (N.S.) 28: 25–31.
1965. On some problems of G. Grätzer and E. T. Schmidt.
Fund. Math. 57: 181–185.
1965a. Neutrality in weakly modular lattices.
Acta Math. Acad. Sci. Hungar. 16: 325–326.
1966. Normal, simple, and neutral congruences on lattices.
Illinois J. Math. 10: 227–234.
1971. On lattices whose lattices of congruences are Stone lattices.
Fund. Math. 70: 315–318.

S. G. Ivanov
1966. Standard subgroups. (Russian.)
Ural. Gos. Univ. Mat. Zap. 5, tetrad' 3: 49–55.
(See also P. G. Kontorovič, S. G. Ivanov, and G. P. Kondrašov.)

T. Iwamura
1944. A lemma on directed sets. (Japanese.)
Zenkoku Shijo Sugaku Danwakai 262: 107–111.

J. Jakubik
1954. On lattices whose graphs are isomorphic. (Russian.)
Czechoslovak Math. J. 4 (79): 131–141.
1954a. On the graph isomorphism of semi-modular lattices. (Slovak.)
Mat. Časopis Sloven. Akad. Vied. 4: 162–177.
1955. Direct decompositions of the unity in modular lattices. (Russian.)
Czechoslovak Math. J. 5 (80): 399–411.
1955a. Congruence relations and weak projectivity in lattices. (Slovak.)
Časopis Pěst. Mat. 80: 206–216.
1957. Remark on the Jordan-Dedekind condition in Boolean algebras. (Slovak.)
Časopis Pěst. Mat. 82: 44–46.
1958. On chains in Boolean lattices. (Slovak.)
Mat. Časopis Sloven. Akad. Vied. 8: 193–202.
1972. Conditionally α-complete sublattices of a distributive lattice.
Algebra Universalis 2: 255–261.
1975. Modular lattices of locally finite length.
Acta Sci. Math. (Szeged) 37: 79–82.

J. Jakubik and M. Kolibiar
1954. On some properties of a pair of lattices. (Russian.)
Czechoslovak Math. J. 4 (79): 1–27.

G. Jameson
1970. *Ordered Linear Spaces.*
Lecture Notes in Mathematics No. 141. Springer-Verlag.

M. F. Janowitz
1964. Projective ideals and congruence relations. I.
Technical report No 51, Univ. New Mexico.
1964a. Projective ideals and congruence relations. II.
Technical report No 63, Univ. New Mexico.
1965. A characterisation of standard ideals.
Acta Math. Acad. Sci. Hungar. 16: 289–301.

M. F. Janowitz

1967. The center of a complete relatively complemented lattice is a complete sublattice.
Proc. Amer. Math. Soc. 18: 189–190.

1968. Perspective properties of relatively complemented lattices.
J. Natur. Sci. and Math. 8: 193–210.

1968a. Section semicomplemented lattices.
Math. Z. 108: 63–76.

1975. Annihilator preserving congruence relations of lattices.
Algebra Universalis 5: 391–394.

J. M. Jauch

1968. *Foundations of Quantum Mechanics.*
Addison-Wesley, Reading, Mass.

J. M. Jauch and C. Piron

1963. Can hidden variables be excluded in quantum mechanics?
Helv. Phys. Acta. 36: 827–837.

1969. On the structure of quantal proposition systems.
Helv. Phys. Acta. 42: 842–848.

Chr. U. Jensen

1963. On characterisations of Prüfer rings.
Math. Scand. 13: 90–98.

M. Jerison

(See L. Gillman and M. Jerison.)

J. Johnson

(See S. D. Comer and J. Johnson.)

B. Jónsson

1951. A Boolean algebra without proper automorphisms.
Proc. Amer. Math. Soc. 2: 766–770.

1953. On the representation of lattices.
Math. Scand. 1: 193–206.

1954. Modular lattices and Desargues' theorem.
Math. Scand. 2: 295–314.

1955. Distributive sublattices of a modular lattice.
Proc. Amer. Math. Soc. 6: 682–688.

1956. Universal relational systems.
Math. Scand. 4: 193–208.

1959. Arguesian lattices of dimension $n \leq 4$.
Math. Scand. 7: 133–145.

1959a. Representation of modular lattices and of relation algebras.
Trans. Amer. Math. Soc. 92: 449–464.

1959b. Lattice-theoretic approach to projective and affine geometry.
In *The Axiomatic Method. With Special Reference to Geometry and Physics. Proc. Int. Symp. at Univ. Calif., Berkeley,* pp. 188–203.
Studies in Logic and the Foundations of Mathematics.
North-Holland, Amsterdam.

1960. Homogeneous universal relational systems.
Math. Scand. 8: 137–142.

1960a. Representations of complemented modular lattices.
Trans. Amer. Math. Soc. 97: 64–94.

1961. Sublattices of a free lattice.
Canad. J. Math. 13: 256–264.

B. Jónsson

1962. Arithmetic properties of freely α-generated lattices.
Canad. J. Math. 14: 476–481.

1965. Extensions of relational structures.
In *Proc. 1963 Int. Symp. at Univ. Calif., Berkeley*, pp. 146–157.
North-Holland, Amsterdam.

1967. Algebras whose congruence lattices are distributive.
Math. Scand. 21: 110–121.

1968. Equational classes of lattices.
Math. Scand. 22: 187–196.

1971. Relatively free products of lattices.
Algebra Universalis 1: 362–373.

1972. The class of Arguesian lattices is self-dual.
Algebra Universalis 2: 396.

1974. Sums of finitely based lattice varieties.
Advances in Math. 14: 454–468.

1974a. Varieties of algebras and their congruence varieties.
In *Proc. 1974 Int. Congr. Math.* I, pp. 315–320.
(See also F. Galvin and B. Jónsson; G. Grätzer, B. Jónsson, and H. Lakser.)

B. Jónsson and J. E. Kiefer

1962. Finite sublattices of a free lattice.
Canad. J. Math. 14: 487–497.

B. Jónsson, G. McNulty, and R. W. Quakenbush

1975. The ascending and descending varietal chains of a variety.
Canad. J. Math. 27: 25–31.

B. Jónsson and G. S. Monk

1969. Representations of primary Arguesian lattices.
Pacific J. Math. 30: 95–139.

B. Jónsson and E. Nelson

1974. Relatively free products in regular varieties.
Algebra Universalis 4: 14–19.

B. Jónsson and Ph. Olin

a. Elementary equivalence and relatively free products of lattices.
Algebra Universalis 6 (1976): 313–325.

J. A. Kalman

1968. A two axiom definition for lattices.
Rev. Roumaine Math. Pures Appl. 13: 669–670.

G. Kalmbach

(See G. Bruns and G. Kalmbach.)

V. Kannan and M. Rajagopalan

1971. On rigidity and groups of homeomorphisms.
In *General Topology and its Relations to Modern Analysis and Algebra* III, pp. 231–234.

L. V. Kantorovič, B. Z. Vulih, and A. G. Pinkser

1950. *Functional Analysis in Partially Ordered Spaces.* (Russian.)
Indat. Tehn-Teor. Lit., Moscow and Leningrad.

I. Kaplansky

1947. Lattices of continuous functions.
Bull. Amer. Math. Soc. 53: 617–623.

D. A. Kappos
1969. *Probability Algebras and Stochastic Spaces.*
Academic Press, New York, N.Y.

D. A. Kappos and F. Papangelou
1966. Remarks on the extension of continuous lattices.
Math. Ann. 166: 277–283.

M. Katětov
1951. Remarks on Boolean algebras.
Colloq. Math. 2: 229–235.

T. Katriňák
1966. Notes on Stone lattices. I. (Russian.)
Mat. Časopis Sloven. Akad. Vied. 16: 128–142.
1967. Notes on Stone lattices. II. (Russian.)
Mat. Časopis Sloven. Akad. Vied. 17: 20–37.
1968. Pseudokomplementäre Halbverbände.
Mat. Časopis Sloven. Akad. Vied. 18: 121–143.
1970. Die Kennzeichnung der distributiven pseudokomplementären Halbverbände.
J. Reine Angew. Math. 241: 160–179.
1971. Relativ Stonesche Halbverbände sind Verbände.
Bull. Soc. Roy. Sci. Liège 40: 91–93.
1972. Die freien Stoneschen Verbände und ihre Tripelcharakterisierung.
Acta Math. Acad. Sci. Hungar. 23: 315–326.
1972a. Über eine Konstruktion der distributiven pseudokomplementären Verbände.
Math. Nachr. 53: 85–99.
1972b. Subdirectly irreducible modular p-algebras.
Algebra Universalis 2: 166–173.
1973. Primitive Klassen von modularen S-Algebren.
J. Reine Angew. Math. 261: 55–70.
1973a. A new proof of the construction theorem for Stone algebras.
Proc. Amer. Math. Soc. 40: 75–78.
1973b. Construction of some subdirectly irreducible modular p-algebras.
Algebra Universalis 3: 321–327.
1973c. The cardinality of the lattice of all equational classes of p-algebras.
Algebra Universalis 3: 328–329.
1974. Injective double Stone algebras.
Algebra Universalis 4: 259–267.
1974a. Congruence extension property for distributive double p-algebras.
Algebra Universalis 4: 273–276.
a. A new description of the free Stone algebra.
Algebra Universalis 5 (1975): 179–189.
b. Construction of modular double S-algebras.
Algebra Universalis 8 (1978), No. 1.
c. On a problem of G. Grätzer.
Proc. Amer. Math. Soc. 56 (1976): 19–24.
d. Congruence lattices of distributive p-algebras.
Algebra Universalis 7 (1977): 265–272.
(See also T. Hecht and T. Katriňák.)

T. Katriňák and P. Mederly
1974. Construction of modular p-algebras.
Algebra Universalis 4: 301–315.

D. Kelly

1972. A note on equationally compact lattices.
Algebra Universalis 2: 80–84.
(See also M. E. Adams and D. Kelly.)

D. Kelly and I. Rival

1974. Crowns, fences, and dismantlable lattices.
Canad. J. Math. 26: 1257–1271.

1975. Certain partially ordered sets of dimension three.
J. Combinatorial Theory Ser. A 18: 239–242.

1975a. Planar lattices.
Canad. J. Math. 27: 636–665.

L. M. Kelly

(See J. G. Basterfield and L. M. Kelly.)

J. E. Kiefer

(See B. Jónsson and J. E. Kiefer.)

N. Kimura

(See A. Horn and N. Kimura.)

M. Kindermann

(See C. Herrmann, M. Kindermann, and R. Wille.)

S. Kinugawa and J. Hashimoto

1966. On relative maximal ideals in lattices.
Proc. Japan Acad. 42: 1–4.
(See also J. Hashimoto and S. Kinugawa.)

M. J. Klass

1974. Maximum antichains: A sufficient condition.
Proc. Amer. Math. Soc. 45: 28–30.

D. Kleitman

1969. On Dedekind's problem: The number of monotone Boolean functions.
Proc. Amer. Math. Soc. 21: 677–682.

D. Kleitman, M. Edelberg, and D. Lubell

1971. Maximal sized antichains in partial orders.
Discrete Math. 1: 47–53.

B. Knaster

1928. Un Théorème sur les fonctions d'ensembles.
Ann. Soc. Polon. Math. 6: 133–134.

S. R. Kogalovskiĭ

1964. On linearly complete ordered sets. (Russian.)
Uspehi Mat. Nauk. 19, No 2: 147–150.

1965. On a theorem of Birkhoff. (Russian.)
Uspehi Mat. Nauk. 20: 206–207.

K. M. Koh

1971. On the Frattini sublattice of a lattice.
Algebra Universalis 1: 104–116.

1971a. On the Frattini sub-semilattice of a semilattice.
Nanta Math. 5 No. 1: 22–33.

1972. On the lattice of convex sublattices of a lattice.
Nanta Math. 5 No. 3: 18–37.

1973. On sublattices of a lattice.
Nanta Math. 6 No. 1: 68–79.

K. M. Koh
1973a. On the Frattini sublattice of a finite modular planar lattice.
Algebra Universalis 3: 304–317.
(See also C. C. Chen and K. M. Koh; C. C. Chen, K. M. Koh, and S. K. Tan; G. Grätzer, K. M. Koh, and M. Makkai.)

M. Kolibiar
1956. On the axiomatics of modular lattices. (Russian.)
Czechoslovak Math. J. 6 (81): 381–386.
1972. Distributive sublattices of a lattice.
Proc. Amer. Math. Soc. 34: 359–364.
(See also J. Jakubík and M. Kolibiar.)

A. Komatu
1943. On a characterization of join homomorphic transformation-lattice.
Proc. Imp. Acad. Tokyo 19: 119–124.

G. P. Kondrašov
(See P. G. Kontorovič, S. G. Ivanov, and G. P. Kondrašov.)

P. G. Kontorovič, S. G. Ivanov, and G. P. Kondrašov.
1965. Distributive pairs of elements in a lattice. (Russian.)
Dokl. Akad. Nauk. SSSR 160: 1001–1003.

A. Kostinsky
1971. Some problems for rings and lattices within the domain of general algebra.
Ph. D. Thesis, Univ. Calif., Berkeley.
1972. Projective lattices and bounded homomorphisms.
Pacific J. Math. 40: 111–119.

A. G. Kuroš
1935. Durchschnittsdarstellungen mit irreduziblen Komponenten in Ringen und in sogenannten Dualgruppen.
Mat. Sb. 42: 613–616.
1943. Isomorphisms of direct decompositions. I. (Russian.)
Izv. Akad. Nauk SSSR Ser. Mat. 7: 185–202.
1946. Isomorphisms of direct decompositions. II. (Russian.)
Izv. Akad. Nauk SSSR Ser. Mat. 10: 47–72.

Z. Ladzianska
1974. Poproducts of lattices and Sorkin's theorem.
Mat. Časopis Sloven. Akad. Vied. 24: 247–252.

H. Lakser
1968. Free lattices generated by partially ordered sets.
Ph. D. Thesis, University of Manitoba.
1970. Injective hulls of Stone algebras.
Proc. Amer. Math. Soc. 24: 524–529.
1970a. Normal and canonical representations in free products of lattices.
Canad. J. Math. 22: 394–402.
1971. The structure of pseudocomplemented distributive lattices.
I. Subdirect decomposition.
Trans. Amer. Math. Soc. 156: 335–342.
1972. Simple sublattices of free products of lattices. Abstract.
Notices Amer. Math. Soc. 19: A 509.
1973. Principal congruences of pseudocomplemented distributive lattices.
Proc. Amer. Math. Soc. 37: 32–36.

H. Lakser

1973a. Disjointness conditions in free products of distributive lattices: An application of Ramsey's theorem.
In *Proc. Univ. Houston Lattice Theory Conf.*, pp. 156–168.

1973b. A note on the lattice of sublattices of a finite lattice.
Nanta Math. 6 No. 1: 55–57.
(See also G. Bruns and H. Lakser; E. Fried, G. Grätzer, and H. Lakser; G. Grätzer and H. Lakser; G. Grätzer, H. Lakser, and C. R. Platt; G. Grätzer, B. Jónsson, and H. Lakser.)

W. A. Lampe

(See G. Grätzer and W. A. Lampe.)

J. Lambek

(See Z. Hedrlín and J. Lambek.)

R. E. Larson and S. J. Andima

1975. The lattice of topologies: A survey.
Rocky Mountain J. Math. 5: 177–198.

K. B. Lee

1970. Equational classes of distributive pseudo-complemented lattices.
Canad. J. Math. 22: 881–891.

L. Lesieur

(See M. L. Dubreil-Jacotin, L. Lesieur, and R. Croisot.)

A. H. Livšic

1962. Direct decompositions of idempotents in semigroups. (Russian.)
Trudy Moskov. Mat. Obšč. 11: 37–98.

J. Łoś

1955. Quelques remarques, théorèmes et problèmes sur les classes définissables d'algèbres.
In *Mathematical Interpretation of Formal Systems*, pp. 98–113.
North-Holland, Amsterdam.

H. F. J. Löwig

1943. On the importance of the relation $[(A, B), (A, C)] < [(B, C), (C, A), (A, B)]$ between three elements of a structure.
Ann. of Math. (2) 44: 573–579.

F. W. Lozier

1969. A class of compact 0-dimensional spaces.
Canad. J. Math. 21: 817–821.

D. Lubell

1966. A short proof of Sperner's lemma.
J. Combinatorial Theory 1: 299.
(See also D. Kleitman, M. Edelberg, and D. Lubell.)

F. Lunnon

1971. The IU function: The size of a free distributive lattice.
In *Combinatorial Mathematics and its Applications*, pp. 173–181.
Academic Press, New York, N.Y.

R. C. Lyndon

1950. The representation of relational algebras.
Ann. of Math. 51: 707–729.

1954. Identities in finite algebras.
Proc. Amer. Math. Soc. 5: 8–9.

1956. The representation of relation algebras. II.
Ann. of Math. 63: 294–307.

G. W. Mackey

1963. *Mathematical Foundations of Quantum Mechanics.* Benjamin, New York, N.Y.

S. MacLane

1938. A lattice formulation for transcendence degrees and p-bases. Duke Math. J. 4: 455–468.

1943. A conjecture of Ore on chains in partially ordered sets. Bull. Amer. Math. Soc. 49: 567–568.

H. M. MacNeille

1937. Partially ordered sets. Trans. Amer. Math. Soc. 42: 416–460.

1939. Extension of a distributive lattice to a Boolean ring. Bull. Amer. Math. Soc. 45: 452–455.

F. Maeda

1951. Lattice theoretic characterization of abstract geometries. J. Sci. Hiroshima Univ. Ser. A. 15: 87–96.

1952. Matroid lattices of infinite length. J. Sci. Hiroshima Univ. Ser. A. 15: 177–182.

1953. A lattice formulation for algebraic and transcendental extensions in abstract algebras. J. Sci. Hiroshima Univ. Ser. A. 16: 383–397.

1958. *Kontinuierliche Geometrien.* Springer-Verlag, Berlin.

1963. Modular centers of affine matroid lattices. J. Sci. Hiroshima Univ. Ser. A–I Math. 27: 73–84.

1963a. Parallel mappings and comparability theorems in affine matroid lattices. J. Sci. Hiroshima Univ. Ser. A–I Math. 27: 85–96.

1964. Perspectivity of points in matroid lattices. J. Sci. Hiroshima Univ. Ser. A–I Math. 28: 101–112.

F. and S. Maeda

1970. *Theory of Symmetric Lattices.* Springer-Verlag, New York, N.Y.

S. Maeda

1965. On the symmetry of the modular relation in atomic lattices. J. Sci. Hiroshima Univ. Ser. A–I Math. 29: 165–170.

1966. Infinite distributivity in complete lattices. Mem. Ehime Univ. Sect. II Ser A. 5, no. 3: 11–13.

1969. Modular pairs in atomistic lattices with the covering property. Proc. Japan Acad. 45: 149–153.

1974. Locally modular lattices and locally distributive lattices. Proc. Amer. Math. Soc. 44: 237–243.
(See also F. and S. Maeda.)

K. D. Magill

1970. The semigroup of endomorphisms of a Boolean ring. J. Austral. Math. Soc. 11: 411–416.

M. Makkai

1973. A proof of Baker's finite-base theorem on equational classes generated by finite elements of congruence distributive varieties. Algebra Universalis 3: 174–181.
(See also G. Grätzer, K. M. Koh, and M. Makkai.)

M. Makkai and G. McNulty
a. On axioms for lattices of submodules.
Manuscript.

A. I. Mal'cev
1966. Several remarks on quasivarieties of algebraic systems. (Russian.)
Algebra i Logika Sem. 5, no. 3: 3–9.

M. Mandelker
1970. Relative annihilators in lattices.
Duke Math. J. 37: 377–386.

J.-M. Maranda
1964. Injective structures.
Trans. Amer. Math. Soc. 110: 98–135.

R. N. McKenzie
1970. Equational bases for lattice theories.
Math. Scand. 27: 24–38.
1972. Equational bases and nonmodular lattice varieties.
Trans. Amer. Math. Soc. 174: 1–43.
1973. Some unsolved problems between lattice theory and equational logic.
In *Proc. Univ. Houston Lattice Theory Conf.*, pp. 564–573.
(See also G. Grätzer and R. N. McKenzie.)

R. N. McKenzie and J. D. Monk
1973. On automorphism groups of Boolean algebras.
In *Coll. Math. Soc. J. Bolyai.* 10. *Infinite and Finite Sets*, pp. 951–988.
North Holland, Amsterdam.

J. C. C. McKinsey
(See A. H. Diamond and J. C. C. McKinsey.)

J. E. McLaughlin
1951. Projectivities in relatively complemented lattices.
Duke Math. J. 18: 73–84.
1953. Structured theorems for relatively complemented lattices.
Pacific J. Math. 3: 197–208.
1956. Atomic lattices with unique comparable complements.
Proc. Amer. Math. Soc. 7: 864–866.
1961. The normal completion of a complemented modular point lattice.
In *Lattice Theory, Proc. Symp. Pure Math.*, II., pp. 78–80.
Amer. Math. Soc., Providence, R.I.
(See also R. P. Dilworth and J. E. McLaughlin.)

G. McNulty
(See B. Jónsson, G. McNulty, and R. W. Quakenbush; M. Makkai and G. McNulty.)

P. Mederley
(See T. Katriňák and P. Mederley.)

R. A. Mena
(See R. D. Byrd and R. A. Mena; R. D. Byrd, R. A. Mena, and L. A. Troy.)

K. Menger
1936. New foundations of projective and affine geometry.
Ann. of Math. (2) 37: 456–482.
(See also L. Blumental and K. Menger.)

G. Michler and R. Wille
1970. Die primitiven Klassen arithmetischer Ringe.
Math. Z. 113: 369–372.

E. C. Milner
(See P. Erdös, E. Fried, A. Hajnal, and E. C. Milner.)
L. Mirsky
1971. *Transversal Theory.*
Academic Press, New York, N.Y.
A. Mitchke and R. Wille
1973. Freie modulare Verbände $FM({}_DM_3)$.
In *Proc. Univ. Houston Lattice Theory Conf.*, pp. 383–396.
1976. Finite distributive lattices projective in the class of all modular lattices.
Algebra Universalis 6: 383–393.
P. Mittelstaedt
1972. On the interpretation of the lattice of subspaces of the Hilbert space as a propositional calculus.
Z. Naturforsch. Tiel A 27: 1358–1362.
A. F. Möbius
1832. Über eine besondere Art von Umkehrung der Reihen.
J. Reine Angew. Math. 9: 105–123.
E. N. Močulskiĭ
1961. Direct decompositions in lattices. (Russian.)
Izv. Akad. Nauk SSSR Ser. Mat. 25: 717–748.
1962. Direct decompositions in lattices. II. (Russian.)
Izv. Akad. Nauk SSSR Ser. Mat. 26: 161–210.
1968. Isomorphisms of direct decompositions in lattices. (Russian.)
Sibirsk. Mat. Ž. 9: 1360–1385.
G. S. Monk
(See B. Jónsson and G. S. Monk.)
J. D. Monk
(See R. N. McKenzie and J. D. Monk.)
J. D. Monk and R. M. Solovay
1972. On the number of complete Boolean algebras.
Algebra Universalis 2: 365–368.
A. Monteiro
1947. Sur l'arithmétique des filtres premiers.
C. R. Acad. Sci. Paris 225: 846–848.
1955. Axiomes indépendants pour les algèbres de Brouwer.
Rev. Un. Mat. Argentina 17: 149–160.
J. T. Moore
(See W. T. Trotter, Jr. and J. T. Moore.)
A. C. Morel
(See T. Frayne, A. C. Morel, and D. S. Scott.)
M. D. Morley and R. L. Vaught
1962. Homogeneous universal models.
Math. Scand. 11: 37–57.
A. Mostowski and A. Tarski
1939. Boolesche Ringe mit geordneter Basis.
Fund. Math. 32: 69–86.
K. Murata
(See T. Fujiwara and K. Murata.)
V. L. Murskiĭ
1965. The existence in the three-valued logic of a closed class with a finite basis having no finite complete system of identities. (Russian.)
Dokl. Akad. Nauk. SSSR 163: 815–818.

P. V. R. Murty and V. V. R. Rao
1974. Characterization of certain classes of pseudo complemented semi-lattices. Algebra Universalis 4: 289–300.

R. Musti and E. Buttafuoco
1956. Sui subreticoli distributivi dei reticoli modulari. Boll. Un. Mat. Ital. (3) 11: 584–587.

J. Mycielski
(See A. Ehrenfeucht, V. Faber, S. Fajtlowicz, and J. Mycielski.)

L. Nachbin
1947. Une Propriété caractéristique des algèbres booléiennes. Portugal. Math. 6: 115–118.
1949. On a characterization of the lattice of all ideals of a Boolean ring. Fund. Math. 36: 137–142.

H. Nakano
1950. *Modulared Semi-Ordered Linear Spaces.* Marazen, Tokyo.

T. Nakayama
(See N. Funayama and T. Nakayama.)

J. B. Nation
1974. Varieties whose congruences satisfy certain lattice identities. Algebra Universalis 4: 78–88.
(See also R. Freese and J. B. Nation.)

E. Nelson
1974. The embedding of a distributive lattice into its ideal lattice is pure. Algebra Universalis 4: 135–140.
(See also B. Jónsson and E. Nelson.)

O. T. Nelson, Jr.
1968. Subdirect decompositions of lattices of width two. Pacific J. Math. 24: 519–523.

W. C. Nemitz
1965. Implicative semi-lattices. Trans. Amer. Math. Soc. 117: 128–142.

A. Nerode
1959. Some Stone spaces and recursion theory. Duke Math. J. 26: 397–406.

B. H. Neumann
1962. *Universal Algebra.* Lecture Notes. Courant Inst. of Math. Sci. New York University.

B. H. Neumann and Hanna Neumann
1952. Extending partial endomorphisms of groups. Proc. London Math. Soc. (3) 2: 337–348.

Hanna Neumann
(See B. H. Neumann and Hanna Neumann.)

J. von Neumann
1936/37. *Lectures on Continuous Geometries.* Institute of Advanced Studies, Princeton, N.J.
1960. *Continuous Geometry.* (I. Halperin, editor.) Princeton Mathematical Series, No. 25. Princeton Univ. Press, Princeton, N.J.

J. von Neumann
(See also G. Birkhoff and J. von Neumann.)

Kung-fu Ng
(See Yau-chuen Wong and Kung-fu Ng.)

E. Noether
1926. Abstrakter Aufbau der Idealtheorie in algebraischen Zahl- und Funktionenkörpern.
Math. Ann. 96: 26–61.

R. H. Oehmke
(See R. A. Dean and R. H. Oehmke.)

T. Ogasawara and U. Sasaki
1949. On a theorem in lattice theory.
J. Sci. Hiroshima Univ. Ser. A 14: 13.

Ph. Olin
a. Free products and elementary types of Boolean algebras.
(See also B. Jónsson and Ph. Olin.)

O. Ore
1935. On the foundation of abstract algebra. I.
Ann. of Math. 36: 406–437.
1936. On the foundation of abstract algebra. II.
Ann. of Math. 37: 265–292.
1937. Structures and group theory. I.
Duke Math. J. 3: 149–174.
1938. Structures and group theory. II.
Duke Math. J. 4: 247–269.
1940. Remarks on structures and group relations.
Vierteljschr. Naturforsch. Ges. Zürich, 85 Beiblatt (Festschrift Rudolf Fueter), 1–4.
1942. Theory of equivalence relations.
Duke Math. J. 9: 573–627.
1943. Chains in partially ordered sets.
Bull. Amer. Math. Soc. 49: 558–566.
1944. Galois connexions.
Trans. Amer. Math. Soc. 55: 493–513.

R. Padmanabhan
1966. On axioms for semilattices.
Canad. Math. Bull. 9: 357–358.
1966a. On some ternary relations in lattices.
Colloq. Math. 15: 195–198.
1968. A note on Kalman's paper.
Rev. Roumaine Math. Pures Appl. 13: 1149–1152.
1969. Two identities for lattices.
Proc. Amer. Math. Soc. 20: 409–412.
1972. On identities defining lattices.
Algebra Universalis 1: 359–361.
1974. On M-symmetric lattices.
Canad. Math. Bull. 17: 85–86.

R. Padmanabhan
1974a. A minimal selfdual equational basis for Boolean algebras. Abstract. Notices Amer. Math. Soc. 21: A–368.

F. Papangelou
(See D. A. Kappos and F. Papangelou.)

D. Papert
1964. Congruence relations in semilattices. J. London Math. Soc. 39: 723–729.

A. Pelczar
1961. On the invariant points of a transformation. Ann. Polon. Math. 11: 199–202.
1962. On the extremal solutions of a functional equation. Zeszyty Nauk. Uniw. Jagiello. Prace Mat. No. 7: 9–11.

A. L. Peressini
1967. *Ordered Topological Vector Spaces.* Harper and Row, New York, N.Y.

P. Perkins
1969. Bases for equational theories of semigroups. J. Algebra 11: 293–314.

R. Permutti
1964. Sulle semicongruenze di un reticolo. Rend. Accad. Sci. Fis. Mat. Napoli (4) 31: 160–167.

R. S. Pierce
1958. A note on complete Boolean algebras. Proc. Amer. Math. Soc. 9: 892–896.
1961. Some questions about complete Boolean algebras. In *Lattice Theory, Proc. Symp. Pure Math.*, II, pp. 129–140. Amer. Math. Soc., Providence, R.I.
1968. *Introduction to the Theory of Abstract Algebras.* Holt, Rinehart, and Winston, New York, N.Y.
1972. *Compact Zero-dimensional Metric Spaces of Finite Type.* Memoirs Amer. Math. Soc. No. 130.
1973. Bases of countable Boolean algebras. J. Symbolic Logic 38: 212–214.

A. G. Pinkser
(See L. V. Kantorovič, B. Z. Vulih, and A. G. Pinkser.)

C. Piron
1964. Axiomatique quantique. Helv. Phys. Acta. 37: 439–468.
(See also J. M. Jausch and C. Piron.)

E. Pitcher and M. F. Smiley
1942. Transitivities of betweenness. Trans. Amer. Math. Soc. 52: 95–114.

A. F. Pixley
(See K. A. Baker and A. F. Pixley.)

C. R. Platt
1974. Iterated limits of lattices. Canad. J. Math. 26: 1301–1320.
1976. Planar lattices and planar graphs. J. Combinatorial Theory, Ser. B. 21: 30–39.

C. R. Platt

(See also H. S. Gaskill, G. Grätzer, and C. R. Platt; H. S. Gaskill and C. R. Platt; G. Grätzer, H. Lakser, and C. R. Platt.)

W. Poguntke and I. Rival

1974. Finite sublattices generated by order-isomorphic subsets. Arch. Math. (Basel) 25: 225–230.

(See also B. A. Davey, W. Poguntke, and I. Rival; C. Herrmann and W. Poguntke.)

V. A. Ponomarev

(See I. M. Gelfand and V. A. Ponomarev.)

E. L. Post

1921. Introduction to a general theory of elementary propositions. Amer. J. Math. 43: 163–185.

D. H. Potts

1965. Axioms for semilattices. Canad. Math. Bull. 8: 519.

W. Prenowitz

1948. Total lattices of convex sets and of linear spaces. Ann. of Math. (2) 49: 659–688.

H. A. Priestley

1970. Representation of distributive lattices by means of ordered Stone spaces. Bull. London Math. Soc. 2: 186–190.

1972. Ordered topological spaces and the representation of distributive lattices. Proc. London Math. Soc. (3) 24: 507–530.

1974. Stone lattices: A topological approach. Fund. Math. 84: 127–143.

1975. The construction of spaces dual to pseudocomplemented distributive lattices. Quart. J. Math. Oxford (3) 26: 215–228.

P. Pudlák

a. Distributivity of strongly representable lattices. Algebra Universalis 7 (1977): 85–92.

b. A new proof of the congruence lattice representation theorem. Algebra Universalis 6 (1966): 269–275.

A. Pultr

(See Z. Hedrlín and A. Pultr; P. Vopěnka, A. Pultr, and Z. Hedrlín.)

R. W. Quackenbush

1972. The triangle is functionally complete. Algebra Universalis 2: 128.

1972a. Free products of bounded distributive lattices. Algebra Universalis 2: 393–394.

1973. Planar lattices. In *Proc. Univ. Houston Lattice Theory Conf.*, pp. 512–518.

1974. Near-Boolean algebras I: Combinatorial aspects. Discrete Math. 10: 301–308.

a. Semi-simple equational classes with distributive congruence lattices. Ann. Univ. Sci. Budapest. Eötvös Sect. Math. 17 (1974): 15–19.

(See also E. Fried, G. Grätzer, and R. W. Quackenbush; B. Jónsson, G. McNulty, and R. W. Quackenbush.)

R. W. Quackenbush and H. C. Reichel
1975. Partitioning bases of Boolean lattices.
Algebra Universalis 5: 148.

R. W. Quackenbush and B. Wolk
1971. Strong representations of congruence lattices.
Algebra Universalis 1: 165–166.

M. Rajagopalan
(See V. Kannan and M. Rajagopalan.)

G. N. Raney
1952. Completely distributive complete lattices.
Proc. Amer. Math. Soc. 3: 677–680.
1953. A subdirect-union representation for completely distributive complete lattices.
Proc. Amer. Math. Soc. 4: 518–522.
1960. Tight Galois connections and complete distributivity.
Trans. Amer. Math. Soc. 97: 418–426.

V. V. R. Rao
(See P. V. R. Murty and V. V. R. Rao; U. M. Swamy and V. V. R. Rao.)

H. Rasiowa
1974. *An Algebraic Approach to Non-Classical Logics.*
North-Holland, Amsterdam.

H. Rasiowa and R. Sikorski
1963. *The Mathematics of Metamathematics.*
Monog. Mat., XLI. Państwowe Wydawn. Naukowe, Warsaw.

H. C. Reichel
(See R. W. Quackenbush and H. C. Reichel.)

B. C. Rennie
1951. *The Theory of Lattices.*
Forster and Jagg, Cambridge, England.

I. Reznikoff
1963. Chaînes de formules.
C. R. Acad. Sci. Paris 256: 5021–5023.

J. B. Rhodes
1975. Modular and distributive semilattices.
Trans. Amer. Math. Soc. 201: 31–41.

P. Ribenboim
1949. Characterization of the sup-complement in a distributive lattice with last element.
Summa Brasil. Math. 2, No. 4: 43–49.

J. Riečan
1958. To the axiomatics of modular lattices. (Slovak.)
Acta Fac. Rerum Natur. Univ. Comenian. Math. 2: 257–262.

L. Rieger
1949. A note on topological representations of distributive lattices.
Časopis Pěst. Mat. Fys. 74: 55–61.
1951. Some remarks on automorphisms of Boolean algebras.
Fund. Math. 38: 209–216.

C. M. Ringel
(See C. Herrmann, C. M. Ringel, and R. Wille.)

I. Rival

1972. Projective images of modular (distributive, complemented) lattices are modular (distributive, complemented).
Algebra Universalis 2: 395.

1973. Maximal sublattices of finite distributive lattices.
Proc. Amer. Math. Soc. 37: 417–420.

1974. Maximal sublattices of finite distributive lattices. II.
Proc. Amer. Math. Soc. 44: 263–268.

1974a. Lattices with doubly irreducible elements.
Canad. Math. Bull. 17: 91–95.

a. A fixed point theorem for finite partially ordered sets.
J. Combinatorial Theory, Ser. A. 21 (1976): 309–318.

b. Finite modular lattices with sublattices of all orders.

c. Lattices with sublattices of a given order $N \leq 9$.
(See also R. Antonius and I. Rival; B. A. Davey and I. Rival; B. A. Davey, W. Poguntke, and I. Rival; B. Ganter and I. Rival; D. Kelly and I. Rival; W. Poguntke and I. Rival.)

I. Rival and B. Sands

1975. Finite weakly embeddable lattices which are embeddable.
Arch. Math. (Basel) 26: 346–352.

F. S. Roberts

(See K. A. Baker, P. C. Fishburn, and F. S. Roberts.)

A. Robinson

1971. Infinite forcing in model theory.
In *Proc. Second Scand. Logic Symp.*, pp. 317–340.
North-Holland, Amsterdam.

1971a. On the notion of algebraic closedness for noncommutative groups and fields.
J. Symbolic Logic 36: 441–444.

H. L. Rolf

1958. The free lattice generated by a set of chains.
Pacific J. Math. 8: 585–595.

1966. The free modular lattice, $FM(2+2+2)$, is infinite.
Proc. Amer. Math. Soc. 17: 960–961.

I. Rosenberg

1967. Maximale Gegenketten im Verband 3^n.
Arch. Math. (Brno) 3: 185–190.

P. C. Rosenbloom

1942. Post algebras. I. Postulates and general theory.
Amer. J. Math. 64: 167–188.

J. Rosický

1975. Sublattices of the lattice of topologies.
Acta Fac. Rerum Natur. Univ. Comenian. Math., Special Number, 39–41.

G.-C. Rota

1964. On the foundations of combinatorial theory. I. Theory of Möbius functions.
Z. Wahrscheinlichkeitstheorie und Verw. Gebiete 2: 340–368.
(See also H. H. Crapo and G.-C. Rota.)

G. Rousseau

1970. Post algebras and pseudo-Post algebras.
Fund. Math. 67: 133–145.

B. R.-Salinas
1969. Fixed points in ordered sets. (Spanish.)
In *Consejo Sup. Investigación. Ci. Fac. Ci. Zaragoza*, pp. 163–172.

S. Rudeanu
1963. *Axioms for Lattices and Boolean Algebras.* (Roumanian.)
Monografii Asupra Teoriei Algebrice A Mecanismelor Automate.
Editura Academiei Republicii Populare Romine, Bucharest.
1964. Logical dependence of certain chain conditions in lattice theory.
Acta Sci. Math. (Szeged) 25: 209–218.
1974. *Boolean Functions and Equations.*
North-Holland, Amsterdam.

D. Sachs
1961. Identities in finite partition lattices.
Proc. Amer. Math. Soc. 12: 944–945.
1961a. Partition and modulated lattices.
Pacific J. Math. 11: 325–345.
1962. The lattice of subalgebras of a Boolean algebra.
Canad. J. Math. 14: 451–460.

V. N. Saliĭ
1972. A compactly generated lattice with unique complements is distributive. (Russian.)
Mat. Zametki 12: 617–620.

Y. Sampei
1953. On lattice completions and closure operations.
Comment. Math. Univ. St. Paul 2: 55–70; and 3: 29–30.

B. Sands
(See B. A. Davey and B. Sands; I. Rival and B. Sands.)

U. Sasaki
1952. Lattice theoretic characterization of an affine geometry of arbitrary dimensions.
J. Sci. Hiroshima Univ. Ser. A. 16: 223–238.
1953. Semi-modularity in relatively atomic, upper continuous lattices.
J. Sci. Hiroshima Univ. Ser. A. 16: 409–416.
(See also T. Ogasawara and U. Sasaki.)

U. Sasaki and S. Fujiwara
1952. The decomposition of matroid lattices.
J. Sci. Hiroshima Univ. Ser. A. 15: 183–188.
1952a. The characterization of partition lattices.
J. Sci. Hiroshima Univ. Ser. A. 15: 189–201.

G. Sauer, W. Seibert, and R. Wille
1973. On free modular lattices over partial lattices with four generators.
In *Proc. Univ. Houston Lattice Theory Conf.*, pp. 332–382.

B. M. Schein
1970. Relation algebras and function semigroups.
Semigroup Forum 1 No. 1: 1–62.
1972. On the definition of distributive semilattices.
Algebra Universalis 2: 1–2.
1972a. A representation theorem for lattices.
Algebra Universalis 2: 177–178.

E. T. Schmidt
1962. Über die Kongruenzverbände der Verbände.
Publ. Math. Debrecen 9: 243–256.
1968. Zur Charakterisierung der Kongruenzverbände der Verbände.
Mat. Časopis Sloven. Akad. Vied. 18: 3–20.
1969. *Kongruenzrelationen algebraischer Strukturen.*
VEB Deutscher Verlag der Wissenschaften, Berlin.
1970. Eine Verallgemeinerung des Satzes von Schmidt-Ore.
Publ. Math. Debrecen 17: 283–287.
1974. Every finite distributive lattice is the congruence lattice of some modular lattice.
Algebra Universalis 4: 49–57.
1975. On the length of the congruence lattice of a lattice.
Algebra Universalis 5: 98–100.
a. On finitely generated simple modular lattices.
Period. Math. Hungar. 6 (1975): 213–216.
(See also E. Fried and E. T. Schmidt; G. Grätzer and E. T. Schmidt.)

M. Schützenberger
1945. Sur certains axiomes de la théorie des structures.
C. R. Acad. Sci. Paris 221: 218–220.

W. Schwan
1948. Perspektivitäten in allgemeinen Verbänden.
Math. Z. 51: 126–134.
1948a. Ein allgemeiner Mengenisomorphiesatz der Theorie der Verbände.
Math. Z. 51: 346–354.
1949. Zusammensetzung von Schwesterperspektivitäten in Verbänden.
Math. Z. 52: 150–167.
1949a. Ein Homomorphiesatz der Theorie der Verbände.
Math. Z. 52: 193–201.

D. S. Scott
(See T. Frayne, A. C. Morel, and D. S. Scott.)

N. Seibert.
(See G. Sauer, N. Seibert, and R. Wille.)

L. N. Ševrin
1964. Projectivities of semilattices. (Russian.)
Dokl. Akad. Nauk SSSR 154: 538–541.

H. Sharp
1968. Cardinality of finite topologies.
J. Combinatorial Theory 5: 82–86.

M. Sholander
1951. Postulates for distributive lattices.
Canad. J. Math. 3: 28–30.

J. Sichler
1972. Non-constant endomorphisms of lattices.
Proc. Amer. Math. Soc. 34: 67–70.
1975. Note on free products of Hopfian lattices.
Algebra Universalis 5: 145–146.
(See also G. Grätzer and J. Sichler.)

F. Šik
a. A note on modular lattices of locally finite length.

R. Sikorski

1964. *Boolean Algebras*. Second edition.
Ergebnisse der Mathematik und ihrer Grenzgebiete, Neue Folge, Band 25, Academic Press, New York, N.Y.
(See also H. Rasiowa and R. Sikorski.)

H. L. Silcock

1977. Generalized wreath products and the lattice of normal subgroups of a group.
Algebra Universalis 7: 361–372.

F. M. Sioson

1964. Equational bases of Boolean algebras.
J. Symbolic Logic 29: 115–124.

H. L. Skala

1971. Trellis theory.
Algebra Universalis 1: 218–233.

1972. *Trellis Theory*.
Memoirs Amer. Math. Soc. No. 121.

L. A. Skornjakov

1961. *Complemented Modular Lattices and Regular Rings*. (Russian.)
Gosudarstv. Izdat. Fiz.-Mat. Lit., Moscow.

M. F. Smiley

(See E. Pitcher and M. F. Smiley.)

M. F. Smiley and W. R. Transcue

1943. Applications of transitivities of betweenness in lattice theory.
Bull. Amer. Math. Soc. 49: 280–287.

F. A. Smith

1974. Projectivity of prime quotients and simple lattices.
J. Algebra 31: 257–261.

R. M. Solovay

(See J. D. Monk and R. M. Solovay.)

Ju. I. Sorkin

1951. Independent systems of axioms defining a lattice. (Russian.)
Ukrain. Mat. Ž. 3: 85–97.

1952. Free unions of lattices. (Russian.)
Mat. Sb. N.S. 30 (72): 677–694.

1954. On the imbedding of latticoids in lattices.
Doklady Akad. Nauk SSSR N.S. 95: 931–934.

V. T. Sós

(See E. Fried and V. T. Sós.)

T. P. Speed

1969. On Stone lattices.
J. Austral. Math. Soc. 9: 293–307.

1969a. Spaces of ideals of distributive lattices. I. Prime ideals.
Bull. Soc. Roy. Sci. Liège 38: 610–628.

1972. On the order of prime ideals.
Algebra Universalis 2: 85–87.

1972a. Profinite posets.
Bull. Austral. Math. Soc. 6: 177–183.

1974. Spaces of ideals of distributive lattices. II. Minimal prime ideals.
J. Austral. Math. Soc. 18: 54–72.

J. H. Spencer

(See N. D. Belnap and J. H. Spencer.)

E. Sperner

1928. Ein Satz über Untermengen einer endlichen Menge.
Math. Z. 27: 544–548.

R. P. Stanley

1971. Modular elements of geometric lattices.
Algebra Universalis 1: 214–217.

1972. Supersolvable lattices.
Algebra Universalis 2: 197–217.

1973. An extremal problem for finite topologies and distributive lattices.
J. Combinatorial Theory Ser. A 14: 209–214.

1974. Finite lattices and Jordan-Hölder sets.
Algebra Universalis 4: 361–371.

A. K. Steiner

1966. The lattice of topologies: Structure and complementation.
Trans. Amer. Math. Soc. 122: 379–398.

1966a. Complementation in the lattice of T_1-topologies.
Proc. Amer. Math. Soc. 17: 884–886.

D. Steven

1968. Topology on finite sets.
Amer. Math. Monthly 75: 739–741.

M. H. Stone

1936. The theory of representations for Boolean algebras.
Trans. Amer. Math. Soc. 40: 37–111.

1937. Topological representations of distributive lattices and Brouwerian logics.
Časopis Pěst. Mat. 67: 1–25.

U. M. Swamy and V. V. R. Rao

1975. Triple and sheaf representations of Stone lattices.
Algebra Universalis 5: 104–113.

K. Takeuchi

1951. On maximal proper sublattices.
J. Math. Soc. Japan 2: 228–230.

1951a. On free modular lattices.
Japan J. Math. 21: 53–65.

1959. On free modular lattices. II.
Tôhoku Math. J. (2) 11: 1–12.

D. Tamari

1951. Monoides préordonnés et chaînes de Malcev. Thése, Université de Paris.
(See also H. Friedman and D. Tamari; S. Huang and D. Tamari.)

O. Tamaschke

1960. Submodulare Verbände.
Math. Z. 74: 186–190.

S. Tamura

1971. On distributive sublattices of a lattice.
Proc. Japan Acad. 47: 442–446.

1971a. A note on distributive sublattices of a lattice.
Proc. Japan Acad. 47: 603–605.

S. K. Tan

(See C. C. Chen, K. M. Koh, and S. K. Tan.)

T. Tan

1975. On transferable distributive lattices.
Nanta Math. 8, No. 2: 50–60.

T. Tanaka

1952. Canonical subdirect factorizations of lattices.
J. Sci. Hiroshima Univ. Ser. A. 16: 239–246.

A. Tarski

1930. Sur les classes d'ensembles closes par rapport à certaines opérations élémentaires.
Fund. Math. 16: 181–304.

1946. A remark on functionally free algebras.
Ann. of Math. (2) 47: 163–165.

1955. A lattice-theoretical fixpoint theorem and its applications.
Pacific J. Math. 5: 285–309.

1968. Equational logic and equational theories of algebras.
In *Contributions to Mathematical Logic*, pp. 275–288.
North-Holland, Amsterdam.
(See also A. Mostowski and A. Tarski.)

S. K. Thomason

1970. A proof of Whitman's representation theorem for finite lattices.
Proc. Amer. Math. Soc. 25: 618–619.

1970a. Sublattices and initial segments of the degrees of unsolvability.
Canad. J. Math. 22: 569–581.

R. M. Thrall and D. G. Duncan

1953. Note on free modular lattices.
Amer. J. Math. 75: 627–632.

J. R. R. Tolkien

1954. *The Lord of the Rings.*
George Allen and Unwin, London.

T. Traczyk

1963. Axioms and some properties of Post algebras.
Colloq. Math. 10: 193–209.

1967. On Post algebras with uncountable chain of constants. Algebras of homomorphisms.
Bull. Acad. Polon. Sci. Sér. Sci. Math. Astronom. Phys. 15: 673–680.

W. R. Transcue

(See M. F. Smiley and W. R. Transcue.)

W. T. Trotter, Jr. and J. I. Moore

1975. The dimension of planar posets. Abstract.
Notices Amer. Math. Soc. 22: A-44.

L. A. Troy

(See R. D. Byrd, R. A. Mena, and L. A. Troy.)

W. T. Tutte

1959. Matroids and graphs.
Trans. Amer. Math. Soc. 90: 527–552.

1965. Lectures on matroids.
J. Res. Nat. Bur. Standards Sect. B 69B: 1–47.

1971. *Introduction to the Theory of Matroids.*
American Elsevier, New York, N.Y.

H. Tverberg
1967. On Dilworth's decomposition theorem for partially ordered sets.
J. Combinatorial Theory 3: 305–306.

A. Urquhart
1973. Free distributive pseudo-complemented lattices.
Algebra Universalis 3: 13–15.

R. Vaidyanathaswamy
1947. *Treatise on Set Topology*. Part I.
Indian Math. Soc., Madras.
1960. *Set Topology*. Second edition.
Chelsea, New York, N.Y.

V. S. Varadarajan
1968. *Geometry of Quantum Theory*. I.
The University Series in Higher Mathematics.
Van Nostrand, Princeton, N.J.
1970. *Geometry of Quantum Theory*. II.
The University Series in Higher Mathematics.
Van Nostrand, Princeton, N.J.

J. Varlet
1963. Contribution à l'étude des treillis pseudo-complémentés et des treillis de Stone.
Mém. Soc. Roy. Sci. Liège Coll. in-8⁰, Vol. 8, No. 4.
1965. Congruences dans les demi-lattis.
Bull. Soc. Roy. Sci. Liège 34: 231–240.
1966. On the characterization of Stone lattices.
Acta Sci. Math. (Szeged) 27: 81–84.
1968. Algèbres de Lukasiewicz trivalentes.
Bull. Soc. Roy. Sci. Liège 37: 399–408.
1968a. A generalization of the notion of pseudocomplementedness.
Bull. Soc. Roy. Sci. Liège 37: 149–158.

R. L. Vaught
(See M. D. Morley and R. L. Vaught.)

O. Veblen and W. H. Young
1910. *Projective Geometry*. 2 volumes.
Ginn and Co., Boston.

V. Vilhelm
1955. The selfdual kernel of Birkhoff's conditions in lattices with finite chains. (Russian.)
Czechoslovak Math. J. 5 (80): 439–450.

P. Vopěnka, A. Pultr, and Z. Hedrlín
1965. A rigid relation exists on any set.
Comment. Math. Univ. Carolinae 6: 149–155.

B. Z. Vulih
1967. *Introduction to the Theory of Partially Ordered Spaces*.
Woltors-Noordhoff, Groningen.
(See also L. V. Kantorovič, B. Z. Vulih, and A. G. Pinkser.)

B. L. van der Waerden
1931. *Modern Algebra*.
Springer-Verlag. First edition, 1931. First English translation, 1949.

H. Wallman

1938. Lattices and topological spaces.
Ann. of Math. 39: 112–126.

Shih-chiang Wang

1953. Notes on the permutability of congruence relations. (Chinese.)
Acta Math. Sinica 3: 133–141.

M. Ward

1939. A characterization of Dedekind structures.
Bull. Amer. Math. Soc. 45: 448–451.
(See also G. Birkhoff and M. Ward; G. Havas and M. Ward.)

A. G. Waterman

1967. The free lattice with 3 generators over N_5.
Portugal. Math. 26: 285–288.

L. Weisner

1935. Abstract theory of inversion of finite series.
Trans. Amer. Math. Soc. 38: 474–484.

H. Werner and R. Wille

1970. Characterisierungen der primitiven Klassen arithmetischer Ringe.
Math. Z. 115: 197–200.

T. P. Whaley

1969. Large sublattices of a lattice.
Pacific J. Math. 28: 477–484.

Ph. M. Whitman

1941. Free lattices.
Ann. of Math. (2) 42: 325–330.

1942. Free lattices. II.
Ann. of Math. (2) 43: 104–115.

1943. Splittings of a lattice.
Amer. J. Math. 65: 179–196.

1946. Lattices, equivalence relations, and subgroups.
Bull. Amer. Math. Soc. 52: 507–522.

1961. Status of word problems for lattices.
In *Lattice Theory, Proc. Symp. Pure Math.*, II, pp. 17–21.
Amer. Math. Soc., Providence, R.I.

H. Whitney

1935. On the abstract properties of linear dependence.
Amer. J. Math. 57: 509–533.

L. R. Wilcox

1939. Modularity in the theory of lattices.
Ann. of Math. 40: 490–505.

1942. A note on complementation in lattices.
Bull. Amer. Math. Soc. 48: 453–458.

1944. Modularity in Birkhoff lattices.
Bull. Amer. Math. Soc. 50: 135–138.

R. Wille

1967. Verbandstheoretische Charakterisierung n-stufiger Geometrien.
Arch. Math. (Basel) 18: 465–468.

1969. Primitive Länge und primitive Weite bei modularen Verbänden.
Math. Z. 108: 129–136.

1969a. Variety invariants for modular lattices.
Canad. J. Math. 21: 279–283.

R. Wille

1970. *Kongruenzklassengeometrien.*
Lecture Notes in Mathematics. 113.
Springer-Verlag, Berlin.

1972. Primitive subsets of lattices.
Algebra Universalis 2: 95–98.

1973. On free modular lattices generated by finite chains.
Algebra Universalis 3: 131–138.

1973a. Über modulare Verbände, die von einer endlichen halbgeordneten Menge frei erzeugt werden.
Math. Z. 131: 241–249.

1974. On modular lattices of order dimension two.
Proc. Amer. Math. Soc. 43: 287–292.

1974a. Jeder endlich erzeugte, modulare Verband endlicher Weite ist endlich.
Mat. Časopis Sloven. Akad. Vied. 24: 77–80.

1975. An example of a finitely generated non-Hopfian lattice.
Algebra Universalis 5: 101–103.

1976. On the width of irreducibles in finite modular lattices.
Algebra Universalis 6: 257–258.

a. Subdirekte Produkte vollständiger Verbände.
J. Reine. Angew. Math. 283/284 (1976): 53–70.

b. Finite projective planes and equational classes of modular lattices.
Atti Conv. Comb. Roma, 1973.

(See also A. Day, C. Herrmann, and R. Wille; C. Herrmann M. Kindermann, and R. Wille; C. Herrmann, C. M. Ringel, and R. Wille; A. Mitchke and R. Wille; G. Michler and R. Wille; G. Sauer, W. Seibert, and R. Wille; H. Werner and R. Wille.)

R. J. Wilson

1973. An introduction to matroid theory.
Amer. Math. Monthly 80: 500–525.

R. M. Wilson

(See T. A. Dowling and R. M. Wilson.)

B. Wolk

(See G. Grätzer and B. Wolk; R. W. Quackenbush and B. Wolk.)

E. S. Wolk

1957. Dedekind completeness and a fixed-point theorem.
Canad. J. Math. 9: 400–405.

M. J. Wonenburger

(See G. Grätzer and M. J. Wonenburger.)

A. Wrónski

1976. The number of quasivarieties of distributive lattices with pseudocomplementation.
Pol. Acad. Sci. Bull. Sect. Logic 5: 115–121.

Yau-chuen Wong and Kung-fu Ng

1973. *Partially Ordered Topological Vector Spaces.*
Oxford Mathematical Monographs,
Clarendon Press, Oxford.

M. Yaushara

1974. **The Amalgamation Property, the Universal-Homogeneous Models, and the Generic Models.**
Math. Scand. 34: 5–36.

W. H. Young

(See O. Veblen and W. H. Young.)

G. Zacher

1952. **Sugli emimorfismi superiori ed inferiori tra reticoli.**
Rend. Accad. Sci. Fis. Mat. Napoli (4) 19: 45–56.

G. Zappa

1952. ***Reticoli e geometrie finite.***
Lezzioni raccolte dal dott.
Giovanni Zacher. Libreria Editrice Liguori, Napoli.

TABLE OF NOTATION

SETS

SYMBOL	EXPLANATION
$\in$	membership sign
$\subseteq$	inclusion
$\subset$	proper inclusion
$\cup$, $\bigcup$	union
$\cap$, $\bigcap$	intersection
$\varnothing$	empty set (void set)
$-$	set difference
$\times$, Π	Cartesian product
A^2	$A \times A$
$\langle a, b\rangle$	ordered pair
$\langle a, b\rangle \in \rho$, $a \equiv b(\rho)$, $a\rho b$	equivalent notations for $\rho \subseteq A^2$, 1, 192
$P(X)$	set of all subsets of a set X

POSETS

SYMBOL	APPEARS ON PAGE(S)
$\leqq$	1
$\geqq$	1
$\leqq_B$	6
$<$	6
$\prec$	10
$\succ$	10
$\preceq$	162
$\succeq$	162
$\parallel$	2
inf	2
sup	2
$\wedge$	4
$\vee$	4
$\cong$	15, 16
$[a, b]$	17
$(a, b]$	17
$[a, b)$	17
(a, b)	17
0	2
1	2
0^b	268
1^b	268
$h(a)$	172
$H(P)$	61
$l(P)$	2

LATTICES

SYMBOL	APPEARS ON PAGE(S)
$\langle L; \wedge, \vee\rangle$	5
$\langle L; \wedge\rangle$	8
$\langle L; \vee\rangle$	7
$\langle L; \wedge, \vee, 0, 1\rangle$	111
$\langle L; \wedge, \vee, *, 0, 1\rangle$	111
$A^b = A \cup \{0^b, 1^b\}$	268
$B(L)$	89
$C(L)$	21
$D(L)$	113
$\mathcal{D}(L)$	19
$\mathcal{D}_0(L)$	19, 42
$I(L)$	18, 80
$I_0(L)$	18, 42
Part(L)	26, 192
$S(L)$	49, 112
Sub(L)	24
L/Θ	21
L/I	151

INDEX

Bei Fragen zur Produktsicherheit wenden Sie sich bitte an:
If you have any questions regarding product safety,
please contact:

Walter de Gruyter GmbH
Genthiner Straße 13
10785 Berlin
productsafety@degruyterbrill.com